P9-CKH-422

architectural inspiration

architectural inspiration

styles, details & sources

Richard Skinulis & Peter Christopher

A BOSTON MILLS PRESS BOOK

First printing

Library and Archives Canada Cataloguing in Publication

Skinulis, Richard, 1947–
Architectural inspiration : styles, details & sources /
Richard Skinulis & Peter Christopher.

Includes bibliographical references.
ISBN-13 978-1-55046-468-9
ISBN-10: 1-55046-468-X

1. Architecture, Domestic—North America.
2. Architecture—North America—Details.
3. Interior decoration.
4. Building fittings industry—North America—Directories.
5. Finish hardware—North America—Directories.
I. Christopher, Peter II. Title.

NA7120.S54 2007 728 C2007-901748-7

Publisher Cataloging-in-Publication Data (U.S.)

Skinulis, Richard, 1947–
Architectural inspiration : styles, details & sources /
Richard Skinulis & Peter Christopher.
[528] p. : col. photos.; cm.
Includes bibliographical references and index.
Summary: A style and source guide that examines residential architecture and design elements from both creative and practical perspectives; includes directory of sources for architectural materials and services.

ISBN-13: 978-1-55046-468-9
ISBN-10: 1-55046-468-X
1. Architecture, Domestic — North America.
2. Architecture — Sourcebook — North America.
I. Christopher, Peter. II. Title.

728. dc22 NA7120.S556 2007

Published in 2007 by BOSTON MILLS PRESS
132 Main Street,
Erin, Ontario N0B 1T0
Tel 519-833-2407
Fax 519-833-2195
www.bostonmillspress.com
Editor: Kathleen Fraser
Design: PageWave Graphics Inc.

IN CANADA
Distributed by Firefly Books Ltd.
66 Leek Crescent
Richmond Hill, Ontario L4B 1H1

IN THE UNITED STATES
Distributed by Firefly Books (U.S.) Inc.
P.O. Box 1338, Ellicott Station
Buffalo, New York 14205

Printed in China

The publisher gratefully acknowledges the financial support for our publishing program by the Government of Canada through the Book Publishing Industry Development Program.

Cover: Main image: The living room of the metal house architect Barton Myers built. The main rooms are essentially post and beam construction. Products left to right: Lionhead fountain carved by Walter Arnold; Espace waterfall tub filler, THG; Hamilton Beaux Arts door set, Rejuvenation Hardware; wall sconce, Hubbardton Forge. Page 1: This staircase borrows from Arts and Crafts traditions. Soft curves follow the natural movement up and down the stairs. Page 2: The stone around this fireplace is from buildings torn down to make way for the Yangtze Valley Dams project in China. The curved beams are a Japanese tradition in which the imperfect beams contrast with the perfect joinery. The rich floorboards are Australian jarrah wood. Page 5: Rudolph Valentino installed the elaborate fountain in this charming Hollywood villa. The stone retaining wall, pea gravel floor and border make a comfortable backdrop for the jungle riot of the garden. Page 9: In this neo-modernist courtyard house, there is as much exterior as interior space. Except for the ipe wood decking, most of the surfaces are limestone with some concrete slabs topped with more limestone. The owners do a lot of entertaining, so the hard surfaces were maximized by Toronto architect Bruce Kuwabara (KPMG Architects). Page 10: You would never know it but this neoclassical mansion is not limestone, but formed cement with a controlled limestone matrix. The cement was lightly sandblasted when the forms came off. Note the decidedly French elements of this house, including French doors and windows and the imposing mansard roof with round windows above the curved-top dormers. The entranceway has a broken pediment, a very neoclassical element. Page 108: The interior of this metal house by Barton Myers combines off-the-shelf materials such as oversized garage doors and steel roofing with concrete floors. Most of the infill walls are stud and stucco plaster, which consists of a skim coat of natural plaster on what is called "blueboard" — a drywall that can take moisture so you can plaster right onto it. Page 472: A modern kitchen by Italian manufacturer Valcucine is all sleek lines, contrasting light and dark woods with stainless steel and glass.

To our wives, Karen and Suzanne,
designers in their own right.

contents

sources

preface

My interest in all things architectural began decades ago when I wrote my first book on log homes. It was given further fuel when I teamed up with photographer Peter Christopher on another book, *Log Homes: Classics of the North*. To our delight, Peter and I found that people who design, build and live in beautiful, well-crafted homes are extremely interesting and a pleasure to be around.

We also discovered that all the things that make a home really work from an aesthetic point of view — in other words, the details — were just as interesting. Why do the arches, columns and wrought-iron work of one house combine together to make it a magnificent showpiece, but on another house look simply awful? How can massive pieces of stone look sensational in one house and unbelievably tacky in the one next door? We wanted to know how the details work and what inspires their use.

In a nice case of serendipity, we ran across some very interesting trends — sea changes, really — that we wanted to communicate and try to interpret. Because we North Americans, and the world at large, are becoming much more sophisticated in terms of how we live and where we do that living, how our homes look and function is becoming very important to us all. Hundreds of billions of dollars are being spent every year by people renovating their homes in North America. Our interest in architectural details is apparently being shared by just about everyone else on the continent. Why not a book that looks at this wave of activity from a uniquely North American perspective?

The passion that we believe infuses this book comes from the countless architects, designers, home owners, artisans and manufacturers whose homes and work grace these pages. Their interpretations are as variable as are their personalities. Many of the quotes you'll find throughout this book come from conversations with these artists, each contributing his or her own take on residential design and decoration. They have explored all manner of extraordinary ideas and have access to sources not all homeowners might find in the course of a trip to a regular neighborhood building center. We are fortunate in enjoying the benefit of these experts' ideas and counsel — in

photographs and in words — plus access to many of the sources where they go to get their materials.

What we have not tried to do is come up with a definitive work on all architectural styles from an exhaustive academic or historical perspective. This book is a source and style guide designed to provide inspiration, ideas and useful information. We wanted to share all of the possibilities that we found as we crisscrossed North America in search of the best and most interesting in architectural details. There are plenty of ideas in these pages to inspire those who want to remodel their home or, dream of dreams, build their own. We also wanted to provide a look at the available products and materials needed to make that dream reality — the best of the best in high-end, hard-to-find or simply unique products. The Products section at the end of each Details chapter contains only a very small sampling of some of the many unique products available. Take inspiration from these and turn to the Sources listing at the end of the book for contact information and web addresses for almost a thousand suppliers, manufacturers, artisans, designers and architects. They will lead you to places you might never have imagined.

It is to celebrate this adventurous spirit that this book was conceived. We hope you will be as inspired in reading it as we were in producing it.

Richard Skinulis
Richmond Hill, Ontario

styles

exteriors

north american styles

All architecture in North America, other than the skyscraper and the single family house, is derived from European precedent and rendered with American technology.

ROBERT STERN, POSTMODERNIST ARCHITECT WITH AN AFFECTION FOR THE PAST

A true restoration of an 1840s Boston Gothic Revival, sometimes called Carpenter Gothic because carpenters would cut out the barge boards and the trim on a band saw. Photographs of the original house helped the architects (Menders, Torrey & Spencer) discover that there were missing finials.

THERE ARE TWO OPINIONS ABOUT NORTH AMERICAN RESIDENTIAL design. One is elegant in its simplicity and final in its view: there is no real North American architecture or residential design; all North American architectural styles are in fact seminal descendents from European and, to a lesser degree, Japanese, styles, and no matter how we have adapted or reworked these styles, they remain derivative.

The other, less dismissive view is that while the immigrants who came to North America did indeed build in the styles they were comfortable with, the circumstances of climate, geography, raw materials and even the means by which people made a living transformed the European prototypes into what are now recognizably North American styles of dwelling.

The simple Cape Cod, the ubiquitous Ranch Style, the California Bungalow, the solid, expansive Prairie Style, the sun-baked adobe Pueblo, the Western Pole house and the log cabin nestled in the woods are among the most recognizable North American styles. These kinds of homes are examples of vernacular architecture. They are characterized by a response to where people live and not according to any existing popular style. Ironically, vernacular styles eventually become popular styles themselves. If we have any claim to North American style, it comes from these humble roots.

Another influence on North American residential architecture is related to private land ownership — the ability to own clear title to land, which is the basis of the settlement patterns of the United States and Canada. In Europe, most people don't have the same expectations of land ownership as do North Americans. In the North American imagination, independence and success are often characterized by the ability to buy a piece of land and put up a family house on it. In North America, a continent of endless forests and wide-open spaces, it has seemed more possible for an ordinary individual to claim a patch of land and build on it. That produces a unique (by world standards, at least) perception of what a dwelling is. It also immediately personalizes a building and suggests it is a reflection of the values and accomplishments of the people who live in it. In contrast, if you live in a 14th-century farmhouse in Tuscany, the architectural style of your home has little to do with who you are other than the fact that you are a continuation of that culture or that family. Over here, your home is almost an instant expression of who you are and your station in life. Cultural geographers have said that the distinguishing characteristic of the North American landscape is that we have always seen it as temporary.

Each of these arguments underlines the immigrant nature of North America — people come here with ideas and customs and then adapt them to this new world.

Oversized buff brick is the cladding for this neo-modernist home designed by Toronto architect Bruce Kuwabara. Its structure of two-story wings joined by a single story middle volume was designed to separate the adult wing from the children's/guest wing with a common space containing kitchen, dining room and other living space. The illuminated fence is Brazilian ipe wood with a zinc liner. The fence also has a roof of prefinished metal so snow or water can't sit on the wood. The windows are milled mahogany with a dark stain. The dark mullions are Scandinavian influenced, and tend to make the openings dissolve because there is less there to catch the eye. To enhance this affect, the sealed units (thermopane) were made of dark brown metal, not the usual metallic silver.

The single family, free-standing house is so much more a common occurrence in the North American landscape than it is anywhere else. If you compare it to farmer's fields in Italy where the guy has plowed around a rock for 500 years, you realize that any Nebraska farmer would have dynamited that rock the day he bought the property. The same thing applies to [North American] houses — they are completely original, completely a reflection of personality.

BARTON PHELPS,
LOS ANGELES ARCHITECT
SPECIALIZING IN MODERNISM

Right: With both Japanese and Adirondack influences, this stone and wood home is made from split-face Ontario limestone from Owen Sound, and Douglas fir. The ledge beneath the wood is the same limestone, only sawn smooth. **Overleaf:** Architect Barton Myers' steel house was designed to make the most of the stunning Southern California scenery and benign climate. The three large sectional doors are "off the shelf" in the sense that they are standard doors made to his specifications. The metal drums over the doors contain rolldown fire shutters. The pool in the foreground — part swimming, part reflecting — is in fact the roof of the guesthouse.

After Frank Lloyd Wright set up Taliesin West, there was a real rethinking of desert architecture that involved natural but also more modern materials like concrete and concrete block, steel and copper because they would last in the desert. There was also a lot of interest in creating shade and orienting and sheltering your building from the sun — a much more intellectual approach.

JOHN DOUGLAS, ARCHITECT

When a new design concept appearing in one country bounces around the world to every other at the speed of light, one might be led to believe that we have now achieved some kind of amorphous global style. But until we truly live in a true "global village," the architectural styles of the world will continue to borrow from and influence each other. One reaction to this march of globalization by residential architects and designers is the concept of critical regionalism, which assumes that one can have a regional fit and that new structures can and indeed should embody some of the same principles as the original models. This is seen as desirable not only on cultural grounds, as a way to keep one part of the world from looking much like any other regardless of history, culture or climate, but also for environmental reasons, as vernacular building styles generally reflect the most time-tested, "green" or sustainable of building methods of a particular region. This common-sense approach may well be the real answer as to whether or not there is an authentic North American architecture.

By the time Columbus sailed into the waters of the New World, there were some 24 million people living in North America and the Caribbean. Except in a few cases, however, the waves of European immigrants who came to North America ignored the building techniques and styles of the indigenous populations. Instead they built what was familiar to them, copying the vernacular building styles of their part of the "Old Country."

Spanish settlers in the Southwest used thick, sun-baked adobe brick (wet mud strengthened with straw, manure or other organic bonding agents) made into walls slathered with mud. This, the most ancient of building materials and methods, was common in villages in Spain, where cheap materials and protection from the sun were just as important as in the deserts of the New World. Settlers from Spain

Opposite: A modern adaptation of a Pueblo style house with Frank Lloyd Wright influences, which can be seen in the Craftsman-style window details and the quarried sandstone with its horizontal banding. The cubist forms of the Pueblo style lends itself easily to contemporary design, making this a good example of a "soft modern" home — one that looks contemporary but lacks the harsh edges of glass and steel that can sap the warmth out of many ultra modern homes. The architect was Eric Brandt. **Above left:** The undulating, sensuous lines of adobe walls (sun dried mud bricks covered in a cement-based stucco), projecting *vigas* (wooden rafters) and traditional kiva fireplace characterize this Phoenix home designed in the 1920s by architect Robert Evans. Note the Mexican sun-dried Saltillo tile forming the patio. **Above:** Walls that look like rammed earth are really a thick cement masonry block plastered with adobe clay, cement, and a little bit of straw. The color comes from the local clay the adobe was made from.

built in the Mission style, but were also influenced by the style of the indigenous population, who had been building round-cornered, vertical adobe walls holding up flat roofs for at least 1,000 years. Either left the color of the local desert or smeared with a coat of whitewash to reflect the sun, adobe was and still is the most natural and efficient of desert building materials.

This massive adobe Pueblo style of house, with its characteristic flat roofs, deep windows, projecting roof beams (*vigas*) and rainwater gutters (*canales*), can still be seen throughout the American Southwest. Adobe styles change from region to region. The true Pueblo-style adobe has rounded corners (to the diameter of a woman's hand, as it was a woman's job to smooth the mud onto the brick). Squaring off the corners in a more angular form is an elaboration known as Territorial style, which can have a gabled roof, pointed arches and a false front. The Spanish Colonial adaptation has windowless, one-story rooms opening onto an enclosed courtyard. The most modern version of this building style is known as Pueblo Revival; it replaces the authentic adobe brick with masonry block covered in emulsion-enhanced plaster. Pueblo Revival usually removes the architectural detailing that most traditional Pueblo style buildings have, such as the projecting roof beams and porch posts, to offer a more stripped-down version with an emphasis on the structural composition. An offshoot of the Adobe building method is the Rammed Earth house, made with moistened soil rammed or tamped into wooden frames that, when removed, leave behind a dense, environmentally friendly wall that can last for centuries.

Spanish Colonial Revival, popular from Santa Barbara, where it is the official building style, to Miami and anywhere with a history of Spanish settlement, gained strength between the World Wars. Its Mediterranean-Moorish roots are obvious in its red roof tiles, stucco walls, arched doors and windows and arcaded and deeply shaded porches. It is more elaborate and decorative than the far older Mission style.

In the American South, lower-class English, Irish and Scottish settlers developed a more flamboyant architectural style designed to awe and impress. The classic Southern plantation mansion took as its model the early two-story wooden homes of the French settlers in the south, which featured long front porches, but elaborated on them with more Georgian-inspired Grecian colonnades.

The first French settlers who sailed to eastern Canada and established New France brought with them the building styles of northern France: houses made of stone with steep gabled roofs and dormer windows. This style is still seen today in Quebec. The English who colonized Upper Canada (now Ontario), the Maritimes and New England at first built in the more medieval style known as half timber.

But by the mid 17th century, the homes of the middle and upper classes were being constructed in the more formal Georgian style (named after a succession of kings George), marked by a rectangular house plan, minimal projections and symmetrical facades. The Georgian home looks comfortable.

Right: Despite a small budget and inexpensive materials, this corrugated steel over wood frame fits beautifully into its Rocky Mountain surroundings of purple sage and pine. The simple rectangle that steps up in the middle was inspired by architect Harry Teageu's love of metal trailers. "Corrugated metal picks up light from different directions than what is hitting the background," he says, "so it always provides contrast. That gives the material a lot of power." **Opposite:** A quaint English cottage. The oversized columns provide a sense of stability, while the rounded window in the gable counters with a feminine softness.

Northern houses...tended to be compact in their massing, making an easy-to-heat, energy-efficient package. The houses were blocky and low-slung compared to Southern examples, where houses were likely to sit over basements or to be built on pilings. High-ceilinged rooms, favored for summer cooking, also contributed to overall building height in the South.

JAMES C. MASEY AND SHIRLEY MAXWELL, *HOUSE STYLES IN AMERICA*

Three different buildings were brought together and built into what historical renovator Tony Jenkins calls "simple structures that were gilded." The two frame buildings on the ends are mid-century Ontario cottage style — the classic story-and-a-half rectangle that is a Georgian derivative. The main stone house in the middle is also an Ontario vernacular Georgian cottage style extended from its original one-and-a-half stories to two by using part of the inside brick wall.

There certainly is an evolved history of North American architecture. But in many ways we are losing our regionalism. We can build glass houses in Chicago that you used to only see in California. People are now very transient, so if they leave one place and go to another they bring their favorite building style with them. But a lot of the richness of regional architecture comes from the regional environment. For instance, Southern houses had the chimneys at either end because they wanted it open in the summer to let the breezes get through. In New England the homes have chimneys in the center to keep the heat central.

JAMES CULLION, ARCHITECT

Except for adding a Juliet balcony, the owner — Toronto designer Dee Chenier — of this renovated Victorian home kept the facade intact to preserve the streetscape. The front doors are a recreation of the original.

"We are sensible people," it says with pride, "And although prosperous, not ones to flaunt our success." Based on the rational classical designs laid out by Renaissance architect Andrea Palladio, the Georgian style is strictly characterized by rows of double-hung windows with either nine or twelve panes per sash, all symmetrically arranged around the central door, flanked by pilasters and with some kind of decorative work above it, often a piedmont.

On the East Coast the Federal style grew out of the Georgian style when, inspired by classical Greek and Roman tradition, architects started adding delicate decorative elaborations such as swags, garlands and elliptical windows to Georgian buildings. Typical Federal-style homes have an arched Palladian window (three windows with an arch over the larger middle window) on the second floor above the front door.

Just as the Federal style became a national style, particularly in the United States, so too formed other national styles as well as vernacular offshoots. Most were derived from European examples such as Greek Revival, which first swept the young nation in the late 1700s when the country needed a link with past greatness. Based on the triangular form epitomized by the Parthenon, it is characterized by low-pitched roofs that are usually gabled, and entry porches held up by columns — usually Doric. Conversely, there was also a great need in the newly independent country to separate itself from the its immediate colonial past; it is no wonder that the Georgian style, so connected to England, gave way

Built in 1898, this stately home has gone through many additions and modifications. Influences of Victorian, Greek Revival and Colonial combine to create a style that can truly be called Eclectic.

to buildings whose classic architectural elements — columns, piedmonts, windows hidden behind porches, dentil moldings and Greek returns — had classical, not colonial ties. (Greek Revival and Federal styles can both fit under the more general neoclassical umbrella.)

Other derivative national styles include Neo Gothic, Neo Colonial, Queen Anne Revival and Italianate, but one of the most important American revivals was Richardsonian Romanesque. Named after Henry Hobson Richardson, who was active in the last half of the 18th century, it was largely based on French and Spanish Romanesque precedents of the 11th century. The style, characterized by massive stone walls, dramatic semicircular arches, deep windows, cavernous recessed door openings and bands of windows, greatly influenced other American architects such as Louis Sullivan and Frank Lloyd Wright.

Swedish settlers brought with them their more rugged tradition of building with logs. The abundance of building material the virgin forests of North America provided, along with the simplicity of construction, made this a building style the settlers couldn't resist. Clear the land with an axe and bucksaw and use the same tools to put up a house with the very trees you have cut down. The practice was soon picked up by the English, Scottish and German pioneers, who took it with then as they pushed on ever westward.

Right: Log house in snow: The classic cabin in the woods. This one-and-a-half story Ontario hewn-log home is the quintessential refuge from the elements. Note the beautifully crafted dovetail joints at the corners.
Opposite: The inspiration for this Colorado house is the Western Pole Barn. The stone used was intended to represent hay bales stored in the barn. The weathered wood and the rusted metal roofing salute the vernacular architecture and materials of the region.

People have romanticized the log cabin, something that was accelerated by the designs of people like R.C. Reamer, which combined Western Mountain log work with the more eastern Adirondack lodge style — the big pole rafters, open spaces and stone fireplaces. Now it has become the style of the mega homes. People come here from somewhere else and want their own Yellowstone Lodge to live in — call it Neo Mountain Revival or Neo Adirondack.

HARRY TEAGUE, ARCHITECT

The rustic log cabin style has undergone two major revivals. The first, often called Camp Style, was developed in the mountains of the eastern United States by architects such as William West Durant, whose Sagamore Lodge combined traditional log construction with the more fanciful and delicate roofs and balconies of Switzerland. The idea of the big rustic log or timber lodge was then taken west by architect R.C. Reamer, who designed the Old Faithful Inn at Yellowstone Park.

The second revival came in the late 1970s when the lure of the big wood caught on with the counterculture "back-to-the-land" movement, which saw log homes as cheap, different and appealing to romantic ideals of history and of self-sufficiency. The look was quickly copied by the wealthy, who wanted the romance of the log house in the woods for their rural getaways and second homes — but with modern conveniences. The result is a multitude of beautifully rendered log homes that would make the pioneers blink in amazement. In keeping with current "anything goes" architectural attitudes, these homes are an eclectic mix of styles and materials, although they generally follow geographical traditions. Most Eastern log homes are made of logs hewn flat with dovetail corners; in the West, the round-log, "Scandinavian-scribed" look with saddle notch corners that project off the ends predominates.

After the Great War ended in 1918, returning soldiers brought back the idea of the Normandy farmhouse, and its feudal arrangement of consolidation of needs: living quarters, livestock, storage, and so on. These compact buildings often had a silo or tower built onto it. Architect Richard Landry was inspired by this design in his Malibu home, which incorporates a reconstructed barn with a metal, silo-like service module that echoes the silos of the past (page 118). Other French country homes moved the tower to the center of the house, encompassing the entranceway. The result was a spattering of mock French chateaus and faux French Country homes that continue to this day.

A timber frame Rocky Mountain ski lodge with an Asian flavor. It was made from local stone and the beam work was done by Colorado log and timber builders Timmerhus, whose principal, Ed Sure, studied in Japan.

A lot of clients come to an architect having already made up their minds about the style they want. They've seen a Bavarian castle or a Tuscan villa and that's what they want. But what I think all architects are waiting for is the kind of client who is willing to explore what would be suitable for the unique combination of who they are, their lifestyle, and the site — as opposed to something preconceived.

PAUL GORDON, A TORONTO ARCHITECT WHO LIKES TO MIX STYLES

A faux French chateau designed by Chicago architect Garrett Eakin. The round tower at the front is influenced by the fortified towers, or sometimes grain silos, that were often built into large farmhouses in Normandy. Wooden shutters, Juliet balconies and a rooster weathervane complete the effect.

Architect Jim Strasman designed this Vermont vacation home reminiscent of an old-fashioned covered bridge. The “bridge” is 120 feet long and made of 12-foot-long Douglas fir trusses. It spans a shallow valley through which runs a creek. Note the two chimneys — one of which is fake for the sake of balance.

At the turn of the 20th century, Colonial Revival styles developed out of a popular interest in America’s past. This interest in the past has been further developed by beautifully preserved and restored historical sites such as Colonial Williamsburg, Virginia, and Lewisburg, Nova Scotia. Williamsburg in particular is a good example of how the architecture of grand public buildings influenced the style of private homes.

But some national styles were truly American. A case in point is the Ranch Style. If one of America’s greatest contributions to world architecture is the single-family house, then the Ranch Style is the main conveyor of that concept. The Ranch Style, a low-slung, one-story home with a shallow roof, usually with an attached garage and often with large “picture windows” or sliding doors opening up to a spacious backyard, was the perfect template for a fast-growing population that wanted a little separation between themselves and the Joneses. Although there are numerous huge, multimillion dollar examples of ranch houses, they became popular with North American masses after World War II when there was a huge demand for houses mainly because of their simplicity of design, efficient use of space, and connection to the outdoors. Pattern books flooded the market, but materials tended to mimic the simplicity of design, so frame and common brick rather than fieldstone or the more expensive clinker brick were preferred. The Ranch Style bungalow was popularized by West Coast architect Cliff May, who was inspired by the sprawling working ranches of the Southwest with their low-slung roofs, wide eaves and central courtyards. He built his first example in Dan Diego, California, in 1932 and from then on there was no stopping the style, which became the prime example of North American informality.

Closely linked to the Ranch Style are the California Bungalow and, in its simple use of natural materials, the Craftsman Bungalow. (The term "bungalow" comes from the Hindi *bangla,* a single-story thatched-roof hut in Bengal province, India, at one time adopted by British colonists as summer homes.) "Bungalow" originally described a one-story building with a central living area around which were arranged other rooms, but it evolved to refer to more complex styles, including a few notable one-and-a-half-story variations. The Craftsman Bungalow was more compactly designed than its spread-out California cousin and, with its exposed roof rafters and bracket work, front porch held up by sturdy columns, dormered roofs and built-in furniture and shelving, was constructed in accordance with the Arts and Crafts movement's main tenet that handmade, artistic work should be available to the common man. (North Americans used machines to mass produce the style, thus bringing it into the reach of the common man, whereas the British — the originators of the movement — tended to use actual craftsmen to create the accoutrements of the style, thereby restricting it to those who could afford it, the upper classes). Among the most important innovators of this North American approach to the Craftsman house were the Californian sibling architects Greene & Greene (Charles and Henry) who contributed the most elegant Craftsman Bungalows, epitomized in the famous Gamble House — a combination of both Mission and Arts and Crafts elements. Designed in 1908 for David and Mary Gamble of Proctor & Gamble fame, the house was truly

One of the best examples of an American Arts and Crafts home and also considered one of the "ultimate bungalows," the Gamble House was designed in 1908 by sibling architects Charles and Henry Greene for David and Mary Gamble of the Proctor & Gamble Company. Along with a very Japanese use of wood (which can be seen in the exposed timbers and shingle siding), the brothers Greene built large, open sleeping porches and terraces along with wide overhangs to take advantage of the southern Californian climate.

684

hand-crafted, but also designed for the benign California climate with its wide sleeping porches, big overhangs and cross ventilation. Another, very local variation is the sturdy brick Chicago Bungalow, with its hipped roof (also one of the markers of the Prairie Style), and porch or sunroom forming an elaboration on the front of the house.

One of the foremost influences on Ranch Style was a national style of equal architectural, if not popular, importance — the Prairie Style. The name comes from the Midwest, and its flat, horizontal solidness. The master of this quintessential American style was Frank Lloyd Wright. It came about in part as a reaction to the vertical, overly ornamented Victorian and Queen Anne Revival homes that rich North Americans were building in the late 1800s. Wright stretched out his two- or sometimes three-story homes (minus basements and attics), accentuating the horizontal with banding and roof lines, sometimes sending a one-story wing off to underline the horizontal even more. The massive, imposing nature of the style is often increased by squat stone or brick pillars holding up porches and balconies. It echoed the Craftsman Style in its use of beautiful detailing in trim, but it did not insist that such ornament be handcrafted. Low roofs and overhanging eaves revealed an Asian influence, but Wright strongly rejected European influences, later developing what he called a Usonian Style (the word derived from the United States of America), in his view, creating "an architecture for democracy."

That's what you get for leaving a work of art out in the rain.

FRANK LLOYD WRIGHT, UPON HEARING A CLIENT COMPLAIN ABOUT A LEAKING FLAT ROOF HE DESIGNED IN OAK PARK, ILLINOIS

Opposite: One of the few Prairie-style homes by Chicago architect E.E. Roberts (built in 1911 and restored by Chicago architect Garrett Eakin), one of Frank Lloyd Wright's major competitors. It has a red brick base and stucco walls with a very deep overhang and a soft 4/12 roof slope to dramatize the horizontal. Eakin added porch extensions right and left and a two-story addition to the right. **Left:** Frank Lloyd Wright himself has said that the Willits House (Chicago) was the first of his great Prairie homes. Finished in 1903, it is a balanced cruciform with a roof line that extends beyond the wings to balconies and terraces that connect the house with the natural world.

Above: Built in 1955, the Louis Penfield House (Ohio) was one of Frank Lloyd Wright's Usonian homes (relatively simple structures that theoretically anyone could own). Made of cinder block, cherry wood and some hardwood plywood, the Penfield house has tall vertical windows to take in the view (right). A simple drive-through carport maintains the clean lines of the design. (For the interior, see page 46.) **Opposite:** A classic Craftsman style home, complete with large overhangs, exposed rafters and porch supports.

Wright also used the Prairie Style to break up the boxed-in Victorian floor plan and create a more spacious, open look, with the emphasis on natural light coming in through rows of casement windows, as well as ornamental brick and a central chimney. Suddenly there were homes with open interiors bathed in natural light. Instead of a jumble of closed-off rooms, there were designated-use areas (dining and living rooms) whose boundaries were merely suggested by elements such as stone fireplaces and art glass panels. Anyone who has wandered the streets of Oak Park, Illinois, where Wright did his first important work, will immediately understand how appalled many of the local residents were at these "modern monstrosities" with their flat roofs and horizontal lines, set down among the turn-of-the-century "painted ladies."

By the 1950s, a continuation of the postwar building boom saw an expansion of the Ranch Style with innovations such as the split-level and the raised ranch (a bungalow with a second floor added). Scorned by the children of the 1960s and 1970s as plebian and ordinary, original one-story ranch homes are now being eagerly snapped up by baby boomers as they come closer to retirement and look for an easy living one-level home with a bit of space around it.

One of the most successful competitors to the Ranch Style, equally visible in 1950s and 1960s-built suburbs, and which has many characteristics in common

(There is an) essential American determination to be protected against the continent's aggressive climate, which is to be achieved by the creation of deeply sheltering, well-heated, well-cooled interior spaces, sheathed in an impenetrable weather seal...the Shingle Style embodies them all.

VINCENT SCULLY, ARCHITECTURAL HISTORIAN, FROM *THE SHINGLE STYLE TODAY*

By using shingles for cladding, adding a kick flare at the base and curved roof lines, the architects (Menders, Torrey and Spencer) came up with this modified Cape Cod style cottage. The shingles are white cedar, treated with a transparent finish to make them last longer. White cedar turns gray — the classic Cape Cod look — instead of the orangey color of red cedar.

with it, is the Colonial Revival Cape Cod. In its original form, the Cape Cod has been around a lot longer than the Ranch. In fact, this simple, wood-clad style, first built by English colonists in the late 1600s, has been the mainstay of New England for 250 years. It was only in the 1930s and 1940s, when an exploding population felt the need for simple, inexpensive housing, that the Cape Cod became the darling of small town and suburb and one of the most popular examples of North American vernacular architecture.

These compact houses are often a story and a half with a higher pitched roof (all the better to provide an attic for storage and to keep the snow from building up). Not surprisingly, the Cape Cod is almost always made of wood — usually wide "clapboards" nailed to a wood frame — because not only has wood always been a plentiful building material in North America, it swells in the winter to keep the icy winds out and shrinks in the hot sun to allow cooling breezes to circulate. Shutters that were originally used to protect valuable window glass from Atlantic gales are now a vestigial stylistic conceit. And just to show how prosaic vernacular architecture can be, some Cape Cod homes have shallow barrel-arch ceilings that copy the arch made by the ribs of wrecked whaling boats used in some old New England churches (pages 60–61).

The classic Cape Cod is also sometimes clad entirely in shingles, not just the roof but the exterior walls and even structural and ornamental details such as posts, dormers and corbels. Everything from fishermen's shacks to the

The shingling of the walls extends to the corbels supporting the eaves of this Shingle Style home.

A lot of bad minimalism is mute, it has nothing to say, it's all about editing. But if you look at the work of the minimalist masters like John Pawson (Britain) and Tadao Ando (Japan), you see that there is a lot of richness of detail even though the spaces appear simple.

JOHNSON CHOU,
TORONTO MINIMALIST ARCHITECT

summer homes of the wealthy have been done in what is known as the Shingle Style, a local manifestation that can now be seen right across the continent. Why did this style catch on so well? In his book *The Shingle Style Today,* architectural historian Vincent Scully argues that this style has "tapped some fundamental items of American belief." These beliefs, he maintains, include bedrock ideals such as intellectual pluralism, where we all come together over shared beliefs that exclude no one; and the democratic approach to housing that takes simple material from local sources that everyone can access. (There is a modern adaptation of this called New Shingle Style. Popularized by architect Robert Venturi, it is recognized by its tall angular forms and mixture of gable and shed roof lines. The cladding is often shingle but it can also be clapboard or board and batten — very popular as a ski chalet.)

One of America's great gifts to the world of early 20th-century architecture was the skyscraper. New ways of building straight up with steel and glass were exemplified by these symbols of exuberant success, and the Art Deco style, with its stepped-back grandeur and zigzag design, was the perfect architectural expression of this form. The world was introduced to Art Deco in 1925 at the Exposition des Arts Decoratifs, held in Paris. Combining a bold, masculine, machine-parts look with Bauhaus, Cubism, Egyptian, Mayan and geometric influences, Art Deco was not only a celebration of the machine age, but also of science and rational thought. Still famous landmarks of New York City such as the Chrysler Building (designed in 1930 by William Van Alen) and the Empire State Building (by Shreve, Lamb & Harmon in the same year) are masterpieces of Art Deco. But the style never really caught on as in residential architecture. Even its Americanized form, known as American Streamline, which did influence the design of everything from radios to locomotives with its look of aerodynamic speed (it was, after all, an expression of America's fascination with machines), wasn't that popular with home owners. Then there is the variant known as Art Moderne, which is a lot like American Streamline but has more in common with the International Style (see below). Sometimes called Depression Modern, it is identified by its smooth, unadorned surfaces, curved windows and walls, metal windows set flush in the wall, steel railings, and semicircular projection over windows and doors.

Although Art Deco and the Prairie Style were huge leaps toward the contemporary look we are familiar with today, the real slide from tradition (any tradition) was first experienced in North America when European architects and designers, fleeing Nazi oppression in the 1930s and 1940s, brought European Modernism with them. Giants of architecture such as Ludwig Mies Van der Rohe and Walter Gropius introduced North America to the International Style. In a good example of architectural synergy, Frank Lloyd Wright greatly influenced the International Style formulators and was in turn influenced by them — in fact, his comeback masterpiece, Fallingwater, has undeniable International influences.

In reaction to most previous styles, with their accent on ornamentation and busy roof lines, the International Style was stripped down and unadorned, with

smooth, industrial cladding often in metal or glass. Roofs were flat, windows were treated as a continuation of the wall rather than mere holes in it, and the use of reveals instead of trimwork added to its minimalist look.

One of the first and perhaps most influential home-grown examples of the International Style, or Modern Movement Style, was Philip Johnson's Glass House, built in New Canaan, Connecticut, in 1948. Greatly influenced by Mies van der Rohe's Farnsworth House, it is the idea of a "home" reduced to little more than a transparent glass box. As Paul Heyer put it in his book, *American Architecture: Ideas and Ideologies in the Late Twentieth Century:* "It is one of the world's most beautiful and yet least functional houses. It was never intended as a home, but as a life-style stage to live with." Another renowned American proponent of the Modern Style was architect Louis Khan, who chose to use simple materials, formal restraint, and a keen awareness of space and light in seeking to create monumental architecture with spiritual qualities.

The International or Modern styles have now become known by the all-encompassing term Contemporary. A residential example of this may still

This 10,000-square-foot Chicago mansion is a blend of Art Deco and Art Moderne, with a little American Streamline thrown in. Built in 1932 (architects Zimmerman, Saxe and Zimmerman), it has a floor plan in the shape of a ship, with the windows on the right the stern and the outside lights simulating the running lights of a ship. Note the repetition of threes in the windows and metal trim, one of the hallmarks of Art Deco.

be a stripped-down building that glories in clean, straight lines, but it also uses new or untypical material such as naturally rusting steel (with brand names such as Corten) and stylistic methods such as reveals (exposing the point where materials meet instead of covering them up with trim). There is also a movement in North America to using off-the-shelf or industrial materials such as corrugated metal, sectional doors, marine plywood and cement board for Contemporary houses — all materials one would normally associate with a commercial building (see the Barton Myers' house on pages 16–17 and 108–109).

Since the 1980s, most new North American suburban housing tracts have been constructed with houses of a style best described as Postmodern Residential or Neo Eclectic. And it is eclectic; just about any style is up for grabs: Neo Gothic, Italianate, Georgian, Shingle, Tuscan, French Chateau, Victorian, Neoclassical (in its third incarnation) and Neo Mediterranean. This flexible approach, which is usually heavy on ornamentation, likes to take certain architectural elements that personify a few particular styles, mix them together and then wrap them around a specific kind of floor plan, usually focusing on the

Opposite: A flat roof is the least expensive style to construct, and that was one reason for the shape of this modern extension that wraps around the back of a century-old home. The other was the simplicity of not having to tie the old and new roofs together. **Left:** An old, unrefinished door and accompanying Victorian screen door are the major elements of this steeply roofed porch. Note the oval window over the door.

Americans commodify everything. Architects and designers get branded. This is helping European manufacturers, who are now seeing North America as an emerging market.

BRUCE KUWABARA, ARCHITECT

big eat-in kitchen opening on to the great room (often incorporating the square footage that would have traditionally been used by the formal living room), and luxurious spa-like master bedroom and bath suites.

The rising cost of both labor and materials is also sending residential designers off in different directions looking for ways to save money. One such avenue, still in its infancy but attracting a lot of attention at the high end of design, is the prefab home. Taking advantage of the economies of building a structure in a factory instead of on-site, with the economies of scale that modern manufacturing and engineering can produce, the prefab market is starting to generate a little bit of noise if you listen carefully. Although the market for prefabs is mainly in the "second home" category of cottages and getaways, and has an image problem (think prefab and you automatically think of slightly tacky and traditional looking cottages), new and very modernist styled prefabs are coming on the market that should make this approach a lot more stylish.

The whimsical, carefree look of designer and antique dealer Michael Trapp's rural Connecticut home combines "a little English, a little Italian and a little French." His backyard entranceway is a pastiche of European and North American antiques, from fluted ionic columns to New England cobblestones and a recycled grain grate as a doormat. The pediment over the door full of bric-a-brac adds to the feeling of artful fun.

This green roof over a guest wing, complete with trees and a garden, shows how much can be done with any flat space. From a practical point of view, the greenery also protects the roof from destructive ultraviolet rays while keeping the home cooler in the summer.

One final area of architectural design that is just beginning to come on strong and promises to grow steadily in the years to come is the "sustainable" or "green" residence. Even more important in Europe, where high energy costs have led to great interest, it is less an architectural style than an ecological lifestyle. The "green" house is usually super insulated and can be made from sustainable materials such as straw bales or rammed earth (moistened earth tamped down inside wooden forms). Green roofs, which consist of plants growing out of soil on a flat roof, are specifically designed to reduce air conditioning bills. More and more architects are designing homes to incorporate active and passive solar systems, which greatly affects roof lines and window placements. And greater attention is being paid to reducing the amount of potentially toxic off-gassing (the tendency of new synthetic materials to give off possibly dangerous gasses) of materials such as carpets, plastics and laminates, by replacing them with more natural and inert materials such as glass, stone, metal and solid woods.

And so it goes. As in any of the arts, architecture never stands still but evolves, advancing and retreating from one style and from one era to another in a constantly morphing creative march. It never stops, nor should it.

interiors

living spaces

If I go into a room thinking it's going to be one way and my client goes in thinking it's going to be another, and when we leave it ends up being totally different, then we have effectively advanced the design.

BARTON PHELPS, ARCHITECT

INSIDE THE SHELTER OF OUR HOMES ARE THE ROOMS WHERE WE live our lives. Sometimes they are designed specifically for one purpose only: places to eat, sleep or shower. But lately we have been combining rooms into larger, more multipurpose areas such as the kitchen, great room, which has gone a long way to reconfiguring the single-family home in North America. The old Victorian floor plan of numerous small, single-purpose rooms originated when there was no central heating. The idea was you would heat only the room you were actually in at the time, usually with a wood, gas or coal fireplace. But with the advent of central heating and air conditioning systems, we now have the luxury of a more open plan. We have also become much less formal in the way we live in our homes.

Dark mahogany columns and pilasters combine for a rich, men's club kind of library.

You can have an eclectic style but all the elements should be connected by some kind of thematic thread.

BOB GREY, ARIZONA ARCHITECT

There are many historic traditions in building design and in architectural adornment, based on principles of art and science, but these are inevitably subject to the whims of popular taste and imagination and dependent on the skills, technology and resources available in a certain time and place. In residential architecture, particularly, the individual imagination is given room to play. Still, within the walls of our own castles, most of us look to the past, and to the successes and examples of architects and designers whose work we admire, historic and current-day, for our own inspiration and interpretation.

We use architectural details in the rooms we live in to give it a certain look: dentil molding and Greek returns for a Classical sensibility; art glass and oak built-ins for the Prairie Style or Craftsman look; the hard straight lines of stainless steel counters, concrete and reveals for a more Contemporary style. But it is an unnecessary restriction to slavishly follow one single style. The reality is you can mix and match any number of styles as long as the overall effect works. Carved East Indian columns can go with Southwestern adobe walls if that's the particular look you are after. Juxtapose hand-hewn beams with carved French limestone if that's who you want to tell the world you are. Anyone with a Contemporary home can entertain their eclectic design instincts with a dizzying array of new products from rediscovered "green" wall coverings to fiber-optic fireplaces.

People want homes they can really live in twenty-four hours a day if necessary. The idea of "cocooning," from a term first coined in the 1990s, has never been more popular. We want comfortable homes that are flexible enough to allow us to throw a dinner part for twenty, but also earn the money for it without having to step outside the front door. We want to be able to do everything from home.

Above: Narrow double doors lead to the sudden expanse of a large open room — a favorite device of architect Frank Lloyd Wright. The living room/dining room of his Penfield House has twelve-foot ceilings and two-story-high windows. The walls and ceiling are brushed plaster with a sand substrate. Note the floating staircase and built-in wooden light fixtures. **Opposite:** Tented by Michael Trapp to look Moroccan, this casual living room is painted the same lapis blue that Yves Saint Laurent painted his garden walls in Marrakech. Napoleon was famous for tenting rooms, perhaps because they reminded him of his many campaigns.

One way to free up your approach to the rooms you live in is in how you go about dividing up space. A monster home with square footage to burn is fine if you can afford it, but increasingly attention is turning to those wanting a smaller home — one that is exceedingly well laid out and carefully designed to have every amenity and comfort, something that is referred to in the trade as a "jewel box." Consider the many possible uses for each room of your home and keep it adaptable.

An aging population is also something to be taken into account when designing a home. The most obvious way to accommodate a less mobile lifestyle is a main floor master suite that makes it easier to ignore the stairs. Hallways and doorways can also be made wider to accommodate wheelchairs

Opposite: The huge expanses and rough construction of a converted barn make it the perfect home for this artist/antique dealer couple. Everything, from the custom balcony railing and blackboard table to the coffee-tree-root chandelier, was designed and built by the owners. Light from skylights, big windows and numerous French doors illuminates the many nooks and crannies. **Left:** A 17th-century Portuguese rendition of *The Last Supper* provides the backdrop to this comfortably cluttered sitting area adjacent to the kitchen.

A ladder that moves around the room on tracks makes it easy to get at the top shelves anywhere in this clothes closet.

and walkers. Walk-in showers are also a good idea, because of their ease of entry (see page 80). Anything that makes life at home more convenient and pleasant should be considered.

There has always been evolution in home design. The combined kitchen, family room, great room concept popularized over the past couple of decades is actually not new but is, in a way, a return to simpler times, when families gathered in a central room or courtyard round the fire for light and warmth. Bathrooms have changed from utilitarian five-by-five-foot water closets to places of luxury and relaxation. Spacious bedroom suites are now the norm. Our closets are huge, compared to what our parents grew up with. But we have more clothes — and other things too. Homeowners are more and more demanding of what a home can provide for them.

If you are just a couple — either young or old — you might want to live in a big open space, which can be very beautiful and spacious. But if you are designing for a family, that might not be a very good way to go because of the noise factor and trouble keeping it clean. Similarly, the living room always served the ceremonial function — a clean, quiet room where you could bring your guests. But that is becoming vestigial. People around the desert are now more informal in their lifestyles.

JOHN DOUGLAS, ARIZONA ARCHITECT WHO DESIGNS FOR DESERT LIVING

Peeled round beams add to the medieval impression provided by this lofty ceiling. Note the metal screen over the high window and the lively play of light allowed by the corner windows.

Family rooms and kitchens have become the focus of every home, so much so that we have abandoned the formal living room for a more informal way of life. But we are at the same time making the family room itself a little more formal, while still remaining part of the kitchen. So we bring a seating or TV-watching area into the actual kitchen, where it has some "family room" aspects to it, but we can also walk away from the kitchen and all that mess and go to the formal family room.

CATHY OSEEKA, CHICAGO ARCHITECT

Kitchen designer Francine Theresa Rodriguez took a photograph of an old arched door and came up with a three-arched opening for this 200-year-old New England farmhouse. Although it is a mix of many styles, the pilasters on either side give it a neoclassical look. The cupboards are distressed mahogany and the island top is a honed rustic marble called "Old Church Yellow."

Despite the proliferation of seeming cookie-cutter housing in new subdivisions, high standards are being demanded of consumers in construction design and detailing. And professional architects and designers have been increasingly in demand in recent years. But today's architect- and designer-shaped home is not necessarily the home with the largest rooms, the grandest furnishings, the most opulent bathtub, or the most expensive, custom-made cabinetry. It may indeed enjoy an unmatched variety and quality of materials, but it is primarily a home in which a wealth of ideas and knowledge, and an ability to tap into great resources and hard-to-find materials, have been put to use to create a useful, comfortable, good house to live in, one that is sheltering, stimulating and pleasing to the occupants.

The headboard wall of this 200-year-old New England farmhouse was made from a sixteen-foot door found in England. Interior designer Catherine Cottrell worked with carpenter Wesley Gerwin to extend the molding of the door to either side. The chandelier and ceiling medallion adds to the grand effect.

People want homes with character. They want those old cracked leather chairs from the 1920s and 30s. You don't want the house to look threadbare, but aged and used — more patina than wear. Whether a house is five years old or 100 years old, you want to have those sign of life — the nicks and scratches and fading and ragged ends of use.

BRAIN GLUCKSTEIN, TORONTO DESIGNER WITH A FLAIR FOR AGED MATERIAL

Designer Brian Gluckstein always wanted a two-story library. By sacrificing an upstairs bedroom and putting in a library office upstairs, he was able to get one. He likes the play of different heights on the main floor because it adds architectural interest to the house. He also likes being able to look down into the lower library where he can display his large collection of books with a certain amount of drama.

the great room

As we have become less formal in the way we live in our homes, less formal room arrangements have become the norm. Many of us, given the choice, don't design a formal living room into our floor plans, or if we already have one, don't use it much. The space that is gained when a formal living room is dropped from the floor plan usually goes right into the connected complex combining kitchen, TV and music room, computer nook and fireplace/sitting area now known as the great room. With everyone in the family working long hours and often on different schedules (at jobs, at homework, at extracurricular activities), today's families need one big room that facilitates simply being together. If that means mom or dad being able to see the kids doing their homework at the counter while dinner is being made, or being able to talk to each other while the TV is on and the dishes are being done, the multipurpose great room serves nicely.

An ideal great room might have a cathedral ceiling with big south-facing windows — the better to let sunlight bathe the heat sink of the stone floor and fireplace. Built-in shelves and media consoles contribute the look of solidity this room demands, as well as help organize its many functions. But over-decorating with lots of molding and trimwork is generally not done, simply because the sheer informality of this room does not demand it. Although the great room is often the TV, games and music center, you can also send this function down to the basement to a "media room" that keeps the noise of the home theater and the Ping Pong table away from the other common areas.

Opposite: Massive exposed roof trusses complete with custom-forged steel hardware, and walls made of little more than a series of sliding glass doors give this long, high-ceilinged great room all the light and air one could wish for. The dry-stack stone fireplace acts as both a focal point and a room divider, and is a good companion material for all that natural fir. Below: The spacious living room/dining room combination of this hewn-log home glows with old wood and stone. A wood-burning fireplace with a simple wooden mantel is a must in such structures, as are historical footnotes — in this case, the fact that the first owner of the pine harvest table at right was reportedly murdered by his son as he ate.

A well-designed small house is like a boat — I learn so much about design by being on well-designed boats.

MARILYN LAKE, ARCHITECT

Opposite: This lofty living room keeps its human scale because the architects (Lipkin, Warner) used a technique perfected by California architect Bernard Maybeck, who brought big beams down to about six feet off the ground. The natural stone of the fireplace as well as the carpeting and the relatively low chandelier add to the warming effect. The fireplace was pulled away from the wall so that light from the high windows would hit it from the back and give it a glow. **Right:** The cathedral ceiling created when a log drive shed was converted to a residence makes for a huge open space that this antique dealer has filled with his best pieces. The wooden coat of arms over the fireplace is from a Quebec courtroom.

People don't live as formally as they used to. I don't know how many people have come to me over the past few years and said: "We would never use a formal living room, but I suppose we have to have one." Often it's because they don't want to hurt the resale value. But I tell them, nine out of ten of my clients tell me exactly the same thing. If you don't want one, spend the money on a really terrific great room that is perhaps a semi-formal room. You can have a TV in it but maybe put in a bar as well and couple of seating groups. This becomes the heart of your house but is near the kitchen, the covered loggia and other essential rooms.

RICHARD LANDRY, ARCHITECT

The slow transition from living room to kitchen in this Newport home is heightened by the continuous ceiling with its gently arched beams. The arch helps tie this large room together, as well as introduce a nautical theme (as in the hull of a ship). To the left is a dining area defined by a bay window overlooking the sea.

the living room

Although the living room is generally falling out of favor, it still has its place. A comfortable, well-appointed room that is sans TV and other distractions, where one can sit with friends and revive the almost lost art of conversation, has great appeal. This is especially true for people without children, or empty nesters who don't have to connect with the family through a great room. The living room can remain what it has always been — a quieter place where your best furniture and architectural design can be on display. We tend to lavish moldings, trimwork, columns and corbels (either real stone, ceramic tile or lightweight composite pieces) on a room that is formal by definition. Even Contemporary homes step up the design in the living room, going for the extra touch that big corner windows, arched ceilings or minimalist fireplaces provide.

Below: Curved concrete walls and alcoves that are dramatically lit with indirect lighting give this basement Art Deco room a decidedly masculine flavor. Note the nautical-style window at upper right. **Opposite:** Sedona architect Aldo Andreoli's barrel vault ceiling, with its metal fastenings and wood-glue laminated beams, creates a soft, informal yet very contemporary room.

Opposite: An 18th-century French mahogany secretary by an oversized oiled-steel door is a perfect example of the successful mixing of modern and antique styles. The mottled finish of the plaster walls is what happens when plaster dries naturally. **Above:** Etched mirrors, chunky trim and steel mullioned windows are the Art Deco touches of this living room. **Overleaf:** The rather formal Georgian architectural paneling in this living room is softened by the curve of the two arched alcoves, which are intended to showcase the two chairs designed by architect Charles Renee Macintosh.

The New York Apartment Houses of
Rosario Candela and James Carpenter
New York Apartment Houses of Rosario Candela and James Carpenter Acanthus Press

ANTIQUAIRES
THE GENIUS OF ROBERT ADAM
His Interiors
THE COMPLETE ETCHINGS OF REMBRANDT
ROOMS
DELANO & ALDRICH

the dining room

Not every new house built today has a separate dining room. Some people have knocked down walls to do away with what they think of as a stuffy, outdated place where guests can't relax and enjoy conversation with the cook. But there are others who prefer the calm oasis of a separate dining room away from the mess of the kitchen, where candlelight and conversation can take precedence. We may be giving up on having a formal living room, but the formal dining room is still considered an essential room by many of us. In fact, instead of disappearing into the great room, kitchen complex, it is becoming even more formal, with larger crown moldings and other classic architectural elements such as columns at the entrance, corbels, chair rails, carved panels and higher ceilings considered essential for this grand room.

Opposite: A definite party room, this dining room has fanciful elements — the lattice work, cove lighting and painted dragonflies — that bring the outside in. **Left:** Art glass corner posts create intimacy and interest in this Prairie Style dining room.

I think dining rooms are all about scale — I don't like being lost in big rooms, because dining is at least somewhat intimate. It's also important to pay attention to the flow, so people are not bumping into each other — going in and out of the kitchen, for example. A fireplace really warms up the dining room, as does being able to look into another room. I like being able to look into the living room and perhaps see another fireplace.

MICHAEL FULLER, COLORADO ARCHITECT

However, as houses become pressed for space because of both the move toward smaller homes and the fact that we use our homes more, the dining room is also being adapted as a multipurpose room. It can double as an office and meeting room, or a place for the kids to do their homework. That means built-in bookshelves — both for their utility and fine appearance. Lining the walls of your dining room with fine china and books can allow it to double as a quiet library away from the bustle of the great room, especially wonderful if it has a fireplace.

One small room that shouldn't be overlooked when making plans, especially by anyone who loves elegant dinner parties, is the "butler's pantry." This is a compact, efficient room with its own sink, countertop and storage areas that lies between the kitchen and the dining room, where the servers can unobtrusively organize the dishes and drinks.

Right: A grand arched doorway, elegant parquet floor, and a gilded light fixture hanging from a painted coffered ceiling create a neo Italian dining room in this Tuscan-style villa. **Opposite:** Glassed in to take advantage of the view, this Regency-style dining porch is both spare and delicate at the same time. Regency (a less formal outgrowth of Georgian) often added top or sidelights to windows to augment their size in an interesting way.

Opposite: Materials from all over the world went into designer Michael Trapp's eclectic dining room. The top of the dining room table is a sidewalk from a 17th-century church in Bath, England, and the base is French limestone designed by Trapp and cut in France. The floor is out of a 15th-century convent in the Pyrenees and the columns are Indiana limestone The big doors and windows — which make for an agreeable inside-outside feel — came from the Capitol building of Rhode Island (designed by McKim), where they were removed and replaced with thermopane. Above: Many a star has imbibed in this dining room in Gloria Swanson's Hollywood home. The ceiling stenciling was done by the present owner, L.A./London designer Martyn Lawrence-Bullard.

the kitchen

The very best kitchen designers always begin by finding out who you are and how you use your kitchen. Even the smallest, most frugally furnished kitchen can be a remarkable place if it is properly organized and designed to suit the people who use it. But while utility is a primary goal in kitchen design, beauty need not be neglected. And even though the kitchen in its most limited incarnation serves a very specific purpose — the preparation and consumption of meals, and the storage of the items needed for those activities — with imagination, a kitchen can be so much more than that. As the kitchens featured in this book demonstrate, the styles and permutations are endless. And because there are so many options available, the ideas you see here can be interpreted and adapted for your own kitchen in your own way and suiting your own budget. But there are some beautiful, one-of-a-kind options also worth looking into.

The materials you choose in finishing your kitchen go a long way to signal its style. The cabinetry especially can set the tone of the room. Some designers are favoring cabinet styles reminiscent of those employed by mid-century modernists such as Cliff May, with sleek clean lines, interesting wood, and no fussy raised woodwork or hardware. Others are harking back to the kitchens of our grandmothers, with stand-alone furniture and open shelving instead of fitted cabinets. See the Kitchen Fittings chapter in the Details part of this book for more inspiration and plenty of product choices.

Below: The owners of this home entertain often, and the sliding screens are designed to seal off the hubbub of the kitchen from the dining room (the food is delivered along the hallway to the left). The shape and feel of both rooms can be dramatically altered by leaving the screens in different positions. The island counter is a slab of granite; the counter is a single piece of stainless steel made in Italy; the cupboards are wenge wood stained dark. Mahogany has been installed everywhere there is wear. Opposite: The faux ruin of this unique kitchen was built to highlight the stone window — a French antique. The butcher's block is also a French antique with over eighty years of use accounting for its smooth indentations. The hand-hewn beams are salvaged square beams that had the hewing marks added on later by hand.

Above: When architect Andrew Kirkowski's took over abstract expressionist Barnett Newman's studio, he wanted a stylish residence that also kept some of the rawness of the Tribeca neighborhood — not an upper East Side apartment only with higher ceilings. He wanted modern but not cold. So he left the large windows unadorned and honest and painted one wall a warm red that can be seen from the street. He accentuated the very clean, Italian style kitchen with walnut cabinetry (from Polyform-Verena) and Louis Polson lights. The huge island also serves as a buffet for parties and is topped with a black granite called jepness. **Opposite:** Rough stone floors and beams, plain cabinetry and a big stone sink contribute to the rustic look of this Connecticut farmhouse. Lots of windows and a high ceiling keep it light and airy.

There are the rooms we live in and want to be comfortable in, and then there are rooms that get a lot of use. The most obvious of these purpose-built rooms is the kitchen. Highly functional, yet increasingly seen as a place to enjoy, relax and even entertain in, the kitchen is also where most of the money is spent. This is where we find a predominant amount of natural (and often more costly) materials, such as stone, wood, glass and metals. Along with inherent beauty, these materials have the virtue of strength and toughness in common. Even glass tile and ceramic tile, which is essentially burnt clay (the literal translation of terra cotta), can be considered natural materials and share the attribute of resistance to wear. Because there is such an amazing variety of tile available — ceramic and otherwise — and because it is so well suited to the rigors of this essential room, we have paid special attention to it in the Floors, Walls & Ceilings chapter in the Details part of this book. There are many materials used in common in kitchens and baths, so when looking through the chapters of this book, be sure to keep an open mind and know that what you might find in one Product section may work just as well in another room.

Kitchens really have become the command center of the house; a place where guests and family congregate. The trouble is that if you are really going to cook, kitchens get messy. When you sit down for dinner you don't want this disaster going on behind you. One solution is to screen the kitchen off from the dining area. Another is to have two kitchens: one a hospitality kitchen and another that is essentially a catering kitchen where all the real work is done and that is tucked away out of sight.

DAVID WARNER, COLORADO ARCHITECT

This well-appointed kitchen is kept from becoming too formal by the beadboard cupboards and trim, which give it a California beach house, country cottage look. Six-foot-wide pocket doors (at the left and out of sight on the right) seal off the living and dining rooms if desired, and make for good traffic flow when open. The oval shape of the room is echoed in the oval island (curly maple top), curved granite countertops (honed to make it look "like rock from the garden") and by curving the molding that runs along the top of the cupboards. The beadboard is made of MDF (Medium Density Fiberboard), a very dense particleboard that doesn't warp and takes paint well.

the bathroom

Bathroom design and furnishings have come a long way since the days of the chamberpot. Today the multitude of choices in taps alone can make a would-be renovator dizzy. Fifty years ago, the average North American homeowner looking to build a bathroom might have two or three tubs to choose from, one or two toilets, and a half dozen sinks. There were only a few manufacturers of bathroom fixtures. It isn't that simple any more. Since the days of ancient Rome, there have always been luxurious baths, but for most of that time, they have been only for those who could afford them. That luxury has today become a middle-class expectation.

Bigger bathrooms include separate showers, his and hers sinks, and the toilet in a separate compartment. Soaker tubs, whirlpools, rain shower rooms....everything you might find in the most luxurious spa. All are created with great attention to detail, materials and finishes — marble being a favorite. Stylistic inspiration comes from all corners of the world, from Italy, Sweden, Japan and beyond. For a wealth of ideas and product choices, turn to the Bathroom Fixtures chapter in the Details part of this book.

Opposite: Representing the values of the client, this bathroom is an elegant but simple, bare-bones bathroom containing Carrera marble, stainless-steel cupboards with stainless-steel mesh, and a long mirror-lined entrance hall. **Left:** A deep and narrow Napoleonic zinc bathtub is the perfect complement to the planned chaos of this draped bathroom. The owner loves the tub because zinc heats up instantly, unlike the old cast-iron tub it replaced, which took forever.

The bathroom has now come into its own as a room, and you enjoy it as you would any room, like the living room or dining room. Today it is much more like a spa, a place where you spend time relaxing. Using materials like stone and glass and falling water adds to that sense of serenity.

LYNN APPLEBY, A DESIGNER WITH A CREATIVE FLAIR IN THE USE OF WATER

This spa-like bathroom is dominated by four pavilions containing the shower, toilet, storage and a change room, based on the look of summer bath pavilions at the beach.

Woven tiles on the floor and knee wall, along with metal vessel sink and clear glass countertop, make for an ancient-looking powder room.

the bedroom

With the great room, kitchen complex bringing the whole family together, harried parents need some space for themselves. At the same time, busy adults with or without families crave a soothing place to decompress from the pressures of long hours of business stress. The solution is the bedroom — or, to be more precise, a multipurpose bedroom suite that has more in it than a bed and a dresser.

The master suite can be designed to contain a fireplace with a small seating area, and perhaps even a media center, wet bar and small office. A small workout area can be set up if you have the room, making this space something more than just a place to hang your clothes and sleep. This approach makes the master suite as self contained and independent of the rest of the house as possible. But keep in mind your own lifestyle when thinking multipurpose: sleep experts suggest that those who have difficulty sleeping keep the bed part of the bedroom reserved for sleeping and adult fun.

Echoing the open plan approach of the great room, the master suite can open up the dressing room, clothes closet area to create a transition space between the bedroom area and the bathroom spa. This transition can be made extraordinary with a two-sided fireplace that warms up the bath and bedrooms simultaneously. Of course, this means more attention has to be paid to dressing room décor, with finely crafted cabinetry and shoe, sweater and T-shirt shelves perhaps made of exotic woods. And why not an island in the middle of the dressing room, with drawers for smaller accessories and a low top for folding or packing?

Below: "Think penitentiary!" was the advice architect Johnson Chou's client gave him regarding his new loft bedroom. He interpreted that to mean he wanted a space in its elemental form; one that was all about the materials in their most natural state without any overt design flourishes. The ceiling is drywall; the floor is slate. The recess behind the bed is a light cove as well as an expanded shelf for books and objects. **Opposite:** Architectural firm The Ideal Environment used oversized roof trusses and exposed V-groove ceiling boards to make this small bedroom look like a cottage.

We find that when people become empty nesters they don't need private space as much in the bedroom.

JOHN DOUGLAS, ARCHITECT

Opposite: An elaborate mahogany headboard acts as a screen to the bathroom and also softens the entranceway — all integrated with lamps, end tables and lighting. This is a very Asian technique — to use something like a shoji screen to separate but not close off — and is a reflection of the Asian background of one of the owners. The walls are smooth plaster and the natural palette sets off the woodwork nicely. **Right:** Big hewn beams and rough floors and ceiling frame this bed made of antique moldings. **Overleaf:** The mezzanine of this artist's studio is kept light and airy with the use of ash floors and perforated steel railing.

ANCIENT ROME
ROME
VENICE
THE ART OF FLORENCE
THE ART OF FLORENCE

GREAT DRAWINGS OF ALL TIME
III
GREAT DRAWINGS OF ALL TIME
IV
HENRI MATISSE

MANAGEMENT
THE AGE OF THE RENAISSANCE
TOM WOLFE
THE HOUSE BOOK
CENTERBOOK
THE VILLA
THE ART PACK
1000 chairs
GANDHI

the home office library

Home offices — sometimes his and hers versions — have become essential new rooms we live in. Often another room has to do double duty and thereby is created a dining room/office, for example, or a library/office. For the same reasons our basements are increasingly filling up with desks and computers, not to mention home gyms and saunas.

Built-in shelves and alcoves are a good way to add charm to a room while helping to stylishly adapt it to different uses. For one thing, they help delineate space in an efficient and beautiful manner — what better way to provide a computer station in the kitchen for the kids, or shelves for the combination dining room, home office; or organize and beautify an expanded dressing room, bathroom? For most people, their home will be their biggest lifetime financial investment; built-ins can only increase its value. And if you are going to be spending more time at home, why not have it look as good and function as well as possible?

If you have the space, and you are going to have one room dedicated as a library, it really should receive special attention. Commonly found on the first floor, but working well anywhere in the house, the library is a perfect excuse to lavish expense and design on the finishing details: the molding, carved wood and finished cabinetry that is used to house books and decorative objects. The sort of cabinetry traditional to a dining room is also suited to a library. If you want to justify the luxury, consider this: your library can be a getaway or informal office, and can also be a good home for a decent stereo so you can enjoy it as a music room as well.

The desktop in this cozy home office designed by architect Buzz Yudell is the same height as the window mullion and one of the shelves in the maple bookcases. Other mullions line up with their corresponding shelves as well for an overall feeling of serenity and simplicity. Although this is not a big space, having the bookshelves go up to the ceiling gives a grand effect to a small room.

Limited space in a ski chalet foyer and an avalanche of boots and gloves combined for a clutter problem. This well-crafted pigeon-hole cupboard provided the solution.

utility rooms

Even the laundry room has seen a little added attention recently. For more convenient placing it may have been moved from the basement up to the first floor, where it shares space in an enlarged mudroom off the kitchen that also doubles as a sewing/quilting/gift wrapping room as well as a prime storage area. Or it moves up even higher in the house, to the second floor, adjacent to the bedrooms where the clothes are stored. The reason for this kind of reallocation is one of convenience — why make extra trips to the basement? But of course the move from the lower realms of the house up into the public areas also means that the dowdy old laundry room has to be spruced up with the features of other "living rooms," such as hardwood floors, ceramics, custom cabinetry and even high-end materials such as granite and stainless steel. Making the floor choices and other finishes of the laundry room or mudroom similar to those in the rest of the house can help keep them from seeming the poor cousins of the other rooms.

hallways

Hallways can be more than a way to get from A to B. They are an opportunity to make a bold design statement, and to connect different parts of the house together thematically. Often the hallway is seen merely as a throughway, as a dumping ground for coats, boots and mail. But hallways and corridors really are a transition from one room to another, often from the public to the private, and can and should contain a certain drama and ambiance all on their own.

The two most common complaints about hallways are that they can be dark and narrow. Good lighting is therefore important. If you can afford the space, a large, inviting hallway that graciously marks the transition from outside to in or from one part of the house to another is always a good idea. However, since hallways are almost always relatively narrow, tables and shelves are best built in and attached to the walls in order to not block traffic flow. Pictures, small plates,

A curved wall covered in a finish called Venetian plaster (a kind of stucco with a mottled effect that looks very deep), with teak wood at the ends. The purpose of this curved entrance piece was to connect the entrance of the house with the spectacular view of Los Angeles at the back of the house. The sculptural shelves in the wall scale it down to a human level, but also provide space for the owner's eclectic art collection. The architect was Aleks Istanbullu.

mirrors or other flat decorative objects are also a good way to brighten up a hall, but the narrower the space the more attention should be paid to keeping pictures above shoulder height to keep them from being knocked off. Having a ceiling that is a lighter shade of the wall color is also an easy way to make the hallway more interesting.

If you're lucky enough to have a hallway that is large, try thinking of it as a room in itself that can be furnished with a couch (a small one) or a writing desk to make the space as useful as possible without interfering with the all-important traffic flow. If you want to make the hallway its own room, the flooring and even the ceiling treatment can differ from those in the rooms the hallway connects. Using the same flooring and or ceiling material throughout, on the other hand, has the tendency to link the rooms together thematically, making it a transition space.

Having one wall of a hallway that is not really a wall — such as one that flanks a two-story high atrium and is delineated with a railing or a colonnade — is the most inviting and spacious approach. The same applies to a wall of windows on one side that looks out on the outside or even into the interior of the house.

Opposite: A floating catwalk over the main hall provides a luminous effect because it doesn't touch the wall on either side. The grid of tiny lights outlines an oiled steel wall while the wire railings provide a high-tech nautical theme. **Left:** This long hallway — called a gallery — is a series of arched doorways that open up onto each other and is typical of Southwestern adobe homes. The terra cotta tiles give a warm look that is cool to the touch.

staircases

Like hallways, staircases are much more than a way to get from one place to another. They are rife with design possibilities and lighting opportunities. Their zigzag nature gives them graphic integrity, and landings offer the perfect place to showcase a carefully chosen vase or chair. Bookcases can be built in underneath them. The architectural details of a carved newel post or baluster become an obvious way to add to the look of a room. Iron railings can add another dimension and element to a space dominated by wood or stone. And no feature of a home is better at making a grand statement as you enter.

But perhaps one of the best things about staircases is that, depending on what level or angle you are looking at them from, they can present several different views.

Right: Called "The Two Bridges," this staircase designed by Paul Froncek was influenced by Japanese bridges. **Opposite:** This graceful, hand-forged wrought-iron railing by the Design-Build Store curves down marble steps to an intricate marble floor.

Most people who design houses don't consider the cadence of the stairs. I hate really steep stairs, but a ratio of six inches and twelve inches can be very gentle and graceful, very easy to walk up and down. I think a lot of attention should go to the stairs — to the rise and the run, to the cadence. What happens when you arrive at the top or the bottom — what do you see — a beautiful ravine out the window or a blank wall?

BRUCE KUWABARA, ARCHITECT AND DESIGNER OF THE "RAVINE HOUSE" ON PAGE 9

Opposite: "Everything in this house, I dragged out of the creek." That's how Sonoma, Arizona artist Doug Andrews came up with this make-it-up-as-you-go staircase and balustrade. The ceiling height was determined by the height of the tree trunk he used for the main roof support. This kind of approach shows what can be done when you discard orthodox thinking and look at materials with a new eye. **Right:** Staircase master Rick Schneider designed this oak and maple staircase so that the steps are wider as you go up (from three feet to six and a half feet), making the sound your feet make deeper as you ascend, like a giant xylophone.

lofts

There is nothing that says urban so well as a loft. Open planned, with big windows offering panoramic views of the city skyline, they are as far from the bourgeois comforts of the suburbs as one can get. They are, in a word, so very "downtown."

The word "loft" suggests elevation, thus, living in the space under the eaves of a tall building. The popularization of the loft as living space grew in 19th-century Paris. The city, at the time full of artists working on the large canvases of the day who needed a cheap, open space with good light, soon saw its attic spaces filling up with the scent of oil paint and turpentine.

North American artists picked up the idea of lofts in a big way right after the Second World War in the garment district of New York's SoHo. When entire industrial buildings started to become available in other areas of New York during the 1960s and 1970s, the idea spread beyond the uppermost floors — and continues to spread — not only in that city but any big city that had a strong

The architects (LOT-EK, New York) turned the cylinder of an oil truck into sleeping pods in this high-ceilinged loft. Not wanting the usual drywall, they gave the pods multiple coats of yellow automotive paint on the inside, but signs of use were left on the outside for an intimate space that complements the industrial feel of the true loft. The doors open hydraulically and the floors are poured epoxy to make it seamless and reflective for a light, airy feel.

Nobody wants to live in a loft like they did in 1971, with a little curtain separating the bathroom from the rest of the space. Nobody wants to always have to look at your stuff all the time. You have to be able to put things away. We try to respond to these very basic needs and yet still let the space breathe as a loft. We try to approach the design with the same honesty that the building has.

GIUSEPPE LIGNANO, AN ARCHITECT WITH THE NEW YORK FIRM OF LOT-EK, WHO SPECIALIZE IN LOFTS

This unusual New York loft includes the basement in a two-story interior courtyard. "We wanted to give the main area a feeling of suspension and displacement," explains architect Sandro Marpillero of Marpillero Pollak Architects. They also wanted recycled material taken from the removed beams to continually show up, as can be seen in the stair treads at left and support beam at the top. The hanging floor is standard angle iron floored with galvanized metal mesh screening hung from a recycled floor joist. The railing of the second floor is constructed of angle-iron frames with both frosted and tempered glass. The floor is maple.

artistic community and large stock of warehouses, especially port cities such as Boston, Toronto, Chicago, Vancouver, Miami, Seattle and San Francisco, to name only a few. Most of these early lofts were in areas zoned commercial and so were technically illegal, something that probably added a little thrill to their déclassé charm. Artists also loved them because their huge windows, high ceilings, exposed raw walls and generally forgiving interiors were ideal as a place to indulge in the sometimes messy occupation of creating art. The original loft was working class and unpretentious, and purposely didn't even try to imitate a residential building. They were idealized as honest, straightforward and heavy duty. And there was (and still is) an expansive lifestyle that these old industrial buildings offered — especially for people who live and work in the same space. The nightlife and fast pace that can make city living so exciting are easily accessed by a downtown loft. All of this, along with the low rent, was more than enough to compensate for the noise and lack of services that are the reality of living in a space not originally designed as a place for people to live.

The urban renewal that took place in most North American cities during the 1980s, along with a freeing up of zoning restrictions, fueled a second round of loft conversions that has only accelerated to this day. (In fact, loft conversions often lead the way in urban renewal.) We have also become more sophisticated about the patina of use in all things residential. For many people, there is nothing new that can match the stories told by a cracked and stained concrete floor, or the romance of years' worth of chips and dings in walls and doors. Rough elm floors, original iron window frames and radiators, concrete or timber pillars, brick walls, even wheezing old cage elevators (properly restored, of course), are what bring a twinkle to a buyer's eye and an extra zero to the price when it comes to lofts.

Nowadays, the look of commercial and industrial conversions is being duplicated in new developments that are essentially loft buildings built from scratch. These oxymoronic new lofts are being referred to as "soft lofts" (or MacLofts) as opposed to the "hard loft" — an authentic industrial building or warehouse. Attempts to approximate the feeling of an industrial building are made by adding materials such as stone, concrete and wrought iron and even in little touches like caged lights and roll-up garage doors. Although lacking the history, use and the downright funkiness of a real loft inside an authentic industrial or commercial building, soft loft buildings are able to avoid some of the messier problems — such as excess noise, weird shapes difficult to adapt and furnish, and lack of services — by dealing with them at the planning stage. In fact, these new buildings can be so perfect that it's hard to tell them from an apartment building with high ceilings, big windows and a couple of columns. You know lofts have arrived as a serious architectural lifestyle when real estate ads start pushing "loft style" or "mezzanine suites" or even the ultimate contradiction we have seen — a "loft-inspired detached single-family home."

Like all residential structures, lofts have strong points and drawbacks. On the plus side, lofts are wide open in terms of how you can furnish them. As essentially big raw spaces, lofts are a blank canvas that can support virtually any

The biggest challenge for most loft buildings is that they were never set up to be lived in. For example, they are much deeper than a standard residential building. The issue of light, therefore, is critical. You have to make sure you can bring light deep into the space. You have to really know how to use translucency, to open up walls.... Otherwise you create deep, dark recesses that are very undesirable.

CHARLES WOLF, A NEW YORK ARCHITECT WHO LIVES AND WORKS A LOT IN TRIBECA

Opposite: Black, hot-rolled steel forms the motif for this loft. The kitchen island and the inside of the doors covering the kitchen wall are made of steel, cleaned with steel wool and oil and given a wax finish. The outside of the doors are made of a cement board material called Eternit. The top of the island is a honed stone called Basaltina and the floor is antique chestnut. The ceiling beams are original, but the soffit was created as a place to put the recessed lighting and air conditioning.

An eight-foot-by-thirty-foot, barrel-shaped Plexiglas skylight fills this artist's home/studio with light; that and the Etruscan fountain give it a sense of being outdoors. A second-floor mezzanine not only provides areas for bed and bathroom, but prevents the space from being boring. Architect Gordon Ridgely designed the space with a sensitivity to how things join and meet each other. For that reason, there were no face welds showing on the steel beams (they look like earthworm droppings, he says). Instead the welds were made inside holes drilled in the steel where they meet the beam.

style, and the bigger the better. Their very elementariness provides the perfect setting for contemporary, even minimalist sensibilities. Their hip yet rough-and-ready qualities make them a great place to live if you work at home, which is why a lot of people involved in small businesses such as communications, publishing, software, photography and architecture are set up in lofts. And they are without a doubt a terrific place to throw a party. Another angle that is only now becoming apparent is that the flat roofs of industrial buildings are perfect for urban roof gardens or green roofs (see the Exterior Details chapter), and preferably both. This is a good thing, especially if you have children, love barbeques or have a love of any kind of nature, as lofts by definition don't have backyards.

The downside to lofts — the real or "hard" variety — is that they weren't originally designed as a place for people to live. Necessary service elements like pipes were not hidden. Storage was not considered — who needs a closet in a warehouse? Noise abatement wasn't given a thought. Privacy was not an issue. Because of their very size and openness, lofts require a non-traditional design sensibility. The trick to designing a loft is to include these necessary comforts and needs while still staying true to the idea that enamored you to lofts in the first place. That means not cutting up the space too much, or lowering the ceilings to make it cozier. And to deaden the noise that goes along with concrete floors and open spaces, soft, sound-absorbing expanses of acoustic materials — such as furniture, screens, rugs or hangings — are usually needed. But what looked good in your old bungalow won't necessarily work in your big new loft.

Then there is the light problem. Any residence has to have some rooms in order to provide privacy. But lofts are often lit by windows along only one or two walls, resulting in dark, forbidding interior rooms. To alleviate this, architects and designers have had to constantly come up with creative ways to let in the light while still delineating interior space. Translucent materials, often on sliding screens, movable drapes instead of walls, clear glass walls and floors, opaque glass, and low dividers are just some of the ways to make your loft still a loft, but one that feels like home.

We are in a period where we are personalizing our residences more than ever, so I don't think there is a strong stipulation of what a loft must be anymore. People are putting together two and three lofts to form one big one, and that kind of luxury and openness is nice. Lofts are definitely tending toward the eclectic and the luxurious with incredible materials being put in such as glass fireplaces and surfaces like back-painted glass and wall coverings that are more like installation art. It's like living in a modern painting.

ANDREW KIRKOSKI

Getting light into the interior of lofts is always a design challenge. These translucent walls are made of steel and a composite material sandwiching a fabric inner layer (the brand name is LUMAsite).

I don't believe in trends. Trends come and go. I tell people to search their own souls, and not do what other people do. When I did my own house I looked at my childhood, my background. Consequently this house is a part of me, so when I come home at night I feel comfortable. Relaxed. I see my culture through the house. It has nothing to do with what the neighbors think.

RICHARD LANDRY, ARCHITECT, WHOSE OWN HOME IS A REBUILT BARN

Simple drywall, big hewn beams and the honey brown color of old worn pine floor boards are all that are needed to show off the artful mélange of books, antiques and artwork in this converted barn. The sculpture in the middle is made from hundreds of burrs.

the reconstructed barn

One good way to capture the idea of a loft, but without the hard edges of an industrial building, is to reconstruct a barn. By that we mean taking down the hand-hewn posts and beams of a barn and reconstructing it on your own new site, usually held together with wooden dowels and in some kind of new cladding. These days the only trick to this kind of construction (in addition to the obvious zoning restrictions in many jurisdictions, which is why a lot of reconstructed barns are in the country) is to find an old barn that is still in good enough shape to use. Usually if the roof is still on, it will be.

A barn is perfect for the kind of person who wants a big, rough space they can do anything they want to. Artists and art dealers, antique dealers and collectors, people who like lots of raw volume or historic atmosphere adore barns. That's because a barn has no history of residential use — so there are no rules. You start with this incredibly strong frame that defines a large but fairly simple space usually based on the rectangle. You're not limited by any tradition or period because barn construction is not of a period in and of itself. About the only restrictions you have are the beams — sometimes you have to work around them to do things like put in windows. See the Floors, Walls & Ceilings chapter in the Details section for ideas on ways to finish the flat surfaces of your reclaimed space.

In a reconstructed barn, you can find a place to hang almost anything. The gleaming pots and pans and stainless-steel hood brighten a space mostly furnished with and finished in wood.

details

exterior details

This is very Japanese thing, but I believe that the design of a house shouldn't shout at you. It shouldn't say it all in one big statement and then leave you out of it. Instead, a house should draw you in to a contractual relationship between architecture and landscape. Because architecture without human experience is nothing — it's too abstract without the patina of living.

BRUCE KUWABARA, ARCHITECT

Emperors and kings of old erected grand palaces and monuments to commemorate their own glories and accomplishments for posterity. Throughout time, wealthy families have bankrupted themselves building show-houses meant to impress the rest of society with their elevated status. The exteriors of our homes present probably the biggest physical statement about who we are that we will ever make. But how we choose to express that statement is changing. The generally conformist attitudes of the last half of the 20th century have morphed into much more individualistic approaches to personal architecture. We associate our own values with architectural styles, borrowing freely from an eclectic variety of historic styles to define our own. Yet we don't so much want to show how we fit in but how we are different.

This extraordinary entranceway reflects the Chinese and American influence of the home's owners. Architects Lipton, Fuller wanted to make a very defined demarcation between the carport and the main level of the house. Instead of having the stairs run directly from the driveway right into the house, they opted to let the energy flow around the gong and the wall, to make it quieter and more separate. The stone on the exterior comes from a local Telluride, Colorado, quarry.

And generally we are much freer, here and now, in how we choose to do that — in the materials we use, the styles we choose and how we mix and match both to dramatic effect.

A noteworthy recent improvement in North American residential design is that we are becoming much more aware of and sensitive to the surroundings into which a new home is going. That means fewer Spanish Mission homes being built in the Rocky Mountains, or Tudor homes in the subtropics, or huge monster homes in settled areas of smaller residences. One reason for this is the increasing sophistication and architectural awareness in North America. Another is a renewed appreciation for the idea of "neighborhood" and our responsibilities to our local community.

Another unmistakable sign that we are evolving in our tastes is our attitude to aged materials. Tarnished metals, painted wood, hewn timbers, rusted steel and tin, wavy glass windows, even found objects such as driftwood and river stones are cherished and used in innovative ways. We are learning from the Europeans that age and use are valuable things. Collectors have always known this, which is why they will come close to tears when they see an antique that has been stripped down to the wood and repainted or refinished. The fact is, old paint and varnish, verdigris and signs of use make a piece worth more than new, even if the finish is coming off in curls. Old rusty steel is more valuable than new steel siding or roofing for the same reason that a thirty-year-old single malt is valued more than a five-year-old blend: because time is the ultimate in priceless things.

Southern California is a dry area and architect Richard Landry wanted this Contemporary home (influenced by the Mexican architect Luis Barragato) to celebrate water. The concrete aqueduct actually divides the house into four quadrants — family, casual, kids and service areas. It spills into a fountain at the front and into a pool at the back — symbolizing the return of water to the earth. When it rains the flat roof collects water, which is sent spilling into the pool and the garden, making the water wall part of the pool filtration system.

The term "Dream House" is accurate because a lot of people are building fantasies for themselves. It's that situation where someone's been on holiday and they love that adobe home they saw or Mediterranean villa, and they get back to the Midwest and want it. Most architects hate this, because there is a certain truth and appropriateness as to why a particular architectural form is in a certain area (or isn't), and when you start pulling it out of its climatic and cultural context just because we can, you can run into problems.

PAUL GORDON, ARCHITECT

Combined themes of glass, steel and water result in a very contemporary industrial feel for this urban reno. Essentially a masonry structure with the first two stories coated in stucco, the building had a pavilion added to a roof that is wrapped in galvanized aluminum siding. An internal stairwell was also added by punching through the masonry wall that runs up the back of the house, creating a three-story curtain wall structure that is galvanized aluminum and glass on the exterior into which a floating steel staircase was inserted. The railings around the balcony are perforated metal set in a metal frame, which creates an impression of lightness — the corners were left open to highlight that effect. The drainpipes are aluminum.

exterior cladding

There are many options to choose from in cladding a home.

STONE Big chunks of stone set in mortar have served as a sign of luxury and solidity for centuries. Dressed limestone, sandstone (and the lesser-known brownstone), fieldstone (a type of sandstone), and even rubblestone all provide a look of timeless beauty and durability when used on the exterior of a home.

But for all of its many attributes, stone is heavy, expensive and hard for anyone but an expert to work with. Pollution can darken stone, and acids and salt can eat away at it over time. (See Kitchen Fittings and Floors, Walls & Ceilings for more on stone as used inside the home.) There are some alternatives to solid stone worth considering.

One relatively inexpensive alternative to solid stone that still maintains that all-important curb appeal is stone veneer. Veneer consists of real stone blocks ranging from ⅜ of an inch to over one inch in thickness, attached to masonry or stud walls. Obviously without structural properties, veneer stone provides the look and feel of real stone without the weight that calls for extensive footings. Veneer products can be used inside or outside the house and are often employed as an accent to more prosaic materials such as wood, concrete or brick.

For a fraction of the cost as well as a fraction of the weight, manufactured stone is showing up on more residential exteriors. Most manufactured stone is made from a combination of cement, stone aggregates and pigment, molded into

Limestone quoins give a Georgian solidity to this stone house and beautifully finish off the corners. The story behind the quoins is that the original owner would take food to the local prison and in return, the prisoners carved the limestone corners.

Left: Santa Barbara rubblestone walls contrast nicely with precast concrete arches. **Center:** The horizontal lines and earth tone of this Arizona flagstone wall are reminiscent of Frank Lloyd Wright designs. **Right:** This heavy limestone entrance has the coat of arms of the family that built it carved over the door. Note the how the gravel drive is separated from the stone walkway of the entrance by a narrow brick border.

a variety of shapes so an obvious pattern doesn't constantly repeat in your wall. They are lightweight, durable and cheaper than real stone, come in a growing variety of options and, like veneer, are not load bearing. Their only drawback is a big one — they aren't real.

You can also buy lightweight polyurethane panels that are molded to look like brick or stone. Molded panels are even cheaper and more lightweight than manufactured stone but look even less like the real thing.

WOOD Wood used to be the most common exterior cladding material in North America, mainly because there was so much of it around. But because of our modern desire for maintenance-free claddings, wood has slowly slipped in popularity. Theoretically, wood must be covered with something to keep the weather off of it — usually paint or varnish or a penetrating stain. These protective coats must be reapplied every few years, and that's where the maintenance comes in. Penetrating stains last longer because they seep deeper into the wood, but you get more choice of color with paint. And stain, because it penetrates rather than covers, does more to highlight and bring out the look and texture of wood grain. But not all wood has to be protected. Highly resinous woods such as cedar and redwood have natural protection from the elements, and many woods weather into a beautiful color and texture with age — think of all the shingle-sided homes facing the salt air of the ocean that have stood up to the elements for decades.

Another wood material designed to stand up to the weather is plywood. Made of layers of wood glued diagonally to increase its strength, plywood is normally used as an invisible sub-skin or subfloor, but weather-resistant grades of marine plywood can be used as a relatively cheap, attractive and easy-to-apply exterior sheathing.

Left: Weathered wood and a local stone on this home mimic the "boomtown" architecture of the North American frontier, in which buildings were thrown up quickly using whatever local materials were plentiful. Note the corrugated metal roof at upper right. Center: Eastern white cedar shingles, stained white and cut square on the bottom half of the wall and scalloped on the top, give this California Beach style house a Mediterranean flair. Right: The white cedar shingles of this hybrid Cape Cod have been allowed to turn gray in the traditional manner. When red cedar is allowed to weather it tends to turn an orangey color.

The most common types of wood siding are bevel siding, boards sawn with one edge thicker than the other; drop siding, patterned upper edge and grooved or lapped lower edge; and vertical siding, square-edged boards with battens over the joints or tongue-and-groove boards. At the extreme low end are siding products that have a base of Oriented Strand Board (OSB) with an exposed-face resin coating that gives color and simulated wood grain texture to the panels.

Whatever siding you use, be aware of the possibilities that trim affords to the exterior of your home. Contrasting colors between trim and siding can make it stand out, which is particularly useful if you have one of the more elaborate period-style homes such as a Victorian or Queen Anne. A look at some of the "Painted Ladies," best exemplified in certain older homes in San Francisco and Oak Park in Chicago, can show just how far you can go with imaginative trimwork and paint schemes. In other styles, the trim is more subordinate. Log homes, for example, use trim in very plain, honest ways, usually consisting of just fascia boards around the eaves and plain wood trim around doors and windows. This is taken to the extreme with Contemporary homes, where the trim is often clean, unadorned bands of metal or painted wood running along the roof lines.

BUILDING WITH LOGS One of the best ways to use wood as an exterior cladding is to build your home out of logs. Solid, rustic and redolent with history, the log house has become the symbol of the North American pioneering spirit. There are a number of ways to build with this renewable and impressive resource.

Hand hewn: Take the logs (usually a softwood such as pine or cedar) and hew them flat with an adze (the modern method is to rip them square at a mill and put the hewing marks on later), then join them at the corners with dovetail joints. The spaces left in between the logs are filled ("chinked") with mortar or the latest in acrylic materials. Neat, trim and easier to deal with on the inside because of the flat walls, this style has all the charm of weathered cabinetry.

Long log or Scandinavian scribe: This method leaves the logs round. Scribes (they are like the compasses you used in geometry class, only oversized) are used to copy the contours of the bottom of one log onto the top of another. This undulating line is then cut with a chainsaw so that the logs sit tight on top of each other with no need for chinking. At the corners where the logs meet, half-moon notches are cut into the top of each log so they lock in place, leaving a few feet of log end sticking out all the way up — like toy Lincoln Logs. This style is massive, with a look so solid is seems nothing could knock it down.

Stackwall: Short, one-to-two-foot-long logs are cemented into the wall for a charming look that is all circular log ends floating in white mortar. Most often used for animal barns, this method has the advantages of using smaller, less costly logs and also making round corners easily for odd-shaped homes. Because the logs shrink but the mortar doesn't, this style usually necessitates a double wall with a vapor barrier in between.

Stockade: Lay the logs vertically instead of horizontally and you have this style, but it is not generally popular for anything other than hunt camps or cottages.

TIMBER FRAME A timber frame house can be almost as rustic as a log home but is built with the energy efficiency of more modern building technologies. If you have seen the inside of a barn, you have seen timber frame — it's all about the massive beams, posts and trusses that compose the self-supporting bones of the structure. That's on the inside. The exterior of a timber frame home often looks like a conventional North American house because they are usually covered in conventional cladding such as shingles or tongue-and-groove wooden siding.

The texture of wooden beams that have been hewn flat by hand with an axe can't be imitated.

This siding is generally composed of panels that are designed to fit into the spaces between the beams. These panels are generally made of an inside and an outside skin of finished material (stone, wood or metal) sandwiching a layer of vapor barrier and insulation. These huge sheets go up fast and are very energy efficient.

VINYL SIDING Vinyl (polyvinyl chloride) gives you the appearance (if you don't get too close) of wood siding. It is available in most styles, including Dutch lap, board and batten, V-groove, shingle and shake, and at a lower cost in both money and maintenance. Vinyl siding comes lots of colors and textures but the cookie-cutter look is unavoidable.

METAL Metal says high tech better than anything else and is therefore well-suited for use as an exterior cladding for Contemporary homes. At one end of the spectrum is stainless steel. Although the material is used all over inside our homes, it starts to grow astronomically in cost when used for exterior cladding in any quantity. A much more interesting and more economical form of steel is the kind that is meant to rust, but only on the outside skin, and is sold under brand names such as Corten. This is the high-end version of the rustic, weathered look that some architects obtain by using authentic rusted steel or tin. Metals such as stainless steel and copper are used most often as highlight or trim pieces on exteriors.

At the other end of the material spectrum is aluminum siding — aluminum with a baked-on color coating. This is the lunch bucket of metal cladding. Like vinyl, it is relatively cheap, very low maintenance and about as stylish as a pink lawn flamingo. Galvanized corrugated steel is also usually seen as the lower end

A reconstructed barn exhibiting a typical timber frame structure made of big hand-hewn beams and bracing members pegged together. The metal roof and the cupolas on top are also traditional architectural elements of the barn. The metal silo-like structure at the right is where the architect put the bathrooms.

Corrugated galvanized aluminum makes for a low-maintenance and stylishly urban cladding. The undulating nature of the corrugation adds visual interest. The awning projecting from the roof is called a brise soleil.

of the materials hierarchy, but its undulating texture can make for an interesting graphic look, especially when used sparingly with other materials. Galvanized aluminum siding can also be used for a stylish and modern effect (see above).

BRICK Brick is a durable and extremely versatile building material, which is why we see so much of it. Bricks are made of clay fired in a kiln, sometimes with a glaze. The nicest, hardest and most expensive brick is called "facing brick," and it is generally used on the most prominent surfaces of a house, as in the front. The next step down is called "common brick" and is used on the sides of homes. A much softer brick known as "backing brick" is used on the inside. Brick veneer is brick in thin slabs, used on frame or masonry walls.

As a general rule, the smoother the brick the more contemporary it looks, and the rougher it is the more antique or historical it appears. Oversized bricks suggest an institutional look that can work very well with a Contemporary home, and different shapes are available for creating structures such as arches and curves.

Varying colors and textures of brick can be mixed in a wall to make interesting designs and patterns, and the laying pattern can be changed to make some bricks project out to create even more texture, a technique used in the Arts and Crafts style.

Bricks and, for that matter, other types of masonry such as stone and block, are relatively energy efficient. It takes, for example, up to eight hours for a 30-degree-Fahrenheit temperature transfer to occur, versus only a few minutes for wood siding, making brick cooler in the summer and warmer in the winter.

Old or salvaged brick can give character to a wall, inside or out. The only caveat is that you don't always know whether or not you are getting interior or exterior brick, or maybe a mixture of the two. Some bricks are so weathered they have the glaze worn right off them and have to be sealed in order to be suitable for exterior cladding.

Among the most interesting and sought-after salvaged bricks is the "clinker" brick. Heavier than normal, they are called clinkers because of the sound two of them make when knocked together. They were originally discarded after the manufacturing process because they had burned too hot in the kiln and their color had gone wrong — usually too dark. They were rediscovered at the turn of the century by American Arts and Crafts designers (notably the California siblings Greene & Greene) and were used to create many a charming detail on Craftsman bungalows.

Ask a brick what it wants to be and it will say "An arch!"

LEGENDARY CONTEMPORARY ARCHITECT LOUIS KHAN

Opposite: Chicago architect Lisa Jaffe used brick to create different planes on this very layered façade. The best facades are three-dimensional because they then have a foreground and a background. This one has three layers: the main volume of the building, then the second layer of brick just four inches in, and finally the bay window. Some of the windows have a double layer of brick, and two lines of limestone profiles run across the front. The steps — which are limestone — start big and then narrow down as they approach the front door. The intention is to greet the visitor at the street. The two rectangular holes above the bay window are scuppers, which drain the roof deck. When it rains, the water comes flying out of the scuppers, over the window and then down to water the garden space below. **Left:** A classic Art Deco home clad mostly in horizontal brick, but with bands of vertical (at top) and angled (long vertical patch under round window at center right) brick, picked out in black. Other Art Deco elements are the semicircular door canopies, the odd-shaped windows, glass brick, and the flat parapet roof.

TERRA COTTA Terra cotta panels used for exterior facing of buildings are referred to as ceramic veneer. This veneer is often plain and flat but the plasticity of the material also allows for sculptured surfaces. The decorative terra cotta used for copings, column capitals, bandings, gargoyles, and other ornamentation is known as architectural terra cotta.

STUCCO Stucco is a type of exterior plaster applied as a two- or three-part coating directly onto masonry, or applied over wood or metal lath. Sometimes referred to as "render," stucco is a lot like wallpaper — it can be used to cover up a multitude of sins when doing a renovation. Used since ancient times, stucco once consisted

The caramel-colored walls, thick wood trim and brackets give this renovated home a South-of-France-meets-Craftsman-Cottage look. A roof garden over the garage was added to provide some interest to the front and downplay the garage. Following the concept that a curve or a wiggle will alleviate something that is too rectangular, the curved stone steps were added to help the somewhat boxy look of the house.

primarily of lime but now is mostly Portland cement, which makes for a much harder material. Stucco is used throughout North America on all kinds of residential and commercial structures, but is seen most on Italianate, "Four Square" and Pueblo style homes, as well as Prairie, Art Deco and Art Moderne, and Spanish Colonial, Mission, Pueblo, Mediterranean, English Cotswold Cottage and Tudor Revival styles. It's cheap and versatile, can be tinted any color and can look like limestone or other dressed stone from a distance. The addition of fake corner quoins and or simulated block lines in the wall can add the faux effect. It's far from maintenance free, however and will crack with age.

Since stucco is put on wet, there are many ways to be creative with the finish to make it smooth, wavy, or pebbly. Some popular finishes include smooth faux adobe and Spanish (often whitewashed), a textured pebble-dashed or dry-dash surface (either introducing real pebbles into the stucco or using a machine to get the texture), fan and sponge texture, reticulated (traced with a pattern), or roughcast (throwing the mortar against the wall with a stiff brush).

There are many stucco products on the market. If you are attempting to repair or otherwise match a historic stucco with a new product, it may prove difficult. Sometimes original ingredients have to be identified and employed; to accomplish that, a visit to your local historical association may be in order.

CONCRETE A mixture of gray powdered cement, water, aggregates and sand, concrete can be formed into any shape and size you can come up with. It can be colored or stained, stamped or formed to resemble just about any material, or mixed with a number of decorative materials on an exterior.

Concrete has always been popular in warmer climes, and near the seacoast where moisture is a problem. For one thing, concrete is easy to mold the decorative shapes that are found in the Spanish Colonial Revival and Art Deco styles that are favored in places such as Florida and Southern California. It is also suitable for the Pueblo Revival style and absolutely perfect for an International or Contemporary style home.

Concrete is resistant to fire, termites and dry rot. Home builders in northern climes use Insulating Concrete Forms (ICF) that have a core of insulation built into a formed concrete panel. It's also a good material for use as a heat sink for trapping and releasing solar heat. Concrete can be tinted in colors, either during or after the construction process and can also be stamped to mimic natural stone, something that is used in paving driveways. One of the most interesting uses of concrete we have come across is on a neoclassical mansion that substituted concrete made with a limestone matrix for dressed limestone. After a little gentle sandblasting, it was impossible to tell it from the real thing (see pages 10–11).

On the down side, concrete is not cheap, and it is estimated that renovating with concrete walls will add from between three and five percent to the cost. It is also susceptible to cracking. The damage can be fixed with a little sandblasting, or it may have to be completely redone.

FIBER CEMENT SIDING Fiber cement siding is made of cement, sand and cellulose fiber (to prevent cracking) that has been autoclaved (cured with pressurized steam) to make it stronger. You can now buy everything from simulated

> You can't change the color of stone and concrete block, but you can always change the color of stucco. In fact the Feng Shui people say the easiest way to change your life is to buy a bucket of paint.
>
> PAUL FRONCEK, AN ARCHITECT WITH AN INTEREST IN FENG SHUI

The best tool in green architecture is design, and often it's just common sense. You start by picking a house design that is suitable for a particular microclimate. If you picked the north side of a mountain that never gets any sun and you put up a huge sheet of glass facing north to get the view, that's not very sensible. You also want to minimize windows facing west. The house often needs heating up a bit in the morning, so east-facing windows are okay, but by the time the sun gets to the west the house is often already getting warm. If you have to, tuck them way back under an overhang, or plant deciduous trees in front (they block the summer sun but let the winter sun hit the house).

HARRY TEAGUE, COLORADO ARCHITECT WHO BUILDS GREEN

brick and shakes to wooden siding and even soffits made from long-lasting, extremely durable and low-maintenance fiber cement. (See the Products section of this chapter for examples of fiber cement siding.)

CONCRETE BLOCK Concrete formed into blocks that contain various aggregates occupies both ends of the architectural spectrum. Because they are cheap and easy to lay, concrete blocks characterize the cheapest, throw-it-up buildings. But because of their smooth, stark look, they have also been used on upper-end Contemporary homes to project a semi-industrial look.

green architecture

The "green" or environmental approach is more a thoughtful use of materials and technologies than it is an architectural style. Thicker walls, more insulation and bigger windows that interact with internal mass to collect and distribute solar heat are some of the best characteristics of the approach.

Building and furnishing materials considered green include adobe, bamboo, straw bale, cork, paper, corn cobs, rammed earth, untreated wood, tile, glass, stone, and lime or mud plasters. Rammed earth is regarded as a good green wall material, as is straw bale construction.

RAMMED EARTH Rammed earth is essentially ordinary earth (sometimes treated with modern stabilizers) that has been tamped down and compacted between wooden forms to make a wall. Employed in the construction of part of the Great Wall of China, rammed earth has been used for thousands of years as a building material. Its big advantages are excellent thermal mass (which needs to be insulated in cold climates), strength, beauty and availability. There is lots of it around. Since rammed earth cures in the wall, it can be built with simple forms and tools, and in a variety of climates. Walls do not need to be plastered and will last for hundreds, even thousands, of years, especially if they are stabilized with raised concrete foundations, overhangs and reinforcing rods.

STRAW BALES Straw bales create a building material much touted by proponents of sustainable living. This renewable material results in super-insulated walls with slightly irregular surfaces. Originally used by the pioneers in Nebraska, who needed something to build with but wouldn't have a tree for miles, straw bales are cheap and easy to use. Most of the time straw bales are used either as infill in a post-and-beam (wood), steel or concrete structure, or as a load-bearing system where the bales themselves support the weight of the roof. The bale walls are commonly wrapped with stucco netting and plastered with mud, lime-sand or cement plaster, but the plaster is sometimes put directly on the bales. Straw bale interior walls are also possible.

INTEGRATED CONCRETE FORMS One of the newest residential construction methods being driven by environmental concerns is Integrated Concrete Forming (ICF). With ICF construction, the walls and foundations are made from polystyrene foam blocks (usually) 12 inches wide, 16 inches high and

Left: These walls are made of straw bales covered in chicken wire, over which stucco is applied. Straw bale construction is relatively cheap and warm and it depends on a renewable resource. **Above:** The "truth hole" — something that is always present in a straw bale home — reveals what the house is made of.

24 inches long, held together with heavy plastic webbing and reinforced with rebar. The 6-inch gap in the center of the block is filled with concrete to add rigidity and strength. The foam blocks snap together like Lego, allowing a crew to throw up an average home in just three to five days. The advantages are a hurricane-proof house with better insulation, fire retardation far superior to wood and even brick, and a building that is 80-percent soundproof. The disadvantages are that the house has to be very well planned, with allowances made for ductwork, electrical and water pipes where they protrude through the wall. And, since you get what you pay for, the cost of building this method is about twenty percent more than conventional wood frame building.

patios and decks

The patio is the transition stage between the comfort and control of the home and the breezy pleasures and uncertainties of the natural world. It is usually hard-surfaced, comfortable and open but always connected in some way — physically or symbolically — to the home itself. Enclosed patios are characteristic of Spanish, Pueblo Revival and Mission style homes, generally built in warmer climates. The original California Ranch homes were designed as low, sprawling buildings with wings embracing a courtyard to provide a private place for families to spend time outdoors.

Decks, which can be thought of as wooden patios, are still popular, but they are evolving beyond the 1980s-style flat, square structure of pressure-treated boards with a railing around it. Many homes now have multiple decks in different sizes and shapes, often with some kind of covering — vines, trellises, pergolas, wrought-iron screens and cloth awnings — as protection from the heat and UV rays of the sun.

"Outdoor rooms" have been prevalent in Europe for centuries. Their sheltered nature makes them perfect for dining al fresco. Essentially an outdoor space connected to the home by sliding glass or French doors, an outdoor room is open to the breeze with a view of the garden but protected in some way from the harsher elements by plantings or gauzy fabrics or awnings. A brick or stone wall can offer a stylish windbreak as well as a place to grow climbing plants, but without completely enclosing the space. Pergolas can provide support for climbing vines as well as shade, but it must be kept in mind that there are exact calculations as well as artistic choices to be considered in building a pergola so that it provides adequate shade without blocking daylight.

Left: A western red cedar deck and trellis with copper banding that was inspired by the California bungalow look popularized by the brother architects known as Greene & Greene (see page 31).The stone of the columns is granite. **Center:** This wooden deck seems to flow into the house. The curved copper roof projects out over the second-story deck. **Right:** This elevated wall of water is made of galvanized aluminum. It not only has stunning good looks but recirculates the water into the pool as well. **Opposite:** This backyard patio in Arizona is all about shade and open air. The small pine poles forming an awning are called lattilas. Because adobe (clay stabilized with cement and a little straw) is so malleable, putting benches (bancas) or carving out alcoves in the wall is easy. The metal lining of the big alcove above the bench was purposefully rusted by spraying it with vinegar and then sealing it. The rust that has dripped down is considered perfectly suitable for adobe. The floor is flagstone.

Partly because of what is going on in the world right now, people have focused more on their home — making it their own personal kingdom. They are focusing inward on areas they can control. The industry is booming as a result. Kitchens, bathrooms, gardens, pools, terraces, are becoming more elaborate and fantasy oriented.

MICHAEL TRAPP, DESIGNER

Outdoor rooms can now offer such amenities as grills, sinks, fire tables, comfortable furniture, heaters and even an internet connection for your laptop. The next logical step — outdoor kitchens — is already here, with many manufacturers making stoves and fridges that can withstand the elements.

porches

A front porch connects us to the street and therefore to the community. The front porch is perhaps the most welcoming of architectural aspects because it is a visible interface between the interior and exterior —the private and public — of a home. There are many architectural styles that historically have incorporated a front porch, including Antebellum, Victorian Gothic and Folk (one story, with elaborate spindlework in the railings), Shingle Style (with stone arches), Queen Anne (wraparound porches), Mission (arcaded entry porch), American Foursquare, and Craftsman Bungalows (with square columns). Greek Revival

The porch on this turn-of-the-century Victorian invites you to sit.

homes may have an entry porch with tall columns that reach to the roof. Some homes simply have a covered entranceway that serves as a small front porch. Others, such as French Creole and Tidewater Plantation style homes, shelter wide porches called galleries that run along all sides and usually on both the lower and upper levels of the house under broad-hipped roofs.

The front porch is a North American favorite, romanticized and idealized as a symbol of a welcoming community. One of the most gracious European-inspired adaptations of the porch is the covered loggia — a colonnaded and roofed patio that in some instances can be closed off with sliding doors, but is normally left open. A loggia provides the feeling of being connected to the outside while retaining the solid, comforting feeling of a real room.

balconies

High above the ground, open to the air, balconies open an upstairs room to the outside. They suspend us over the terrain and are therefore the most private of the outside rooms. The scale and style of your balcony will depend on the style of your home: Beaux Art style buildings are usually massive and grand (as in the New York Public Library) with balconies of equal grandeur. An Italianate home might have a porch over which a balustraded balcony provides a pleasant place for a container garden. The wrought-iron balconies of New Orleans are Spanish in origin but Victorian in character. You might find your inspiration in the famous Juliet balcony in Verona, Italy, for an ornamental and functional balcony off an upper-story master bedroom suite. So decorative are balconies that they don't even have to be functional, and even a faux balcony or intricate wrought iron can add interest to windows and French doors.

Left: Steel railings and ladders add to the "ocean liner" look of this Art Deco home. **Center:** The design inspiration for this semicircular deck was taking your arm and saying "we have a view from here to here" and then making a deck that literally points at the view. The wood is ipe (a type of mahogany) and the handrails are untreated teak. Both woods have been allowed to weather naturally. **Right:** Smaller even than a Juliet balcony, the metal screen nevertheless finishes off these windows. Note the hardware at the bottom of the shutters.

One powerful idea that has been developed in North American architecture is the unified exterior envelope — a roof and walls that vary in an overlapping way almost like a garment or a blanket. This has resulted in some very complicated homes that have these continuous roofs, with hips and valleys and sheds that enclose the whole structure. A newer idea is Sculptural Fragmentation — you actually break a house apart into its components and combine them into the most sculptural, formal, energetic ways you possibly can — something that Frank Gehry did early on.

BARTON PHELPS, ARCHITECT

roofs

A roof is much more than a weather shield. Architecturally, roofs are great indicators of what style a house actually is. They will also often reveal the age of a home. Roofs vary in pitch and construction from style to style and region to region. As a general rule in North America, except for in the Southwest, where flat roofs are the norm, the steeper the roof the older the building style.

Not much has changed in roof styles in the past half-century or so. However, one recent innovation is that large eaves overhangs, which had begun to disappear with the advent of mass air conditioning, especially in warmer climes, have seen a bit of a resurgence as people become more concerned with energy costs and other environmental issues.

The materials used on roofs also have not changed dramatically, except that the colors and surfaces that reflect light and thereby reduce energy costs have increased in importance. In the same vein, roofs that function as solar collectors are becoming more common. But recent technological innovations have incorporated the solar panels into the roof itself, so they become almost invisible. This has removed the spiky and for the most part unattractive look that environmentally "green" houses used to have, when solar panels were cantilevered out or propped up on the roof.

When looking at roofs, consider the climate — do you have snow all winter, rain for months on end, hot sun beating down most days? Traditional styles described below are often common to particular geographical areas.

FLAT The simplest roof construction, it is most popular in dry climates where rain runoff and high snow loads are not a problem. Flat roofs are typical of the Adobe and Pueblo style but also Art Deco, Art Moderne, International and many Contemporary homes. They are also suitable for green roofs (see below).

Surrounded by dry hills, the roof of architect Barton Myers' home always has standing water in it (there are about 30,000 gallons of water on all three of his roofs). The water acts as an insulator, collecting heat in the daytime and keeping the house from losing too much heat at night. He also uses the water for fire prevention, and likes "the Islamic idea of surrounding yourself with beautiful reflecting pools of water as a psychological relief to the dryness." The drums over the windows contain steel shutters that act as fire, sun and security protection.

The hipped roof on this tower gives it a medieval appearance. Note the shed dormer in the center. The house itself, which is concrete block covered in golden yellow stucco, is called Sogno Dorato or Golden Dream, a phrase Italian mothers use when saying good night to their children.

GABLED The most common roof style in North America. Based on the Greek temple of antiquity, it has triangular ends (the gables) and a clean, simple look that is often echoed in the dormers and door treatments. Having two gabled roofs run into each other is called a cross gable roof and results in valleys and interesting planes that can give an otherwise boring house style a complex look with more depth.

SALTBOX A New England style in which the main (two-story-high) roof extends to cover a one-story addition at the back, often for a summer kitchen or storage area. Modern homes may have the single story at the front.

A-FRAME An engineering marvel that encloses space in the cheapest way possible — albeit with some loss of headroom at the sides — the A-frame is all roof. Introduced to North American in 1957 by architect Andrew Geller, it is most often used for cottages and remote getaways.

GAMBREL A gable with a slight bend on each side, with the upper roof steeper in pitch than the lower. Often used for barns and in Dutch Colonial homes.

MANSARD Flat on top and sloping down steeply on all four sides. The style is named after French architect François Mansart (1598–1666), who used it to provide extra living space in the attic. In North America, you will find mansard roofs on everything from Second-Empire and French Chateau style to Contemporary homes (see page 168).

HIPPED A hipped roof slopes down to the eaves on all four sides; the line where the slopes meet is called the hip. It may form a perfect pyramid with a single point at the ridge. Hipped roofs are generally found on French-inspired American Foursquare, and on everything from Prairie Style to Colonial and Victorian styles.

roofing materials

ASPHALT SHINGLES Asphalt shingles are the sensible shoes of the roofing material world — inexpensive, long lasting and common. Although they don't offer the dimensional, architecturally interesting aspects of the more expensive shake, tile or metal roofs, asphalt shingles are easy to install, usually last from between twenty and thirty years, and are fire retardant.

At the bottom of the ladder is the "three tab" asphalt shingle. These are the shingles you see on two out of three homes in North America. They have a core of either felt or fiberglass that is impregnated with asphalt, which is then covered with a protective layer of ceramic granules that are resistant to ultra violet rays. They come in various colors, but nothing you would call "designer."

There is a form of asphalt shingle that is about twice as thick and is known as an "architectural" or "dimensional" shingle. Laid out in random patterns, these specialty shingles can be bought in patterns that resemble wood, tile or slate but at a fraction of the cost of the real thing. Because of their thickness, they give a bristly appearance that also helps mimic the look of cedar shakes.

You can also buy asphalt in long rolls that are usually three feet wide. Sometimes known as "rolled roofing," it is even cheaper than shingles and is usually only used to cover sheds and other utility buildings. However, we have

Our company has a rule that the shingles on the roof have to be western red cedar and on the walls they must be eastern white cedar. When you let them go their natural color they bleach different colors — the walls go a soft fawn gray, like the Cape Cod saltboxes, and the roof goes a dark charcoal gray. But it's a mistake to put western red on the walls because they streak. If the building is exposed on all sides, we will let the shingles age naturally, but if the building has shady parts or areas where the snow builds up in the winter, we will accelerate the process by putting a bleaching oil on the shingles, otherwise you can get uneven ageing.

MARILYN LAKE, ARCHITECT WITH THE FIRM THE IDEAL ENVIRONMENT

seen creative use of rolled roofing — making what is essentially a large shingle out of it to create an exaggerated and (if carefully done) interesting-looking roof.

The crudest form of asphalt roofing is known as "hot mopped." Used exclusively on flat roofs, it involves the slathering on of hot, liquid asphalt onto the roof surface. Extremely cheap and more or less effective, it is also unhealthy to apply, bad for the environment, and about an ugly a roof as you can imagine. This kind of roof sometimes has a layer of pebbles put on top to make it last longer.

SHAKES AND WOODEN SHINGLES A shake is a wooden shingle that has been split off of a block of wood, is rougher than a shingle on at least one side and usually has no taper. (Untapered shakes are sometimes called "bard shakes.") A wooden shingle is sawn on both sides and has a taper that varies from ½ inch to one inch in thickness, depending on the product. Both can come in varying widths. Wooden roofing has more than twice the insulating value of asphalt shingles, four times that of fiber-cement composites, and five times that of slate. Generally speaking, shakes last longer than wooden shingles.

Among the big pluses of using wooden shingles — especially cedar — is that pound for pound it is one of the strongest materials you can use on a roof. Yes, materials such as concrete and ceramics have great strength, but they are also heavy. Wood is relatively light but also adds more structural strength to a roof than any other material, something that is vitally important in areas prone to earthquake. Wooden shakes and shingles also have very good insulation properties, something ceramics, concrete, slate and metal obviously do not.

Of course, you can use wooden roofing shingles on the sides of your house (shakes are considered too thick), but to get a more flat, tailored look you should look for a specially made sidewall product called "R&R" or "Rebutted and Rejointed."

Historically, shakes were split by hand using a sharp-edged tool, a cross between a knife and an axe called a "froe," and a mallet to hit it with. Although this is still done today, the expense of hand-made shakes has driven most people to the machine-made variety.

Wooden shingles and shakes are made from various woods, but mainly cedar (western red, eastern white and Alaskan yellow varieties) and, to a lesser extent, pine, which doesn't last as long. (Although we have heard of a pressure-treated southern yellow pine that does a good job of imitating cedar for less money.) The great thing about cedar is that the resin that permeates the wood is highly rot and insect resistant. Red cedar is basically a rainforest tree that produces resin to protect itself from mold and damp, which is why it doesn't rot and is resistant to UV light. However, most cedar will naturally turn gray as it weathers. Many people like this look, but if you wish to keep it from turning gray you can treat it with a preservative fire retardant, or buy pressure-treated material that does the same thing. Wooden roofing that does not have a fire retardant on it is banned in some areas.

Whatever type you choose, selecting a lighter color will keep your roof, and your attic, cooler in the summer.

Left: Cedar shakes — like these — are cut straight, giving them a bristly look. Cedar shingles are cut with a taper, which helps them lay flatter. **Center:** Clay tiles work perfectly with this Southwestern style of home. Note how the tiles are doubled up on the bottom and set in cement. Clay tiles can last forever but their excessive weight means you need a strong roof system. **Right:** The tiles on the gable end are original clay tiles over 100 years old. The tiles on the roof are new and were matched by going to the original manufacturer — Ludowici Roof Tile (see Sources) — who still makes the product.

CERAMIC TILES Ceramic tiles are made of clay fired at high temperatures. They are the oldest of roofing tiles, having been used on roofs all over the world for centuries. They are durable, resistant to rot, fire and insects, and their shape promotes good air circulation on the roof. Ceramic tiles definitely give a house a classic Mediterranean flavor and are especially suited to southwestern, Italianate, Spanish Colonial or Spanish Mission homes. However, they are fairly expensive, add a lot of weight to the roof, and low-quality tiles (those not fired hot enough) are susceptible to frost damage. Although these tiles are breakable, they are also reusable, and we have heard of tiles being used for centuries on different roofs.

There are two basic types of ceramic roof tiles — overlapping and interlocking. Barrel tiles (sometimes called pantiles) are the most common form of the overlapping variety. Shaped like a cylinder cut in half, they are laid on a bed of mortar with one side up (forming a trough) in one row and the other side down in the row beside it, all across the roof for a continuous cover. It has a Mediterranean look to it because that's where it originates from. This is the most authentic and ancient way but also the most labor intensive. Some other forms that ceramic tiles can take are flat, shake-like rectangles and thinner versions that look like slate shingles. These tiles are usually nailed to the roof through special holes added during the manufacturing process, but they can also overlap, usually on one side and at the top. This flat tile has a heritage more English than Mediterranean.

Originally ceramic tiles were fired only in the color of the clay they were made from, and traditionally had the classic look of terra cotta. Today you can get ceramic tiles in all kinds of colors; sometimes the color is mixed in and sometimes it is painted on the outside.

SLATE Slate shingles are actual pieces of rock and as such have most of the same properties as ceramic tiles: they are strong, long lasting, fire-, rot- and fungus-resistant, and aren't affected by acids. They are always flat and come in natural slate colors that range from shades of green, gray and black but also include purple and red. Much of North America's slate comes from Vermont.

CONCRETE Concrete tiles are the poor man's ceramic. Popular since the early 20th century, concrete tiles are made from Portland cement and various aggregates and are extruded under pressure. Cheaper, just as durable, resistant to rot, insects and fire and somewhat lighter, concrete tiles lack the subtle color variations and pigments of their ceramic cousins. Like ceramic, however, concrete is heavy and needs an extra-sturdy roof system. Some of the newer types of concrete tile are lighter, have cellulose mixed in to add strength, and come in various colors. Concrete tiles that simulate wooden shakes and shingles are also available.

METAL Occupying the high end of the roofing world, metal roofing is making a bit of a comeback. Zinc, copper and lead roofs were all the rage a few hundred years ago, at least for the wealthy. Today they are admired for their architectural beauty, relative light weight, high strength-to-weight ratio and durability. In fact, metal roofs are so lightweight that they can be laid right over existing roofs, if desired.

Using metal as a roofing material can have a positive ecological impact as well. It is not only recyclable, but the Florida Solar Energy Center claims that its research shows metal absorbing 34 percent less heat than asphalt shingles, and homeowners switching to metal roofing reported saving up to 20 percent on their energy bills. Of course, metal itself has almost no insulation properties.

Standing seam metal roofing, in this case made of Galvalume — a very durable galvanized aluminum. This is more expensive than a shingle roof but not a lot more, plus it has the advantage of being fireproof. The seams give it good detail with a crisp edge.

The most popular application for metal roofs is called "standing seam." In this method, the upturned edge of one sheet of roofing is attached to the upturned seam of the sheet adjacent to it. This produces the distinctive long vertical ridges that give the roof that old-fashioned, highly architectural look, something that particularly suits high-pitched roofs. The style is often used with steel roofing but can be used with any metal roofing and is one of the easiest roof systems to install.

Metal shingles are also available, as they have been for centuries. Usually made of steel or aluminum, they are either sold as individual shingles or four-foot-long panels made to look like rows of shingles. They can be made to resemble wooden shakes and shingles or antique Victorian metal shingles. For intricate or curved roofs, individual tiles are the easiest to use.

Steel is the most commonly used of the metal roofing materials, and it is almost always galvanized (steel coated in zinc). However, a new product called Galvalume, which is coated with both zinc and aluminum, is even more weather resistant. Some steel roofing now comes with an acrylic covering with the same kind of ceramic granules that cover asphalt shingles, with a clear acrylic overcoat.

Aluminum is similar to steel in terms of its properties. In fact it is even more lightweight and is also cheaper. Its resistance to weathering makes it suitable for coastal areas. And aluminum does not rust.

The queen of metal roofing is, of course, copper. Strong, weather resistant and absolutely gorgeous, copper is also the most expensive. It usually comes in sheets but can also be obtained in individual shingles. Left unprotected, copper oxidizes into the greenish-blue manifestation known as verdigris, seen on cathedrals and historic buildings around the world. Not only does this verdigris produce a good-looking patina, it actually protects the copper from further oxidation. To prevent oxidation from happening in the first place the copper must be coated, either with an acrylic or with some thin coating of another metal, usually lead. The acrylic keeps the copper color while the lead gives it a dull, pewter-like patina.

Zinc is similar to copper in its anticorrosion properties, and is even more malleable. Like copper it also oxidizes, but with zinc the color is more of a bluish white.

dormers

In a nutshell, a dormer is a way to introduce light and air into the upper story of a house. It also gives some added dimension to the roofline. It is a window with its own roof and sides that projects out of the roof. (For more on windows, and how they can affect the façade of homes, see the next chapter, Doors & Windows.) The dormer's much rarer cousin, the wall dormer, is set flush into the wall plane, usually bisecting the roof line at the eaves (see page 184).

Dormers are often associated with traditional-style, one-and-a-half-story homes, but as you will discover there are several types of dormers adaptable to many styles.

SHED The simplest of all dormers, in which the flat roof of the dormer comes straight off the roof of the house. It is used often on Shingle Style homes, sometimes with a very steep pitch, to great dramatic effect.

The early settlers came [to Arizona] with wooden Victorian houses that fell apart because intense sun destroys wood. The indigenous houses used heavy timber when using wood, so it has resistance, and they embedded it in the wall. We like to use permanent materials like concrete, concrete block, steel, stone — things we know will hold up and not require a lot of maintenance. For example, we like to use unpainted copper gutters or steel and then leave it.

JOHN DOUGLAS, ARCHITECT

The face of the dormer should always be the window and the window only — no front wall, just a little peaked roof. This keeps it visually light. But if you put a structure on each side of the window, it looks much heavier. Sometimes you have to do that because of interior considerations — if you need more height inside, for example. Framing the extra sides as blind sidelights, all done in the same trim, will give you somewhat the same effect.

TONY JENKINS, HISTORICAL RECONSTRUCTION CONTRACTOR

The shed dormers across the back of this elegant home prove that this simplest of dormer styles can work anywhere.

Green roofs are excellent at reducing air conditioning, because there is moisture inside the soil and, as it evaporates — just like the human body sweats — it wicks the energy into the air, not the house. Plus the membrane underneath that keeps the moisture off the roof doesn't get affected by ultraviolet radiation, so you get a longer life out of your roof. It isn't a form of insulation, however, so there is not much benefit in the winter, which is why most homes that have a green roof also have high levels of insulation in the roof.

MARCO VANDERMASS, ARCHITECT AND PARTNER WITH THE GREEN ARCHITECTURAL FIRM BREATHE ARCHITECTS

GABLED Essentially a square box with a triangular roof on it. It is typically seen on Cape Cod, Craftsman Bungalow, Colonial Revival and raised-roof ranch homes. It is the generic dormer, at least in North America.
PEDIMENTED Inspired by classical Greek temples, this triangularly-roofed dormer can be found on Georgian, Federal and Colonial Revival styles in North America.
HIPPED Take a gable dormer and put a bend in it at the peak. It is typical of Prairie Craftsman, and Shingle styles.
ARCHED An essential element of many French architectural styles, including Beaux Arts and Second Empire, but also Italianate, Richardson Romanesque, Spanish Colonial and Mission style. When an arched dormer comes to a point in the middle, it becomes Gothic.
EYEBROW Originally seen on the thatched roofs of English cottages. Today seen mostly on Craftsman Bungalow, Shingle, Romanesque and Queen Anne styles. (Some eyebrow dormers contain no window at all, but are merely shallow architectural elements or roof vents.)
INSET A gable dormer that is pushed back, or recessed, into the roof.

green roofs

The "green roof" concept has been around ever since the Hanging Gardens of Babylon were constructed in the 8th century B.C., but it really started to catch on in the modern world in Europe in the 1980s as a result of rising energy costs and concern about the environment.

There are some myths about green roofs that should be put to rest. First of all, a green roof is not a garden on your roof, although it can be thought of as one. In fact, green roofs vary from simple, relatively low-cost operations with some hardy plants or grass that is not readily accessible, to elaborate, expensive affairs with deep soil and a variety of plants including trees and shrubs that are easily accessible and meant for recreational as well as ecological uses. This last type may call for irrigation systems and a very strong roof system.

Essentially a green roof has a job to do — it's there to protect your home (and the roof itself) from heat and ultraviolet rays. The plants used on a green roof have to be tougher than usual. They must be resistant to elements such frost, drought and high heat. In other words, they have to be low-maintenance, usually alpine varieties that are able to look after themselves for the most part.

Having some potted plants or big planters on your roof doesn't qualify. A green roof is a flat roof covered in wood, concrete, metal or any strong, weather-resistant material. This in turn has to be covered in some kind of impervious material. A growing medium — either soil or some other kind of growth material — is then planted with various plants. It is quite simple in theory, but the roof has to be strong enough to take the weight, and proper drainage and waterproofing are essential.

Flat roofs are fine, but a gentle slope is not only possible but often desirable because of the need for drainage. In fact a slope of 1.5 to 2 percent is considered ideal, and slopes of 30 and even 40 percent are possible.

Left: Hardy, drought-resistant plants are perfect for this green roof. Center: The plants on this green roof keep the sun's ultraviolet rays off the roofing material and the soil absorbs most of the rain that would normally go into the storm sewers. Right: Ornate clay chimney pots from Superior Clay Works (see Sources) top a brick chimney.

Green roofs capture rain that would normally have to be dealt with by storm sewers. This is especially important during heavy storms when storm drainage system can be overwhelmed. They help filter the water that does enter the storm sewer system, meaning less in the way of pollutants in the water table. They reduce the amount of heat that a roof captures during the day and releases at night, resulting in decreased air conditioning costs. And they last longer.

chimneys

Whether brick, stone, steel, ceramic or even log, chimneys add the crowning touch to your roof. To work properly, your chimney must be enclosed by the heated envelope of the house. This gives you, in essence, an insulated chimney and therefore an efficient draft that pulls the smoke out of the house. Unfortunately, a lot of builders, especially those who build big subdivisions, run the chimney up the outside of the house, surrounded by a brick or wooden "chase" in order to have more room on the inside. It takes a while for this kind of chimney to warm up, and that means a potential for backdraft and smoke in your home every time you light a fire.

The most primitive way to build a chimney is out of short round logs that taper in size as you go up. This can work well with a log cabin, as would a fieldstone or rubblestone chimney. Today, most of the invisible part of a chimney (which goes up the inside of the house) is made of insulated stainless steel or brick, often with a clay liner. Even if your chimney looks like brick, if it's relatively new, chances are it's really stainless steel surrounded by real or fake brick in order to comply with building codes.

The aesthetic part of a chimney is the cap — the part that emerges through the roof, or at least the decorative finish to the functioning tube that connects to

We often use wire mesh (held up with re-bar) coming down the corners of a house as a "rain leader." It's more or less a fun thing, a playful way of dealing with water, because in the winter it freezes into fantastic shapes. But it is also a way to channel water that is not about a pipe. Runoff is used to water the garden and is therefore not connected to the storm sewers.

MARCO VANDERMASS, GREEN ARCHITECT

Left: Rain water goes from an upper stone-filled drainage box to the lower through this carved-stone scupper (the facing material is sandstone), a creative way of dealing with water and the kind of architectural detail that makes a house. **Center:** An elaborate metal rain chain. **Right:** A stainless steel drain spout, out of which hangs the drip chain.

the fireplace or stove. The chimney cap is the crowning glory of your roof and as such, a golden opportunity to finish your home off. You can make a chimney cap out of any non-flammable material, but the most common are clay, brick and metal. At the higher end of metal chimneys is copper, with stainless steel a popular and less expensive alternative. Most chimney caps come with a small roof to keep the rain out and a mesh spark arrester, something that is particularly important if you have a wooden shake or shingle roof. The mesh will also keep birds and animals from nesting in your chimney. Some chimney caps are of complex design in order to deal with tricky cross winds to increase draft.

drainage

In theory, drainage is simple: water hits the roof and drains off to be collected by the gutters that run along the edge of the roof. The gutters funnel the water into pipes at the corners of the house that divert it away from the foundations. The most environmentally responsible way of dealing with the water is to keep it away from the storm sewers by aiming it directly into the garden, pond or swimming pool (see below).

These gutters and troughs can be made of virtually any metal, although aluminum is the most common because it doesn't rust. Copper is the high-end choice, but we have also seen them made of wood and even stone. The drainage conduits that get the highest marks, however, are "drip chains" — various pieces of metal connected into a chain that hangs from gutters to the ground. We have even seen wire mesh used to the same effect. An attractive architectural element in their own right, these chains allow the water to slide down from the roof, generally into ornamental beds of stone or gravel.

products

exterior details

Limestone, Shenandoah
Genesee Cut Stone & Marble

Limestone, Mayfair Blend

Limestone, Round Hill Blend

Limestone, Colonial Stone, Mixed

Minnesota Granite Split Face

Georgia Cherokee Creole

Michigan Cobblestone

Colonial Limestone, Buff

Limestone, Castle Rock, Buff Gray

Brown Country Limestone

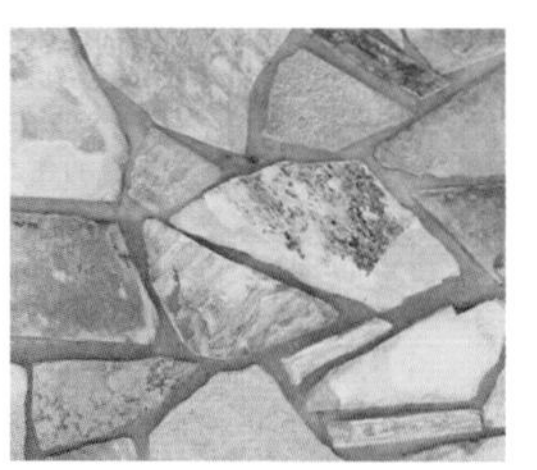

Limestone, Pine Log

Michigan Split Redstone

Granite, Aqua Grantique
Krukowski Stone Co.

Sandstone, Autumn Sunburst

Sandstone, Chestnut Boulder

Slate, Processed Riven Rock, Desert Pearl
Universal Slate

Slate, Processed Riven Rock, Multi *Universal*

Slate, Processed Riven Rock, Charcoal *Universal Slate*

Indiana Limestone
Genesee Cut Stone & Marble

Limestone, Brian Hill

Limestone, Tennessee Ledge

Limestone, Bluestone Split Face

Wisconsin Weathered Edge

Sandstone, Apple Creek
Krukowski Stone Co.

Sandstone, Carmel Crown

Sandstone, Glacier Bay Ashlar

Sandstone, Omega Blend *Krukowski Stone Co.*

Sandstone, Sandy Creek

Sandstone, Sawn Heights

Sandstone, Prairie House

Sandstone, Ashwood Palace

Sandstone, Beaver Creek

Sandstone, Ashwood

Sandstone, Confederate Gray, Ashlar

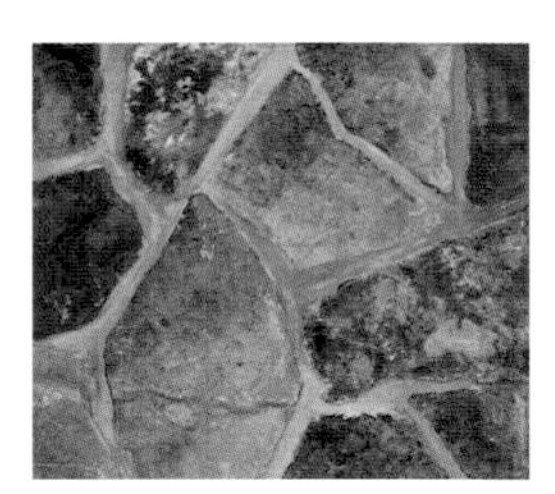

Sandstone, Confederate Gray, Wall

Sandstone, Chestnut Ashlar

Sandstone, Vail

Sandstone, with Granite Cobbles

Sandstone, Glacier Bay Fortress

Sandstone, Palace Blend

Robinson Rock, Arapaho *Robinson Brick Company*

Robinson Rock, Blueriver

Robinson Rock, Goldrush

Robinson Rock, Sierra

Granite, Processed Castle Rock, Black Glitter *Universal Slate*

Limestone, Processed Castle Rock, Desert Pearl

Limestone, Processed Castle Rock, Green Glitter

Old Brick Originals, Cigar Factory, tumbled
Robinson Brick Company

Old Brick Originals, English Pub, tumbled

Old Brick Originals, French Quarter, tumbled

Old Brick Originals, Peppermill, tumbled

Old Brick Originals, Seawall, tumbled

Old Brick Originals, Smoke Stack, tumbled

Old Brick Originals, Train Station, tumbled

Brunswick Queen *The Henry Brick Company*

Breckenridge

Brunswick

Hartford

Madison

Madrid *Carolina Ceramics Brick Co.*

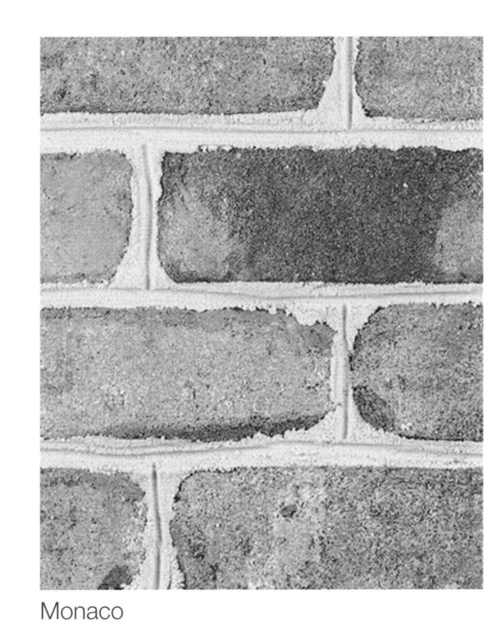
Monaco

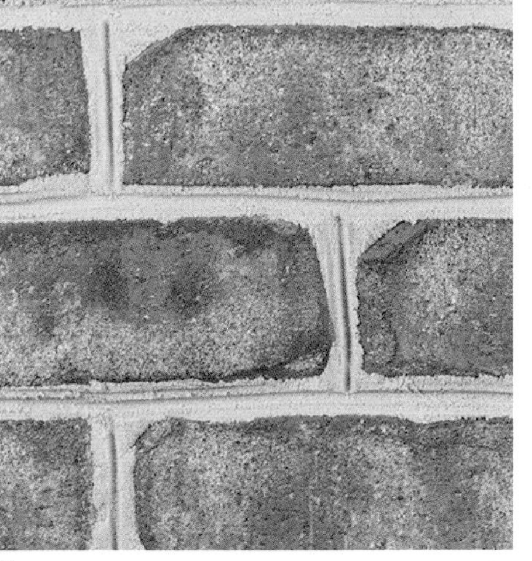
Sierra

Madison Queen
The Henry Brick Company

Old Cahaba

Old English

Princeton

Providence

Providence Queen

Virginian

Virginian Queen

Old Cahaba Queen

Chestnut Smooth *Carolina Ceramics Brick Co.*

Blue Black

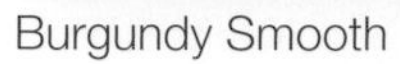

Burgundy Smooth

Burgundy Iron Smooth

Burgundy Velour

Camelot Iron Spot Smooth

Chestnut Velour

Colonial Grey Velour

Colonial Grey Smooth

Crimson Iron Smooth

Crimson Iron Velour

Dogwood Smooth

Dogwood Velour

Empire Ivory Smooth

Iron Spot Smooth *Carolina Ceramics Brick Co.*

Heritage Smooth

Heritage Velour

Rosewood Smooth

Rosewood Velour

Empire Ivory Velour

Sable Smooth

Shadow Grey Smooth

Shadow Grey Velour

Teakwood Smooth

Teakwood Velour

Terra Cotta Velour

Topaz Smooth

Sable Velour

Arboretum, pine *Cape Cod Finished Wood Siding*

Cape Cod Gray, pine

Cape Cod White, pine

Chesapeake Blue, pine

Driftwood Gray, pine

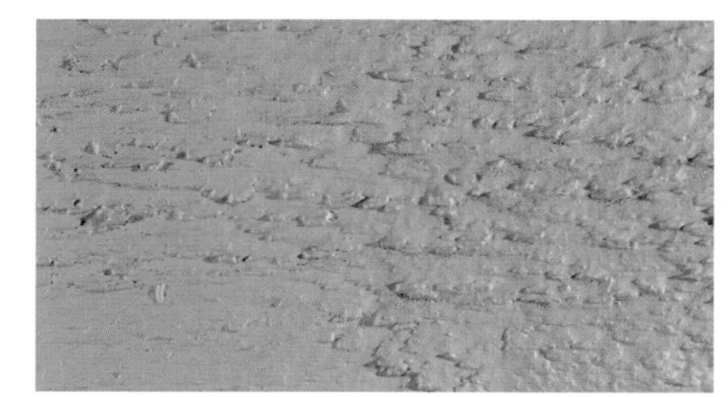

Sandalwood, pine

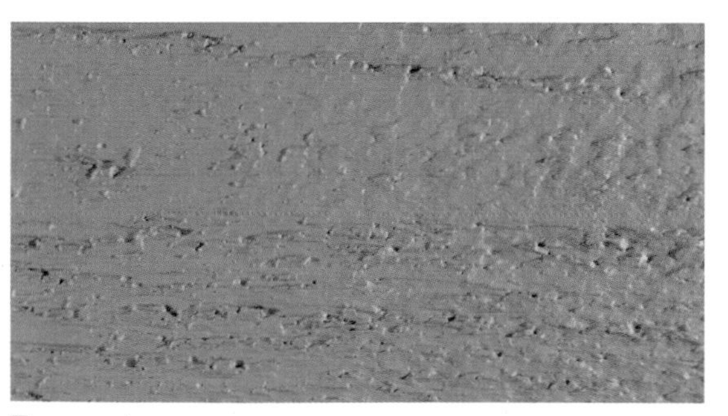

Taupe, pine

Victorian Gray, pine

Both fish-scale (scalloped) and square-cut shingles were used as siding on this house, and are made of eastern white cedar stained white. The chimney bricks have been painted to match.

HardiePlank Lap Siding (all), 011
James Hardie

Lap Siding, 002

Lap Siding, 003

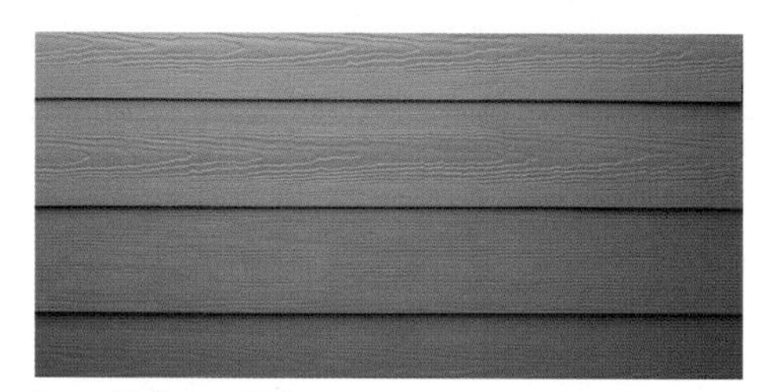

Lap Siding, 004

Lap Siding, 005

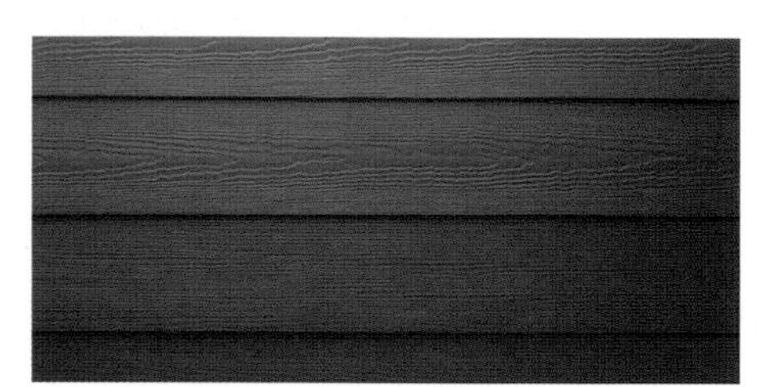
Lap Siding, 006

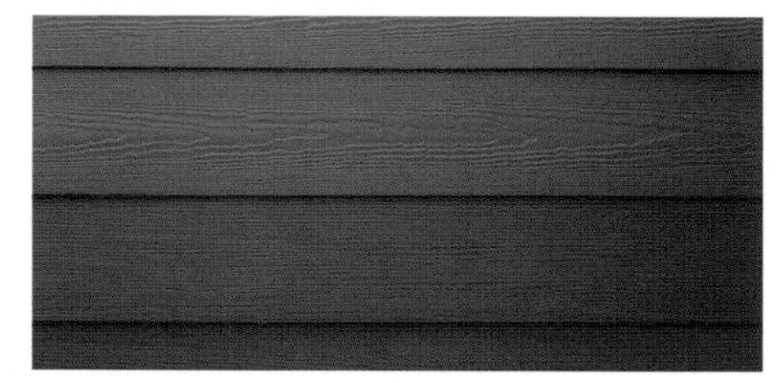
Lap Siding, 007

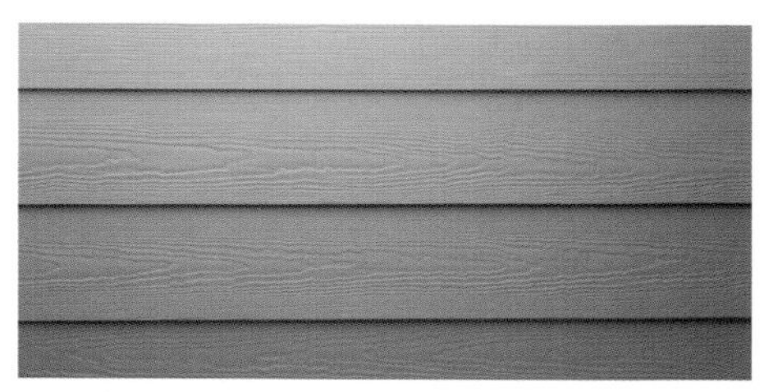
Lap Siding, 008

Lap Siding, 009

Lap Siding, 010

Lap Siding, 016

Lap Siding, 019

Lap Siding, 021

Lap Siding, 023

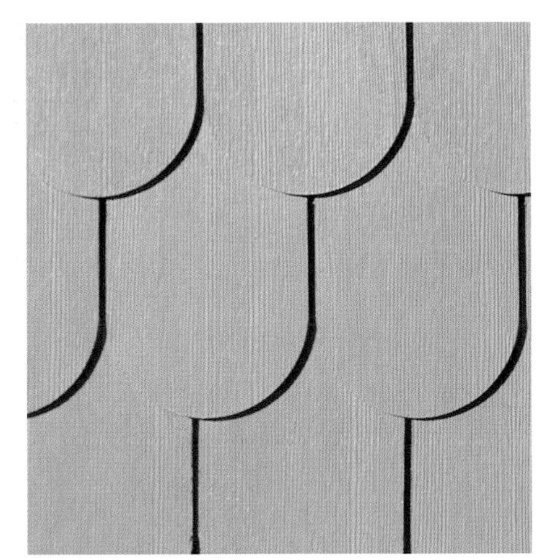
HardieShingles,
Half Round Notched Panel, 101

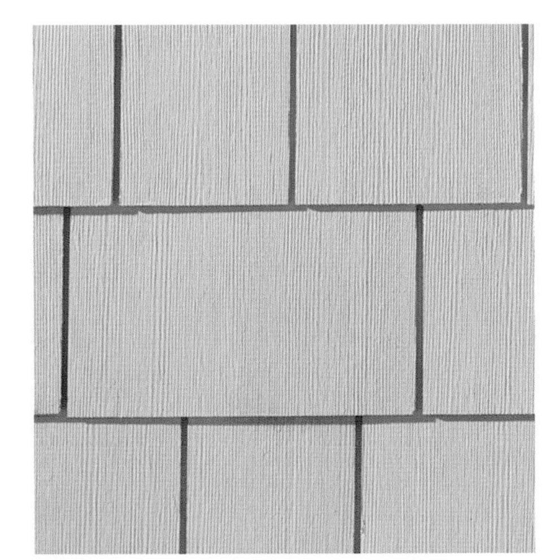
HardieShingles, Straight Edge Panel, 92

HardieShingles, Staggered Edge Panel, 97

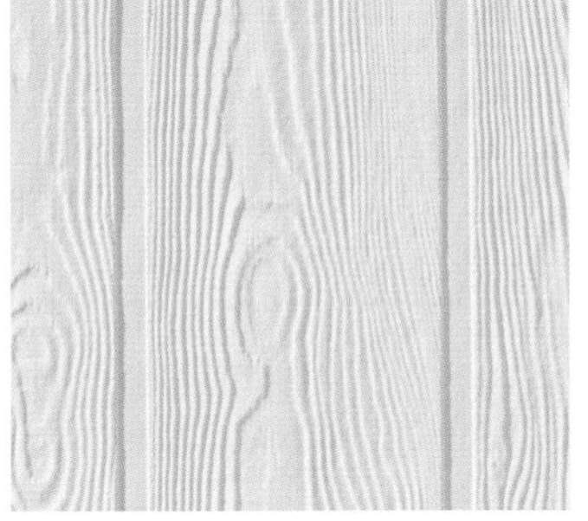
HardiePanel Vertical Siding, 125

HardiePanel Stucco, 340

HardieShingles, Individual, 88

HardieShingles,
Straight Edge Panel, 87

Cedar Groove
CertainTeed Siding

Cedar Stain

Mahogany Stain

Maple Stain

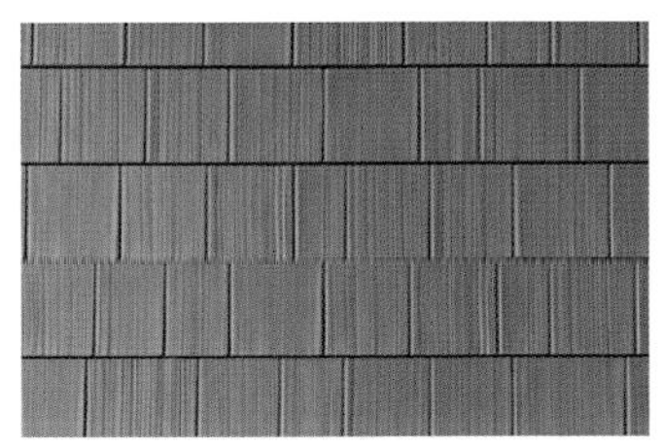
Perfection Shingle

Random Square Staggered

Stucco

Dutch Lap

cedar shingles

Western Red Cedar, Green semi-transparent oil stain
Shakertown Cedar Shingles

Western Red Cedar, Gray semi-transparent oil stain

Western Red Cedar, Redwood semi-transparent oil stain

Unfinished Western Red Cedar

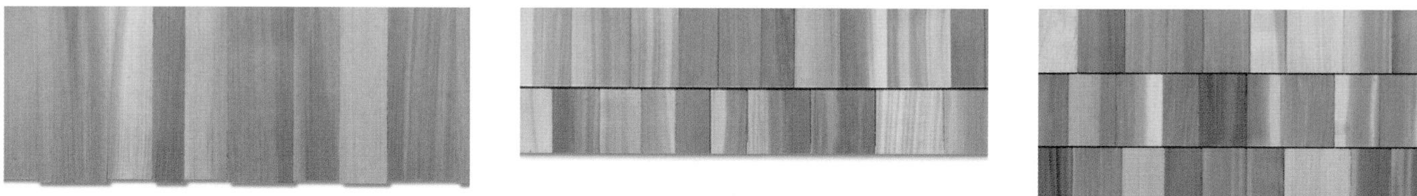

One course, staggered panel

Two course keyway panel

Three course keyway panel

metal

Stone-coated Steel (all),
Granite Ridge Thunderstorm
Gerard Roofing Technologies

Guardian

Canyon Shake Chestnut

Stone-coated Steel (all), Tile Terra Cotta

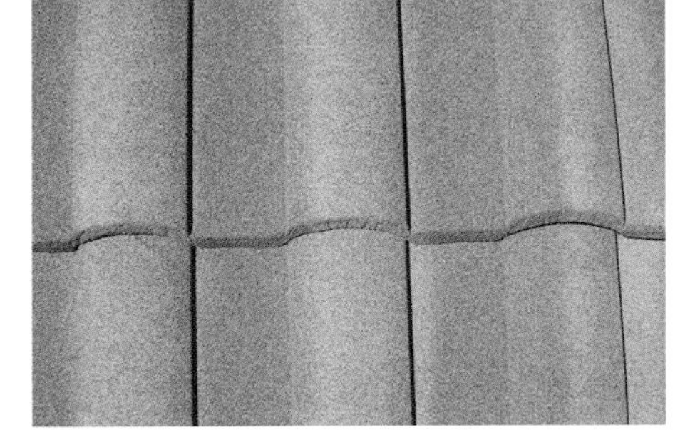

Barrel Vault Sunset

Canyon Shake, English Suede

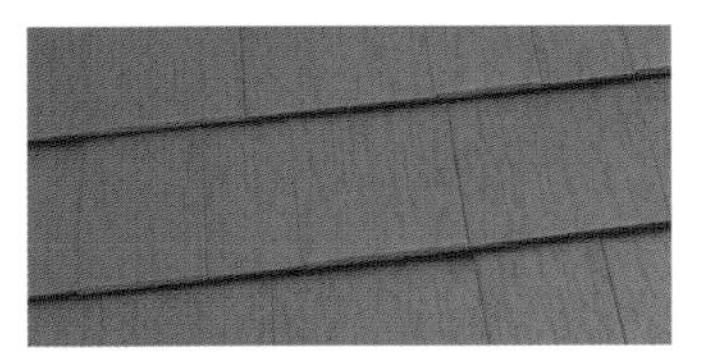

Aluminum (all),
Oxford Aged Bronze
Classic Metal Roofing Systems

Oxford British Red

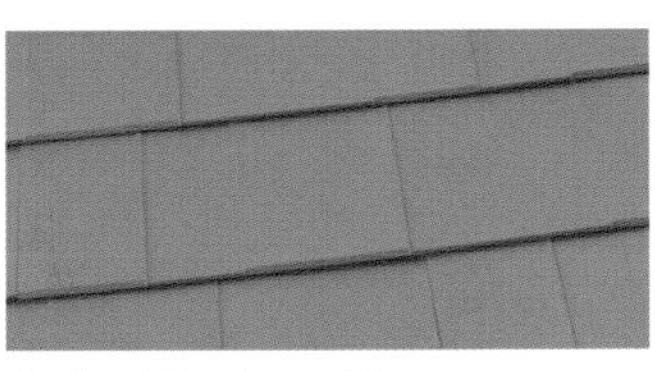

Oxford Buckeye Tan

Oxford Copper

Oxford Copper Patina

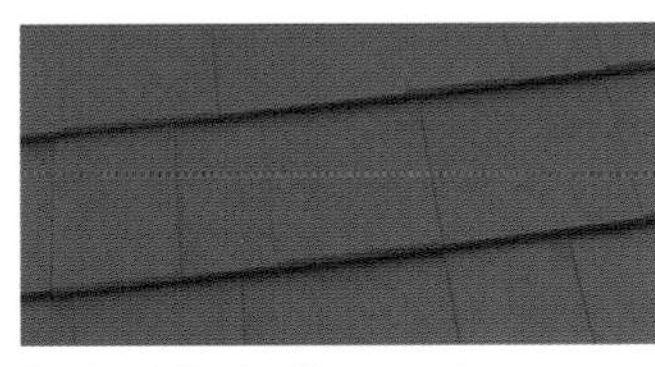

Oxford Dark Charcoal

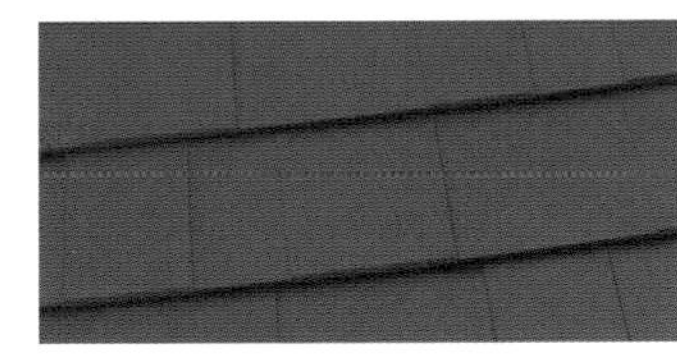

Oxford Fairway Green

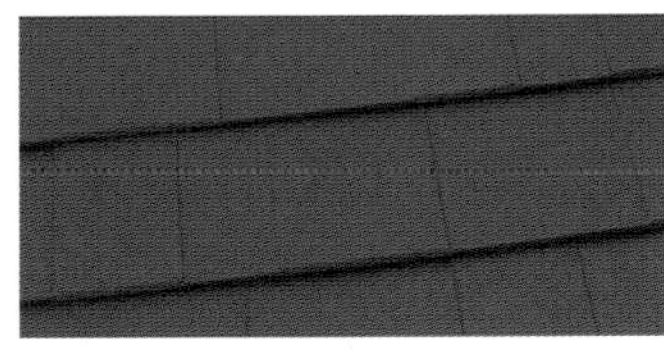

Oxford Musket

Oxford Tile Red

Oxford Weathered Wood

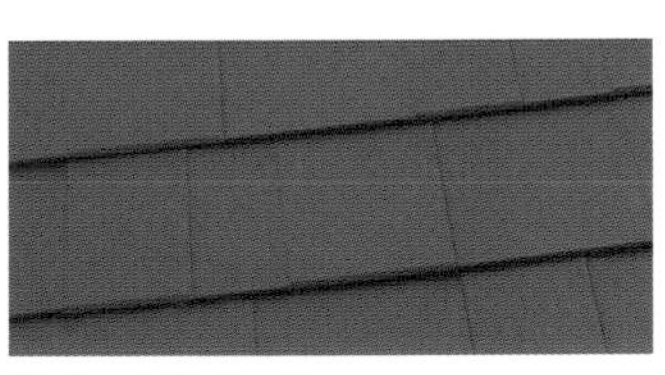

Oxford Slate Gray

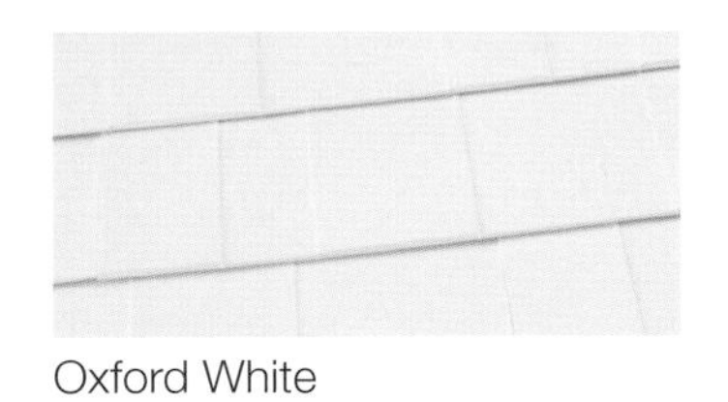

Oxford White

metal

Steel, Standing Seam, Buckskin *Classic Metal Roofing Systems*

Steel, Standing Seam (all), Caramel

Deep Charcoal

Forest

Mustang

Pearl

Shake Gray

Terra Red

White

Recycled Metal (all), Rustic Black

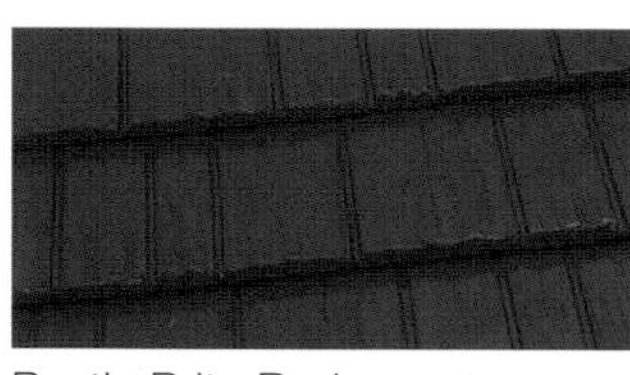
Rustic Brite Red

Rustic Deep Charcoal

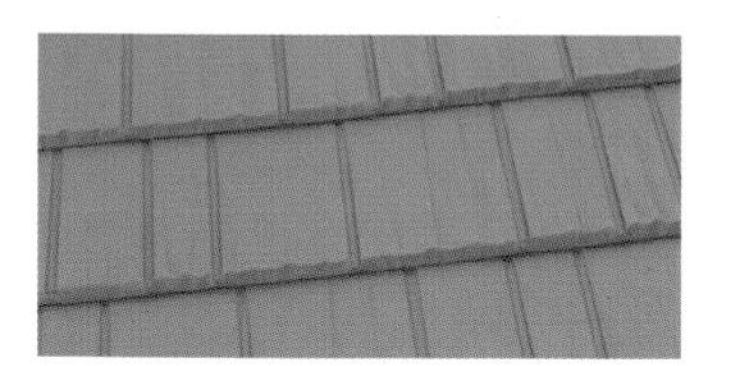
Rustic Buckskin

Rustic Caramel

Rustic Charcoal

Rustic Mustang

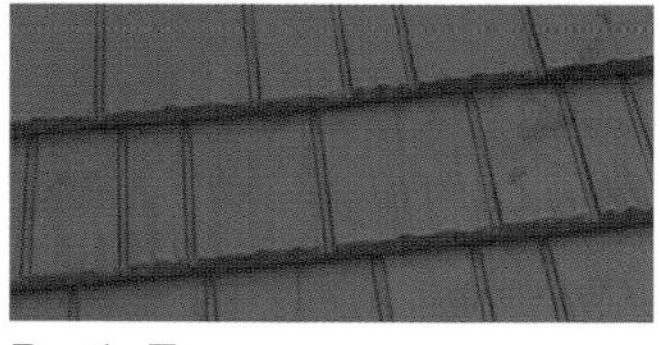
Rustic Terra

Rustic White

Aluminum, Country Manor Shake (all), Slate Gray

CMS Aged Bronze

CMS Buckskin

CMS Copper Patina

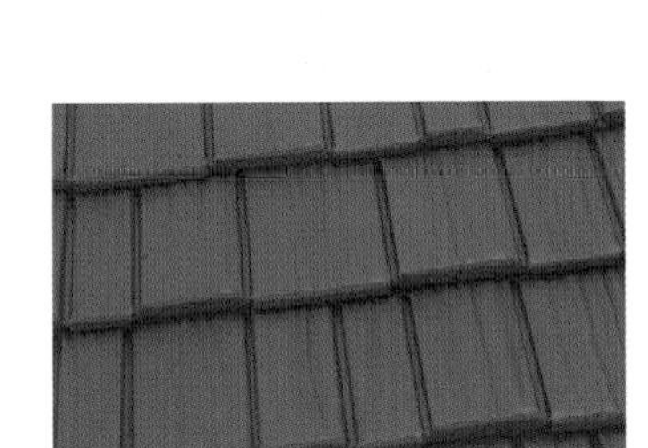

CMS Dark Charcoal

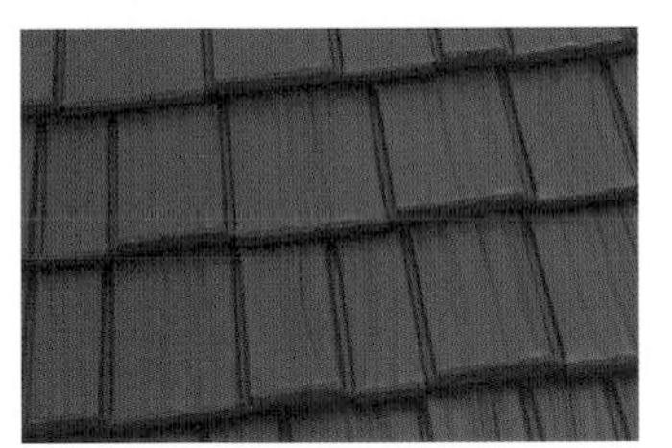

CMS Fairway Green

CMS Musket

CMS Tile Red

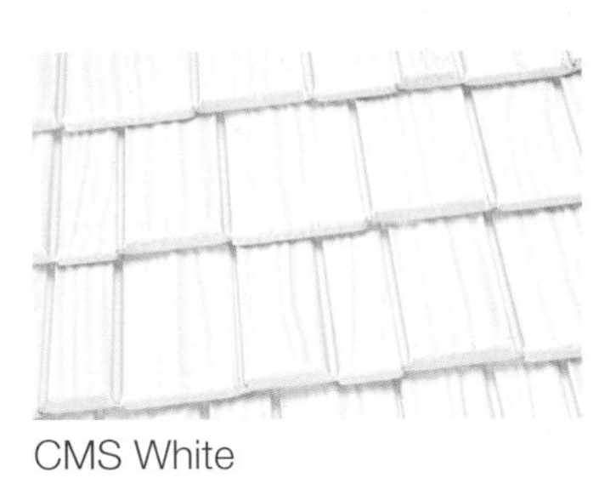

CMS White

Steel (all), TimberCreek Beech

TimberCreek Mahogany

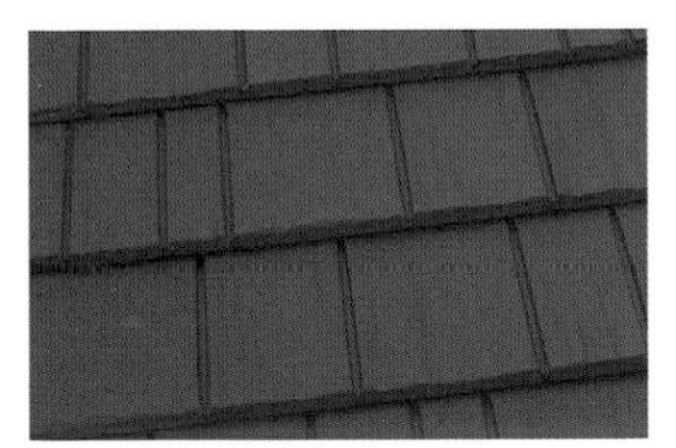

TimberCreek Spruce

TimberCreek Maple

TimberCreek Terra Cotta

concrete

Riviera Sequence *Vande Hey Raleigh*

Costswold Stone Sequence

Custom Brush Sequence

New England Sequence Slate

Shake Sequence

High Barrel Spanish Sequence

Tapered Turret shingles rising to a glass point, from *Vande Hey Raleigh.*

Single concrete shingles

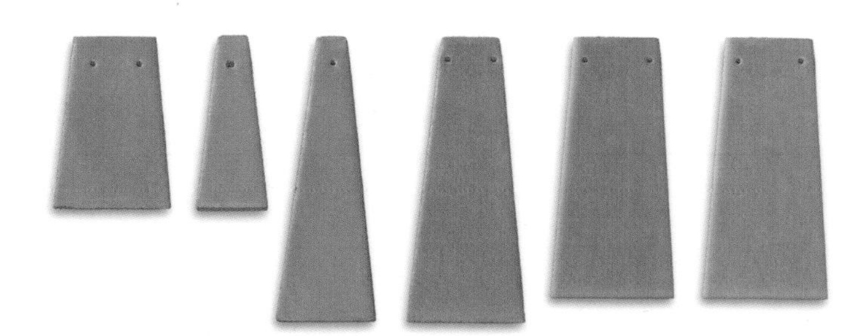
Single Tapered Turret concrete shingles

Riviera Shingles

Concrete Shingle

Turrets Shingle

Cotswold Stone

Custom Brushes

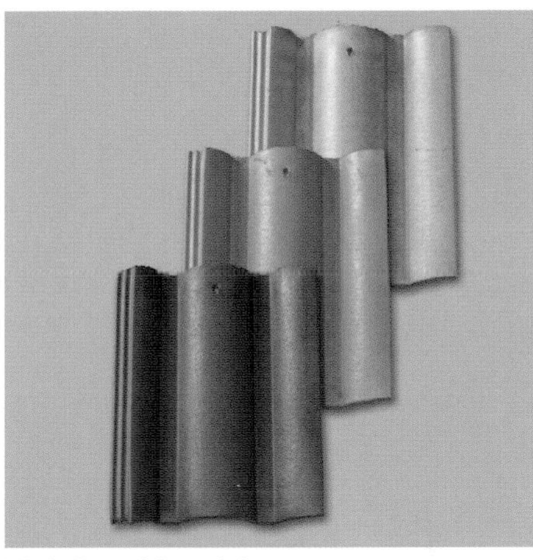
High Barrel Spanish

English Shingle

Modern Slate

slate

Non-weathering Gray Green *Greenstone Slate*

Non-weathering Mottled Purple

Non-weathering Purple

Semi-weathering Gray Green

Some slate changes color naturally and gradually when exposed to the elements. Non-weathering (or unfading) slates will show the least amount of color change, weathering slates will show the most, and semi-weathering are somewhere in between.

Vermont Gray Black

Royal Purple

Vermont Strata Gray

Vermont Clear Gray

Unfading Black, Semi-weathering Gray Black, Semi-weathering Vermont Gray, Unfading Purple, Unfading Mottled Purple and Green, Unfading Green, Custom Blend, Custom Blend, Custom Blend *North Country Slate*

These complex roofs show off the rustic beauty of slate roofing tiles.

Unfading Purple, Unfading Gray, Unfading Mottled Purple and Green

Semi-weathering Sea Green

rubber

Seneca Cedar, stained gray *EcoStar*

Seneca Black

Seneca Cedar

Seneca Chestnut

Seneca Green

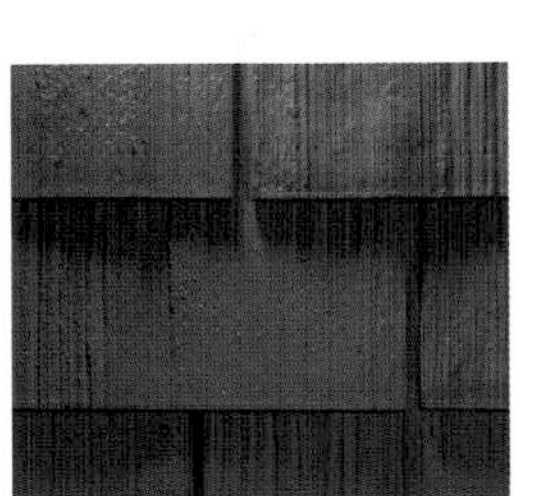
Seneca Midnight

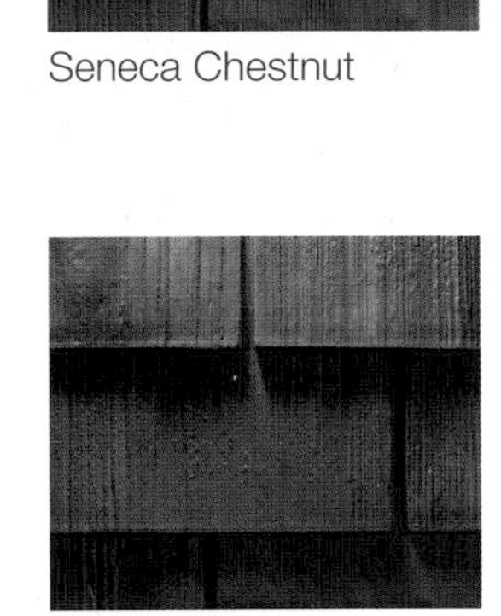
Seneca Red

Majestic Smoke Gray

Majestic Earth Green

Majestic Traditional

Majestic Beaver Tail

Majestic Bevel

Majestic Chisel Point

Titus

Majestic Chisel

clay

Traditional Red *Northern Roof Tiles*

Traditional Brown

Arbois

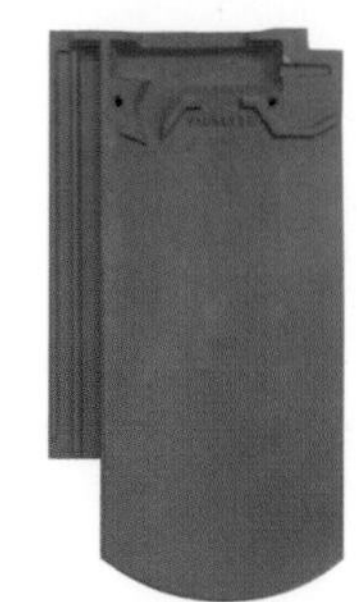

Vauban Beavertail

Vauban, Champagne

Victorian, 441

Victorian Grey, 451

Actua Champagne

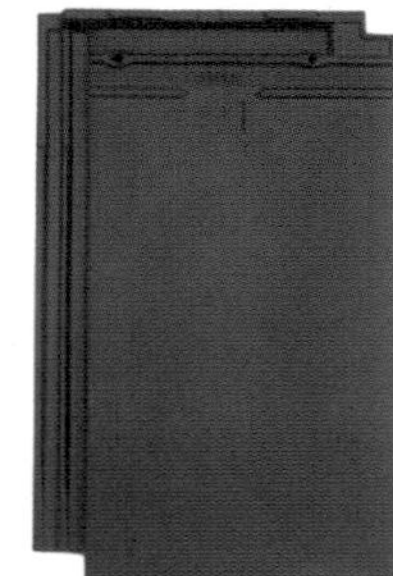

Actua Slate Grey

Altusa

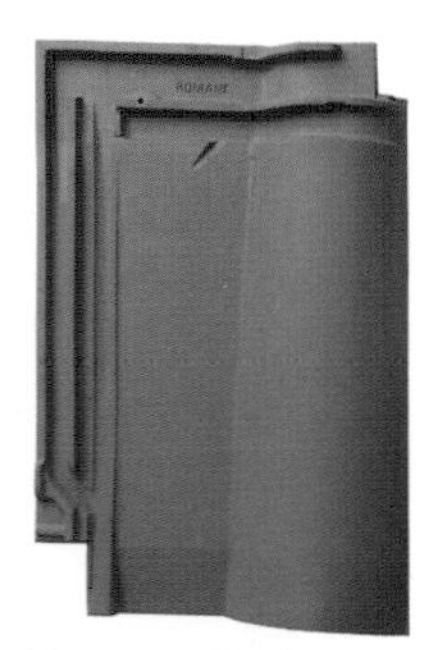

Romane Red

Romane Antique

Romane Straw

Arboise Beavertail

ceramic

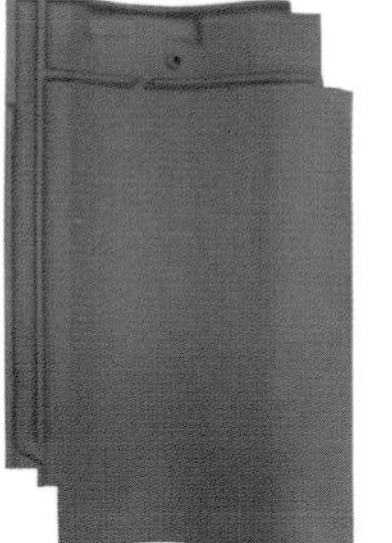

Madura Natural Red

Romane Ocra

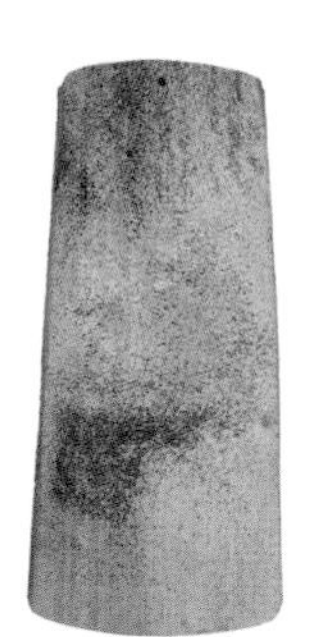
Mission, Mixed Tone

Tudor

French Lichen

French Red

French Vineyard

Frontier

Goxhill

St. Foy Red

St. Foy Chevreuse

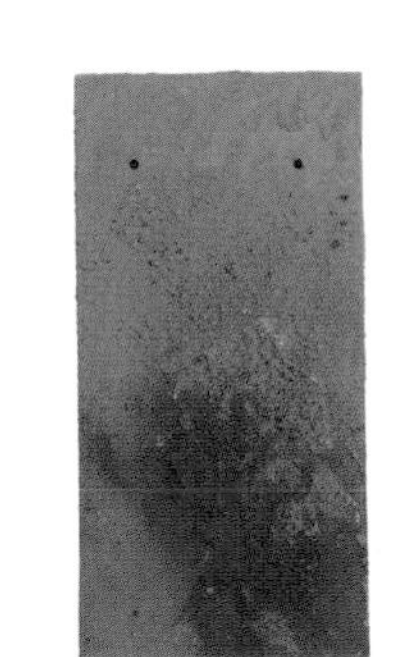
Pilgrim Granular Weathered

Glazed ceramic roof tile
Purple Sage Ceramics

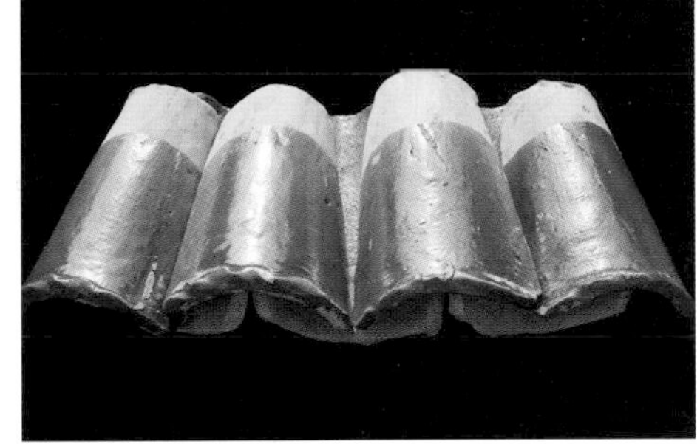
Verde handmade terra cotta
Saint Tropez Stone Boutique

chimney pots

Jumbo Federal Spike, glazed terra cotta *ChimneyPot.com*

Halifax Windguard

Federal Covered Short

Medium Brushed Windsor

Magnum, Edwardian, Super Magnum, Mini Edwardian, Edwardian *Superior Clay*

Large Governor, Victorian, Kent, Governor

Bishop chimney pot, copper *Jack Arnold*

Cathedral, Small Birdcage, Diamond, Birdcage *Superior*

Two Victorian glazed terra cotta chimney pots, Halifax Windguard (left) and Halifax Register (right) from *ChimneyPot.com.*

drainpipes

Bacinellas Minoletti Collector
Park City Rain Gutter

Bacinella Roma Collector

Copper gutters, collector and downspout.

Square Collector Box

Vent Cap

Bochetta Collector

Standard Outlet

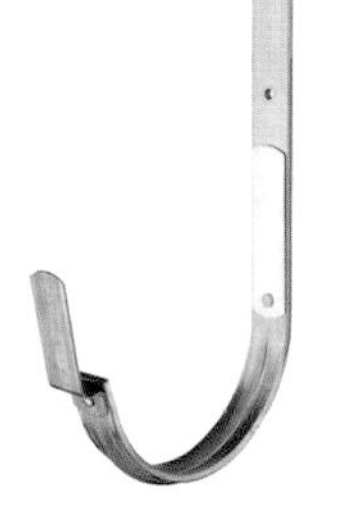

Gutter Bracket Roof Mount

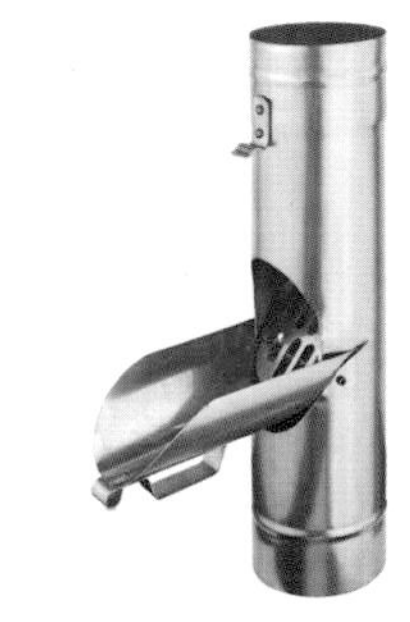

Deflector (Cleanout)

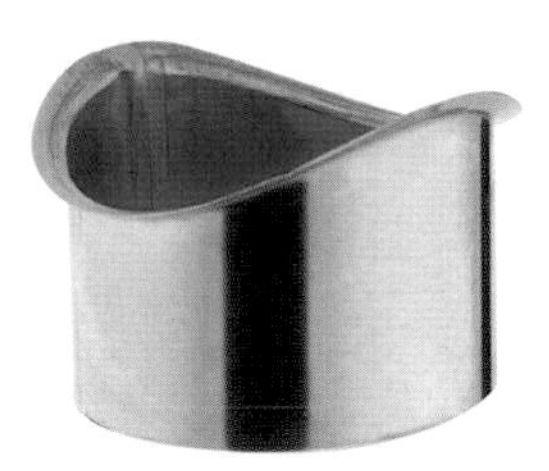

Mini Outlet, Shaped

Round End Cap

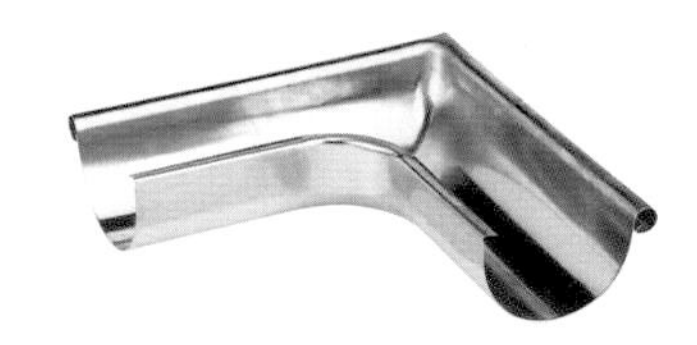

Outside Miter Box

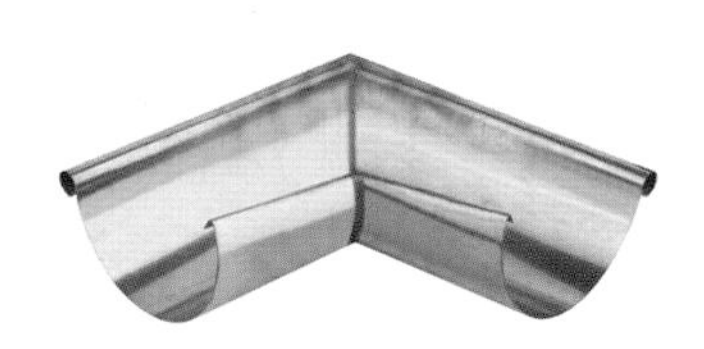

Inside Miter Box

weathervanes

Dragonfly weathervane
Westcoast Weathervanes

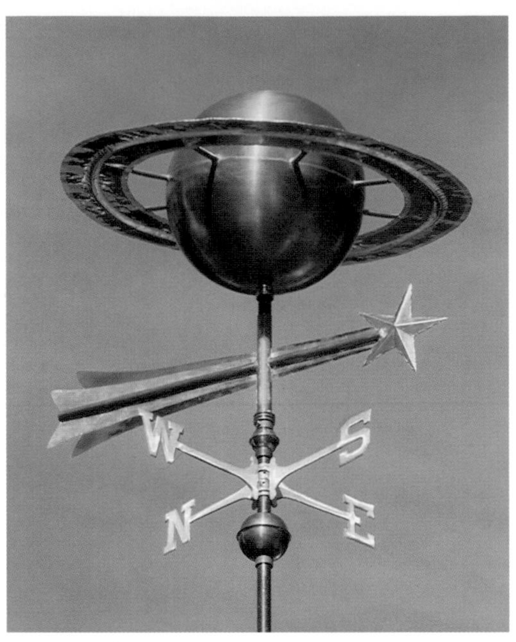

Saturn & Comet weathervane

Gallo *American Rain Gutter*

Cavallo

Drago

Strega

Banda

ridge caps

Elaborate iron ridge caps provide the finishing touch for this beautifully shingled roof. Ridge caps are the kind of detail essential in remaining true to a period style. Note the snow guards at lower center that keep ice and snow from falling.

Balmoral Cresting
The Architectural Iron Company

Empire Cresting

Regency Cresting

Gramercy Cresting

Empire Snowguard

Regency Snowguard

Chelsea Finial

doors & windows

I like to do corner windows, especially in the city because you can look down the street instead of just across the street.

PAUL FRONCEK, ARCHITECT

NEXT TO THE BASIC ARCHITECTURAL STYLE, YOUR CHOICE OF DOORS and windows has the most impact, inside and out, on how your home actually looks. Whether you have European-influenced arched Tuscan doors and windows, multipaned Old English style casements or double-hungs, as in Georgian or Federal style homes, a classic Palladian window or the sleek metal reveals of a Contemporary sliding door, you should be aware that whatever you choose for these inherently functional elements will define the style of your house. The use of glass in a home is all about manipulating light: this purpose can extend beyond windows and insets in doors to include glass in a multitude of other incarnations. More than most architectural elements, doors and windows combine form and function to make a visual statement.

The broken pediment over the porch, as well as the columns, beams and entablature are neoclassical, reflecting Greek and Roman precedents, while the arched windows, mansard roof and French doors add a Gallic flavor to this stately mansion.

When we design an entranceway, we start by trying to understand how a client wants you to feel when you come up to their front door. Do they want people to think: "What a beautiful, cozy little space," or come through the door and say: "Oh my god! Look at this!" The entrance should be appropriate to the rest of the house. When you walk through the front door, you shouldn't be disappointed. That means getting the scale and proportions right.

RICHARD LANDRY, ARCHITECT

entranceways

Some entrances are underplayed and humble, others anything but. Generally speaking, your front door is the first part of your home a visitor will see up close and touch and is therefore the real first impression whether they realize it or not. A front door speaks volumes about its owners: will the sound of its closing be the thin *clank* of a cheap used car or the solid *thunk* you get when you close the door of a Rolls Royce? Then there is the hardware. Door knobs and latch plates are often overlooked details that can make or break the overall impression of your entranceway. And don't forget a matching metal kickplate along the bottom exterior of the door, a practical necessity that can also add beauty.

These stunning wooden doors with art glass windows cap the ends of this vaulted entranceway (made of Chicago common brick), which was inspired by the long, narrow passageway in Frank Lloyd Wright's home in Oak Park, Illinois, that leads unexpectedly into the high-ceilinged "Children's Room." This one leads to the foyer seen on page 334.

The copper-shingled interior of this porch was designed and created by Colorado artisan Steven Stein. Elaborate care was given to this porch roofing — at the end of a long walkway — so that visitors to the home would be rewarded after the long walk.

Mailboxes are another entranceway detail that shouldn't be ignored. Instead of an afterthought, your mailbox can add ambiance as long as it is complements the rest of the entranceway: for example, a nicely tooled wrought-iron box for a Craftsman home, or a sleek stainless-steel contrivance on a Contemporary home. The same possibilities arise in choosing mail slots, doorbells and door knockers. While providing security, security grilles over small windows at eye level in your door can give an Old World look; if they are backed by a small hinged door, they can evoke the speakeasies of Prohibition. House numbers can fit into the overall scheme but are also an excellent way to add a distinctive touch for relatively little expense. Numbers that are ceramic, worked metal, carved wood or stone and, for the more contemporary-minded, fiber optic lights are all available, the only caveat being that they should be visible from the road and sidewalk.

Right: Narrow steel columns add a touch of delicacy to the cantilevered mass of this discreet entranceway. The door is mahogany finished in a dark stain. **Opposite:** Although essentially a Prairie Style home (with wide overhangs and horizontal banding of brick), this entranceway has elements of Greek Revival (dentil molding) and Art Nouveau (the strong organic curves of the floral design cut into the stone around the door). Note how welcoming the low brick walls are as they seem to embrace the visitor. Also note how the staircase framed by the window at left seems to float.

Porches and overhangs do more than keep your callers protected from the rain. They collaborate with other architectural details of your façade to make a design statement: a curved-metal half circle says Art Deco, while a flat roof supported by thin metal pipe is contemporary. And the neoclassical porch, with its columns and portico, is a direct link to the ancient Greek temple.

All of the details contained in the entranceway are enhanced by what leads up to it. Walkways and plantings are a inviting way to make the transition from the street to the home, the more curving or undulating the more interesting. For a charming Old World feel, have your entrance path pass under a trellis. Whatever you choose to do to enliven your entranceway, avoid the dreary ordinary if you want to create a feeling of welcome.

219

doors

Doors provide passage in and out of your home, and from one room to another. Although they come in an infinite number of styles, generally, doors can be classified into just a few types:

SOLID These can either be paneled (having raised and indented sections) or flush (slab). These common doors can have lights (glass) in the doors or in sidelights (beside the door) or fanlights (above the door). Generally speaking, panel doors have a more classic look and suit traditional home styles, and flush doors are more appropriate with contemporary homes. Georgian panel doors usually have both side and fan lights, sometimes so extravagant they become the visual focus of the façade. Victorian doors are known for their decorative, often stained glass. The Craftsman home is usually entered by a simple, squared off and symmetrical door that can have paneling of angled tongue and groove. A grilled "speakeasy" hole in the door that opens up to view callers gives a taste of the Old World.

FRENCH Glazed panels that go from floor to ceiling with both doors opening, usually into the interior of a room, but in some cases, outward. (Some manufacturers call French doors in which only one door is active garden doors or terrace doors.) Elegant and light in appearance, French doors are perfect for opening and closing off space — such as the breakfast room/patio or living room/dining room. The drawbacks are expense (especially if you want thermopane), and the fact that they are hard to screen.

SLIDING GLASS Offers the best view and ventilation but is visually not that interesting (and not appropriate for period homes). Usually only one side slides open, but more expensive models can be found in which both sides open. In addition, you can have large doors custom made that pivot in the middle. On the

Opposite: Framed by simple stone, plaster and wooden beams, this door (the design comes from Tibetan Buddhism) was hand carved in Mexico. Left: Substantial wooden Gothic doors are perfect for the heavy and formal limestone exterior of this home. Center: Delicate art and acid-etched glass on French doors provide both light and privacy for this grand bathroom entrance. Right: A Douglas fir pocket door. Pocket doors are particularly good for small rooms that don't have space to accommodate swinging doors.

interior of homes, the sliding door is often in the form of a pocket door, which slides efficiently out of sight into the wall — perfect for opening up and closing off rooms. Some manufacturers offer "French" sliding doors, which mimic a traditional French door in the size and shape of the framing but work as a sliding door.

SLIDING GARAGE-TYPE Long used in Europe and elsewhere to close up shops for the night or open up cafes to the street in fine weather, these segmented doors slide up into metal channels in the ceiling. Now they are being used as a convenient way to open up a room — or even an entire wall — to the outside. They can be used as a stock size or ordered custom from a garage door manufacturer.

INTERIOR These are usually less elaborate than exterior doors, except for the door trim, which can be almost as decorative. Contemporary designs favor flush mounted doors and reveals are used for a more minimalist look. Most custom doors are either solid (smooth, flat) or paneled (built up with moulding). The least costly doors are hollow or foam core, which have the benefit of being lighter than solid wood. The heavier the door and more solid the construction, the more satisfying it feels when you close it. Metal-clad doors, while providing a little more security, are also appropriate for more modern styles, from Art Deco and American Streamline to the International style.

Interior doors come in various incarnations to suit their functions. For example, pocket doors slide out of the way into the wall when space is a problem. And French doors, with their many panes of glass, flood the room with light as well as a sophisticated ambiance. Swinging doors, which are useful in the kitchen, can be given a professional air with the inclusion of a circular window around eye level as are seen in certain New York apartments.

Left: Steel fire doors descend from drums above the door to seal off these sectional and sliding glass doors, partly for security but primarily as protection from fire from the surrounding dry Californian hills. **Center:** Designed by Canadian sculptor Edward Falkenberg to look like a painting framed by the end of a hallway, this plank door was stained black — the eccentric red handles and swiped of red paint look like brush strokes. **Right:** Knotty pine boards, simple forged hinges and rope door pull prove you don't have to spend a lot of money if you have imagination and taste.

Windows are only good for three things: one, view; two, daylight; three, ventilation. The double-hung window in the middle of the wall doesn't do any of those things. You can't look out of it because it's in the middle of the wall and the wall is dark on both sides so there is too much glare. It doesn't light very well because the light doesn't hit anything and reflect off of it. And it's a lousy ventilator because it's hard to adjust and its not in a good position to set up air currents.

BARTON PHELPS, ARCHITECT

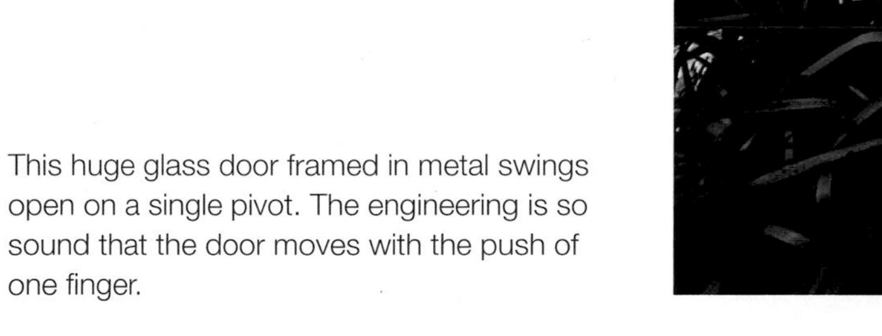

This huge glass door framed in metal swings open on a single pivot. The engineering is so sound that the door moves with the push of one finger.

The positioning of interior doors is important because it governs traffic flow. For example, if you have doors that open up in the middle of walls, the traffic flow will be funneled into the middle of the room. You probably don't want a prime view of your powder room toilet, no matter how elegant, from the head of your dining room table. Choose carefully, when you have the opportunity, which way a door will swing and where it is placed in the wall, and you won't regret the choices you make later.

Opposite: Acid-etched, tempered glass was installed from floor to ceiling inside the front door of this narrow home to make a foyer without cutting down on light. By using a single sheet of glass and not a frame wall, an additional six inches of space was created for the breakfast area to the left by Toronto designer Dee Chenier. **Left:** This renovated TV studio had a huge back door to bring in equipment. The oiled steel door was put in as a stylish replacement.

What separates a good house from a great house is how it engages the environment. For example, even a small fence and gate in front of a modest house will create a sense of entry. If your gate is the same shape as the front door, you are being prepared for the aesthetic experience.

GARRETT EIKEN, ARCHITECT

gates

In a courtyard or garden wall, gates are an opportunity for whimsy and extravagance. They are the intermediate portals between the street and your home. They can latch with a simple looped string or swing open electronically between stone pillars complete with intercom speaker. Gates can be light and airy when constructed of metal or wooden spindles or even wire mesh, or heavy and secure (not to mention private) when made of solid material. They can be primitive, as are those made of scrap wood or thin saplings, or sleek and modern in copper or stainless steel. And always think "more" with gates — more elaboration and more detail than what you would find on a regular house door or entrance. You can and should see your gates as an excuse to be creative, both in materials and design.

In a courtyard, fence or garden wall, gates are an opportunity for whimsy and extravagance. Framing the entrance to your property, gates also provide your first line of security and, as such, are usually substantial. This of course doesn't mean gates should lack beauty. One way around this is with the slender steel bars of the wrought-iron gate, something that fits in very well with an adobe, stone or brick wall. Wrought-iron gates also have the ability to let in light and show off the house in a very stately way. To shut out the world and increase privacy, a solid wood door works best. Elaborately carved doors (see facing page) can add elegance to an entranceway, but a gate of simple dowels and boards can perform just as well at a fraction of the cost and works particularly well for a Cape Cod or Southwest style residence. The style of your gate will usually be compatible with whatever fence, lattice, pergola or other structures you use to define your landscape. Generally, the grander your home, the more imposing your gate can be.

Left: Carved plank driveway gates are remotely controlled from the house. **Center:** A Craftsman gate with antiqued hardware. **Right:** Wooden gate in a Spanish design set in an adobe wall that has been whitewashed. **Opposite:** Antique doors from India are set in this adobe wall, made of engineered adobe block (from Clay Mine Adobe in Tucson) covered in adobe clay and cement.

garage doors

Historically, the word garage is derived from the French *garer* — to shelter or protect from the elements. Although they are one element of residential design that many people gloss over, garage doors are more than simply a utilitarian front for a storage area. For one thing, if your garage fronts the house, as is usually the case, you are looking at about half of your actual building frontage. Think of that when you think of "curb appeal." Even if you have a garage at the back of the home, it will still form an integral part of those behind-the-house courtyards.

In fact, the first garages were the stables in which the horses and carriages were kept (hence the popularity of "carriage house" garage doors). Garages were originally separate buildings. Many of those older homes in which the chauffeur or carriage driver once had quarters above the garage now have offices or guest rooms in the floor above. Eventually, however, as land became more valuable,

This garage door was hand-painted to resemble the rough stone walls that surround it.

Left: Textured stucco with a modified Greek key design. Center: A good example of carriage house doors, made of mahogany. Right: Panels of copper treated with acid are a fine addition to these wood paneled garage doors. They are flanked by a fieldstone column and Craftsman light fixtures.

the garage was more often incorporated into the overall envelope of the main house itself. In either case, garage doors are in important design element, although they seldom call for as much detailing as, for example, an entranceway. The only exception to this is when there is a desire to make the garage doors blend in more, in which case you see elaborate doors that mimic the entranceway, or even doors artfully disguised as the surrounding walls (see the painted door in the photograph at left).

Theoretically, unless you are going to keep your car in an open carport, just about any kind of door arrangement will do. There are, however, three common kinds of door that serve particularly well for garage doors.

HINGED Consisting of two large doors, hinged at the sides, that open out like French doors. This is the oldest style of garage door and complements a more traditional exterior, as in a Craftsman, Colonial or Victorian home. Because they usually have several small, multipaned windows, they let in more light than many modern doors. The downside is that they usually have to be opened manually, they take up space once they are open, and they can be obstructed by objects in the drive, snow or even cars parked too close.

SECTIONAL Two or more panels, generally of wood or metal, hinged together so they can roll straight back into the top of the garage on rails. This door has the advantage of working well with electronic garage door openers, plus it provides more headroom than the one-piece swing door.

ONE-PIECE SWING This door swings up on a pivot. It has less headroom than a sectional but at lower cost, and looks almost the same from the outside. One advantage to the one-piece swing door is that, since it doesn't have to hinge or bend, it can be more easily made of material that complements the rest of your exterior such as clapboard or shingles.

With older, original windows it is important to not juxtapose them too much with newer windows. The idea is to let them have an identity of their own.

RICHARD TRAMAGLIO, BOSTON ARCHITECT

Left: An ornamental urn sits impressively in front of these wooden double hung windows. **Center:** A large gabled dormer with four multipaned lights splits the eaves of this roof. **Right:** Casement window deeply set in a flared cedar shingle wall. Note the cherry wood band over the top. **Opposite:** Antique French windows open up into a greenhouse and provide a country courtyard atmosphere.

windows

Windows do more than just let in light, the view of the outside and fresh air. Yes, they keep out the weather and other potentially harmful aspects such as ultraviolet rays and even intruders. They open and close, tilt and slide, but more than that, they determine your interior and exterior views. There are areas in which you can you can afford to skimp when planning your building budget, but windows is not one of them. There are questions of shape and size and materials to be considered, and a look at the standard window configurations described here can be helpful. From these traditional and modern options, you may discover the perfect window with which you can create an exterior façade and interior views to suit your home. They can be finished to match your exterior trim on the outside and your interior trim stain on the inside. Or you may be inspired to venture into the world of custom windows, with your own signature divided light patterns, designed to complement other existing architectural features of your home. A good place to begin is by considering the types of window available.

DOUBLE HUNG This classic look — essential for the Georgian style, which usually has five double-hung windows evenly spaced across the second floor facade —which has been around for centuries consists of two sheets of glass, the upper sash able to slide down (and in some versions, tilt in for cleaning) and the lower one free to slide up. One benefit is the minimal amount of space it needs in order to open, making them ideal for areas where space is a priority, as in hallways, porches and small rooms. Another is that it allows you a choice in ventilation — from the top or the bottom of the window.

CASEMENT A one-piece window hinged at the side so it can be cranked open either to the outside or inside. Casements provide a clean, solid surface of glass

You can't go to a mass commercial window manufacturer and get a properly detailed window. We used to order them without trim and then take the sills off and completely rebuild them, but that's too expensive. Now I go to a small craft shop and get them made custom.

TONY JENKINS, HISTORICAL CONTRACTOR

This bay window has been clad in copper, in part for its stylish look but also to match the copper gutters and drainpipes that grace the rest of the house. The copper has been coated with a heavy-duty wax (used on bowling alleys) to retard verdigris.

and are said by some manufacturers to provide a better insulated seal, because there are no gaps between sashes (as in a double hung) for cold or hot air to get in. This style works well with Contemporary homes as well as Craftsman, Tudor and Mission styles. Two casement windows side by side form a French window. A casements is convenient over a sink because it's easier to crank it open than to lean forward and lift a sash. (The crank handles can interfere with the operation of window treatments, however, so look for handles that fold into themselves.)

SLIDERS Usually have one side that is fixed glass and one side that slides past it (like a horizontal version of the double hung). Often found in Contemporary homes.

AWNING Hinged at the top and opens to the outside. An awning window is a good choice for a room in need of ventilation but also protection from rain. Awnings are often used in combination with (flanking) picture windows.

TRANSOM A transom window is like an awning window in that it is hinged at the top. It is sometimes set over picture windows, double-hungs or casements, or over French doors to increase the sense of height and provide further ventilation.

JALOUSIE Narrow plates of glass that can be opened and closed like a Venetian blind. A jalousie is one of the defining elements of the Florida room and other tropical styles. Because it doesn't seal very well, it is usually used where you don't need the airtight seal achieved by thermopane.

BAYS AND BOWS A bay window is traditionally made up of a series of three windows joined together at angles (often a large "picture" window flanked by narrower side lights that open) to form a protruding space or "bay." A bow window is similar to a bay, only made of curved glass and forming a curved bay.

The balcony of this home "pushed" the windows over in the design, so architect Barton Phelps crinkled the windows until they became what he calls "a kind of architectural act."

Chicago has a lot of gray-sky days, so I do a lot of skylights in which you can't actually see the sky — their only function is to let in the light.

PAUL FRONCEK

One elaboration on the bay window is the oriel window, which projects from an upper story and is supported by decorative brackets. The oriel is most often seen on Victorian homes or public buildings. A bay or bow window is a good way to bring the outside in and a natural location for a window seat.

THERMOPANE Two, or sometimes three, panes of glass sealed together with either air or argon gas trapped between them to act as an insulator.

SKYLIGHTS A great way to get light into dark interiors, especially tightly packed city homes. They have a reputation for leaking, however, so get them installed by a reputable dealer who offers a warranty. The old problem of heat buildup during the day has been solved by the use of sliding blinds that can be used to easily cover them up, as well as low-e glass coatings. One of the most interesting elaborations is the tube skylight that essentially "pipes" in light along a reflective tube (somewhat like fiber optic lighting) into any space in the house you want. The only drawback is you don't get to see the sky.

This glass block floor is connected to a tunnel skylight that brings light down from the roof to a prism that lights the glass block floor on the third and second floor. At night, when the lights are on, it provides a diffused ambient light.

Left: A round-topped door set in a stone wall covered in foliage. The delicate iron railings taper in toward the door in a way that sweeps the visitor into the home. **Center:** Antique metal-grilled windows set in a modern adobe wall. The neutral look of adobe makes it *simpatico* with antiques. **Right:** The pointed arch of this window marks it as Gothic in style.

door & window styles

When considering the kind of windows and doors you want for your new home or your remodel, keep in mind that you probably have more scope than you think. Obviously the basic style of your doors and windows should fit in with the style of your house, but that doesn't mean you have to be a slave to it. As long as you keep the essential form of the style in your window and door treatments — the various grilles, bars, glass pane size and trimwork — you can play around with the size in order to take advantage of a nice view or to let in more light if you want. This is especially true of the (more private) sides and back of the house. No matter what style you are using, try to keep the tops and sides of the windows lining up with each other in order to maintain symmetry. On the inside, the window and door trim work best when they are the same as the wall and ceiling trim.

That said, there are some window and door styles that are more flexible than others. Craftsman-style windows and doors are often a good match for most high-end homes, not just Craftsman bungalows.This is partly as a result of the timelessness of Arts and Crafts look, but also because of the style's insistence on quality and creative appearance. The double-hung window has always been synonymous with the Craftsman bungalow, but awning and casement windows are also permissible, as are rows of windows in a dormer, or a picture window with a transom over and a double hung on either side. Another Arts and Crafts approach is to create a combination of a picture window (with a transom) and double-hung flankers. Metal-framed windows, usually multipaned casements, have always been used in the Art Deco and International style of Contemporary house and their appearance says "modern" more than anything else. Another

These steel columns are here so that the corners could be open enough to accommodate the view of a ravine. Because there are also steel beams in the ceiling, only one column was needed but architect Bruce Kuwabara split the columns into two so they would be more slender. The two columns are also a reference to contemporary Italian architect Carlo Scarpa, who had great success in pairing elements. The effect is one of tranquility.

style to consider, no matter what you are building, is the rustic elegance of the Southwestern look, epitomized by rough, wide boards and hand-wrought iron trim on doors. Wrought-iron screens, sometimes in breathtaking elaboration, provide architectural elegance and interest along with the obvious security benefits.

Arched windows, the hallmark of styles such as Gothic (when pointed) and Spanish Colonial, Italianate and Palladian (when rounded), can be more difficult to find window treatments for than rectangular windows. Round windows present the same problem, although they are particularly good for accenting a façade. The eyebrow window, perfect for the English Cottage look, is even harder to dress. Dormer windows, typically found in Cape Cod style houses, are explored in the previous chapter, Exterior Details.

With windows, we try to not just punch an opening in the center of the wall, but to use the opening to wash its surface. If it's in the center of the wall you can often have a glare condition because of the contrast between the window and the solid wall. But if you slide that window over so its edge is meeting the perpendicular of the wall, you get a wash of light against that edge, which softens the glare and gives a much better quality of light into the room.

BUZZ YUDELL, ARCHITECT

The sun porch on the back of this house softens the stone. The style is Regency — a more relaxed outgrowth of the Georgian period. The railing looks Craftsman-like because the Arts and Crafts style borrowed from the Regency period 100 years before — only usually in a heavier style. Notice the top lights over the windows and the transom over the door — another Regency favorite.

A wall or screen of windows provides maximum transparency and the feeling of being outside. But this often gets used in the wrong setting. Like in a ski chalet that has a huge wall looking out at the slopes — if they are only on one side of the space they can produce so much glare you can't see anyone on the other side. One remedy is to have windows on other walls. The other is to use less glass.

RICHARD TREMAGLIO, AN ARCHITECT WHO WORKS WONDERS WITH SMALL SPACES

Before deciding on window placement, spend some time on the site if you can with a compass and a light meter in order to see where the sun hits and with what intensity at different times of the day. You don't want the sun to hit your TV screen at any time, but sunshine in the kitchen or breakfast nook in the morning is a real joy. Bedroom windows should be positioned to let in cool breezes at night, but installing walls full of north-facing windows in a cold climate can result in higher heating bills.

Like a picture frame around a painted landscape, this platform draws our eye to the view by framing it, even though there are already floor-to-ceiling windows behind it.

door & window materials

Windows and doors are generally available in just a few basic materials.

WOOD It's the most expensive but it also looks the best. There is a huge visual difference, for example, between plastic and real wooden mullions. With wood mullions, the wood has to be milled, something that becomes even more costly with thermopane because you need three sets of mullions (one on the inside surface, one on the outside surface and one sandwiched in the middle to give it a continuous look). The only other drawback to wood is that it has to be protected from the elements and is therefore relatively high maintenance, although using naturally preserved resinous wood such as redwood and cedar can minimize this.

WOOD CLAD Essentially wooden windows clad on the outside with either aluminum or vinyl. These have the benefit of the look of wood on the inside and easier maintenance on the outside.

VINYL Although it doesn't have anywhere near the charm of wood, vinyl (polyvinyl chloride or PVC) lasts forever, doesn't have to be recoated with anything periodically to protect it, and most vinyl windows have hollow cavities inside them that protect against heat loss and condensation.

ALUMINUM Has the same low-maintenance qualities as vinyl but none of the class and beauty of wood and is said to transfer heat and cold more quickly.

One choice you should consider when buying windows or glass doors is low-emission (or low-e) glass. A coating on the glass blocks certain long-wave radiant heat from entering you home, which helps to keep it cool. This works very well with sunrooms, sliding-glass doors and glassed-in cupolas, especially if you don't want to have to block the view with curtains or blinds. Another option is argon gas and yet another is storm-proof glass, especially valuable in hurricane-prone areas.

Left: A natural wood door with a finely worked wrought-iron gate in this adobe home. **Center:** Layers of moldings and very thick window posts set off these wooden mullioned windows. The transom lights at the top add a finishing touch and are often used in Georgian, Regency and Arts and Crafts homes. **Right:** Frosted glass in a stainless steel door.

The use of architectural glass is all about the manipulation of light in and around the home. One of the obvious attributes of glass in residential settings is that it allows light to pass through while still providing acoustic and visual privacy. This makes it perfect for large, open spaces such as lofts and converted barns and schoolhouses, where the desire to introduce light into the interior can conflict with the needs of privacy and room delineation. It also works well in the bathroom, where an easy-to-clean surface that can obscure sightlines is essential.

Right: This glass brick wall extends down to the floor below. **Opposite:** A glittering corner wall of recycled, lead-paned windows lets the light into the second floor of this Craftsman home.

glass block

Glass block has been around since the early 1800s, when crude squares of solid glass were used to let light into the holds of ships. The first true hollow glass blocks were made in 1902, but it wasn't until the 1930s that the manufacturing process was perfected, just in time for the Art Deco and Art Moderne styles. The look has never really gone away, and can be seen today especially in homes that cultivate a retro International Style. Keep in mind, if you are interested in untraditional shapes and motifs, that glass brick is easy to lay in curves. And, because of their obscure nature (clear glass block lets in 80 percent of the available light), they are great in bathrooms and other settings where you want to maintain privacy but also let in the light. Etched, pebbled or obscure glass can produce a soft glow of diffused light.

architectural glass

Durable, easy to clean and of unlimited variety in appearance, architectural glass is becoming one of the most popular of architectural materials — it is now used on countertops, walls, floors, and even sinks. Glass can be etched or sandblasted to give it a pattern or make it obscure. Textured glass can be remarkably interesting to look at, for its own sake as well as what the texture does to light throughout the course of day.

A lot of architectural glass is tempered glass — glass that has not only been made stronger but shatters into thousands of tiny, harmless pieces instead of dangerous shards of glass. This is essential when installing glass walls or floors and is the reason shower and sliding doors are tempered glass.

Left: Glass doors and screens provide acoustic privacy without blocking the light, which is particularly important in smaller homes. The square etched pattern also provides definition to the space, which alleviates one of the dangers of opening up an old space. **Center:** Oak leaves are the theme of this sandblasted window. While looking good, the obscure glass also doubles as a privacy screen. The sandblasting was done by Toronto artist Jeff Jackson. **Right:** Half-inch sandblasted glass was put up in front of this window on a busy street in order to provide privacy and to avoid putting on any kind of window covering. Serendipitously, the glass acts as a projection screen, so the lights of the cars play on the glass at night in a form of kinetic sculpture that pleases the artist who owns the home.

products

doors & windows

exterior doors

Rustic Collection, solid wood
Therma-Tru Doors

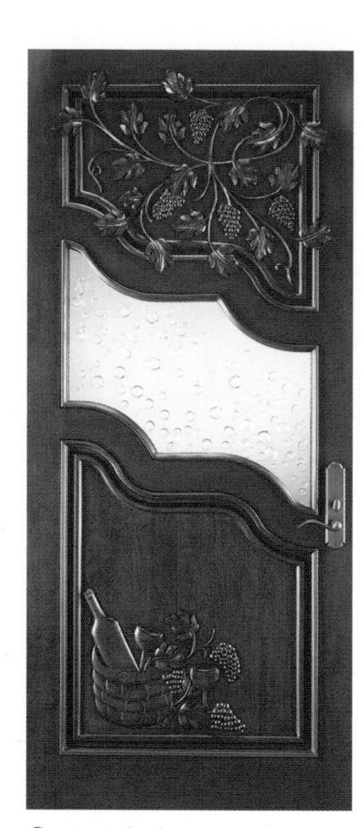

Grapes design, wood
Jeld-Wen Doors

IWP wood panel
Jeld-Wen

Aurora door with rustic speakeasy grill *Jeld-Wen*

Premium pine panel door
Jeld-Wen

Custom eight-panel Aurora, fiberglass

Premium single, fiberglass

Premium double, fiberglass

doors from reclaimed material

Door made of antique wood and grillwork from India. *La Puerta Originals*

Door made from English library panels and carved wood from Pakistan.

Antique Colonial doors from Egypt and Astragal from India.

Door made from rusted tin roofing and a plow disk.

Antique doors and other elements from Pakistan.

Courtyard gate made from antique Indian doors and reclaimed Douglas fir.

Mesquite garden gate made from antique Mexican door and reclaimed timbers.

La Puerta Originals' New Mexico yard contains hundreds of antique doors.

entrance doors & sidelights

Aurora Craftsman Style, fiberglass
Jeld-Wen Doors

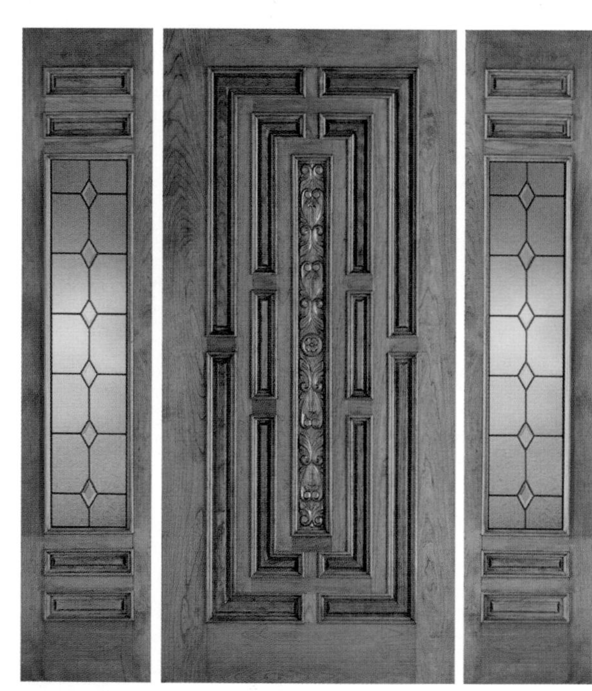

Custom IWP with sidelights *Jeld-Wen*

Raised panel with beveled glass,
Classic *Marvin Windows & Doors*

Prairie with Arts and Crafts glass,
Craftsman *Marvin*

Wooden panel with curved lights,
Luminary *Marvin*

Italian Renaissance grillework
with rain glass, Old World *Marvin*

Raised panel with black patina
grillework, Rustic *Marvin*

Double storm entrance
Pella Windows & Doors

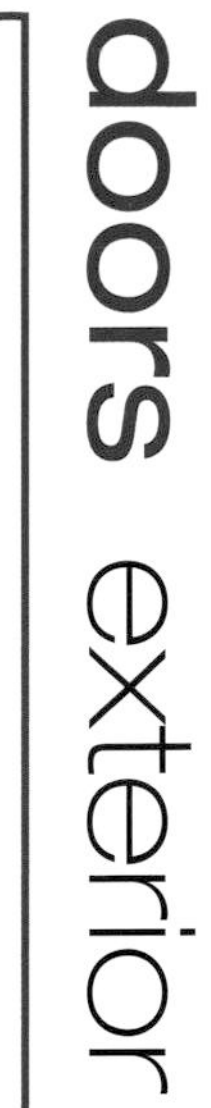

Three-panel, 10-foot mahogany
The Design-Build Store

Arched mahogany with sidelights
Upstate Doors

Solid panel A-18 *Amberwood Doors*

Fiberclassic with Sedona Glass, fiberglass
Therma-Tru Doors

Mahogany with triple-glazed glass *Upstate*

Mahogany with hand-forged ironwork
Design-Build

Curved-top wood door, A-1
Amberwood

Craftsman with art glass *Therma-Tru*

Pineapple pediment, header and keystone *Balmer Moldings*

interior doors

Quarter-sawn white oak *Upstate Doors*

Coped capitals with frieze and pilasters *Balmer Moldings*

Wood with sunburst glass, A-17 *Amberwood*

Acroter frieze and fluted pilaster surround *Balmer*

Wood panel, B-5 *Amberwood Doors*

B-1 wood panel *Amberwood*

Mahogany with crotch mahogany panels *Design-Build Store*

Arched three-panel mahogany with crotch mahogany panels *Design-Build*

Arched door in poplar *Design-Build*

Cherry French doors with five light-transom *Upstate Door*

Wood with Prairie-style grilles, Architect Series *Pella Windows & Doors*

Wood sliding doors with Prairie-style grilles *Pella*

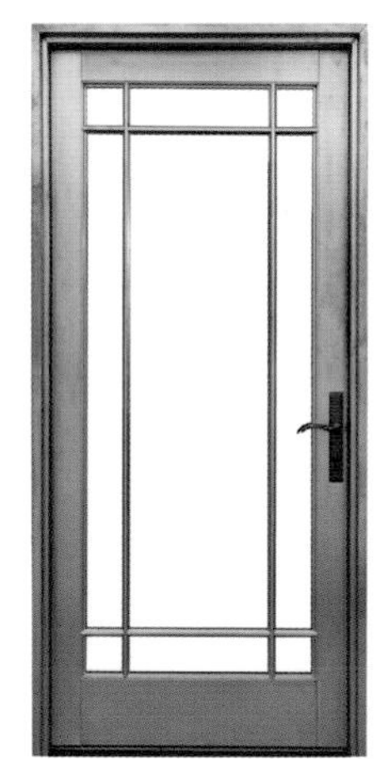
Single with Prairie-style grilles *Pella*

Mahogany pocket doors *Upstate*

carved wood

Buffalo Portrait *Richard Cornelius*

Alaskan Bull Moose *Cornelius*

Single Carriage House *Jeld-Wen Doors*

Short Panel Traditional *Jeld-Wen*

Classic Line 3, Z-brace *Real Carriage Doors*

Classic Line 1, Z-brace *Real Carriage*

Craftsman Traditional 1 *Real Carriage*

San Clemente insulated western red cedar *Garage Doors Inc.*

Gold Coast insulated western red cedar door with wrought-iron hardware and vents from *Garage Doors Inc.*

Pasatiempo insulated paneled western red cedar with dentil molding *Garage Doors Inc.*

Estate Series 86 with knockers *Jeld-Wen Doors*

house numbers

Brulee White *Clay Squared to Infinity*

Grotto
Clay Squared

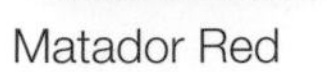

Matador Red

Marigold

Federal Blue

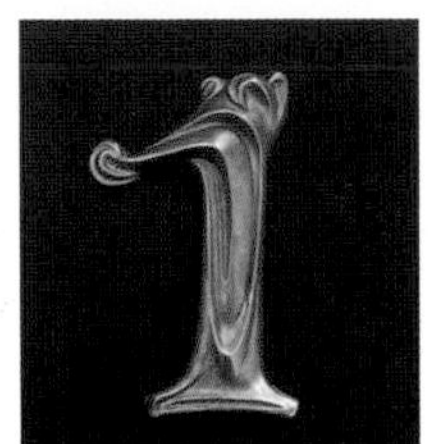

Art Nouveau house numbers, cast bronze
Historical Arts & Castings

Frank Lloyd Wright-designed house numbers, bronze and nickel. *Historical Arts & Castings*

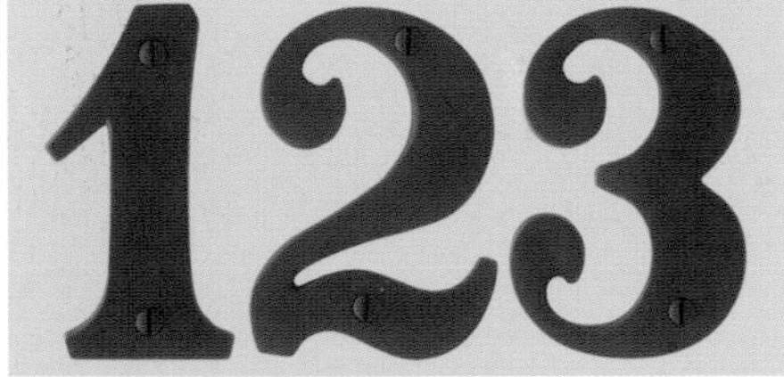

Brass 8238 *Rejuvenation Hardware*

Brass 4097
Rejuvenation

Wrought Iron
Acorn Manufacturing

Vine motif, ceramic
Rocheford Handmade Tiles

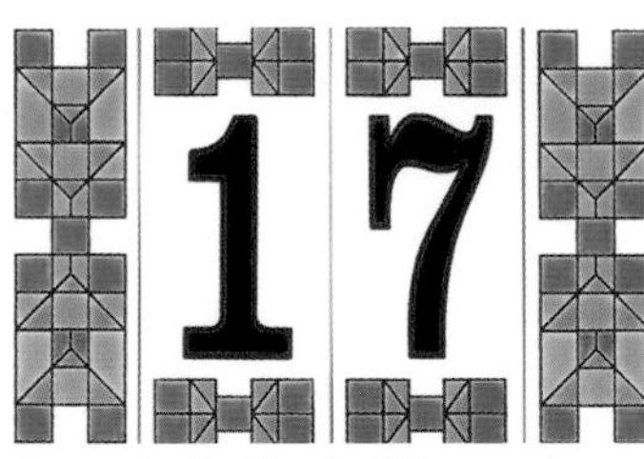

Ceramic *DuQuella Tile*

Hand-carved stone *Walter Arnold*

Arts and Crafts, yellow and gold glaze *Rocheford*

Fleur de Lis, turquoise glaze *Rocheford*

mailboxes

Antares Contoured
Galaxy Mailboxes

Arcturus Contoured

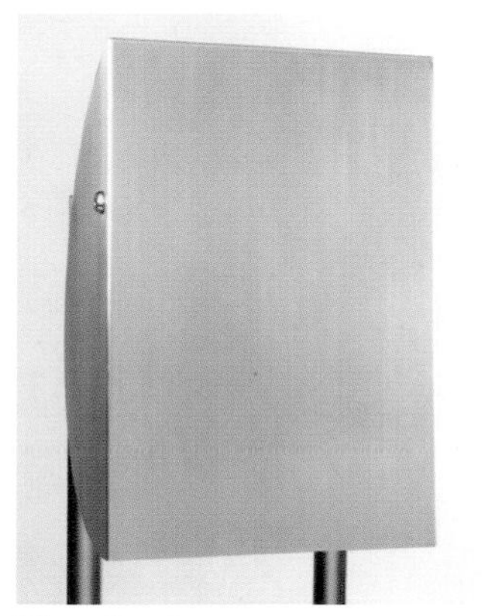
Plate Studio

Capella

Vega

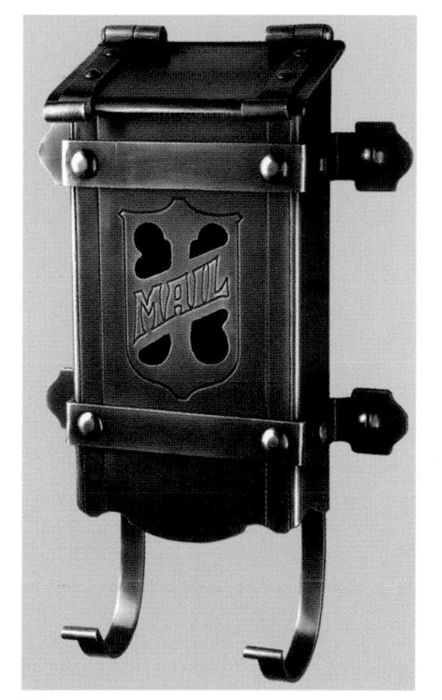

Harmon
Rejuvenation Hardware

Roland

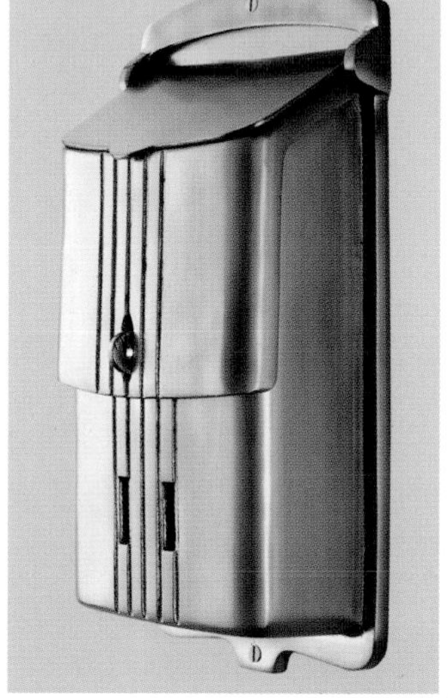
Shelton

Venting picture *Marvin Windows & Doors*

In-swing French Casement *Marvin*

Deco bay *Marvin*

Ultimate double hung *Marvin*

Casement with curved transom *North Star Vinyl Windows & Doors*

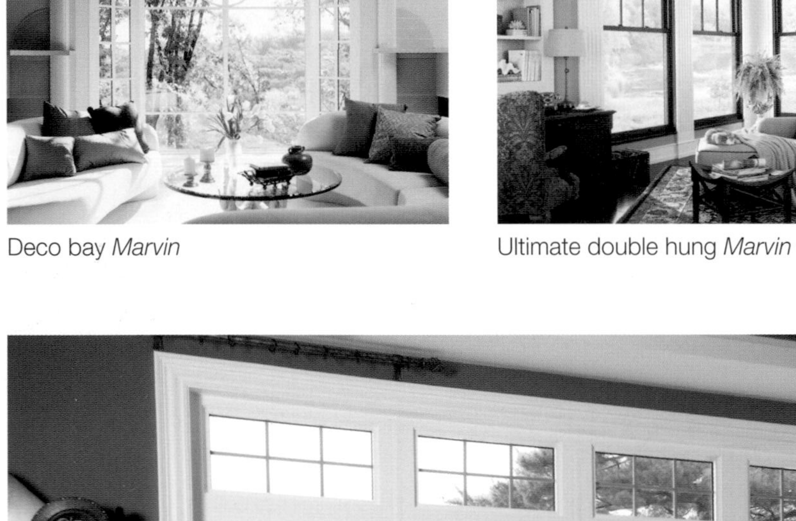

Casement with transoms *North Star*

Double hung with twelve lights *Marvin*

Anodized aluminum *Kolbe Windows and Doors*

Custom clerestory *Pella*

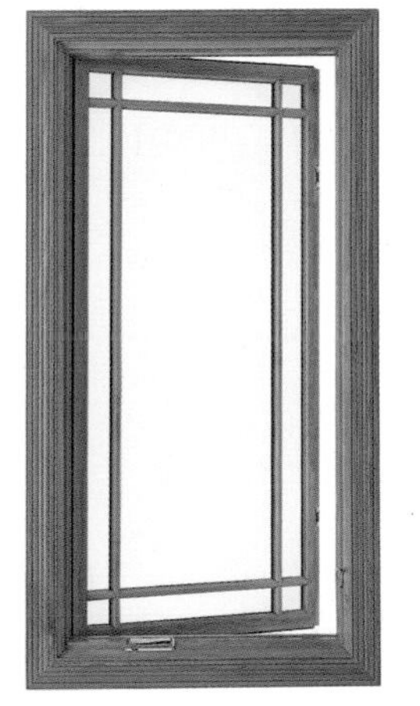

Casement, Architect Series *Pella Windows & Doors*

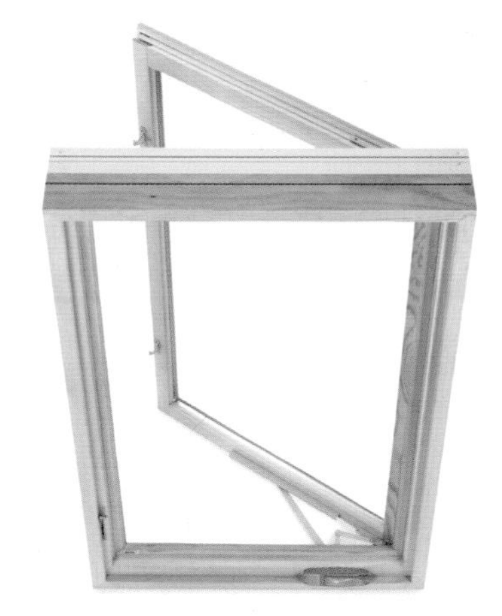

Casement *Marvin*

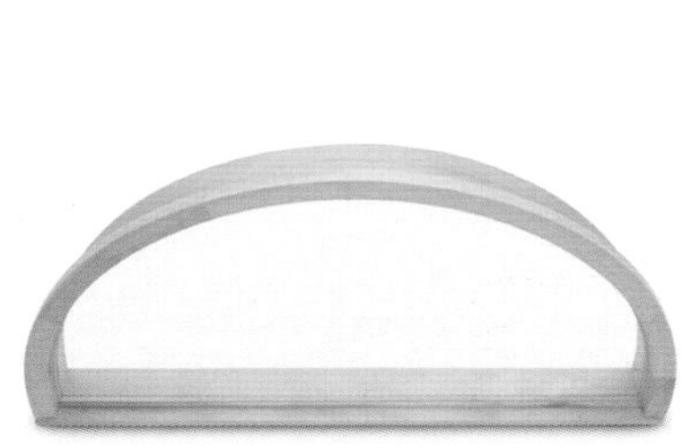

Round-topped fixed *Marvin*

Venting picture *Marvin*

French casement *Marvin*

Push-out casement *Loewen Windows*

Double access awning
Loewen Windows

Copper clad Umbra

Half round

French Chateau casement

Push-out casement

Double-hung

Custom bay window *Pella Windows & Doors*

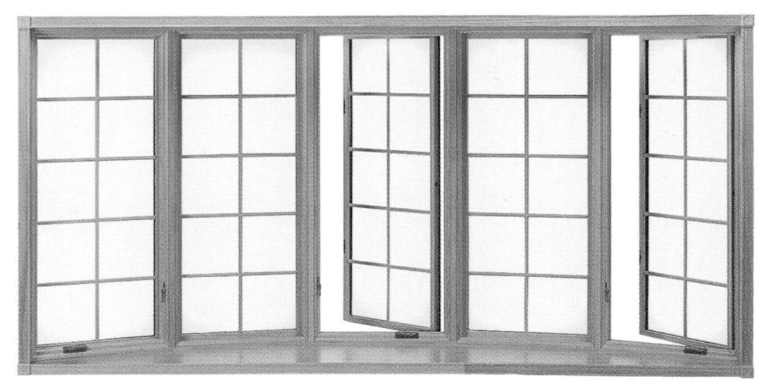

Bow window with five opening casements *Pella*

Bay window with fixed center and flanking double-hungs *Pella*

Awning, wood *Marvin Windows & Doors*

Glider *Marvin*

Bow *Marvin*

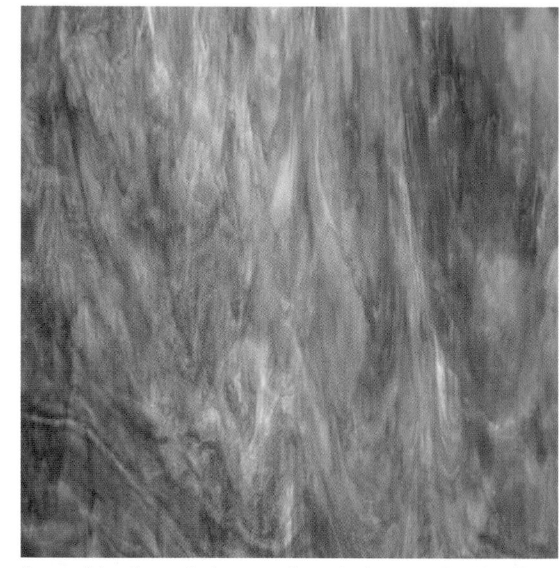

A combination of clear and opal glasses, 01467-S, *Armstrong Glass*

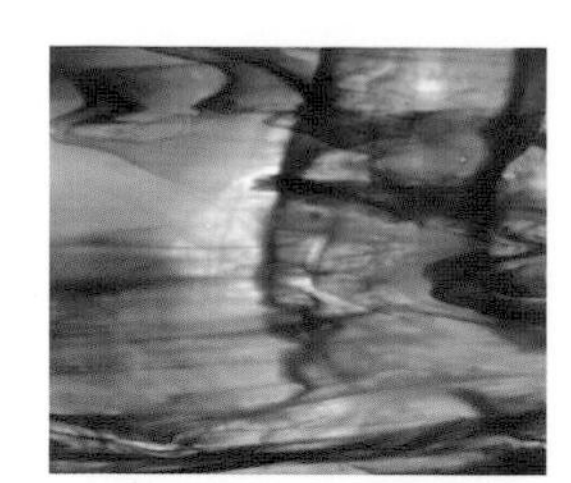

Copper red, smooth 5017-S

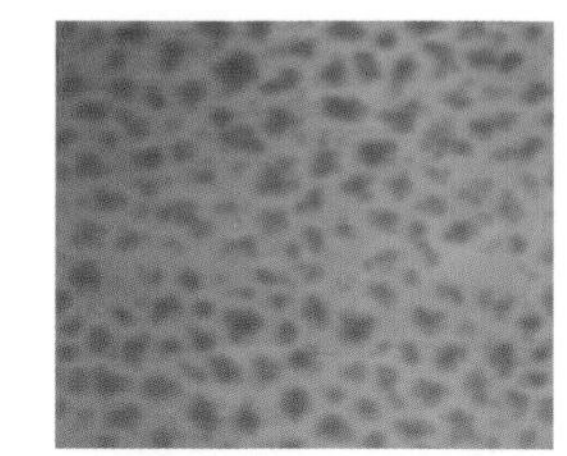

Flesh opal, 010-W

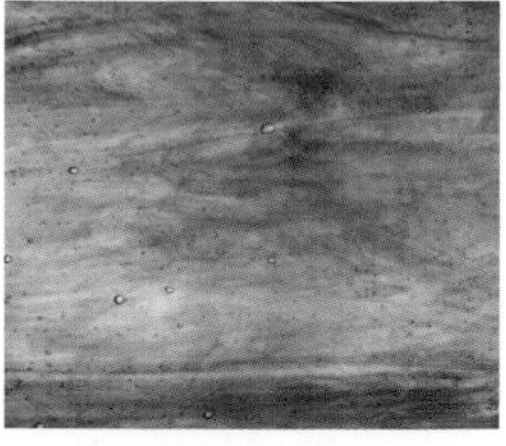

Opal, 436-C

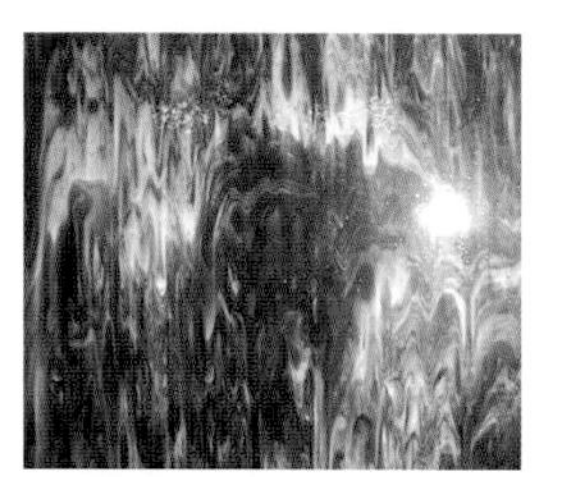

01325-S

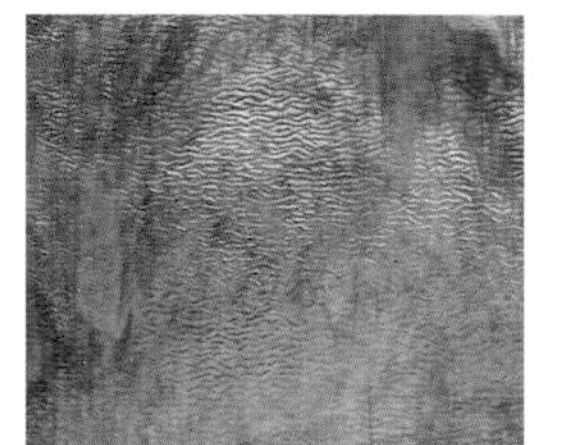

Textured ripple

436-P194-SR

Blue opal, 1042-S

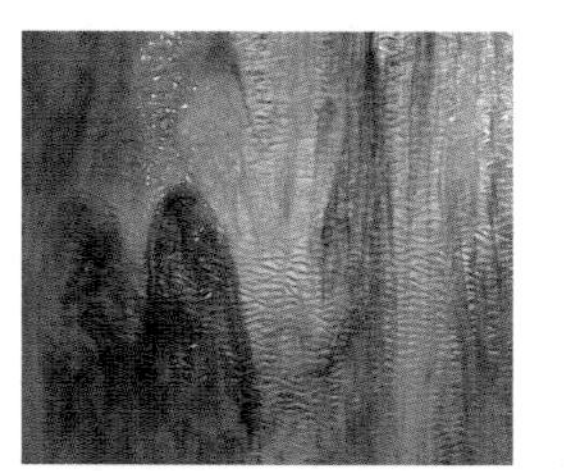

Textured Ripple, 106-P

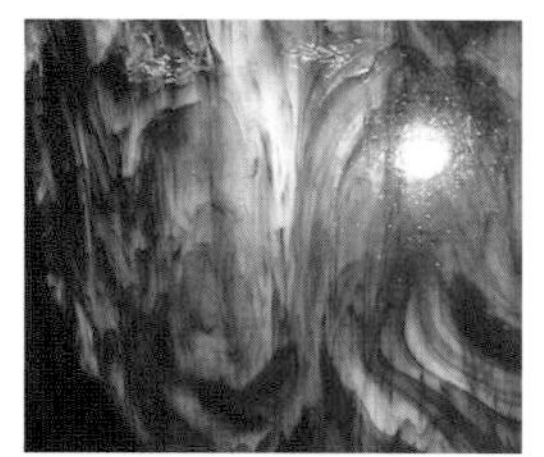

Clear opal, 05-CS

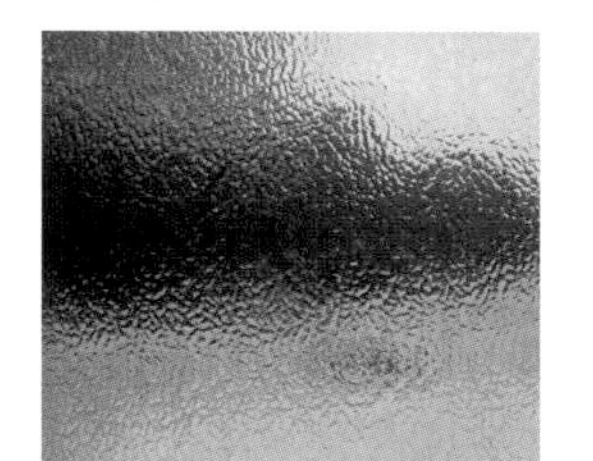

Cobblestone, 34-C

Brilliant Sunrise, 87-SO

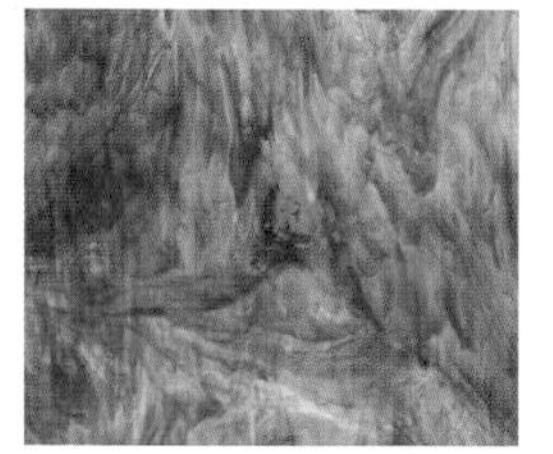

Wispy orange, smooth, 894-S

Iridescent, 00-WR

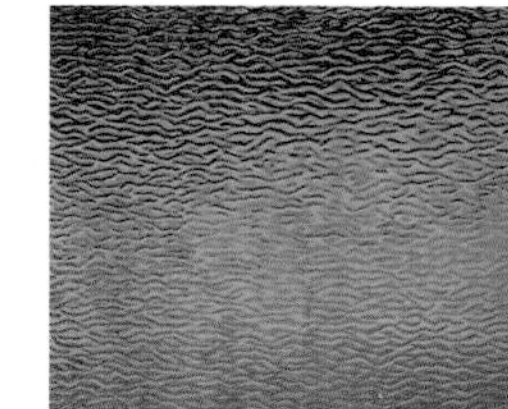

Textured Ripple, 90-P

Iridescent Granite Clear (00-GR) seems to shimmer when the light hits it.

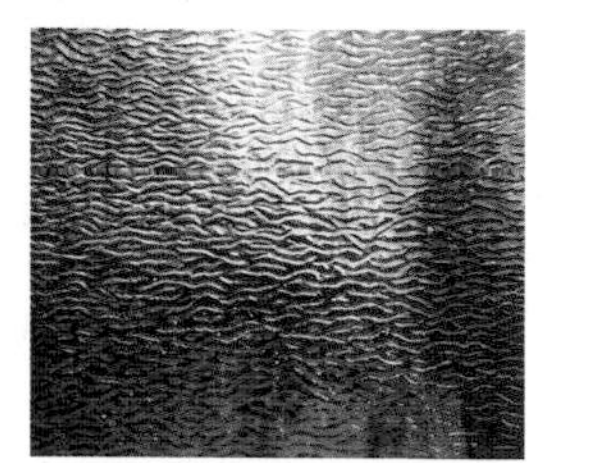

246-P, textured ripple

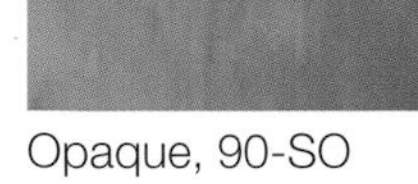

Opaque, 90-SO

Iridescent, 494-S

Float Fire fusing glass can be fused with standard clear window glass. Float Fire 1013 *Armstrong Glass*

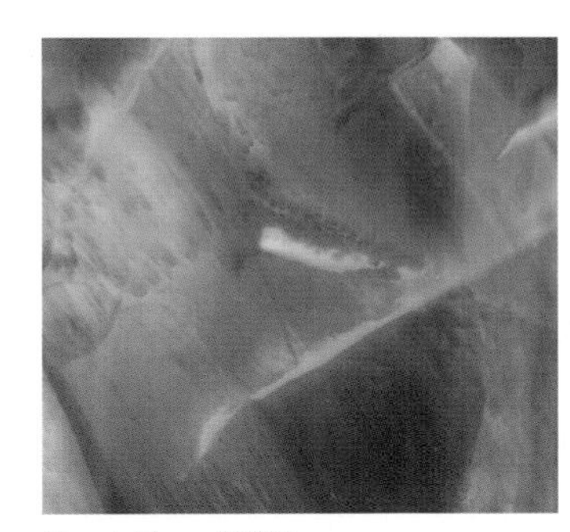

Float Fire, 1005

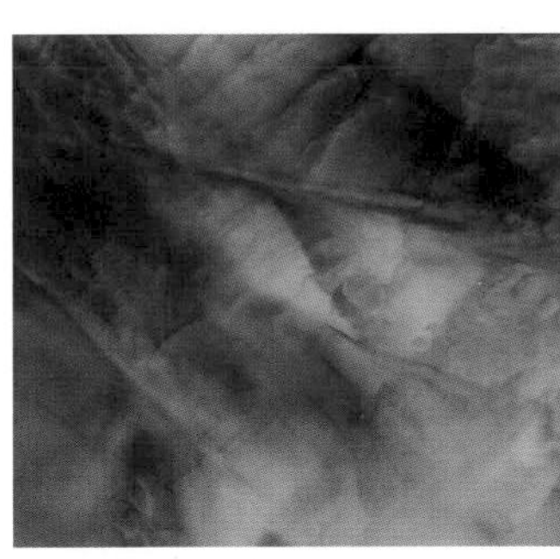

Float Fire, 1007

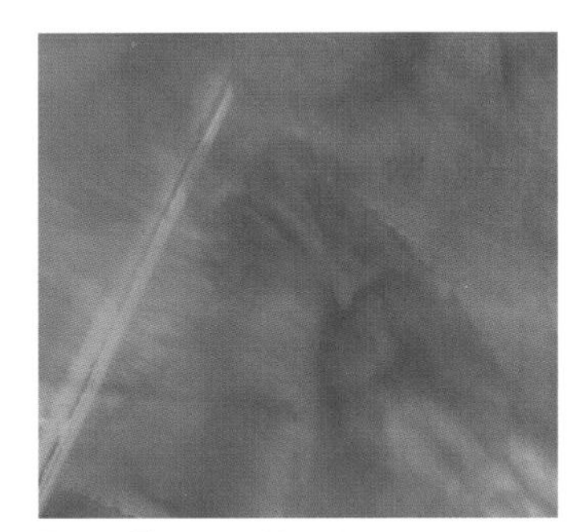

Float Fire, 1008

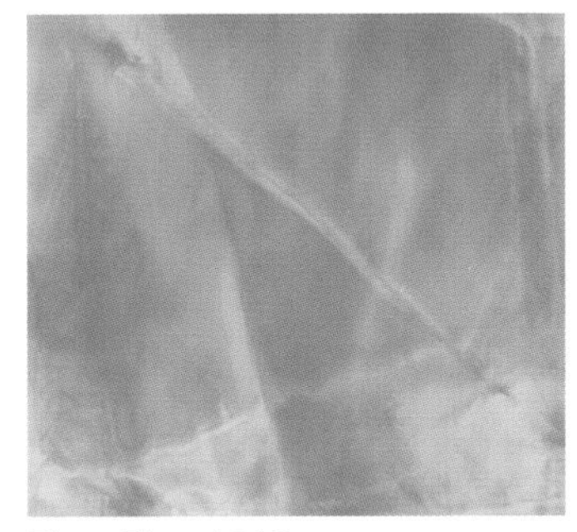

Float Fire, 1010

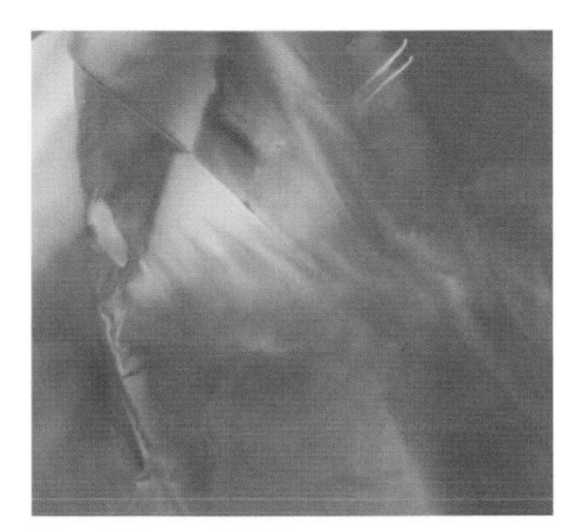

Float Fire, 1017

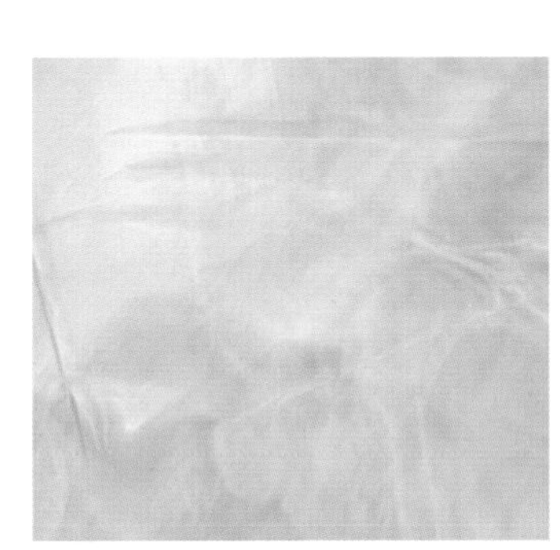

Float Fire, 1021

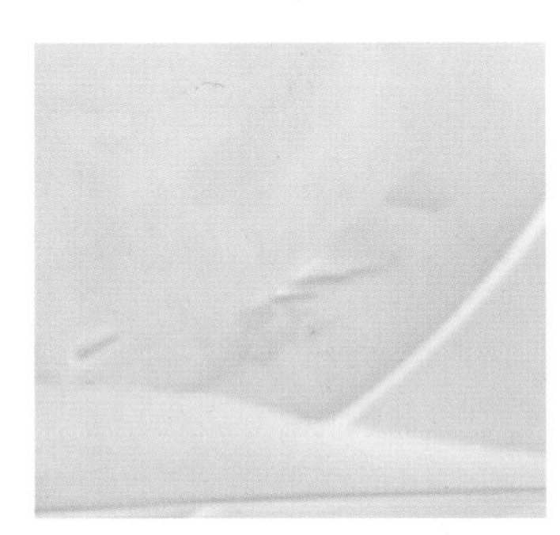

Float Fire, 1000

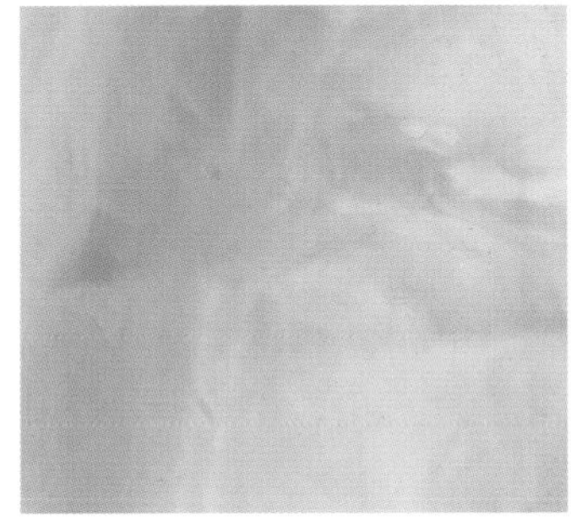

Float Fire, 1018

Float Fire, 2002

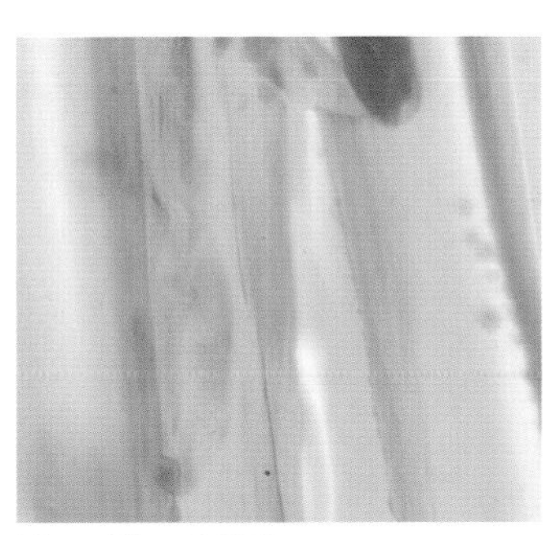

Float Fire, 2005

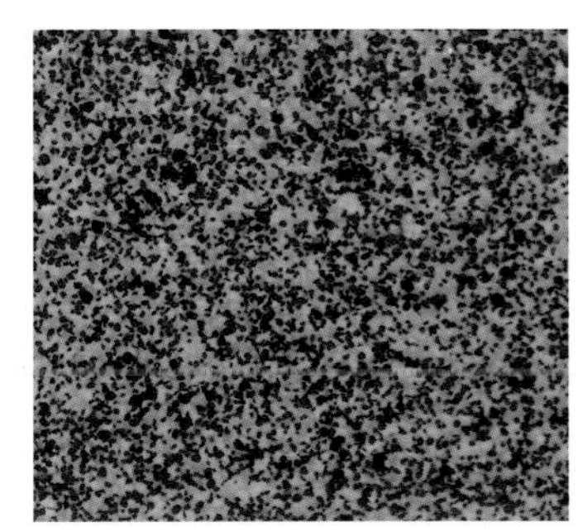

Float Fire, 1250

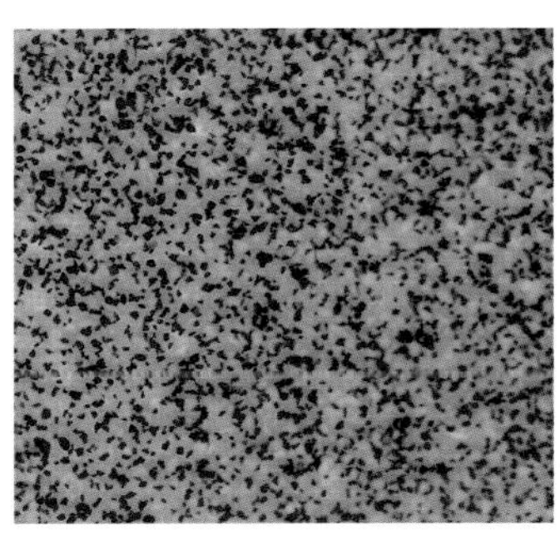

Float Fire, 1210

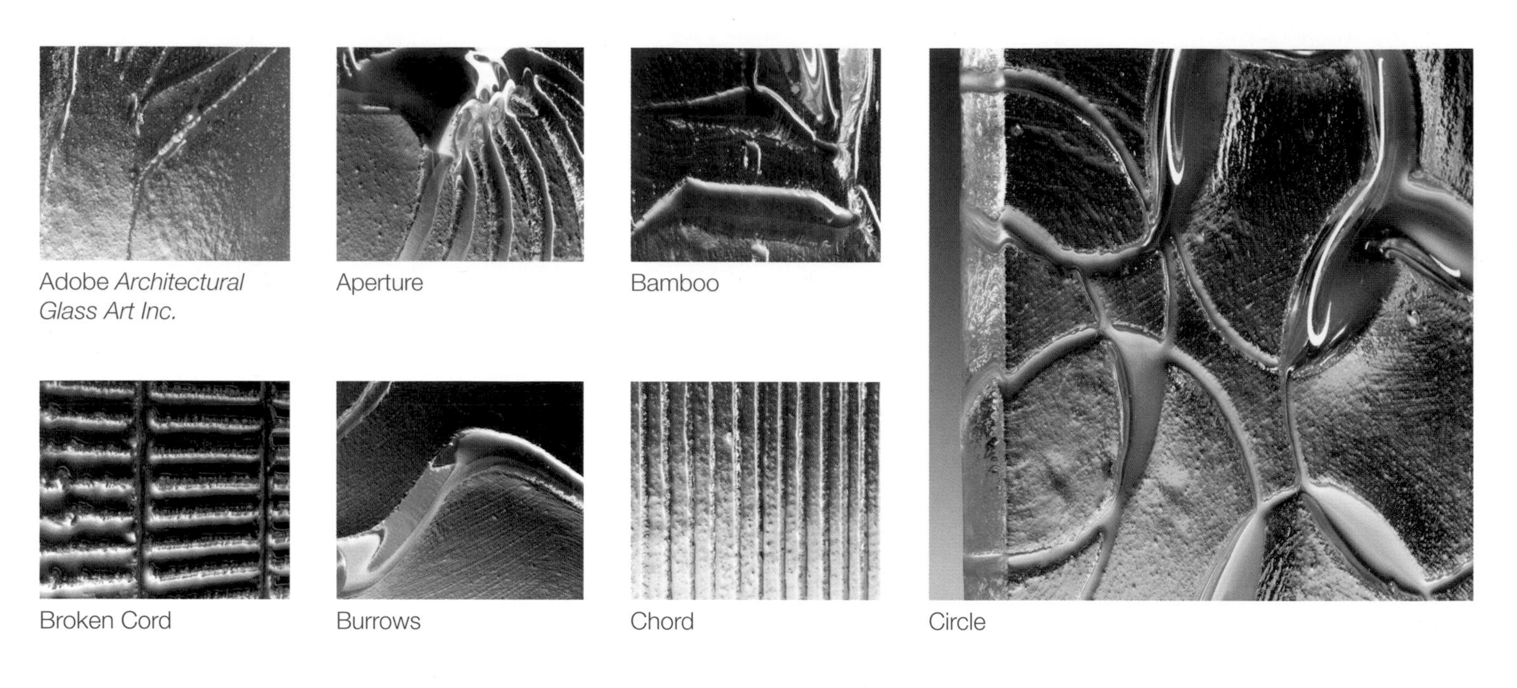

Adobe *Architectural Glass Art Inc.*

Aperture

Bamboo

Broken Cord

Burrows

Chord

Circle

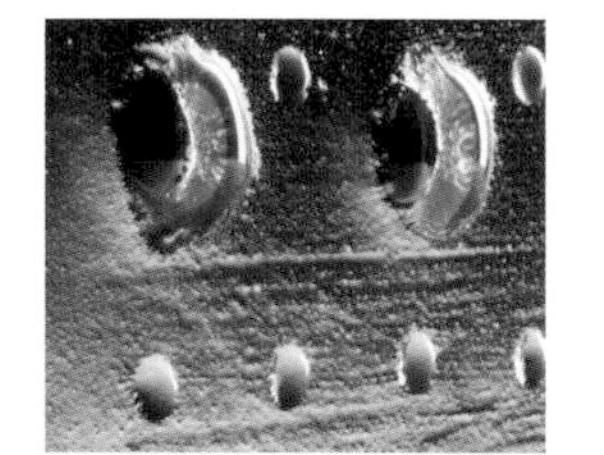

Dot Shift

Embossed Leaves

Metro

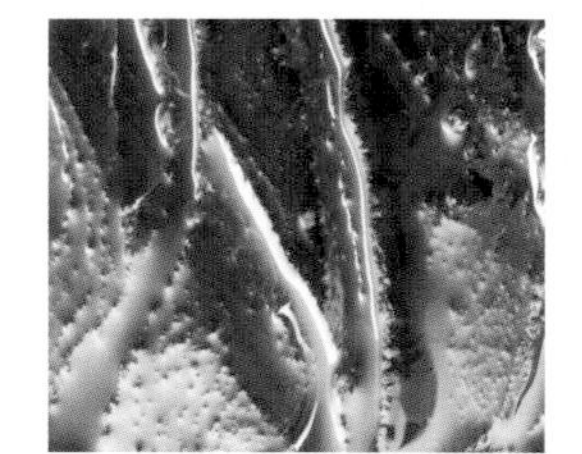

Parched

Scallop

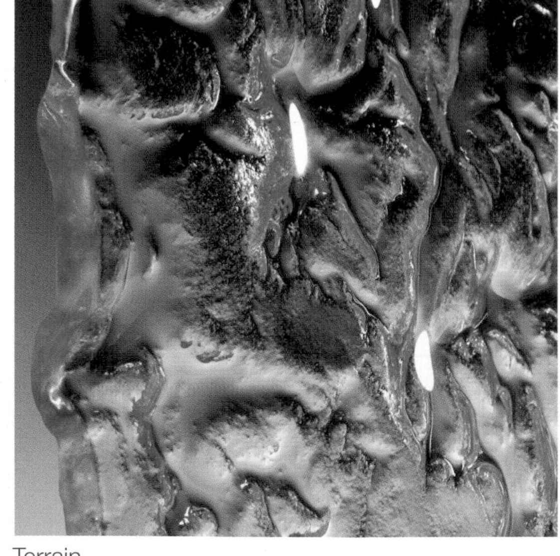

Terrain

Squares

Staccato

Stipple

Strand

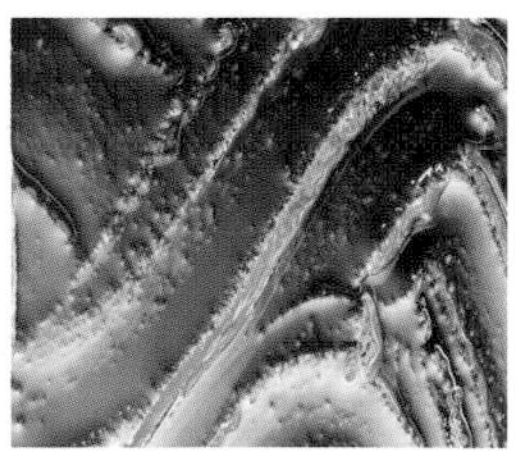

Tide

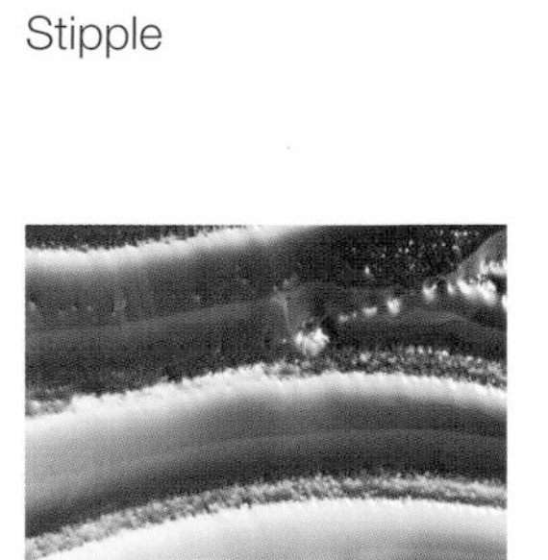

Wavy

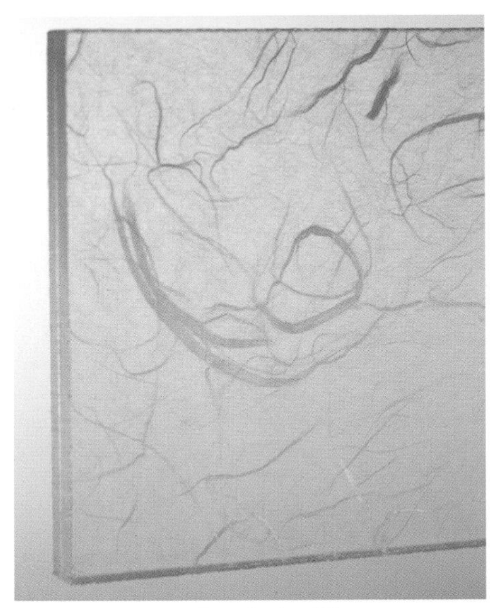

Fiber-imbedded *Bendheim Glass*

European Clear Corduroy *Bendheim Glass*

European Clear Lineum

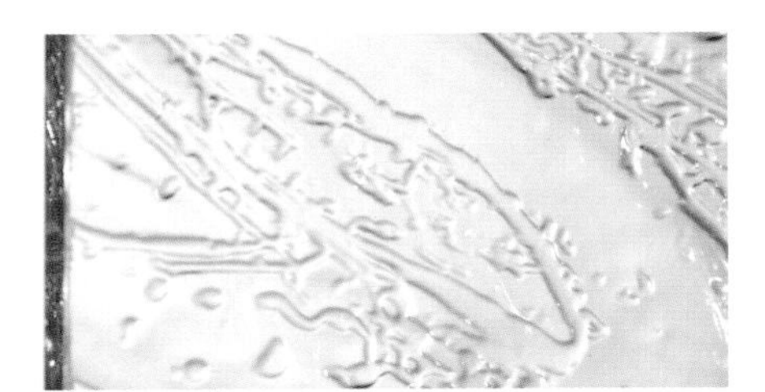

Clear Ice Fern

European Clear Etch

Float Fire, 0602
Armstrong Glass

Float Fire, 0603

Float Fire, 0604

Float Fire, 0605

Float Fire, 0606

Float Fire, 0607

Float Fire, 0608

Float Fire, 0609

Float Fire, 0617

Float Fire, 0618

Float Fire, 0622

Float Fire, 0624

Leaded Arts and Crafts tree *Jay Curtis*

Leaded etched and beveled underwater scene

Leaded etched and painted grapes

Leaded etched and painted abstract
Jay Curtis

Mother Earth, etched *Raynes & Co.*

Leaded etched and painted poppies
Jay Curtis

Rusholme Window *Art Zone*

Experience

Landing window

Vortex

Weaving

Erasing the Grid

Family Photo Album

Northern lights

Icescapes *Pittsburgh Corning*

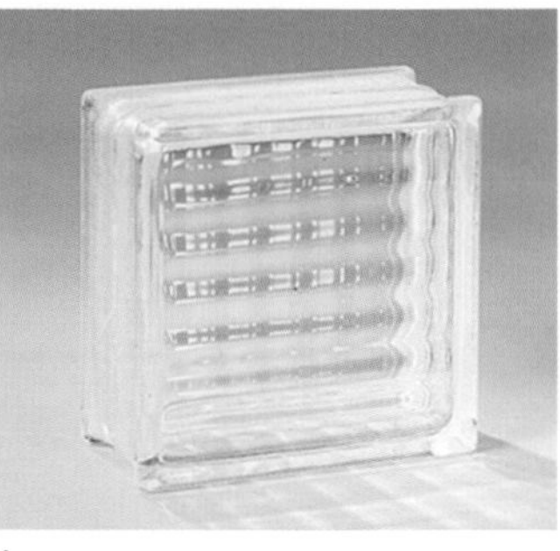
Argus

Argus Parallel Fluted

Arque Decora

Decora LX Cut

Decora Endblock

Decora Premiere

Decora Thinline

Delphi

Decora Endcurve

Endblock 6 x 8 (Decora)

Endblock (Icescape)

Endblock 8 x 8 (Decora)

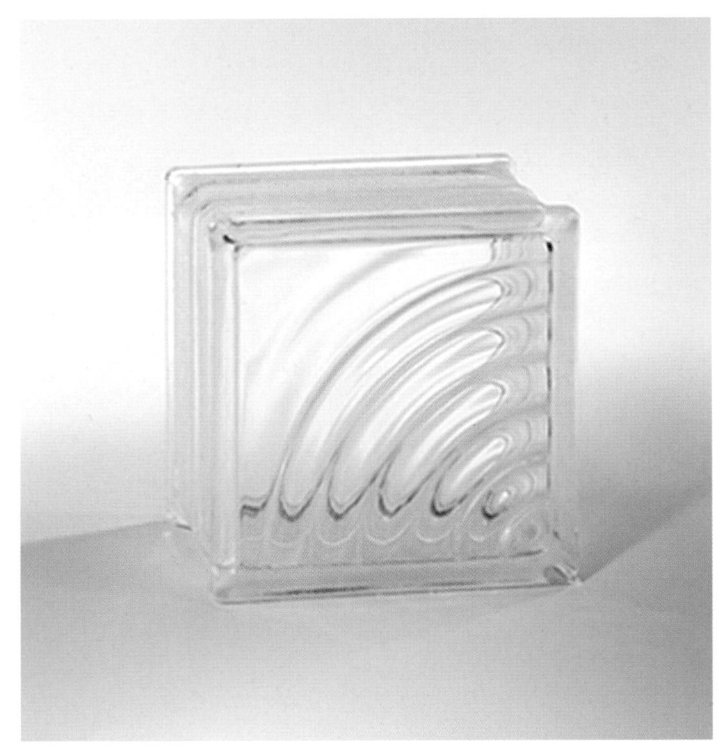
Spyra

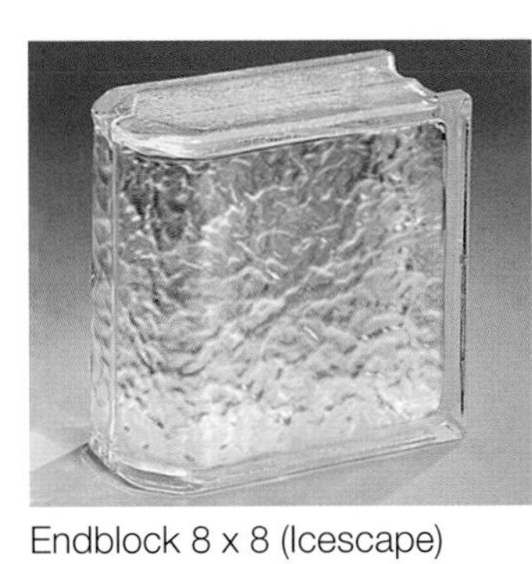
Endblock 8 x 8 (Icescape)

Essex AA

Hedron LX Cut

Hedron Corner (Decora)

Hedron Corner (Icescape)

Tridron (Decora)

Tridron (Icescape)

Vistabrick stippled

Icescape Thinline

Vistabrick Paver

Vistabrick Corner

Vue

floors, walls & ceilings

We are using a lot more reclaimed antique floors. In modern spaces it gives a kind of loft look, and in traditional spaces it gives the feeling that it has been there forever.

BRIAN GLUCKSTEIN, A CANADIAN DESIGNER WITH A LOVE OF OLD MATERIALS

EVERY ROOM IN YOUR HOME IS BASICALLY A BOX, FRAMED BY the floor, walls and ceiling. Take one step past these practical essentials and the fun begins, because now we enter the world of detail: material, color, texture, line, even the juxtaposition of light and shade come into play. The rich glow of a jarrah wood floor, the luminescence of a translucent resin wall panel, the way the light plays off the facets of a pressed tin ceiling or makes shadows on a Venetian plastered wall… These are all elements that speak volumes about our environment. A coffered ceiling and a parquet floor can conjure up visions of Versailles just as oiled steel and laminated glass convey a sense of cool sophistication. When you approach walls and ceilings as the blank canvases they are, you open up a world of possibilities.

Different shades of French limestone and black slate create a foyer floor that is a dramatic example of parquetry that seems three dimensional.

The range of choice, durability and sheer beauty of tiles make them an almost essential architectural finishing material. Ceramic tiles especially are widely available employed — and we have gathered a large and inspiring selection of ceramic tiles in this chapter. Tiles made of other hard materials such as glass, stone, metal and concrete are also featured here and in other chapters of this book, including Exterior Details.

The choices in color and texture offered by paint, wood, wallpaper and other surface treatments present infinite options as well. For inspired choices in materials and architectural elements to add interest to walls, floors and ceilings, consider not only the ideas and products showcased in this chapter, but also the decisions made by designers and architects as evidenced in photographs throughout this book. The potential for architectural detailing exists everywhere, but nowhere more than these in these interior planes.

This inch-thick laminated glass floor not only creates interest, but lets light get to the first floor kitchen underneath.

floors

Floors take more punishment than any other interior surface of your home. Your floor has to be durable, attractive and, ideally, acoustically appropriate so that every sound is not amplified. Polished stone floors such as marble provide a cool elegance. Stone tiles and wooden parquet floors set in a pattern create visual interest. Concrete — polished or roughcast, colored or plain — is forgivable and easy to maintain. Wood floors, whether solid or laminate, new or recycled, offer a solidity and warmth. And a ceramic tile floor with its parallel grout lines adds a graphic component that can be striking and space-altering. Renovators and new builders today often use the same flooring throughout the main living spaces of a home, to help create a feeling of spaciousness and link adjoining rooms visually. This is increasing the demand for floors that perform well under all circumstances.

Whether you favor a high-tech approach that looks to the future with engineered floors, laminates, and sustainable materials such as bamboo or turn to the past with reclaimed and salvaged material, you have many choices in flooring materials.

SOLID WOOD Oak is still the perennial favorite. For a departure from the usual, consider other kinds of wood: chiefly maple and ash but also exotic woods such as Australian jarrah, a hard, heavy and close-grained wood with a rich dark mahogany-red color. It is so durable, the streets of Berlin and London were once paved with blocks of jarrah. Or use exotic wood borders to frame the more prosaic oak or pine, or even ceramic tile or brick — see page 233.

Some experts discourage the use of solid wood floors in kitchens because of the danger of water damage and swelling (see Engineered floors for alternatives). And darkly stained wood floors tend to show up every speck of

Left: Wide plank pine floors paired with an old cook stove. **Center:** The rich tones of a parquet de Versailles style floor. **Right:** The very formal and luxurious look of an oak parquet floor. Oak, especially red oak, is a popular wood for flooring because of its durability and tight, intricate grain.

An old subfloor painted in a faded checkerboard tile pattern.

The deeply burnished look of reclaimed elm sets off this parquet floor and built-in bench.

dust and dirt. But if you have your heart set on solid wood, go ahead — just be sure to deal with a good supplier who guarantees the quality and wear of your floor.

Real parquet floors (short, narrow pieces of wood laid down in patterns like tile, rather than the ready-made ones), while visually interesting, are laborious and tricky to install and require on-site finishing.

RECLAIMED WOOD Salvaged or reclaimed wood comes from various sources: sunken logs, demolished houses and warehouses, barns and even railway bridges. The results can be nothing less than spectacular. That's because most reclaimed wood is from old-growth forests, meaning they offer denser, harder and more stable wood and in much wider widths and with a tighter grain. Reclaimed wood also shows marks of history and use: nail holes, plane marks, faded paint, bolt holes, and even personal marks such as carved initials — something that gives it lots of character. There is a growing industry in hand-scraped and other intentionally stressed wood flooring products. Hand-scraped means just that — wood planks that have been scraped and otherwise aged. Less costly is distressed hardwood flooring, which has marks and wear applied by machine.

Reclaimed wood is usually thought of as a "green" product because it doesn't require the cutting down of existing trees, and it saves the used material from being consigned to landfill. Check if the wood has a SmartWood certification. The SmartWood Rediscovered Wood Program (part of the Rainforest Alliance) evaluates forest products operations that are reclaiming or using reclaimed, recycled, and/or salvaged wood materials. Those that meet the SmartWood standards are awarded certification for Rediscovered Wood (RW).

PAINTED FLOORS Painting a pattern on your floor, especially as an alternative to replacing an old, worn floor, is definitely an option to be explored. In fact, people have been painting designs on floors for centuries. Faux black and white tile motifs and rug patterns were popular in Colonial times, but today's designers and renovators have taken this quaint craft and opened it up. Everything from fake Persian rugs motifs to Jackson Pollack splatter paintings can be used, especially on the funky old floors found in lofts or underneath age-old linoleum.

The best way to repaint old floors, if you want them to look aged, is to use multiple layers: a hard, durable oil-based paint for the base coat and soft, less durable water-based (latex) coats on top. Over time the tops coats will wear down in high traffic places so that the base coat begins to show through. You now have the appearance of centuries of wear along with the old texture of the floor.

A traditional pattern painted on the original floor of an 18th-century New England farmhouse. Note the checkerboard tile pattern painted on the floor in the background.

ENGINEERED FLOORS Durable, low maintenance and available in a wide range of wood veneers and widths, engineered floors are a good choice for a surface that takes a lot of pounding and abuse. In the manufacturing of engineered hardwood floors, the wood plies are stacked on top of each other in opposite directions to make a plank. Having the plies reversed helps counteract the natural tendency of wood to expand and contract with different levels of humidity and moisture — making this kind of floor good for kitchens. Engineered wood planks are much more dimensionally stable than solid wood and can be used over wooden subfloors, old vinyl floors or even concrete slabs. And the top ply (or finish layer) of an engineered wood floor can be a different (and much better) wood than the layers underneath. This allows manufacturers to offer surface layers in a wide variety of both domestic and exotic hardwood species without driving the costs out of sight. Plus, because an engineered floor is factory prefinished with a polyurethane finish, once it's installed that's it — no on-site sanding or urethaning is needed. Another advantage of engineered floors is that the widths of the planks available vary from 2¼ to 7 inches. This allows for the alternating of various widths for a more interesting floor.

It's not easy to tell the difference between a high-quality engineered floor and solid wood, but it is with the cheaper varieties. The best is cut in slices that follow the grain like real wood slabs. The alternative is to cut off thin veneer from the log in one continuous sheet, which employs more of the log but gives the repetitive grain look that plywood has. Also, the more plies the floor has underneath the veneer, the more stable (and expensive) it is. If you buy inexpensive engineered flooring that has a very thin veneer (¼ inch thick), you can't sand it to fix any blemishes that occur over time. Good-quality engineered floors (veneers up to 9⁄16 inch thick) can be sanded once or twice, compared to solid oak floors, which can be sanded as many as seven times.

LAMINATE Laminates are a bit like engineered flooring, in that they are made of a visual top layer that is glued to a multi-ply panel underneath. The main difference is that with laminates, the top layer is a photographic image of wood or anything else you would like, for that matter — clouds, tile, grass or bamboo. It is more scratch resistant than engineered or real wood floors, but once it is scratched it can't be repaired. It is however, extremely easy to clean and maintain. A floating laminate floor is one that is installed over a high-density foam or cork cushion. This helps reduce noise and is softer and therefore easier on the feet.

BAMBOO Bamboo is a grass, not a wood, and as such it can grow to maturity in four to five years, not the decades it takes most hardwoods, making its yield up to 25 times higher and thus much more sustainable. It can also be grown without fertilizers and pesticides, making it environmentally correct. It's just as hard as maple and oak, and is considered 50 percent more stable — resistant to expansion and contraction due to changes in the weather — than even red oak.

Bamboo comes in a natural golden blond color (it takes stain quite well) but can also be purchased in a smoked/carbonized variety that has a deeper, richer tone reminiscent of caramel or amber. Both varieties have very distinctive grain

Warm looking, durable and relatively soft on the feet, cork flooring is increasing in popularity.

patterns and joint marks (the joint marks are visible in the horizontal-cut products, while the vertical-cut variety shows a more even color and grain). Bamboo usually comes prefinished, can be nailed or glued down like engineered flooring, and is just as low maintenance. Bamboo is also available in door and window moldings, wall panels and stairway treads.

CORK Used since the early 20th century, particularly in Europe, cork is finally available in North America as well. Like bamboo, cork is a renewable resource — it's the bark of a Mediterranean tree that is harvested every nine years, so the tree is not actually cut down. Cork's excellent sound absorption, anti static and insulating properties plus its durability and resistance to common pests have increased its share of the flooring market. Its variation in color (from honey brown to green, red, chocolate and black) and its variegated texture give it a stylish richness. Its antimicrobial, mold, mildew and water-resistant properties also make it popular as a kitchen floor, but only if it is sealed. One of the drawbacks of cork is that is will fade in direct sunlight and sometimes yellows with age. And, although it is water resistant, cork floors are not immune to moisture and humidity and for that reason wet mopping is not recommended as it may cause the seams to swell. Similarly, you shouldn't do things like allow a moisture-saturated floor mat to remain on the cork floor.

You can buy prefinished or unfinished cork tiles and planks. One innovation is a product with a vinyl topcoat that protects the cork and makes it easier to clean. Other products have a non-toxic oil finish that can be walked on immediately and can still be refinished. Cork floors that do not have a surface finish on them can be color stained, and then sealed. You can glue your cork floor or float it over a wide variety of subfloors, including wood, concrete slabs and some types of existing floors.

It is said that cork has a "memory" and therefore recovers well from compression (think of a cork being pulled from a bottle), but pads are still recommended beneath table legs and other heavy objects. Cork floors can be repaired by grinding a bit of cork tile into dust and mixing it with polyurethane to form a paste to fill the cracks or blemish.

CERAMIC TILE The popularity of ceramic tile floors in America was given a boost early on by the noted architect Andrew Jackson Downing. In his book *The Architecture of Country Houses,* published in 1850, Downing raved about ceramic floor tiles for residential use because of their practicality and decorative possibilities, especially for high-traffic areas such as vestibules and entrance halls.

Ceramic floor tiles come in two distinct categories: "once fired" (monocottura) or "twice fired" (bi-cottura). The once-fired variety is fired at a hotter temperature and therefore usually has a stronger and more break-resistant base, while the twice-fired is given the opportunity to have a finer and more interesting glaze.

Real leaves were taken to a ceramic artist, who copied them to make tiles for this powder room floor.

One common form of floor tile is quarry tile (the name comes from the fact that they were originally made from quarried stone); these are extruded while wet and then fired to make a rustic square or rectangular tile that is often used outside. The extrusion process leaves a compacted surface that becomes very hard when fired. These tiles are unglazed so there is no top layer to wear off, which makes them extremely durable and slip resistant. (Floor tiles need a certain degree of roughness to make them safe.) Unglazed ceramic tile has no fused, or glazed, surface, so the natural color of whatever clay was used shows through. Unglazed tiles offer a rich, earthy look to a room and minimize things like scratches and wear patterns. The downside is their lack of color choices. It can also be difficult to find a grout to match the color. They can be allowed to become wet, but they stain easily because they have no glazed protective layer, and so it is very important that they be sealed for use in kitchens. A paver is a simpler, cruder form of quarry tile that is usually a little thicker and made from the dust pressed method. Satillio is a hard, decorative floor tile with Spanish origins.

Another common floor tile is terra cotta. From the Latin for baked earth, terra cotta tiles are glazed or unglazed fired clay tiles, made either by hand or by machine. The handmade tiles have more character; in fact, the ones from Mexico are renowned for bearing the occasional child's footprint or paw print. The machine-made tiles are said to be denser and therefore more frost resistant. For interior floors, real terra cotta tiles offer a warm, antique look. Their relative softness means they should be sealed very well and may need to be refinished periodically.

Ceramic floor tiles are thicker and are fired at higher temperatures than wall or ornamental tiles, making them much stronger. They also come in larger sizes

Left: A ceramic tile border enlivens a plain wood floor. **Center:** A round mosaic tile pattern set in a limestone tile floor. **Right:** Because of its impermeability, ceramic is an ideal material for use in bathrooms. **Opposite:** The cool, Mediterranean look of terra cotta tiles.

People are now understanding that you cannot get that wonderful feeling you get when you walk into a perfectly done old house without using reclaimed and antique materials that have the patina of age.

BRIAN GLUCKSTEIN

A limestone floor surrounds a beautifully complex marble inlay patterned to look like a rug. The hand holding the crystal ball on the newel post is a cast of the architect Joe Brennan's own hand.

Left: Seventeenth-century French stone floor tiles with a chestnut border. **Center:** An Indian slate island in an oak floor, including border. **Right:** Square marble tiles with a round center piece make the transition between two different kinds of wood floor.

(the rule of thumb is the larger the room, the larger the tiles should be), but small mosaics are also common. Also, the surfaces are not as highly polished as wall tiles, and can have a mat or even unglazed finish.

The extreme inflexibility of ceramic floor tiles means you have to have a completely rigid, smooth subfloor to stop them from cracking. One of the criticisms of ceramic floors is that they are hard, and that can mean hard on the back. They are also fairly unforgiving when dishes or glasses are dropped. And unless properly sealed, the grout can pick up stains. All of these qualities make some people leery about using ceramics for kitchen floors. On the plus side, ceramic floors will outlast just about any other floor material you can think of. Because ceramic tile may break, make sure you order more tiles than you need for the job so you have some spares on hand.

For more about ceramic tiles as used on walls and other surfaces, see the Tiles section later in this chapter.

STONE TILES Natural stone tiles provide a level of obvious luxury to a home. Because it is a natural product, no two stone tiles look the same. The flip side of the coin is that all natural stone is more or less porous and has to be sealed with a penetrating sealant to prevent staining. A good rule of thumb is that honed surfaces take wear better than polished — for one thing, a polished stone will be etched by acids like juice or wine and lose its polish, while it won't show up as much on a honed surface.

Marble is one of the most expensive, if beautiful and hard, stone flooring materials but it is even more porous than most, and is usually used only inside the home. Granite is the hardest, but slate, with its dark, waxy sheen, is also popular. Both can be used inside or outside the home. For a very interesting form of slate, which often includes rusty sandstone-like and other colorful patches, try Indian Slate. Both can be used inside or outside the home. Travertine, a kind of pitted limestone, is often "tumbled," a process where tiles are intentionally roughed up to give an antique quality. And limestone itself, although it comes in a more restrained palette (usually in shades of off-white, bone and mellow yellow shot through with darker spidery lines and flecks), has a very old, almost ancient look about it that makes it perfect for the things like the French Country House style. (For more about the properties of natural stone, see the Kitchen Fittings chapter.)

VINYL Not exactly the hippest flooring material around, vinyl flooring (sometimes called resilient flooring) is still one of the most popular. The reasons are simple: dollar for dollar it's the cheapest to buy, easiest to install and keep clean, and the most resistant to water (which is why it's a favorite for kitchens and bathrooms). On the down side, it cuts and scratches easily. If you are going to have a problem with your vinyl floor, it will probably be around the seams.

Vinyl is essentially a plastic (polyvinyl chloride, or PVC). Today's vinyl tiles do a pretty good job of replicating the look and textures of ceramic tile, stone and wood grains. It comes in continuous sheets of 6-foot and 12-foot widths, and in vinyl tiles of various widths and thickness. The durability of a vinyl floor depends on the thickness of the wear layer — the top layer. The thicker this wear layer — which can range from .005 to .025 of an inch — the more durable, and expensive, the flooring. The actual pattern or color is sandwiched between the wear layer and the backing. Most resilient floors also have a foam undermat that gives a certain bounce and comfort to the floor.

You can lay a new vinyl floor over an old vinyl floor with only slight complications from the rise in height. If the old tile is embossed, you must first prepare it with a leveling material. That's one reason not to use embossed tile; another is that embossing attracts dirt much more than smooth flooring.

Although vinyl tiles will eventually wear through and expose the different colorations underneath, this wear and tear has acquired an attractiveness that is being sought after. That means there is a market for antique vinyl tiles just as there is for any other salvaged or distressed material. And if vinyl tiles are laid in an interesting mix of colors and patterns by someone with a good eye for design, they provide a certain retro look (see photo opposite).

LINOLEUM Linoleum is primarily made of all-natural products, such as linseed oil (from flax), wood powder, limestone, resins and some colored pigments. The backing is made from jute, which is a natural grass. Although linoleum floors have been around for more than a century, many people confuse it with vinyl floors. The fact is the two floors are very different and should not be grouped together. Because linoleum is made from natural ingredients, its colors tends to fade under natural light, and it is more susceptible to decay and disintegration when exposed to moisture, foods and beverages, cleansers and common household chemicals. Although not as popular as it used to be, there are several manufacturers still producing linoleum floors for residential use, including Armstrong with Marmorette, Forbo with Marmoleum, and Domco Tarkett with Linosom.

The artists who own this kitchen had more taste than money, so they used ordinary materials in extraordinary ways. Working with architect Harry Teague, they had perforated metal cupboards made, and used a sense of balance in laying inexpensive vinyl tile in a pleasing pattern.

TERRAZZO Terrazzo is a tile formed of marble (or stone) chips and dust, all bound together with cement and has its roots in ancient Roman mosaics. It is durable and can be used just about anywhere, especially in high-traffic areas. The finish is a "honed" finish, that is, smooth but not polished. A penetrating sealer should be applied to new terrazzo floors to seal the pores in the cement and thus retard stains absorption. Although terrazzo is considered easy to take care of, it is best to use a commercial cleaner made especially for terrazzo. All-purpose household cleaners, soaps, detergents and wax removers usually contain one or more alkalis, and so should not be used on terrazzo.

Although it has been around for centuries, the sleek, smooth, clean look of terrazzo made it a favorite of Art Deco homes and commercial buildings. For the same reason it you see it in many Contemporary homes.

AGGLOMERATE TILE Agglomerate is like terrazzo, only the matrix is bound with either epoxy or other resins instead of cement. This kind of stone tile goes by different names: artificial marble, faux marble and faux stone. It is usually made of a mix of marble or granite chips of various sizes, and the appearance will vary greatly depending on the size of the chips and the color of the resin or other binding agent. It can also be suitable for outdoor use in patios and walkways.

RUBBER Because of its extreme durability, stain and scratch resistance and cushioned resiliency, rubber has been primarily confined to industrial applications — but not anymore. Rubber floor tiles are almost indistinguishable visually from commercial vinyl tiles, but some designers are using industrial rubber sheet flooring, often ribbed or studded, in surprising places and in hip pastel and bright primary colors. Using textured rubber is especially popular in wet areas such as kitchens and bathrooms because rubber can become slippery when wet. It is also good for softening hard-edged spaces, deadening sound, and insulating, so it is perfect for industrial loft conversions.

Rubber flooring used to be made from rubber trees, but most now comes from synthetic sources. A significant number of rubber flooring products on the market are made from recycled tires, making them a renewable resource. Whatever the source, rubber flooring is a durable, if costly, flooring alternative.

CONCRETE The durability, elementary nature and inherent imperfections of concrete have moved it to the fore of the "less is better, and the more natural the better" movement. A properly sealed concrete floor is about as worry-free a material as you could wish for. Acid etched, stained, tinted, painted, stamped,

Left: A beautiful, if oversized, example of terrazzo in which small stones are embedded in concrete and polished smooth. **Center:** A purposefully haphazard painting of concrete in a tile pattern achieves a certain sloppy elegance. **Right:** A wooden heat register cover painted to fit in with the wood floor around it.

scoured or channeled, concrete is an extremely flexible material visually. The more adventurous can even add other materials to a freshly poured concrete floor — anything from shells, computer chips and bits of glass and marble to tiles, river stone and wood. Washing away some of the top layer of concrete to expose the aggregate below before it dries can also make for an interesting and very rugged texture. Polishing can also reveal whatever has been seeded into the mix, but in subtler ways. (See Kitchen Fittings for the use of concrete in countertops.)

Many industrial and warehouse buildings converted to residential use have concrete floors that have been lovingly burnished and polished, sanded and sealed by their new owners to provide a beautiful and appropriate floor. Concrete is a chameleon material in that it can be made to look like almost anything from brick and slate to polished stone and textured fabric (but it is also inert, which is good if you have allergies). It can be scored to make tiles of any shape or size, stained to look like terra cotta and polished to give the high luster of lapis lazuli or granite. Concrete's versatility even extends to the next tenant, who could, if they wish, lay new flooring using the concrete as a more than stable subfloor.

Acid etching is also a way to jazz up the look of concrete floors. Despite the name, acid doesn't have much to do with the process. It consists of using metallic salts in an acidic, water-based solution that reacts with the hydrated lime (calcium hydroxide) in hardened concrete. What you get by using one of various commercially available chemical stains are permanent colors embedded in the concrete. The colors are variations of three basic color groups: black, brown, and blue-green.

One warning: With concrete's high weight-to-mass ratio, it's no surprise to find this kind of floor in industrial buildings, either on the first floor or resting on steel girders on the higher floors. This weight must be taken into account for any residential use on anything other than an on-ground slab — something that works particularly well in southern climes.

Concrete is inert and impervious (if sealed properly) to snow and sand. It can also be a delightfully cool medium in hot climes, but the converse is also true. For that reason, new concrete floors are often poured with radiant heating coils embedded in it, all the better to counteract concrete's penchant for sucking the heat out of both the air and your feet.

heating register covers & grates

Even such details as heating register covers can add to or subtract from a room's looks. Common sheet-metal vent covers can be upgraded to intricate cast-iron grates, wooden grates that match the grain of the wood (or are at least painted in a faux grain to blend in). And certain ceramic grates are available in various colors and patterns (see the Products section on page 280 for examples.) Antique salvage companies and reproduction hardware manufacturers are good sources for period-style heating register covers.

walls

When we think of walls, our attention is usually drawn to their surface coverings — paint, wallpaper, wood and plaster. But the size, placement, shape and even the angle of the walls of a room are the greatest — if not obvious — elements that affect how a room makes us feel. A simply clad, straight-angled wall will make any architectural details or adornments in the room stand out. But a curved, angled or odd-shaped wall will become the focal point of the room itself, especially if it is clad differently then the other walls of the room.

Concrete is a good choice for odd-shaped and curved walls (nothing is more welcoming or better for gently directing traffic than a curved wall) because it is so easy to shape and has integral strength. Concrete as a building material was first popularized in North American by architect Bernard Maybeck, who began using concrete inside and out after thirteen of the homes he designed were destroyed in the Berkeley (California) hills fire of 1923. Concrete walls, as exemplified by Maybeck, project a feeling of solidarity and offer every texture from highly polished to rough plastered and may even incorporate imaginative protuberances.

Below: A cork floor, walls paneled in elm and horizontal slats allowing light in from the first floor give this wine cellar entrance a calm, quiet feel. Opposite: This carved and varnished cedar wall was sculpted by Edward Falkenberg. Notice how the beveled edges of the vertical boards make them stand out in relief.

The New York architectural firm LOT-EK likes to deal with lofts by introducing one or more partitions that are not made of typical wall material, in order to redefine the space without chopping it up. In this case they used the side of 40-foot shipping container, whose sides don't touch the floor or the ceiling of the loft. This, and the fact that lettering was left on makes it identifiable as an object. They then installed the typical kitchen functions — stove, sink, cupboards — in hatchbacks that open and close so that any mess could be quickly closed off when company appears. The bedroom is treated similarly (see page 100).

Wood paneling can be rustic or refined, depending on the wood chosen and the way it is used. Wainscoting is common architectural wall treatment that provides a strongly traditional look. Wainscoting (which shouldn't be higher than three-quarters of the wall height) is also a great way to tie a room together, especially when it is used on other elements such as cabinets or counters. Usually made of beadboard or tongue-and-groove wood, it can be extended to cover the entire wall for a warm, country home style. Of course, the ultimate in wood walls is one made of logs, either the sensuous, mammoth look of round log construction (beautiful, but its undulating surface creates difficulties when installing cupboards and trim or even hanging pictures) or the clean, flat look of hewn logs with its defining chinking lines in between.

Walls that move or are translucent are extremely useful if flexibility is a concern. This is particularly important in small homes but also in lofts or other industrial buildings in which you are dealing with raw space. Walls that slide along a track or fold up accordion style allow rooms to have more than one function. A cozy study can metamorphose into a large airy dining room simply by moving a wall out of the way. A moveable wall can allow you to increase privacy to create, for example, an office or guest space, and then open up the space when the need for privacy is gone. Similarly, being able to let in light (a process designers call "daylighting") while still retaining acoustic or visual privacy is important anywhere a dark interior is a problem. Everything from rice paper (the classic shoji screen) and etched glass to translucent resins and other plastics — sometimes in elaborate patterns — can be used to make a virtue of necessity by creating a wall that glows with light (see Products section of this chapter). And transparent interior glass walls offer acoustic privacy while still making a visual connection between different spaces possible — for example, cooking while still keeping an eye on your exuberant kids.

Partial walls — "knee" or other forms of truncated walls — are great for delineating space while still letting in light and keeping sight lines open. This is particularly useful in open plan homes where, for instance, a low-wall suggestion of where the kitchen starts and the family room ends is more appropriate than a complete wall. And you can think of cutting out arched walkways or even glass-less windows into a wall to connect rooms, such as a bedroom–dressing room, or a living room–dining room.

For a high-tech look, consider a metal wall. The ultimate in cool, metal is durable, low maintenance and comes in a variety of finishes from hard shine to a dull, warm pewter-like glow. It can be polished, burnished, brushed, embossed or honed to any number of patterns and textures in materials such as copper, stainless steel, zinc and aluminum (see the products section of this chapter).

And for a very finished look, built-ins can't be beat. Not only do they provide a place for everything, built-in bookcases, cupboards, niches, armoires and wall units provide interest, solidity and complexity to a room and they do it without taking up additional floor space.

Left: A soft, inexpensive industrial material called homasote (made of recycled newspaper) that has been sanded down was used on this "soft loft" wall as sound deadener. The orange material to the right is fabric, put on to hide a bathroom door. The floor is ebony-stained red oak. The lighting in the indented pillar is LED lighting, computer-programmed to change to any color. **Center:** Simple cast-concrete steps fit right in with the adobe-like walls. Note the wrought-iron handrails and the adobe over the radiator at bottom left. **Right:** Built-in cupboards provide a finished, custom look.

tiles

Tiles have been employed on wall, floor and ceiling surfaces for millennia. From the intricate mosaics of ancient Pompeii, still impressing tourists, to today's glass and metal backsplashes, tile takes an infinite number of forms and serves an equally endless number of purposes both decorative and useful. Tiles come in square, rectangular, hexagonal, even (occasionally) odd shapes, but also in different materials. The sheer hardness of stone (limestone, marble, onyx and more) tiles make them perfect for floors but also for walls, especially in bathrooms or as a backsplash in the kitchen. Semi-precious stones such as jade and jasper also make beautiful — even spectacular — tiles, as does the pearly luminescence of seashells. And metal tiles, introduced when Art Deco got hot as a style in the 1920s and 1930s, is also being used today, especially for hard-edged industrial look.

Because they are not expected to bear much weight or do anything other than look good, wall tiles are manufactured with a higher water absorption rate but a lower breaking strength than floor tiles. Wall tiles are where you find the most creative expression of the ceramic art. Among the tiles included in this chapter you'll find many decorative and field tiles of different materials, styles and production techniques, including ceramic majolica, delft, porcelain and encaustic tiles, mosaics, glass, metal, hand-carved and hand-painted tiles, stone and gemstone tiles, plus borders and pictorials. For more on floor tiles, turn to the Floors section of this chapter.

The bulk of wall or floor tile are composed of "field tiles" — those that fill up the field or general area of a surface. Field tiles are generally monochromatic or at least subdued in decoration. More elaborate designs are usually reserved for the border tiles that delineate the field around the outside, usually in the same style. This exuberance can also include three-dimensional shaping such as sloped, bull nose or other curved edges.

Once exception to the solid field tile look is the pictorial. A pictorial is a tile that constitutes one small piece of a picture (an object, scene, design or even a phrase). When installed together they form a homogenous whole, often with border tiles to tie it all together are used as a focal point for a wall (both interior and exterior), as part of a fountain or incorporated in a backsplash. The look is very Old World, and pictorials are often (but not always) mosaics.

CERAMIC TILES In this book we have given special attention to ceramic tiles, as there is such a vast choice available today in this age-old material. We have been combining the elements of earth, water and fire to make ceramic tile for over 4,000 years. What was once only available to kings and high priests is now available to us all in an astonishing array of materials, colors, sizes and applications. The art form has seemingly been perfected.

Today, the biggest producer of ceramic tiles is Italy, followed by other European countries such as Holland, Germany, Spain and Portugal, and Japan and Morocco, but Mexico, the United States and Canada all have healthy tile industries.

Rustic decorative ceramic tiles contrast with the sleek copper-topped countertop in this butler's pantry between kitchen and dining room.

Details cost money. Angles and curves cost money. Changing materials while you are building cost money. Complicated roofs cost money. So it's best to have a simple roof and a simple foundation to keep costs down. Instead, put your money and elaboration into the small things you can touch and feel: moldings and trim — things that are on a human level, that will make sounds or have light impinging on them.

MARILYN LAKE, AN ARCHITECT WITH AN EYE FOR DETAIL

Glass tiles of different shapes and sizes introduce some dazzle to this bathroom.

Throughout history, the look and composition of ceramic tile has been borrowed from cultural and religious motifs. The ancient Romans perfected the art of using small tiles in mosaics, and today we still often see mosaic floors and wall borders laid out in traditional Roman or Greek designs. Many Islamic countries were forbidden to portray the human form, so they became masters at producing intricate geometric designs rendered in vivid colors. As the world became smaller through exploration in the 16th and 17th centuries, Chinese influences became popular in the West. One result of this was the familiar blue and white Asian-influenced designs of the wall tiles made by the Dutch potters based in the town of Delft, which became the name for these distinctive tiles. Another European leap in the ceramic arts was the production of Majolica tiles. Majolica was the first modern European pottery to achieve bright, rich colors. This was accomplished by applying opaque colors over the red clay (terra cotta) tile. The term Majolica has now become widely used for any

These Mexican tiles, applied with big grout lines to a kitchen island, combine with the big wood of the beams and the wide antique fir floor to create a rustic space, but not one lacking in refinements.

pottery with a thick, three-dimensional feel due to the thickness of the glaze. They also have a brightly glazed surface that is usually painted freehand.

The United States found its start as a major tile producer during the glory days of the Arts and Crafts movement, from the 1850s to the mid 1900s, mostly in the Midwest. Influenced greatly by ceramic artist William de Morgan, who was a friend of Arts and Crafts founder William Morris, this style grew in popularity and greatly increased the market around the world and in North America as well. Many small- and medium-sized North American tile makers got their start then and continue to this day, having created the base for a vibrant and diverse indigenous boutique tile industry. The handmade tile industry for which Mexico is so famous began back in the 16th and 17th centuries, when Spanish immigrants brought their knowledge of pottery and tile making to Mexico. Their particular style of rustic and irregular handmade, handpainted tiles are known as talavera, after the pottery town of Talavera de la Reina, in Spain.

Recent technological advances in ceramic tile production, from high-tech glazes to super-efficient production methods, have made the most beautiful and durable ceramic tiles attainable for almost every household in North America. You can get ceramic tile that looks like marble, porcelain, granite, limestone, slate, and even wood, rusted steel and leather. Even the metallic-looking "oil-slick" tile that was so popular — and expensive — during the Victorian era has been made affordable today due technological advances.

Motifs and glazes from every century are either made on these shores or imported. This abundance of ceramics has coincided with, or has at least been partially fueled by, our increasing interest and sophistication in all things architectural. The enthusiasm for domestic decorating has also been a boon to the scores of boutique tile makers, as much artist as artisan, who now have a market for their unique, small batch and often custom product, many of which appear on these pages. These smaller tile makers are known not only for their own particular shapes and designs, but for the unique glazes they come up with.

PROPERTIES OF CERAMIC TILES Quite apart from their intrinsic beauty, ceramic tiles have mechanical qualities that make them ideal for a host of applications from countertops and backsplashes to floors and walls. They are resistant to stains caused by almost all acids, alkalis, and organic solvents. They are not affected by oxygen, so they don't oxidize or deteriorate in the air. They are water resistant. They are extremely hard. There are tiles laid down in ancient Rome that are still in use today. Beyond that, a dizzying array of shapes, sizes and finishes are available. (Note: Color shades of ceramic tiles may vary from batch to batch. This is often seen as one of the more charming aspects of ceramics. If you don't share this view, be sure to look for the same lot number and shade number to insure uniformity.)

There are three basic ways to make ceramic tiles: Hand cutting the shape out of pressed clay; extrusion (pushing the wet clay out of a shaped tube; and the dust pressed method. The dust pressed method (which includes most commercial ceramic tiles) involves the mixing of clays, talc, and other ingredients, which are then pressed into a mold at extremely high pressures to form the tile.

A stunning example of Majolica tile work by Matthias Ostermann, one of the North American masters of this old style of tile making.

The term "vitreous" refers to a ceramic tile that is hard, dense and impervious. Most tiles for residential applications are semi-vitreous, while the least dense, or non-vitreous tiles are used primarily for decorating walls in dry areas. Glazing is usually used to protect this last kind of ceramic tile.

Ceramic tiles that have a coating (usually a mixture of ground glass and pigments) put on them that is fused on the surface by firing in a kiln, are called *glazed*. With the exception of quarry and encaustic tiles, most tiles are glazed.

An *unglazed* tile has no glaze coat baked on so it will be the color of the clay that the tile was made of, ranging from a light sandy brown to a red brick shade. Since they have no relatively delicate glaze to worry about, unglazed tiles are more durable and are therefore used in high-wear areas such as floors. Unglazed tiles are usually sealed.

Encaustic tiles have their patterns embedded or stamped right in the clay itself during the manufacturing process. They are not glazed and are traditionally used as decorative flooring tiles. Geometric tiles are often used as borders.

PORCELAIN All ceramic tiles can be considered either porcelain or non-porcelain because ceramic tiles are either made by firing clay or by the porcelain method. Made using essentially the same dust-pressed method as fine china, porcelain tiles can be fired at higher temperatures than ordinary ceramic tiles and are therefore tougher, have a lower water absorption rate (.5 percent or lower) and are more stain resistant while at the same time displaying a finer grain. They are used for floors, walls and countertops and often have the color go all the way through the tile, which greatly reduces the consequences of scratches and chips. Porcelain tiles also have a very high freeze-thaw rating, making them suitable for exterior use.

MOSAIC From one to two inches square, their small size means that mosaic tiles can be used to make patterns or pictures and are perfect for contours like rounded or odd-shaped shower stalls or counters. They are used on both floors and walls, depending on what they are made of and what their strength and surface is, but the fact that mosaic floors have more spaces between more tiles increases their slip resistance. Mosaics can be glazed or unglazed, porcelain or non-ceramic (stone, seashell, and glass, for instance), and are often attached to paper or mesh in pre-arranged patterns that can be immediately applied.

GLASS As well as being extremely durable and easy to clean, glass tiles offer a visual depth unavailable with ceramic tiles. Depending on thickness and texture, glass tiles do things to light that can enliven a room like no other material. Glass tile surfaces fall into three main categories: clear, frosted and iridescent. They are relatively expensive but they can also be environmentally friendly, if you choose tiles made of recycled material.

One caveat: glass tiles are slippery. Some manufacturers apply a non-slip surface for just that reason. You can also use small glass tiles in a mosaic for a more gritty, non-slip surface. Of course the smaller the tiles, the more grout you must use and the grout is where most of the maintenance is. The grout may stain, so it has to either be sealed or cleaned with a bleach solution. One fairly new (and expensive) answer is to use a stain-proof epoxy grout.

Left and Center: The hearth tiles of this Jerusalem limestone fireplace are from a photograph of a stone façade and carved-oak leaf detail from an ancient Italian church. The image was then transferred to tile by Imagewares.
Right: Unpainted wood and unglazed terra cotta tiles go comfortably together.

METAL TILES The high-tech clean appearance of metal tiles — mainly stainless steel but also copper, zinc and various amalgams — makes them popular for contemporary settings, particularly as accent tiles. This is particularly important to keep in mind, because a large field of metal tiles can be overwhelming. Their strength makes them popular for floors, especially in ultra modern kitchens. And their durability makes them perfect for backsplashes and even countertops. The myriad of finishes available, from pewter dull to blindingly high gloss, adds to their versatility. Also, since metal tiles are impervious, they don't need to be grouted and therefore present a more homogenous look devoid of grout lines.

STONE TILES Stone tiles have many of the same attributes of metal in terms of strength and durability although they often have to be sealed to prevent staining. Unlike glazed ceramic tiles, stone tiles are homogenous throughout and therefore chips and scratches are not nearly as noticeable. Stone can also be manhandled in different ways to achieve different textures and looks, from the antiqued appearance of tumbled tiles to the roughed surfaces of "blazed" (treated with heat) tiles. Rather than metal's cutting-edge look, stone has a solid, timeless feel.

At the outer edge of the stone palette are gemstone tiles. Made from semi-precious materials such as amber, jade and jasper, they are naturally expensive. This and their innate flashiness confines them more to use as accent tiles, but they can provide a beautiful richness when used with taste. The same can be said for seashell tiles, which use the pearly luminescence and striations of shells to make a bold decorating statement. Examples of all of these tiles can be seen in the Products section under Tiles, as well as in the Kitchen Fittings and Bathroom Fixtures chapters.

ceilings

Sometimes known as the "fifth wall," ceilings are without a doubt the most ignored area of a room. This is unfortunate because a large unadorned ceiling can make any room seem cold, cavernous and drab. Ceilings with moldings, murals, medallions or beam work, on the other hand, can give the impression of spaciousness even when of standard height. You can do much with an existing ceiling, with paint, moldings, medallions and novel materials, and with treatments similar to or differing from adjoining rooms, to either draw attention to them or, in the case or truly irredeemable surfaces, to direct the eye to other, more appealing views.

When planning a ceiling, think first about going up. High ceilings give a sense of grace and grandeur to a room. You can simulate this look by lowering

Opposite: Architect Richard Landry's clients eschewed a formal living room, but wanted a formal sitting area, so they combined the living room with the library and came up with this elegant space by the fire. The coffered ceiling is a series of elongated octagons that are lined with mahogany molding. **Left:** This soaring exposed timber frame ceiling made of Douglas fir beams is in the casual "Coastal Living" style in which most interior surfaces are painted white. The vaulted ceiling, massive supports and open plan make this home feel like a loft. The floor is made of very hard Brazilian ipe wood, the counters are honed Algonquin limestone, and the fireplace is made of split Algonquin limestone.

Right: Nothing suggests the rough frontier more than smooth round beams. Sometimes the bark is left on for an even more deliberately rustic look. **Opposite:** This antique brick dome over a games table is a good an example of Old World construction and a pleasing and now uncommon way to crown the space. The roughness of the brick minimizes the echo you get with smooth domes.

your wall trim. Apart from that, using trim around the edges and light fixtures gives a sense of depth and visual interest, as does interesting ceiling treatments such as stamped or embossed tin, patterned wood strips, plaster, corrugated or oiled steel, lattice work, or hewn beams.

Of course, the beams don't have to be hewn, and can merely consist of the exposed rafters. Some beams — called "box beams" — are fake constructs that have no structural use and are there just to provide interest and atmosphere. In any case, beamed ceilings open a room up because they draw the eye up. This also works with paneled ceilings and even those made of beadboard. Board made of rough-cut pine and cedar, especially of the knotty kind, can provide a particularly rustic feel. And some kind of paneling or beam work is often used on cathedral ceilings to maintain interest in what can look cavernous if there is too much flat surface.

Ceiling medallions, usually made of plaster but also available in synthetic materials such as MDF and PVC, are also a great way to liven up a ceiling, especially around light fixtures. They can be a fairly simple affair of centric circles for an elegant look, or intricately patterned geometric or leafy, floral moldings (as in the popular "acanthus" style) for period authenticity. Crown molding can also be an effective means to draw attention to a ceiling. See the Architectural Details chapter, especially the Products section, for examples of ceiling medallions and crown molding.

The elliptical, barrel-arch ceiling in this bedroom continues past the wall into the closet and then the bathroom, with a pane of glass for acoustic privacy. A lip around the bottom ties it all together. Barrel-arch ceilings usually take advantage of unused attic space.

The shape of the ceiling is also a great way to introduce interest to a room. Barrel arch ceilings, which were very popular in the 1930s, can look especially dramatic, especially in kitchens. A true barrel arch is a full half circle, but you can also have very shallow versions in the living room or even the bedroom that provide a very sophisticated and calming affect, especially when lit with cove lighting (see above). But really any shape can work if done using pleasing proportions, and any material — everything from oiled and corrugated steel to simple plywood or rough-sawn boards — can look stunning if it suits the surroundings.

A coffered ceiling is also a good way to make a dull ceiling interesting while also making it look higher. Either in a honeycomb or diamond pattern, coffered ceilings can be shallow or deep, subtle or extraverted. They can be made custom or be purchased as a system that combines the larger beam elements (usually of veneer) with acoustic inserts made of plaster, wood or steel.

I had a friend whose parents were architects and he thought it was weird that all the furniture belonged to someone else — the Eames chair, the Mies day bed…

ANDREW KIRKOSKI,
NEW YORK ARCHITECT

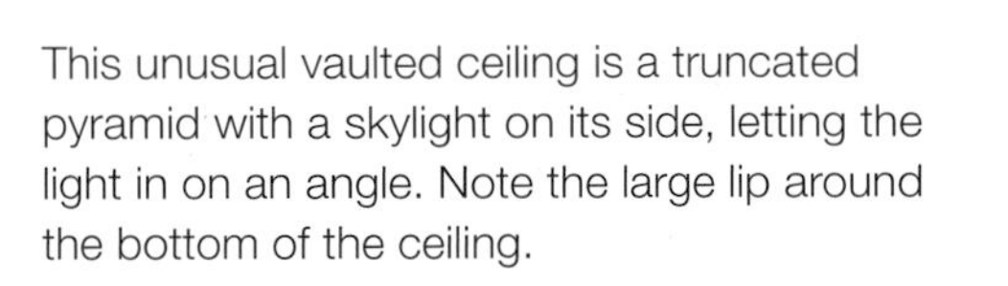

This unusual vaulted ceiling is a truncated pyramid with a skylight on its side, letting the light in on an angle. Note the large lip around the bottom of the ceiling.

Left: Exposed plywood ceiling and joists (with fairly elaborate iron hardware) are a good example of a style in which the honesty of the construction and materials is the main design element. **Center:** A carved wooden ceiling, the undulating curves of which are echoed in the slightly fluted shades of a wrought-iron light fixture. **Right:** Taking out the drywall ceiling and adding steel beams allowed for another foot of ceiling height in this Contemporary home. The exposed beams and oiled steel plates add a high-tech-looking edge to the space and draw the attention up.

Domed ceilings are the most dramatic, and can be made even more so with creative cove lighting. Stone or brick domes are extremely dramatic and give an unmistakable aura of age and solidity to a room. We know of one domed ceiling that had the stars in the sky — exactly as they were on the night the architect was born — painted on it.

In the kitchen, some good alternatives to the ubiquitous Florida ceiling — which consists of sheets of pebbled plastic concealing banks of cold and buzzing fluorescent lights — are ceilings that are barrel arched, coffered or domed, with recessed lighting backed up by task lighting over work areas. A kitchen ceiling that differs from an attached great room is also a good way to define the spaces while still linking them. For more ideas on ceiling light fixtures and fans for use throughout the home, see the Heat & Light chapter.

PRESSED TIN TILES This kind of fancy patterned ceiling was all the rage in the later part of the 18th and early 19th centuries, but then fell out of style and homeowners began to cover up their "old-fashioned" pressed-tin tiles. This led to many a happy surprise for renovators who cautiously looked under their dropped ceiling to see what was underneath, only to find a hidden treasure that had beautifully survived the decades.

Pressed tin ceilings are made by pressing (embossing) a pattern into the metal, which used to be tin but these days is usually steel. The patterns available range from Victorian to Art Deco and to what is known as "Americano," which relies heavily on a star motif (see pages 298–9). Pressed tin can be painted, is very durable and provides an instant traditional look for ceilings, walls and backsplashes. It comes in sheets and tiles, but also cornices and fillers (borders), and is often available prepainted.

products

floors, walls & ceilings

solid wood

Chateau, Autumn Maple *Shaw Industries*

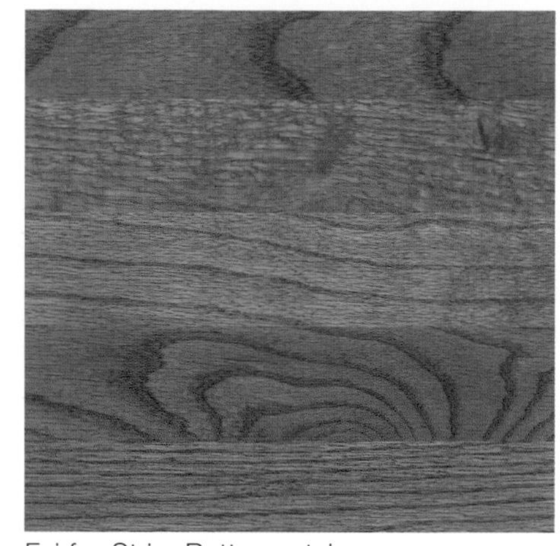

Fairfax Strip, Butterscotch

Arts & Crafts, Molasses

Sentry Plank, Latte *Shaw*

Sentry Plank, Cappuccino *Shaw*

Brazilian Cherry
Launstein Hardwood Products

Cherry, Wide Plank
Launstein Hardwood Products

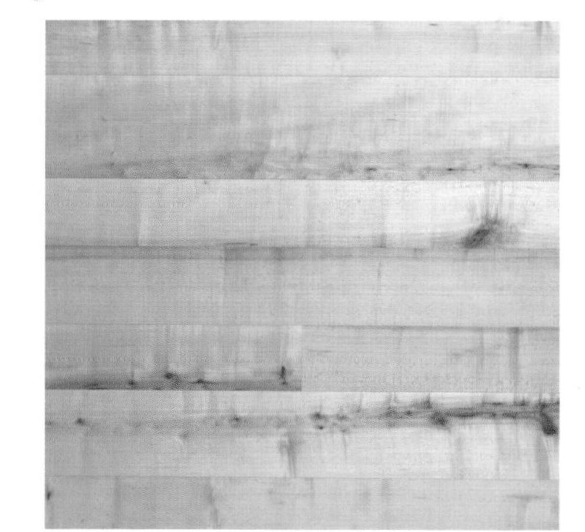

Hard Maple, Wide Plank

White Oak, Wide Plank

Hickory, Wide Plank

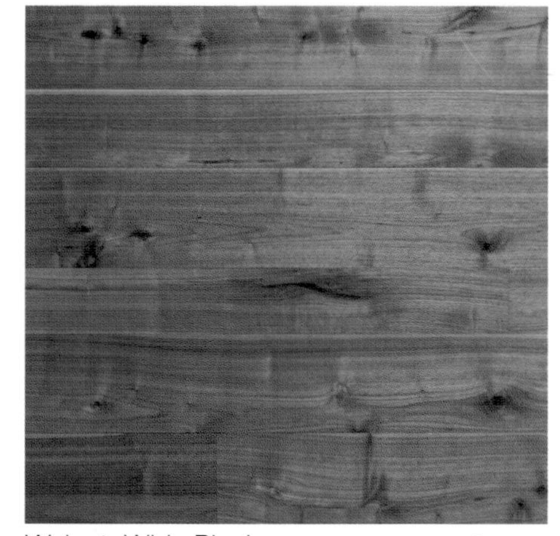

Walnut, Wide Plank

Red Oak, Wide Plank

Hickory Country

Red Oak Common

Red Oak Country

Red Oak Select

White Oak Country

Ash Country

solid wood

Walnut Common
Launstein Hardwood Products

Walnut Select

White Oak Common

White Oak Select

Cherry Country

Hard Maple Country

Hickory Common

Hard Maple Common

Hard Maple Selelct

hand-scraped solid wood

The worn texture of hand-scraped wood from *Homerwood Hardwood Floors.*

Amish Black Walnut

Amish Cherry

Amish Hickory Natural

Amish Hickory Saddle

Amish Oak Saddle

Amish Red Oak

Amish White Oak

Hickory Saddle

Bordeaux

Butter Rum

Espresso

Red Oak

reclaimed wood

Foot-worn Pine *Longleaf Lumber*

Chestnut Blend

Hardwood Forest Blend

Heart Pine, Clear Flat-sawn

Heart Pine, Naily

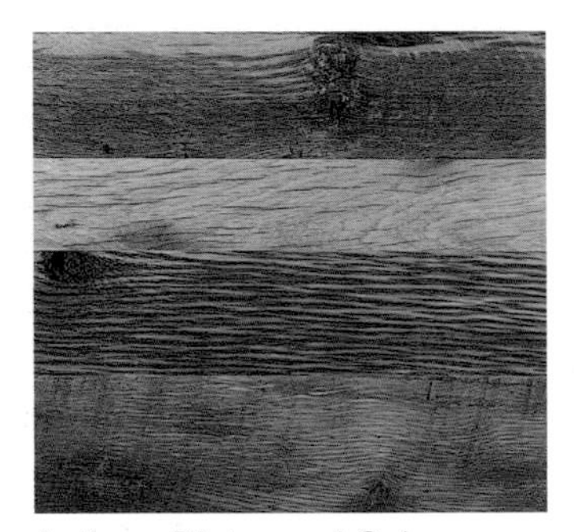

Antique Distressed Oak
Aged Woods

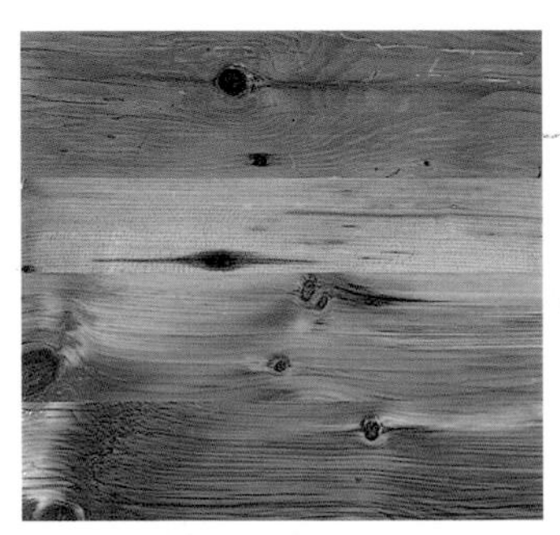

Antique Distressed White Pine

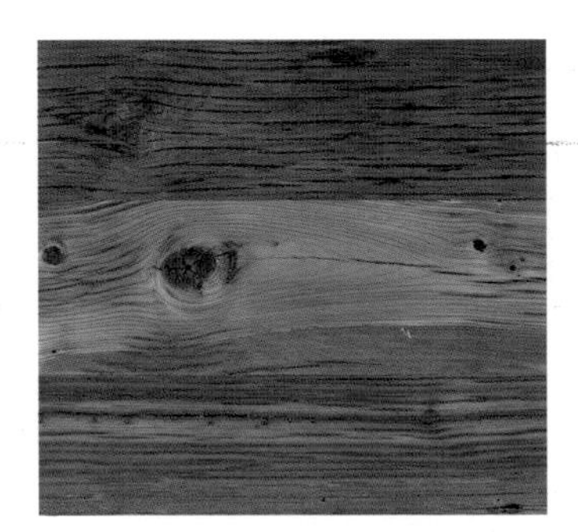

Antique Distressed Yellow Pine

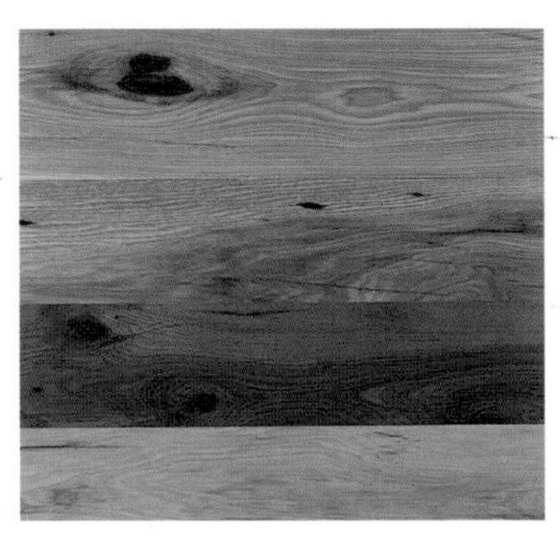

Antique Distressed Hickory

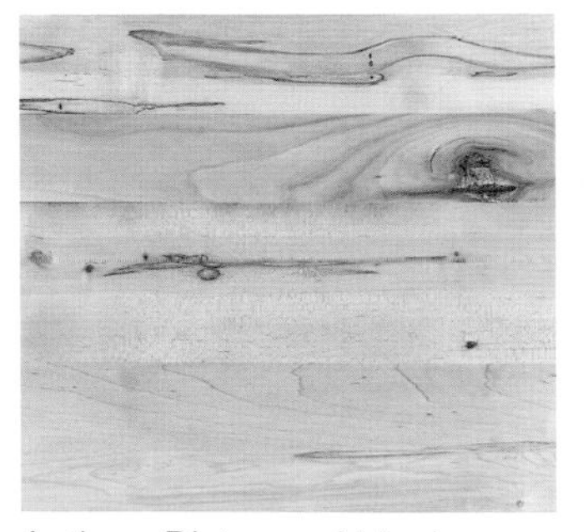

Antique Distressed Maple

Bunkhouse Plank Oak

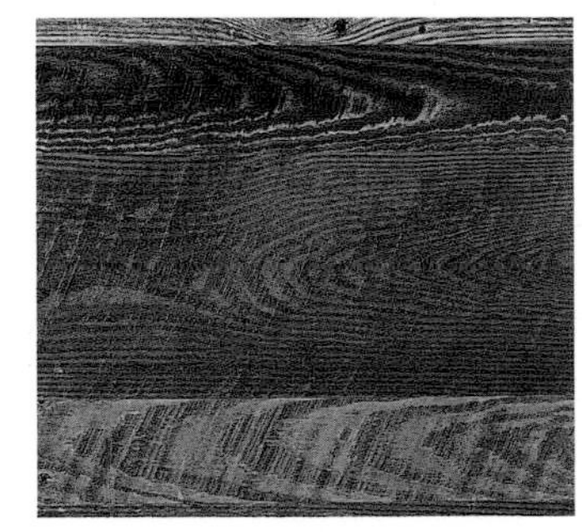

Bunkhouse Plank Pine

Beech (as is) *Longleaf Lumber*

Heart Pine Rustic
Longleaf Lumber

Heart Pine Select, Flat-sawn

Heart Pine Select

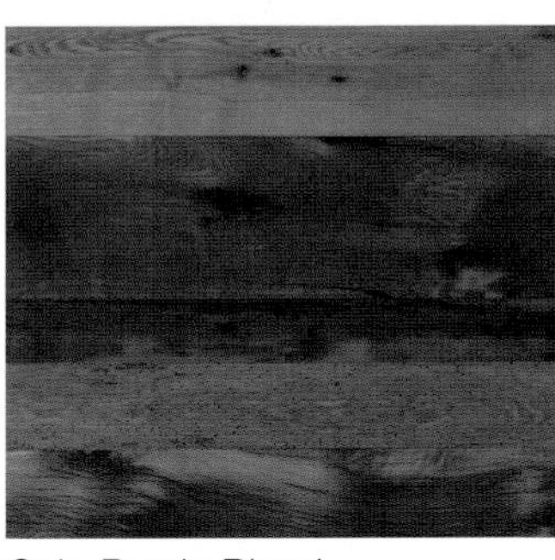

Oak, Rustic Blend

Pumpkin Pie

Red Oak

White Oak

White Pine

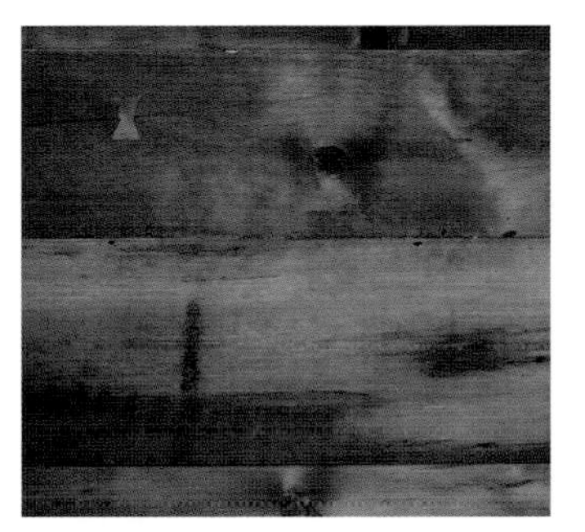

Wide Oak

Ye Olde Yankee Blend

The plain, honest look of Foundry Maple.

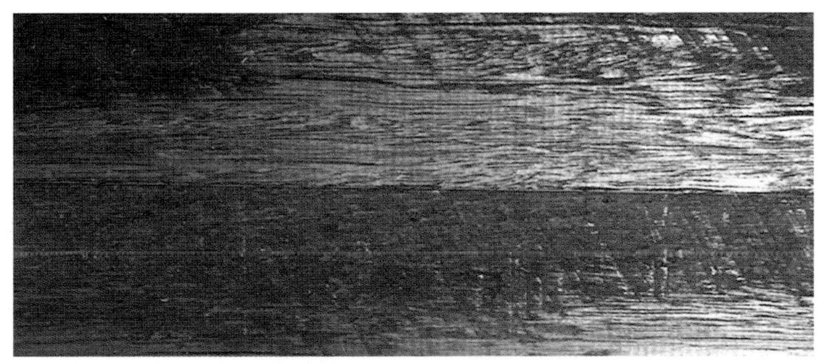

Walkabout Character Jarrah

engineered wood

Santos Mahogany *Pergo*

Cherry Mocha

Milan Maple

Lancaster Oak

Northam Oak

Wynwood Oak

Beacon Hill Cherry

Windham Cherry

English Black Walnut

Hawthorn Maple

Bordeaux Cherry

Springhill Maple

Dark Oak

Smoked Oak

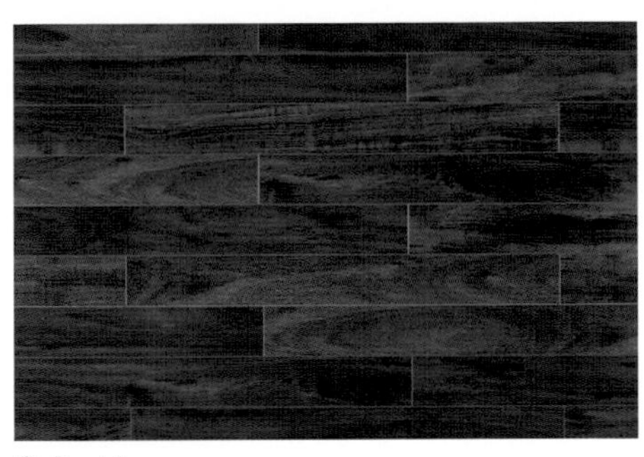
Salted Lapa

Chalked Oak

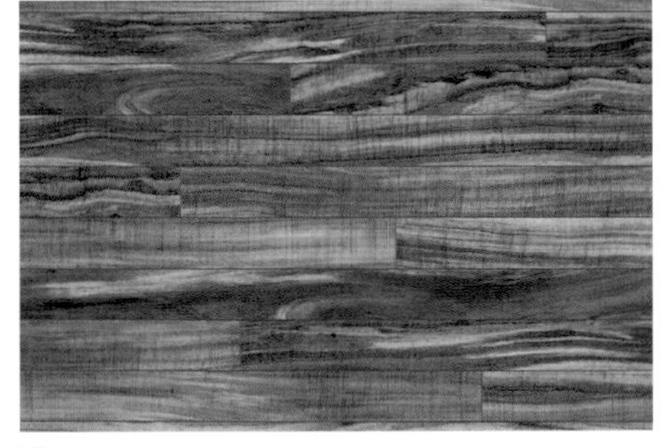
Koa

Ironwood

Jatoba

Merbau

Indian Tigerwood

Australian Eucalyptus

Dolce Mahogany

African Padauk

engineered wood

Parkway Natural Oak *Shaw Industries*

Parkway Honey Oak

Providence Autumn

Parkway Country Oak

Providence Caramel

Wilmington Victorian

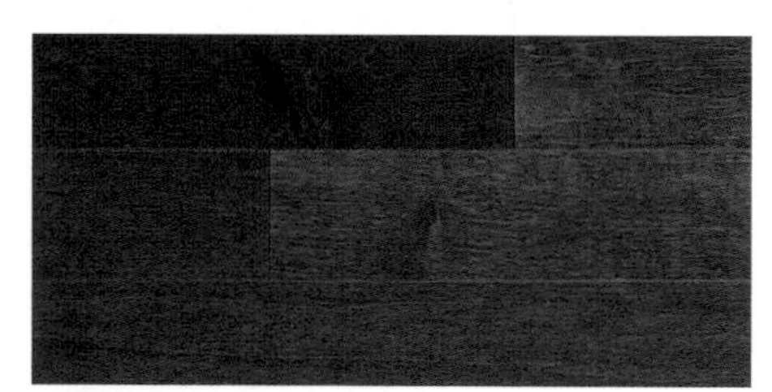

Savannah Maple Merlot

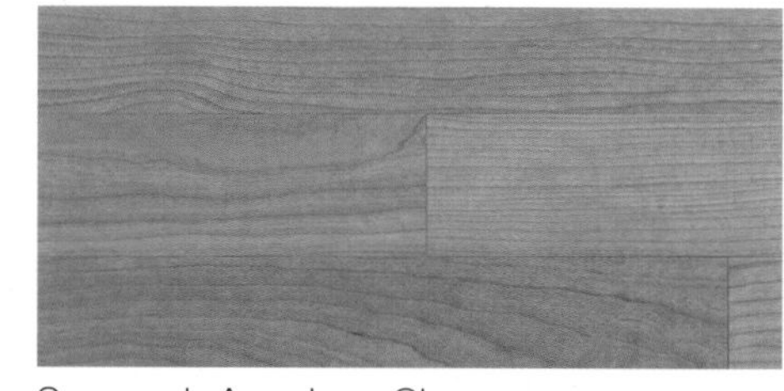

Savannah American Cherry

Savannah Brazilian Cherry

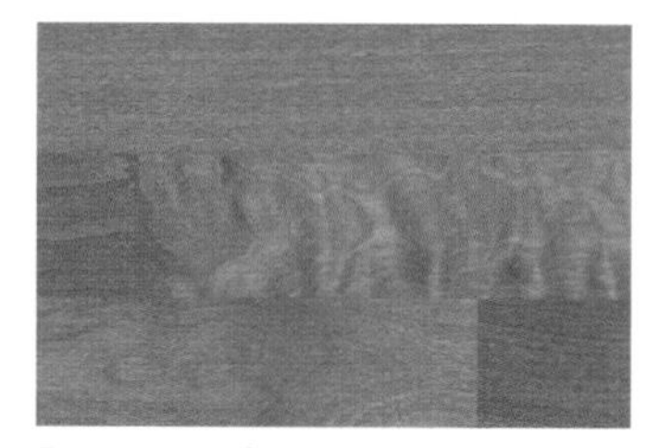

Discovery Cherry Plank

Discovery Chestnut Plank

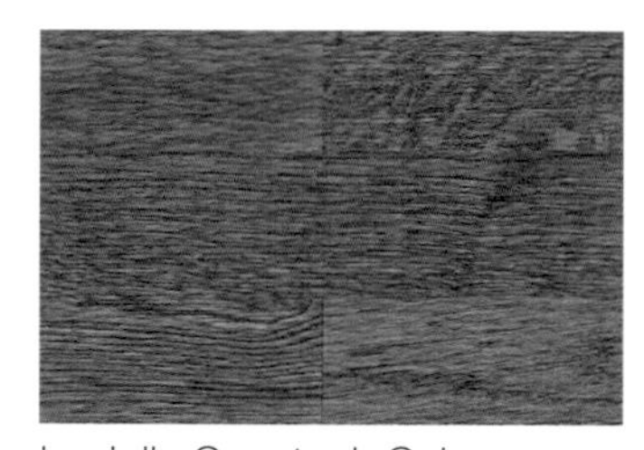

La Jolla Gunstock Oak

La Jolla Sable

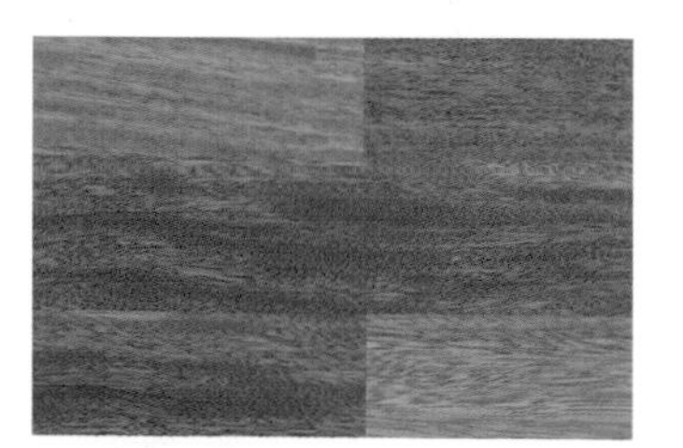

Highland Ridge Royal Mahogany

Highland Ridge Brazilian Walnut

Highland Ridge Natural Maple

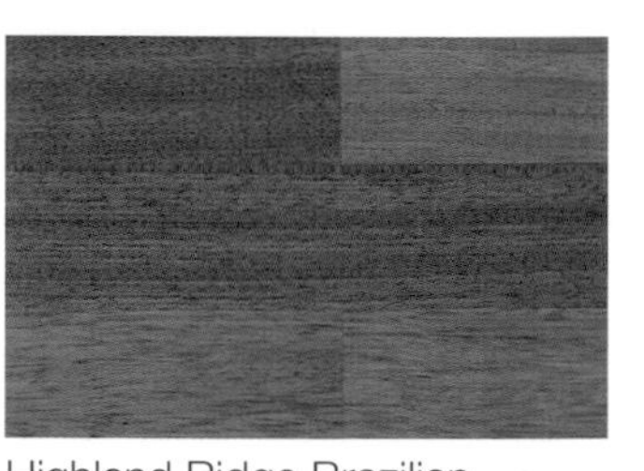

Highland Ridge Brazilian Cherry

hand-scraped engineered wood

Ventura Antique Brass *Shaw Industries*

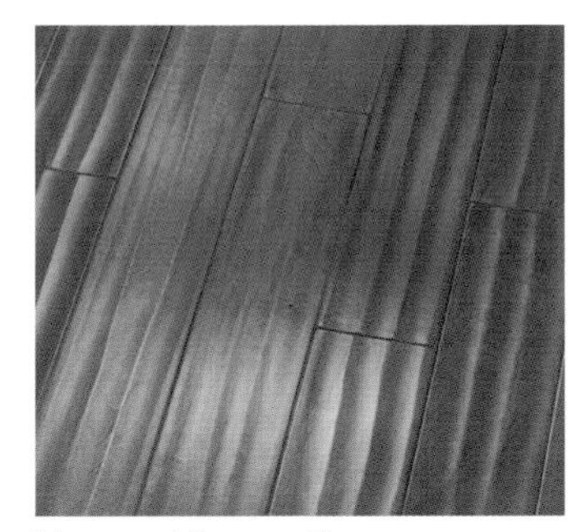

Ventura Vintage Sage

Ventura Cinnamon Spice

Ventura Aged Merlot

Ventura Manhattan Mocha

Palm Beach Mandarin Cherry

Palm Beach Imperial Walnut

Palm Beach Australian Rosewood

Palm Beach Merbau

Vicksburg Maize

Vicksburg Cider

Vicksburg Harvest

Vicksburg Espresso

linoleum

Marmoleum (linoleum) floor by *Farbo* in two contrasting colors.

Marmoleum 53030
Farbo

Marmoleum 53038

Marmoleum 53123

Marmoleum 53125

Marmoleum 53127

Marmoleum 53139

Marmoleum 53146

Marmoleum 53164

Marmoleum 53173

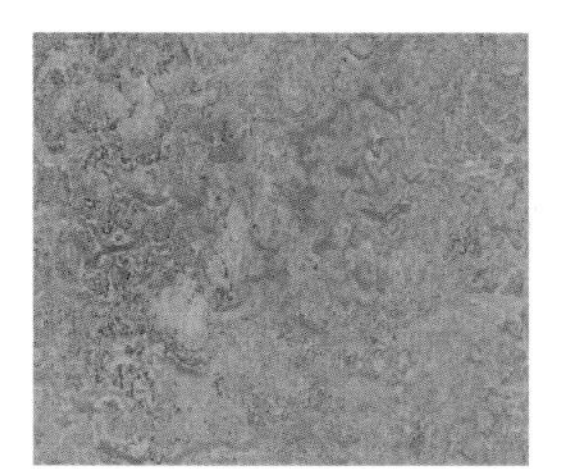

Marmoleum 53174

Marmoleum 53825

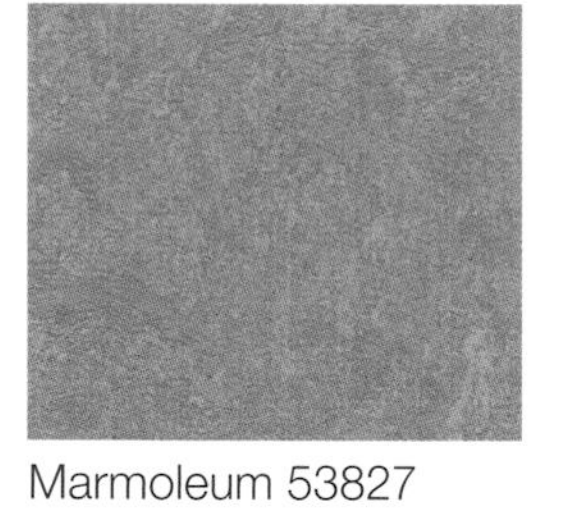

Marmoleum 53827

Marmoleum 53846

Marmoleum 53855

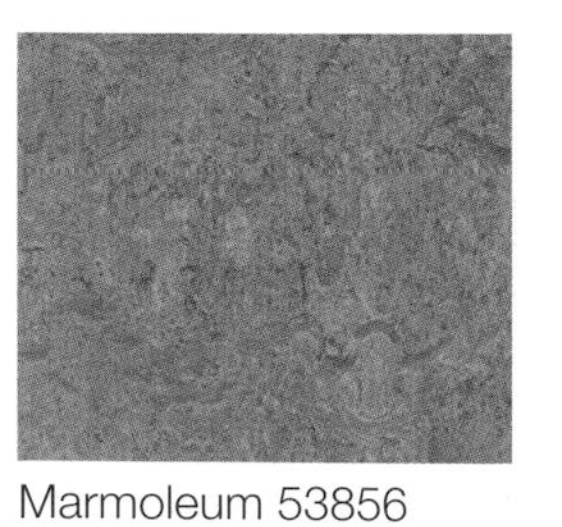

Marmoleum 53856

Marmoleum 53858

Marmoleum 53860

Black and white Marmoleum tiles in a bold zigzag pattern make a dramatic hallway.

Earthpath Golden Clay *Congoleum*

Victorian Plank Nutmeg

American Slate Arizona Rust

Earthpath Baked Clay

Quartz Stormy Greige

Rapolano Desert Chimney

Solitaire Earthen Brown Stone

Autumn Mist Earthen Teal

DC Options Sierra

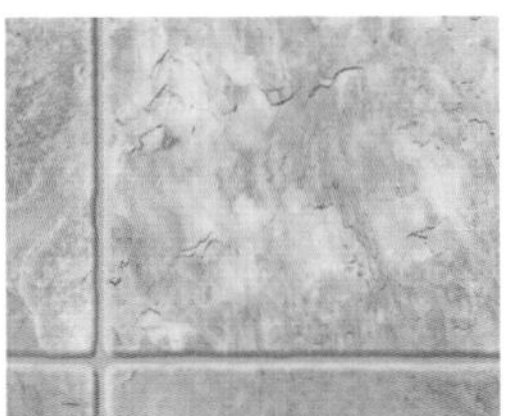

Montego Slate Amber

Sahara Warm Taupestone

Spirit Clay Bluestone

Rock Garden Seaside

Random Paver Plummit

Colonial Cherry Natural

Danbury Oak Colonial

Brentwood Cherry

stone

Blue Limestone *Paris Ceramics*

Tumbled Bourgogne

Tumbled Limestone

One-off reclaimed floor

Antique Blonde

Antique Jerusalem Limestone

Bourgogne Mosaic

Cotswold with Cabs

Honed Copper Slate Bricks
Stonehenge Slate

Honed Copper Mosaic

Sandblasted Limestone, Lilac

Polished Sandstone Parquet
Paris Ceramics

Autumn Border, Honed Slate

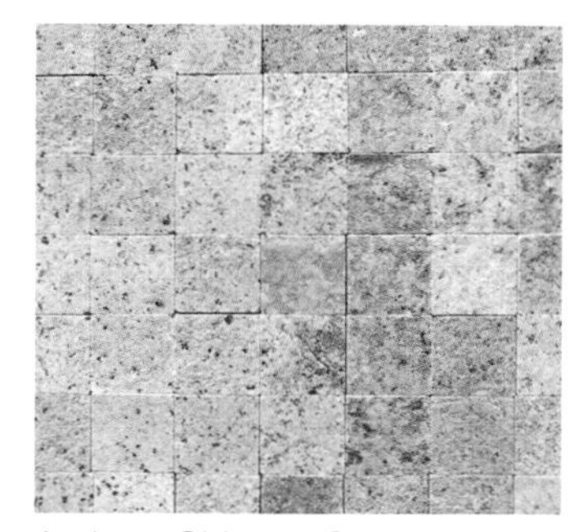

Antique Chinese Granite
Rhodes Architectural Stone

New Jade Ink Limestone,
Point Stalk Finish

slate

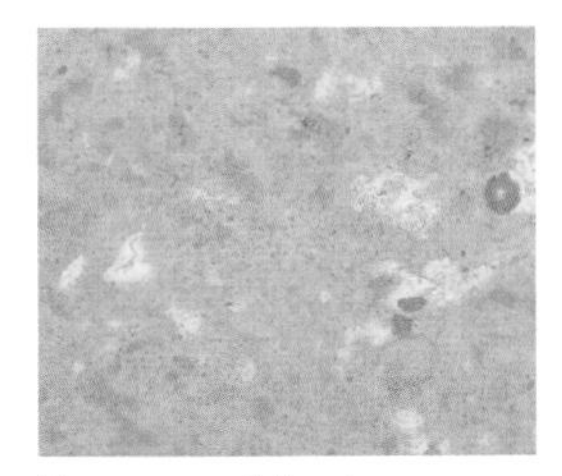
Processed Azul Martimo *Universal Slate*

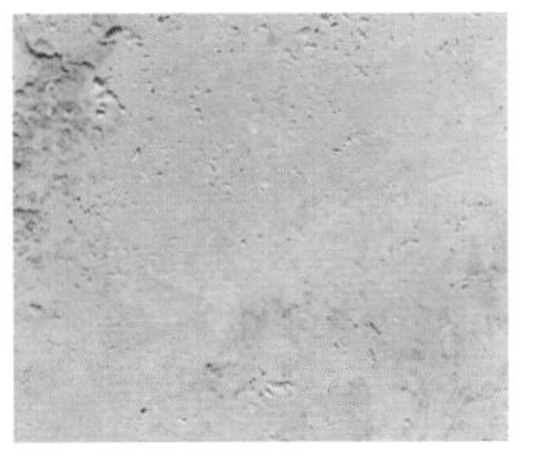
Processed Certina Cobbled

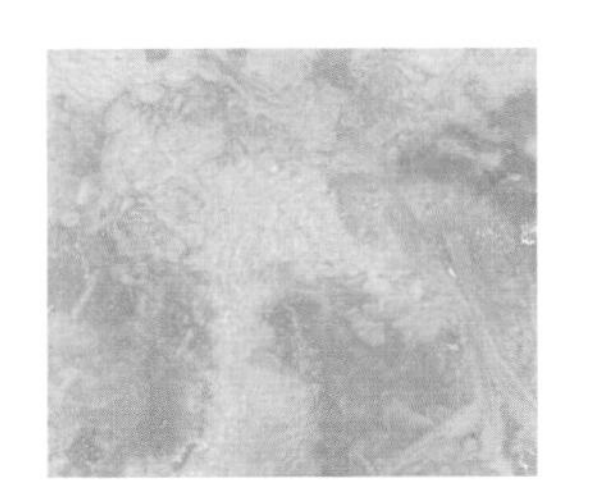
Processed Ivory

Processed Morocco

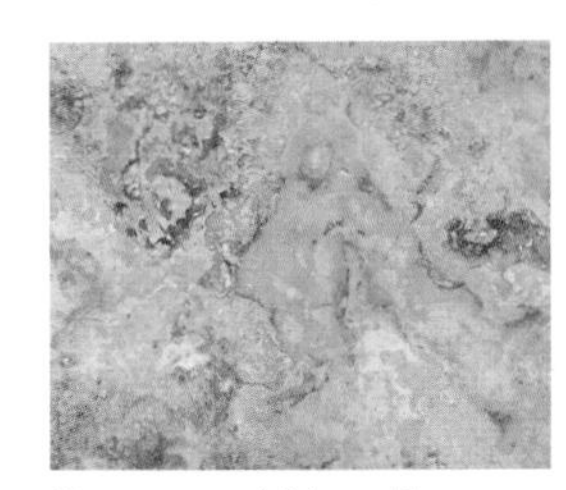
Processed Napolina

Processed Roma Antiqua

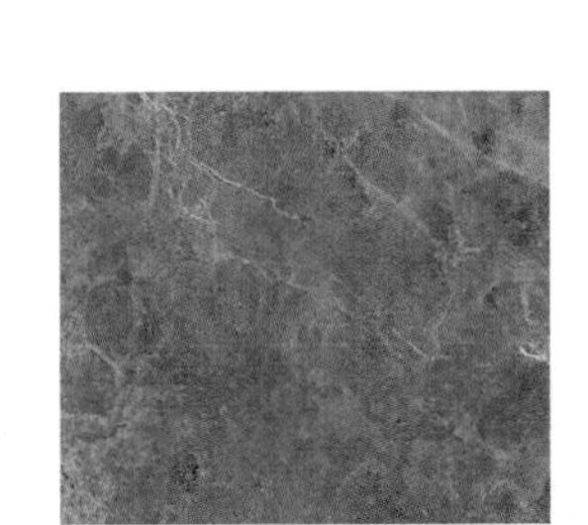
Processed Tobacco Dark

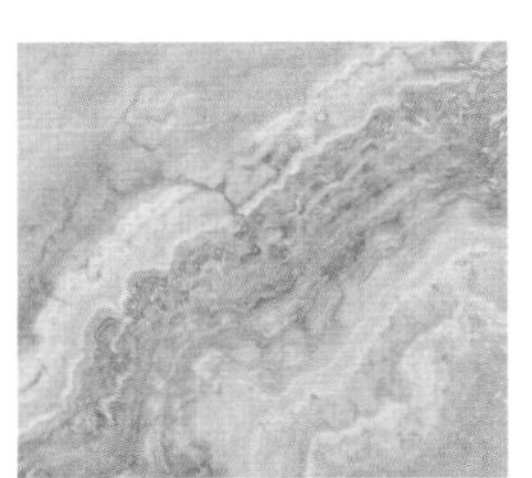
Processed Roma White

Processed Silk

Processed Walnut

Processed Toffee

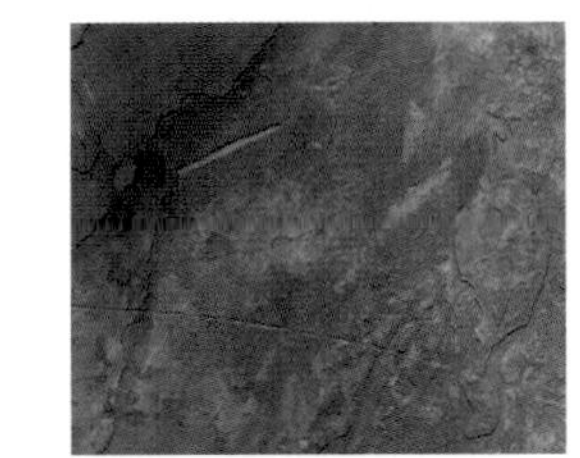
Processed African Gold

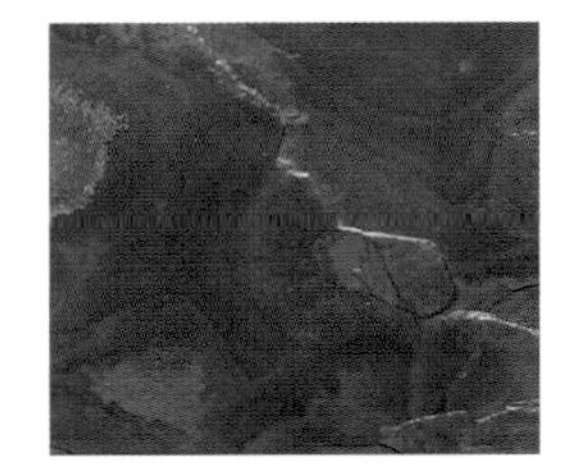
Processed Autumn Gold

Processed Black Cloud

Processed Black Pearl

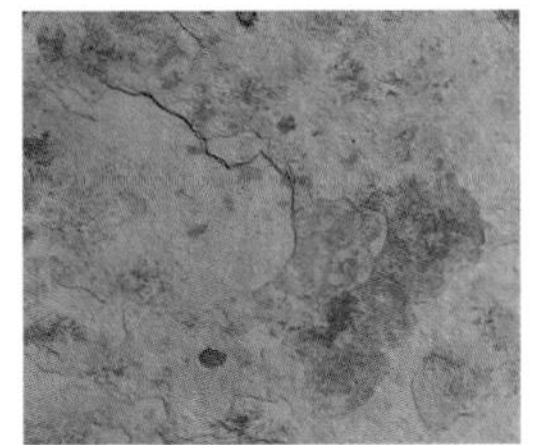
Processed Butterscotch

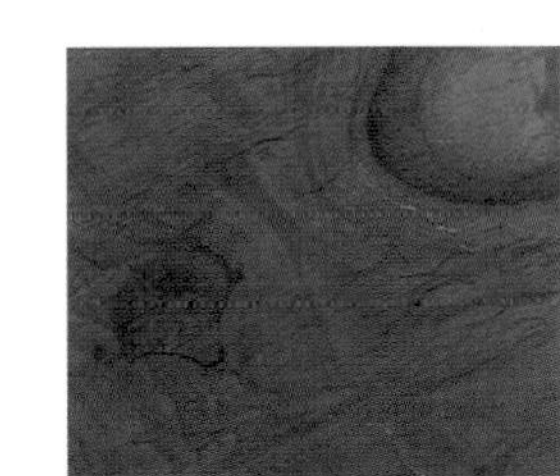
Processed Mountain Twilight

Processed Evergreen

Processed Indigo

Processed Platinum

Processed Sea Green

Processed Sunburst

rubber

Faux Leather rubber floor *RobinReigi*

Faux Leather rubber floor

Repel Rubber *Flexco*

SpexTones Rubber

terazzo

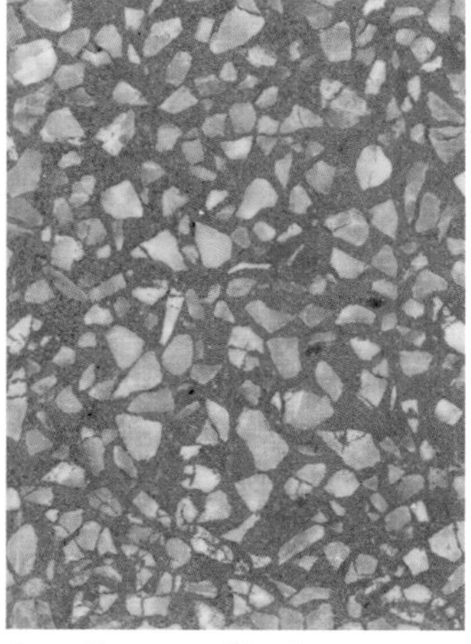

Grey Terrazzo *Rhodes Architectural Stone*

Green Terrazzo

Olive Terrazzo

Antique Salmon Granite Terrazzo

resin

Riverstone, marble pebbles in polyester matrix *RobinReigi*

metal & glass

Recycled Aluminum *Eco-Friendly Floors*

Recycled Brass

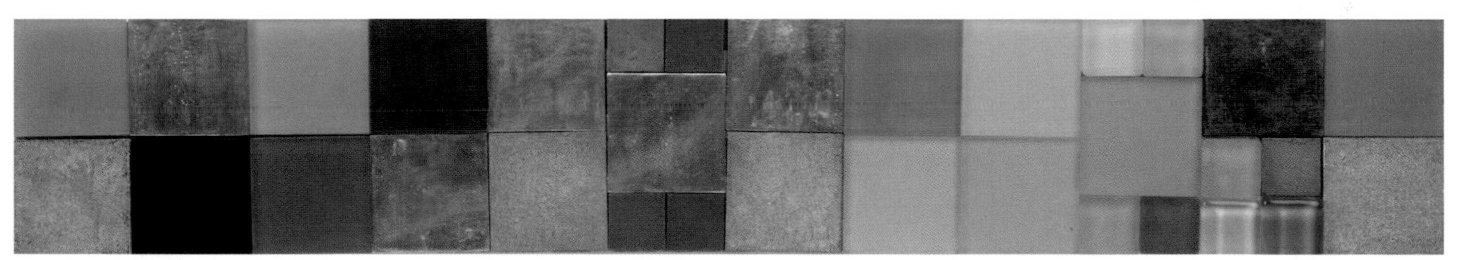
Mixed Recycled Metal and Glass

Stainless Steel *Rigidized Metal*

Recycled Aquavida and Granny Smith Glass *Eco-FriendlyFloors*

porcelain

Green *Crossville*

Almond

Beige

Black

Moka

Palais Versailles

Palais Louvre

Palais Elysee

Palais Luxembourg

Processed Old Stone Antra *Universal Slate*

Processed Old Stone Gris

Processed Arke Coke *Universal Slate*

Processed Ambienti Avorio

Processed Ambienti Brown

Processed Ambienti Grigio

Processed Ambienti Nero

Processed Ambienti Salvia

Processed Ambienti Arke Grey

Processed Arke Silk

Processed Arke Tan

Processed Caesar Di Barge

Processed Caesar Sintra

Processed Easy Slate Avorio

Processed Easy Slate Rosso

Processed Pastorelli

Processed Pastorelli Nero

Processed Uno Cromle

Processed Suburbia Soho

Processed Suburbia

Processed Uno Cremle

Processed Suburbia Queens

Soho Supergres Sub

Tribeca Supergres

porcelain

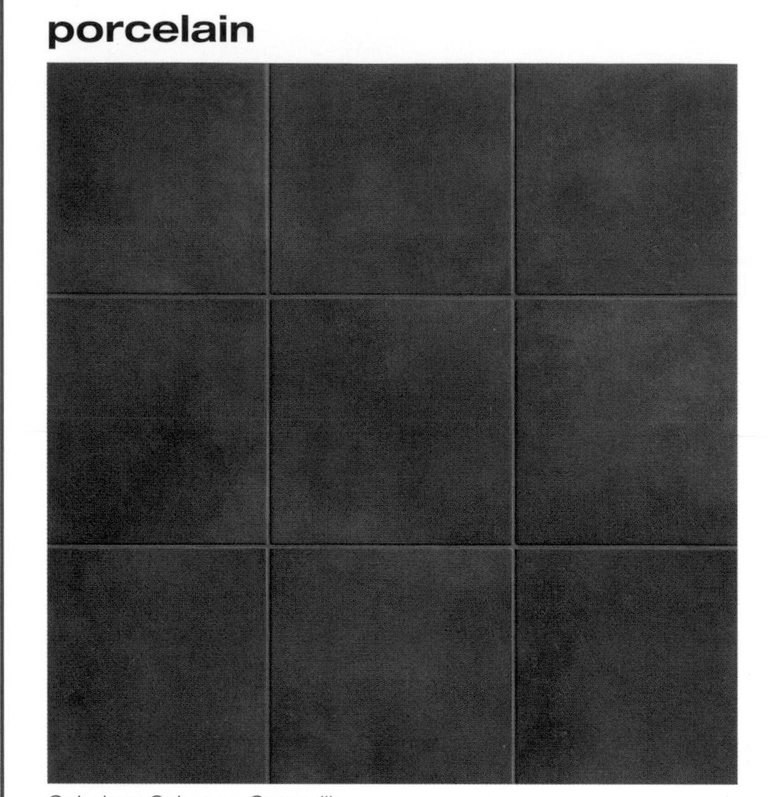
Colorbox Caboose *Crossville*

Colorbox Marshmallow

Colorbox It's a Boy

Colorbox Treehouse

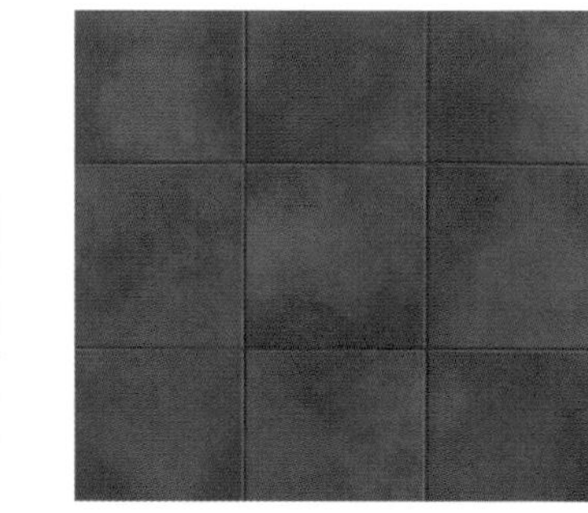
Colorbox Camping Out

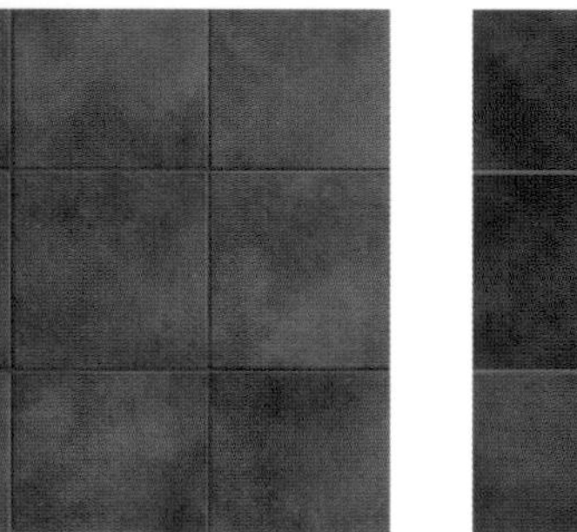
Colorbox I See The Moon

Colorbox Chocolate Candy

Colorbox All Night

Colorbox Grape Jelly

Chemistry Boron

Chemistry Argon

Chemistry Krypton

Chemistry Rubidium

Empire, Crown Dore

Empire Laurel Green

Empire Napoleon Blue

Empire Empress Silver

Empire Persian White

Empire Elba Night

Abisko Ebano, faux wood in porcelain *Country Floors*

Biorgogna Digione, porcelain

ceramic

Kennet, Crossings collection, floor or wall tile *Country Floors*

Westminster, Old English collection, 16-tile pattern

Sheherazade Crossings collection, floor or wall

Saxon Diamond, Crossings collection, floor or wall

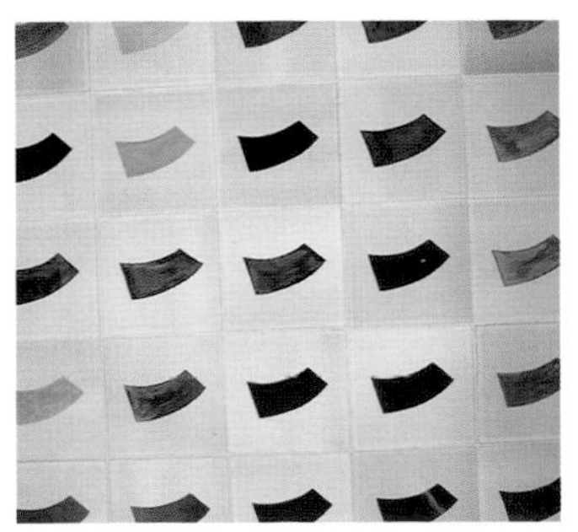
Scimitar ceramic *Paris Ceramics USA*

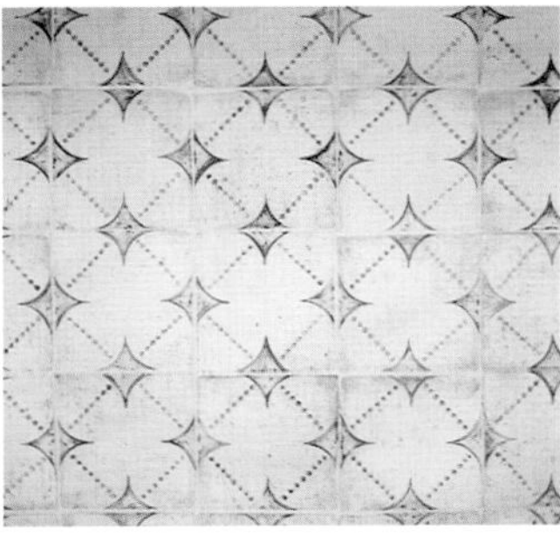
Therstar ceramic

terra cotta

Terra Cotta octagons with Deco inserts
Malibu Ceramics

Terra Stone octagon paver with decorative keys

Santa Barbara Red Mission paver with decorative key

Custom Decorative Glazed Star and Cross

Terra Stone picket paver with decorative keys

Terra Cotta with Deco inserts

Handmade Moroccan Terra Cotta
Saint Tropez Stone Boutique

Handmade Moroccan Terra Cotta

Hexagonal Terra Cotta tiles with square glazed filler tiles

photographic imaging

The Street Smart line of ceramic tiles from *Imagine Tile* directing traffic in a bathroom.

Manhattan, Street Smart
Imagine

Crosswalk Center tile, Street Smart

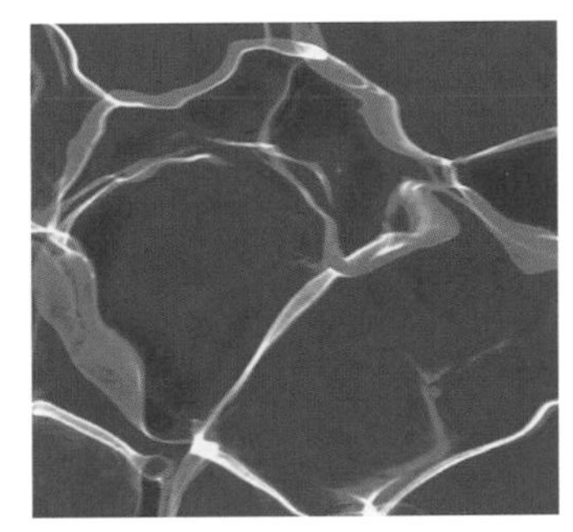

Water

Desert

Water tiles from *Imagine* on bathroom floor.

Hearthstone tiles copied from a photograph of an antique textile by *Imagewares.*

ceramic

Barrett ceramic vent *Handcraft Tile*

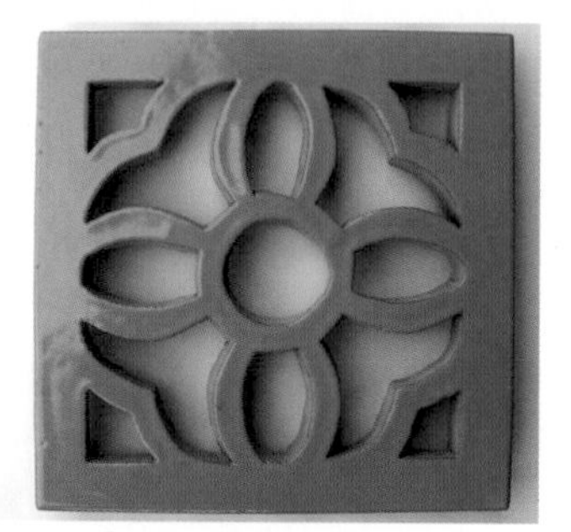
Camellia ceramic vent

Camellia

Delarose

Delarose

Delarose

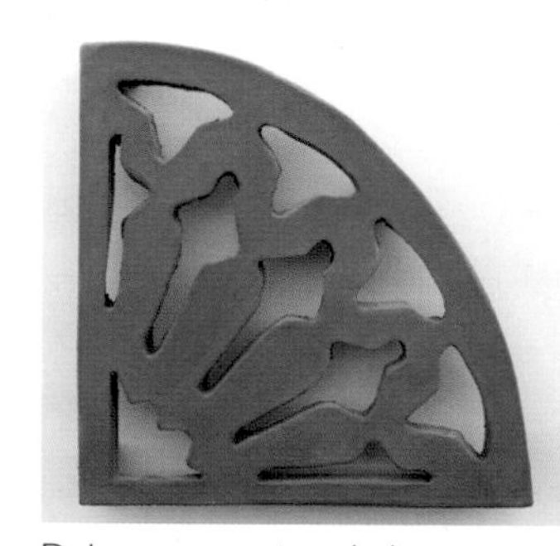
Delarose quarter circle

Lattice

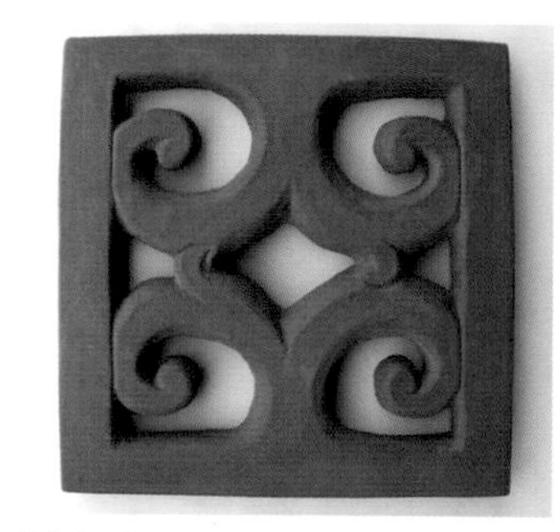
Maleck

Zen

Spiderweb

Spiderweb

Maya

wood & metal

Adjustable damper, Trimline insert
Launstein Floors

Base vent

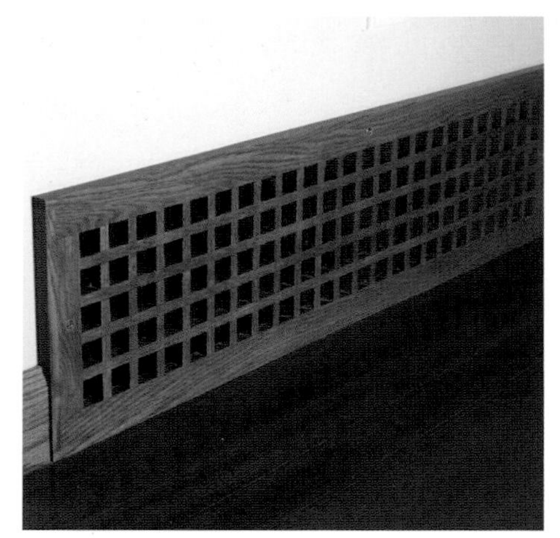

Egg Crate vent

Trimline Flush

Trimline Insert

Vent Flush

Black metal grilles *Acorn Manufacturing*

acrylic & resin

Varia resin panel, Via by *3form*

Varia resin panel, Play, crushed turquoise

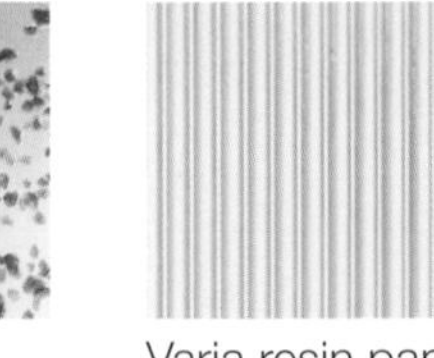

Varia resin panel, Vertu V Mondo

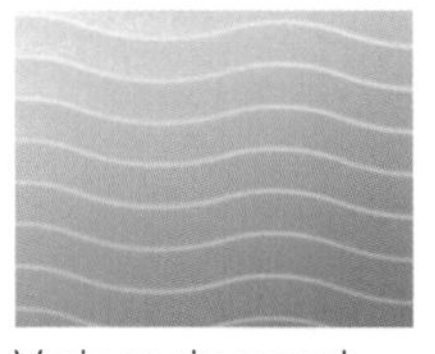

Varia resin panel, Quo Mondo

Varia resin panel, Genoa

Screen-printed panel, Varia Cirque

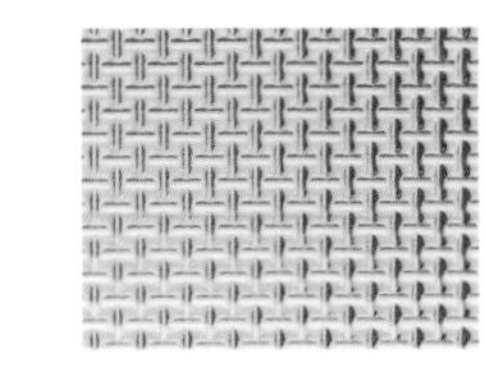

Screen-printed panel, Varia Futura

Screen-printed panel, Varia Eclipse

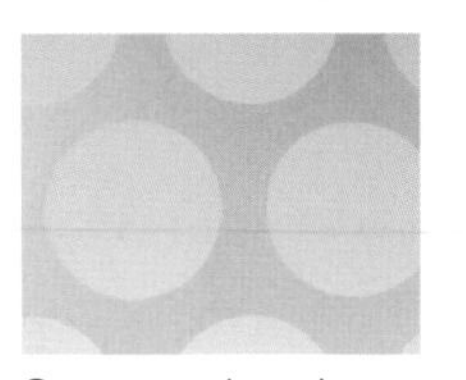

Screen-printed panel, Varia Mega Cirque

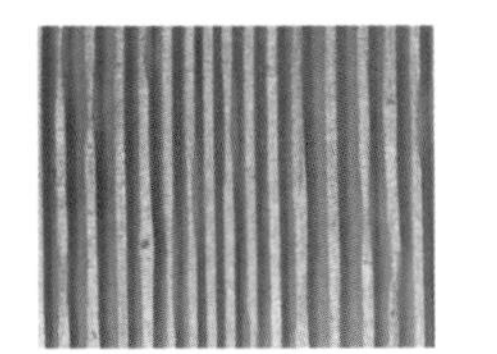

Varia resin panel Moderna, Tempo Ice

Varia resin panel Moderna, Linea Ice

Varia resin panel, Organics, Bear Grass

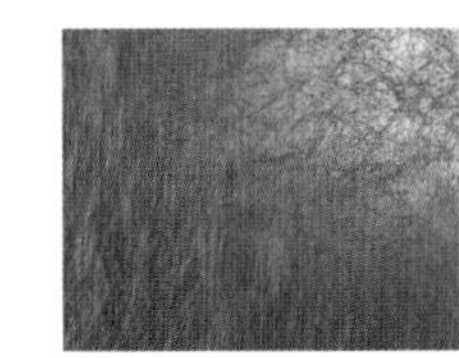

Varia resin panel, Organics, Ion Green Tea

Champagne Silver, silver leaf tiles *Maya Romanoff*

Varia resin panel, Organics, Fossil Leaf

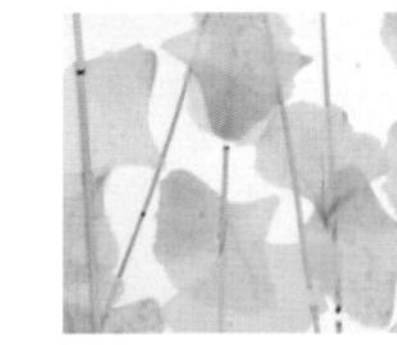

Varia resin panel, Organics, Gingko Thatch

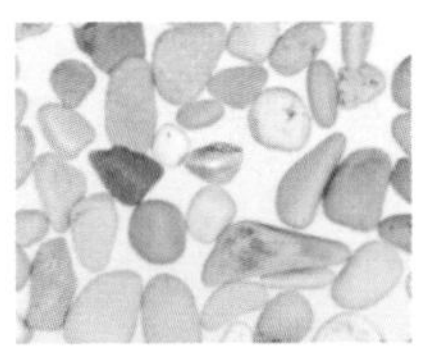

Varia resin panel, Organics, River Rock

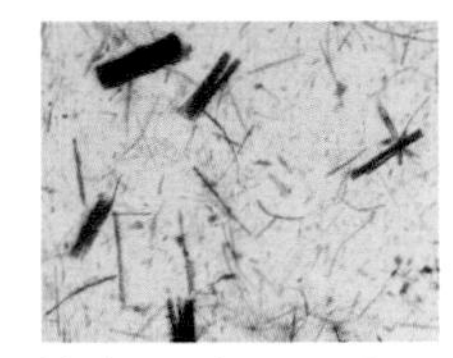

Varia resin panel, Organics, Tabako

Eggshells embedded in lacquer *RobinReigi*

Cast rubber resin, gel resin *RobinReigi*

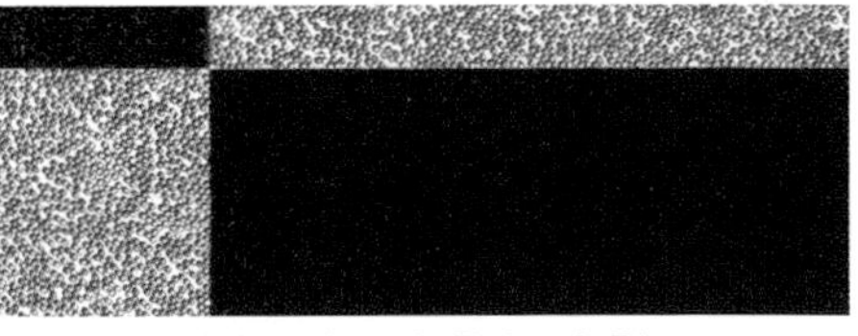

Bedazzled glass bead, Ruby & Bianca *Maya Romanoff*

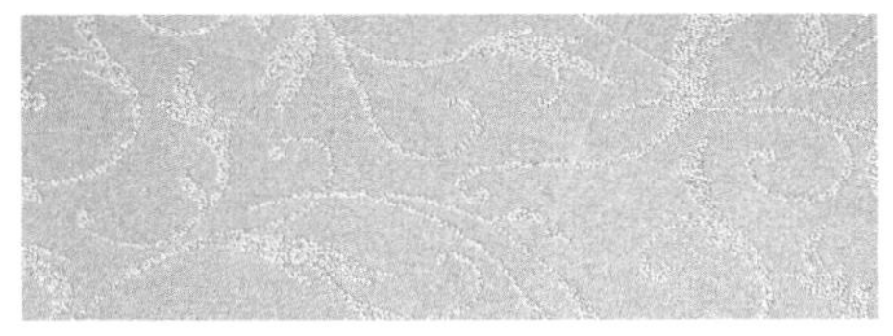

Relief Marquetry glass bead *Maya Romanoff*

metal

Mirro Maze *Spectra Metals/Pacific Crest*.
All decorative stainless steel this page available in 4 x 8 and 4 x 10-foot sheets.

Basketweave

Braided Tubes

Circles

Clouds

Currents

Dancing Vines

Egyptian Tile

Flaming Mirror

Herringbone

Jungle Vine

LA Style

Mist

Rolling Dunes

Shimmering

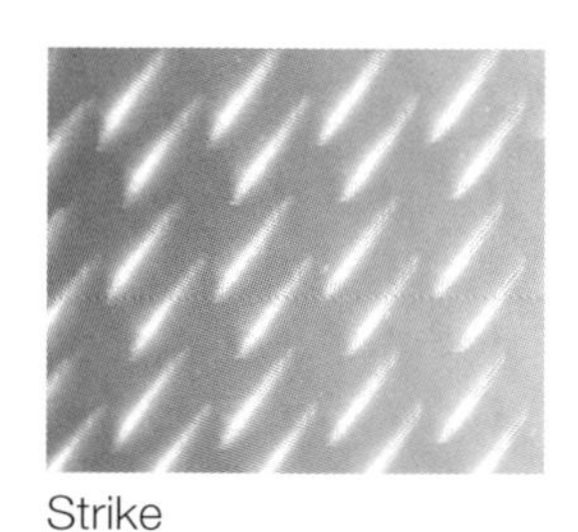

Strike

Square Reflections

Vegas Clouds

Twister

decorative painted

Branch, hand-painted Majolica *Levy Larocque*

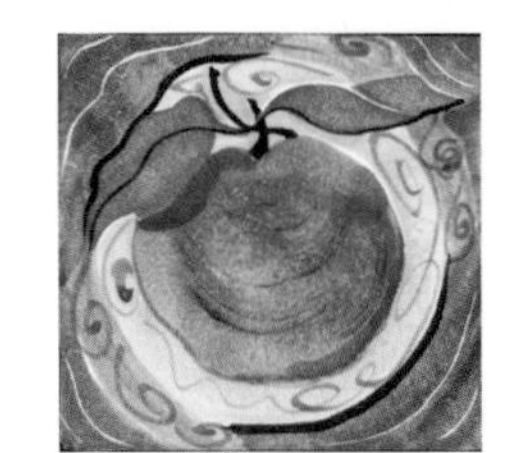

Apple

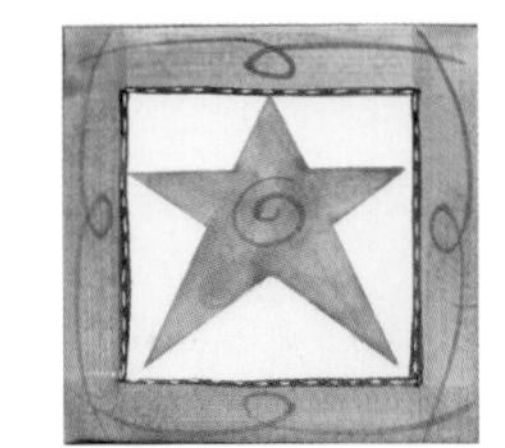

Star

Big Fish

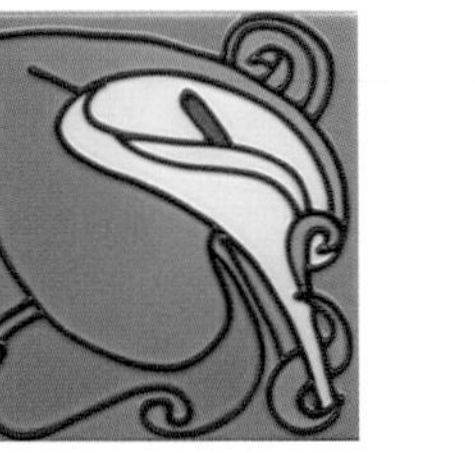

Stylized Calla Lily
DuQuella Tile

Mackintosh Glasgow Rose

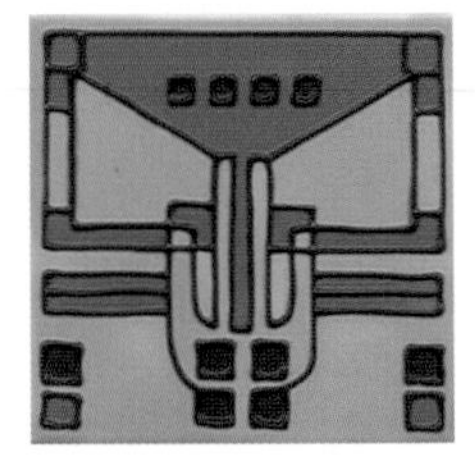

Geometric Design

Angelfish, hand-painted
Dy Witt

Glyph

Hieroglyph

Monkey Bird

Warrior

Benecio Décor
Jeffrey Court

Diego

Mojave

San Louis Rey
Jeffrey Court

Sierra Stop

Clara Décor

Fish Art *Majolica Mosaics*

Yellow Lily *Majolica Mosaics*

Orange Sofa
Majolica Mosaics

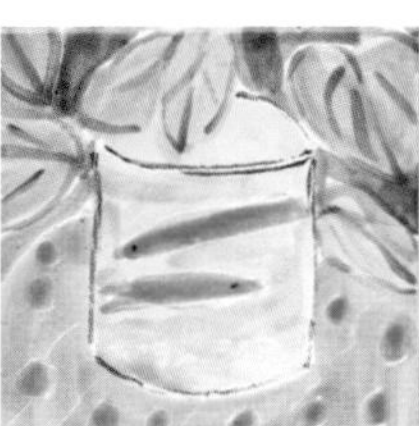
Fishbowl

Purple Lilly

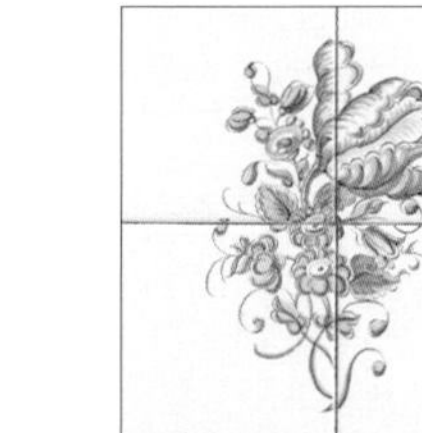
Peony Bouquet, Delft
Jeffrey Court

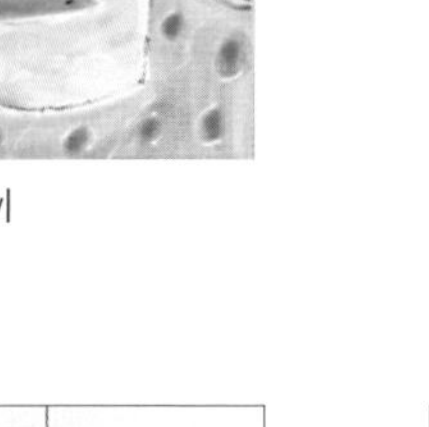
Tulip Bouquet, Delft

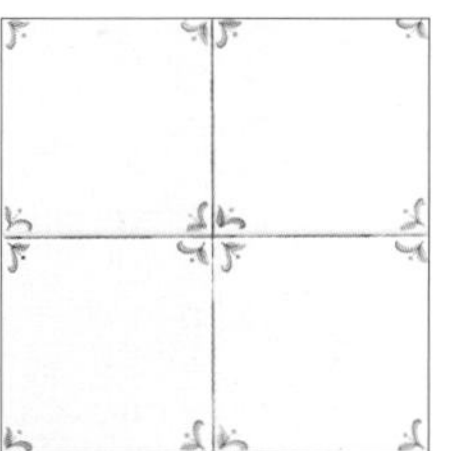
Leaf and Dot, Delft

Ship mural *Motawi Tile*

Tropical Palm mural *Elsner Artistic Tile*

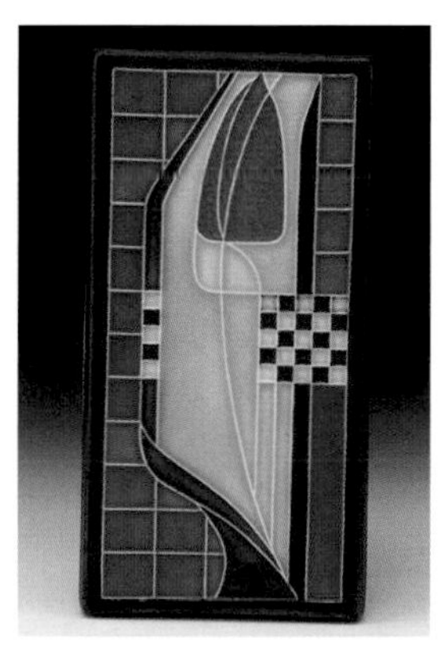
Rennie's Lattice (based on a textile pattern by C.N. Mackintosh) *Motawi Tile*

The Reader pictorial *Motawi Tile*

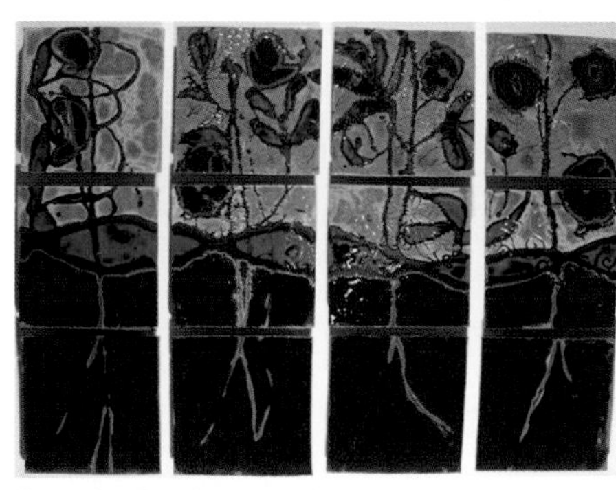
Cosmic Garden mural
Clay Squared to Infinity

Blosfeld Barley *Clay Squared to Infinity*

decorative raised

Lily *Earth Marks*

Oak

Daisy

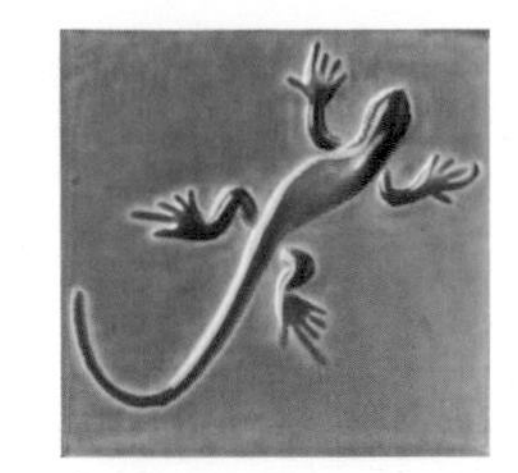

Gecko

Cloisonne Craftsman *Whistling Frog Tile*

Cloisonne Landscape

Cloisonne Cat

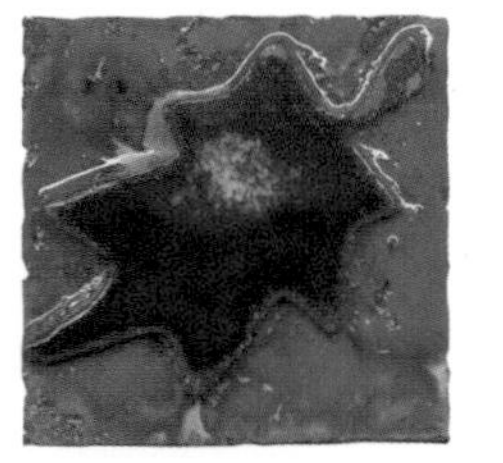

Three Inch Leaf *Clay Squared to Infinity*

Antique Patina

Night Gold

Matador Silver

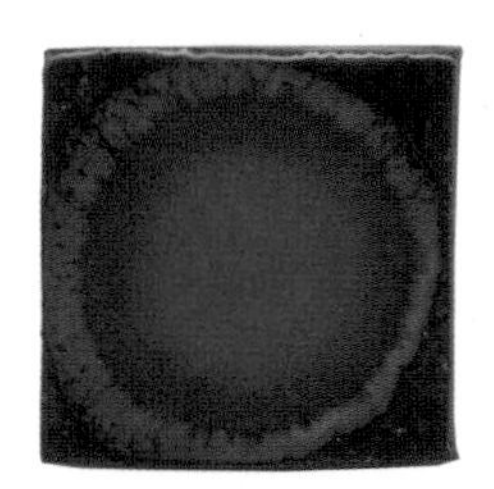

Matador Gold

Oriental Fish *Urban Jungle*

Sun

Egyptian Sun

Fancy Bird

Modrian

Etched Morocco with gold *Purple Sage*

Etched Morocco with gold

Thistle polychrome *Motowi Tile*

Dard Hunter Rose

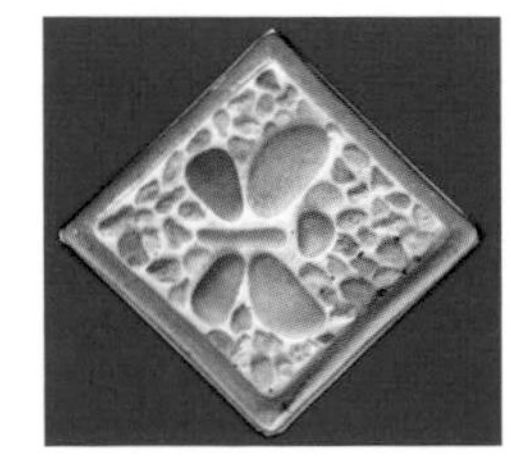

River Rock *Manet Tile*

Rounded Celadon *Tactile Geometrics*

Four Diamond Azure
Tactile

Textile Marigold

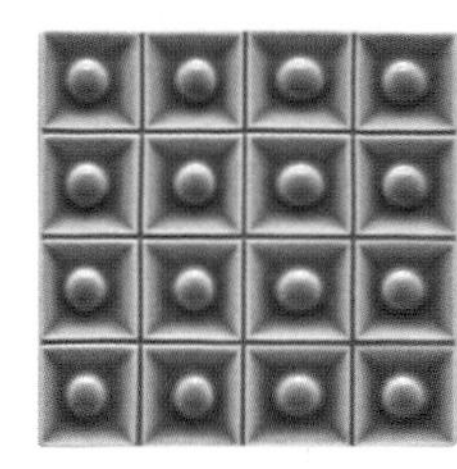
Pomo Ivy

Relief, Catalina
Jeffrey Court

Relief, Cherries

Relief, Apple

Basket Cuenca
Jacobs Tile

Angel Straw
Bambino Tile

Bobcat
Terrapin Tile

Meadow Vine
Talisman Tile

Daisy Cow Straw
Crackle *Bambino Tile*

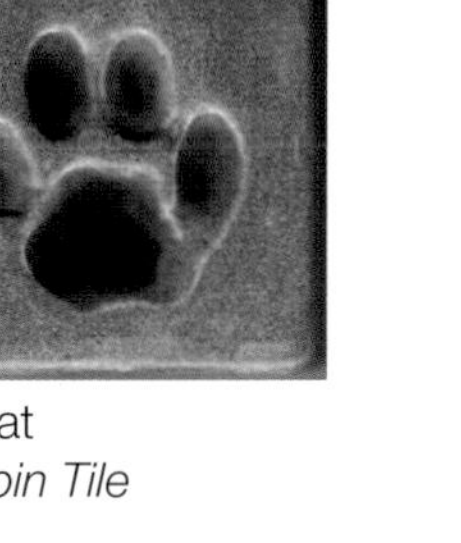
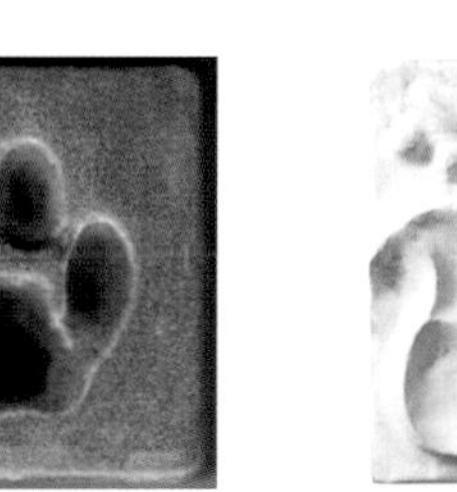

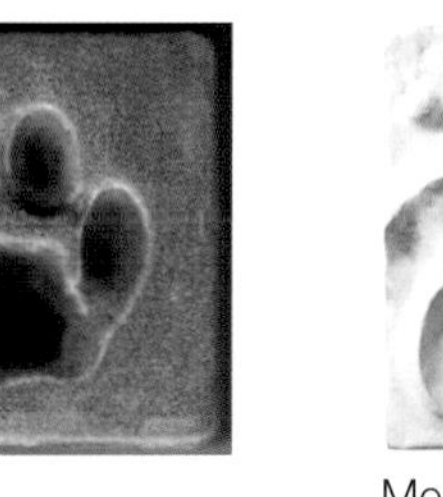
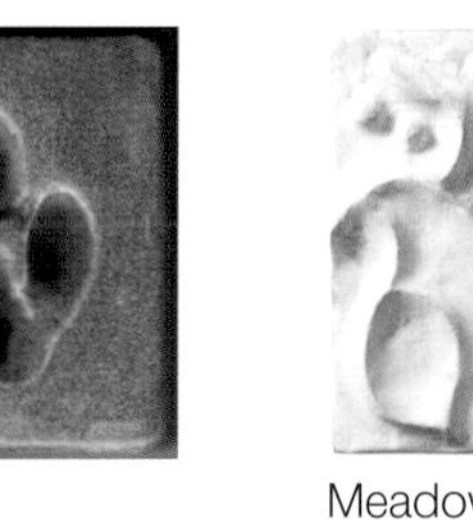
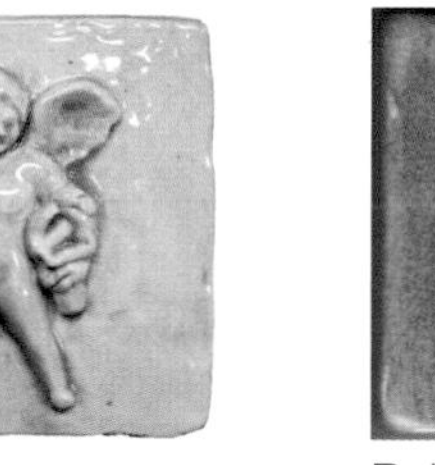

Circles *Jacobs Tile*

Griffin Green
Bambino Tile

Scallop *Terrapin Tile*

Spiral Wave
Talisman Tile

Fish *Bambino Tile*

Apricot Blossoms
Jacobs Tile

The Hunt
Bambino Tile

Ginkgo *Terrapin Tile*

Shadow Square
Talisman Tile

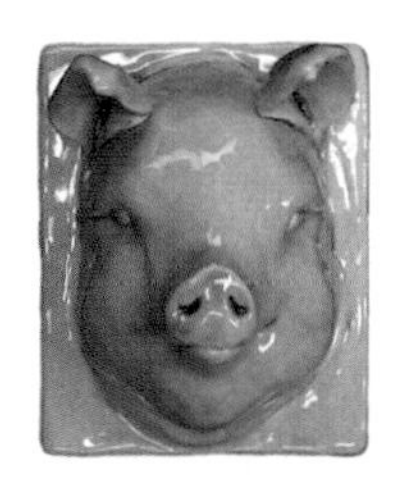
Pink Pig
Bambino Tile

ceramic tiles

Rishi *Helen Weisz Architectural Ceramics*

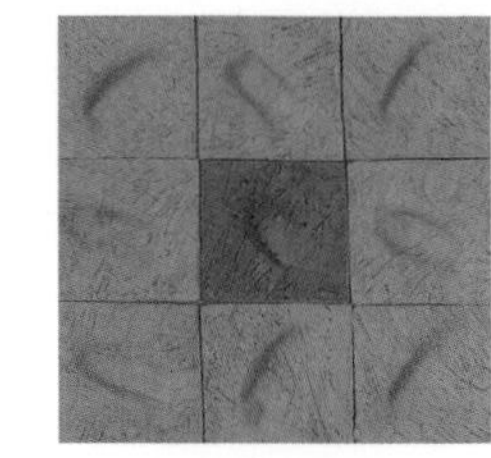

Whirling Squares

Dati

Hobi

Grapevine and Olive
Fraser Tile Works

Olive Baxter Hill

Flowerpot
Revival Tile Works

Pinecone

Pinecone Dot

Freestanding wall
Eric Pilhoffer

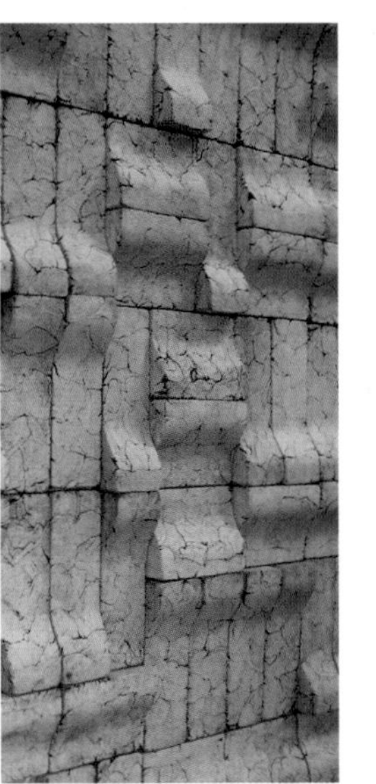

Curved wall

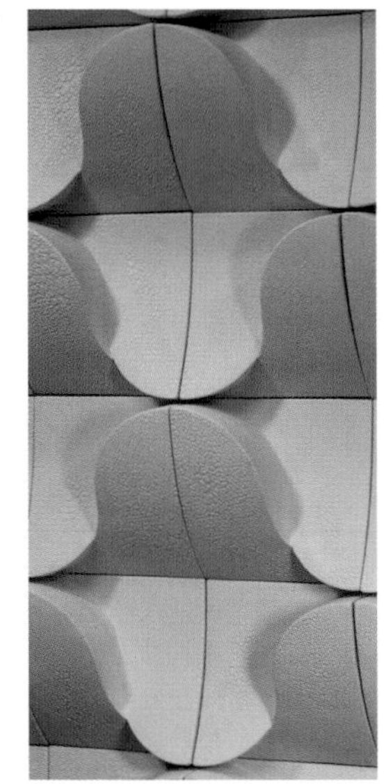

Wave Module

Tile Brick kitchen from *Imagine Tile.*

ceramic borders

Floral Wave *Bambino Tile*

Chestnut

Poppy

Twin Chimeras

Meadow Vine *Talisman Tile*

Shadow Key

Spiral Wave

Diamond

Astrogal

Ophelia

Twareg handmade terra cotta
Saint Tropez Boutique

Kasbah Moroccan terra cotta

Shell *Bambino Tile*

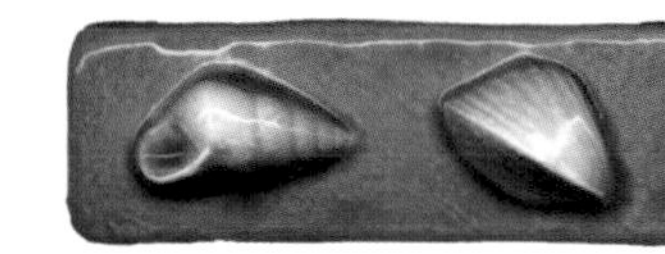
Shell Seaspray

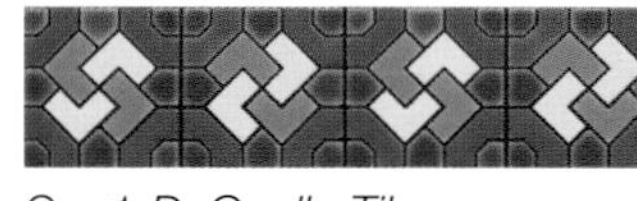
Geo1 *DuQuella Tile*

Geo2

Delft Maritime
Elsner Artistic Tile

Geometric

Blue & Yellow

Tassle

Cornice Molding
Revival Tile Works

Fleur De Lis *Jeffrey Court*

Two Piece Adobe

BDRMG *DuQuella Tile*

field

Color samples *Lavabo*

Cobblestone *Sonoma Cast Stone*

Clouds

Bamboo

Field tiles, Old California Palette *Revival Tile Works*

Slickstone 1 *Sonoma Cast Stone*

Slickstone 2

Softstone

Pebblestone

Slatestone

Field tile 1 *Talisman Tile*

Field tile 2

Olive Drab *Manet Tiles*

Peru

Monochrome *Sonoma Cast Stone*

Traditional Batchelder, Claycraft palette *Revival Tileworks*

Bronze *Paris Ceramics*

Squares *Sonoma Cast Stone*

Baton Bricks *Ann Sacks*

Ceramic 50010 *Manet Tiles*

field & basketweave

Hand-made field tiles
North Prairie Tile Works

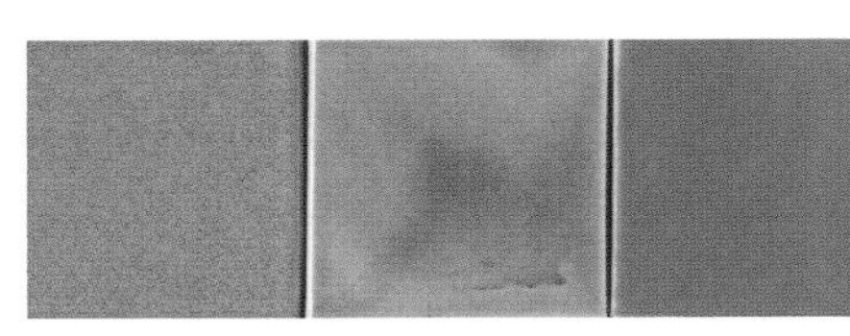

Mosaic field tile 402 *Manet Tiles*

Mosaic field tile 405

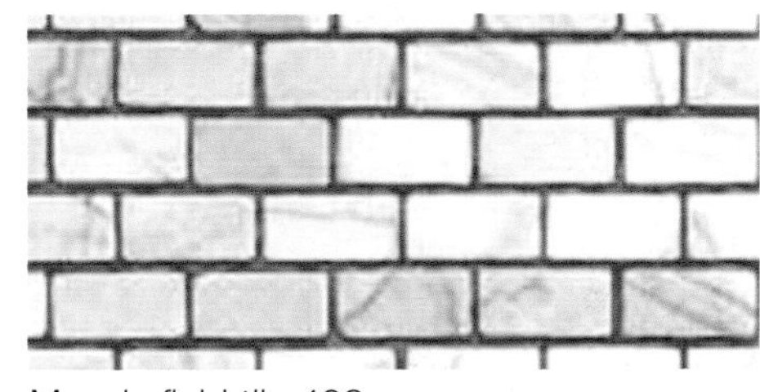

Mosaic field tile 408

Basketweave Casablanca
Jeffrey Court

Ida

Haven

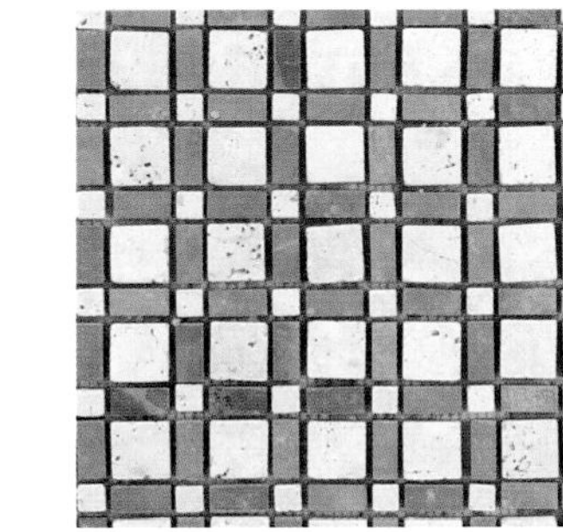

Dot Dash

Rectangular *Jeffrey Court*

Carrara Wicker *Ann Sacks*

Giallo Basketweave

Triple Basketweave

glass

Nike Trapunta
Lavabo

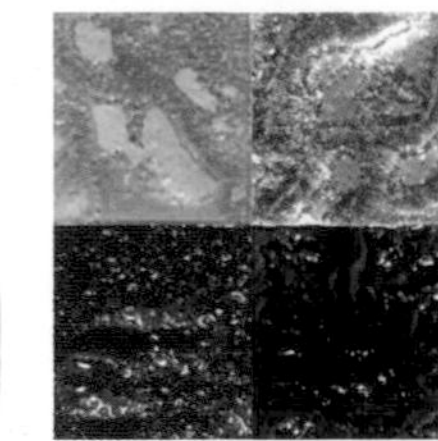

Nike Luxor

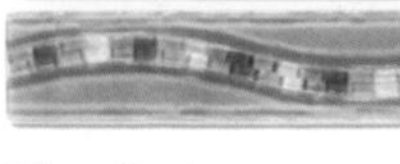

Nike Onda

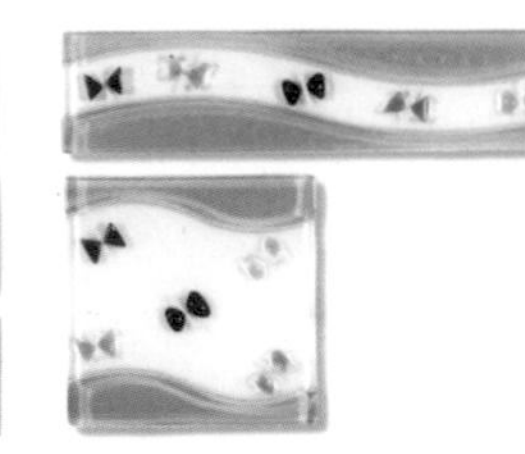

Nike Papillons

Nike Sirio

Textured Bronze
Crossville

Textured Cinnamon

Textured Blue Foil

Textured Crystal Green

Textured Mica

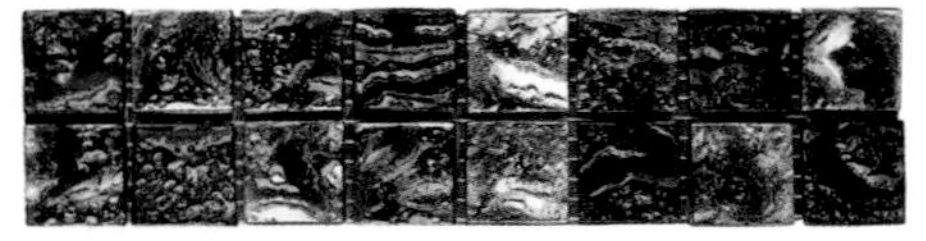

Blue Foil Mosaic Liner Bar

Silver Leaf Mosaic Liner Bar

Citrine/Topaz/Bronze Liner Bar

VG 120 Blue Foil Liner Bar

VG 100 Mica Liner Bar

VG 130 Citrine Liner Bar

Bamboo *Ann Sacks*

Ice *Ann Sacks*

Lavender

Lemonade

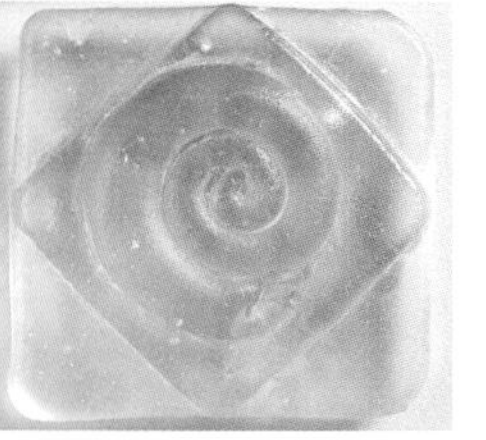

Snail Deco *Ann Sacks*

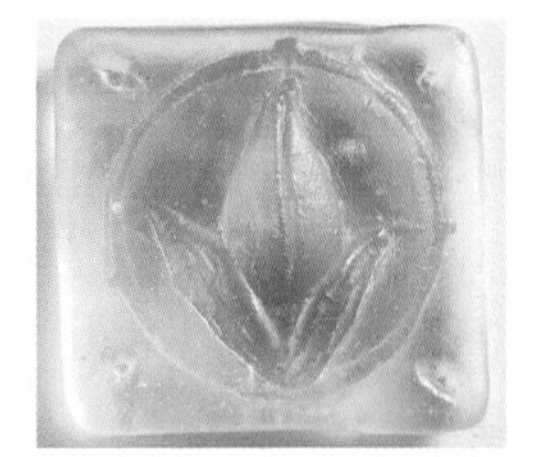

Bud Deco

Bird Deco

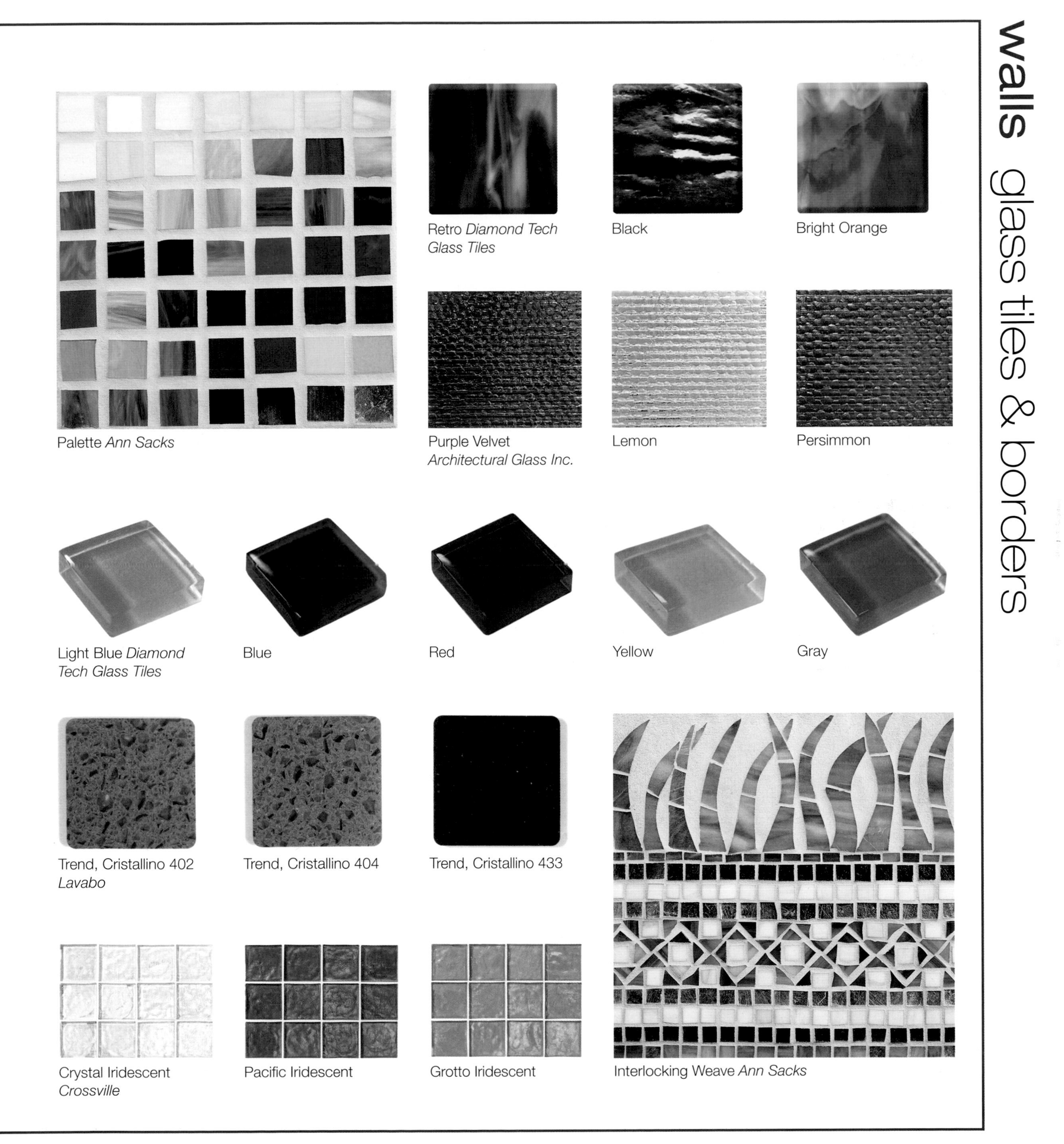

Palette *Ann Sacks*

Retro *Diamond Tech Glass Tiles*

Black

Bright Orange

Purple Velvet *Architectural Glass Inc.*

Lemon

Persimmon

Light Blue *Diamond Tech Glass Tiles*

Blue

Red

Yellow

Gray

Trend, Cristallino 402 *Lavabo*

Trend, Cristallino 404

Trend, Cristallino 433

Crystal Iridescent *Crossville*

Pacific Iridescent

Grotto Iridescent

Interlocking Weave *Ann Sacks*

metal

Sun, bronze *Talisman Tiles*

Sun white bronze

Presidio B, iron rustic
Jeffrey Court

Palm, bronze
St. Gaudens Metal Arts

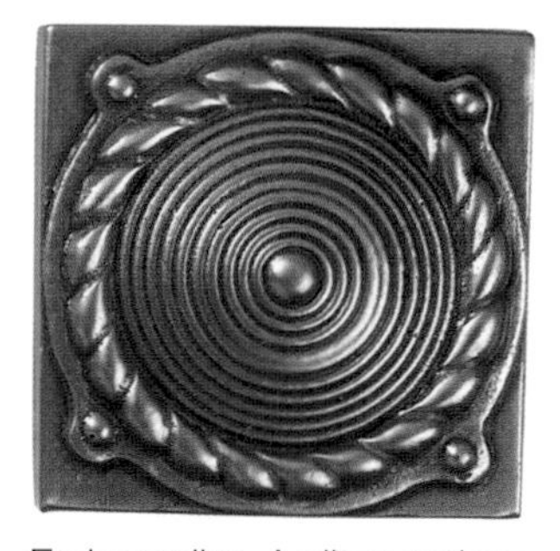
Embacadiro, A silver antique
Jeffrey Court

Embacadiro B silver antique

Ashbury, bronze satin

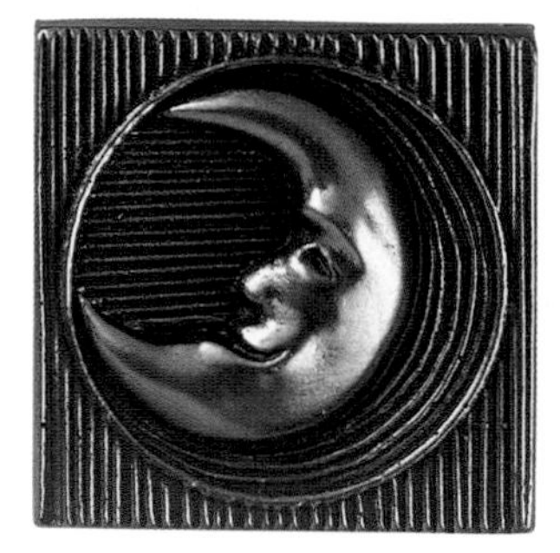
Haight, bronze satin

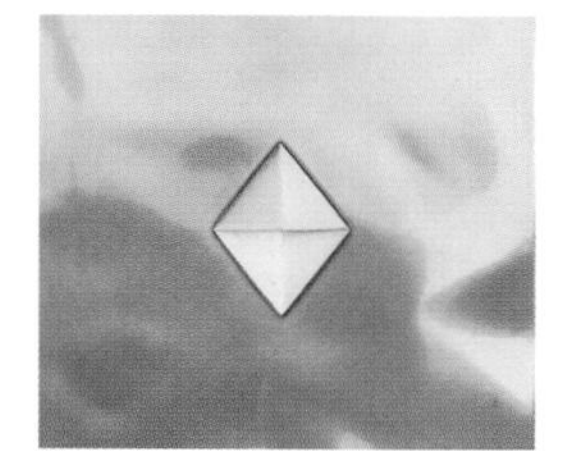
Flutes and Diamonds, nickel silver *Crossville*

Tuscan Vineyard nickel silver

Country Orchard bronze

Tomatoes, bronze

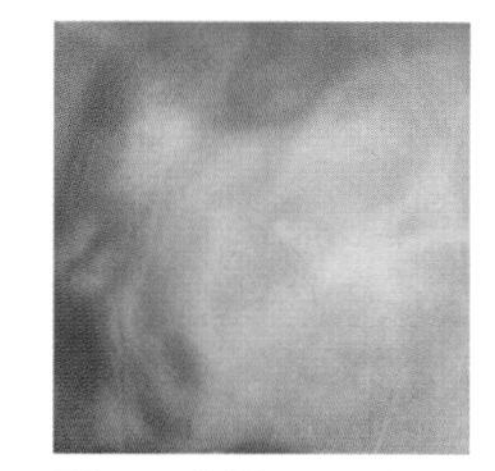
Riveted Plates nickel silver

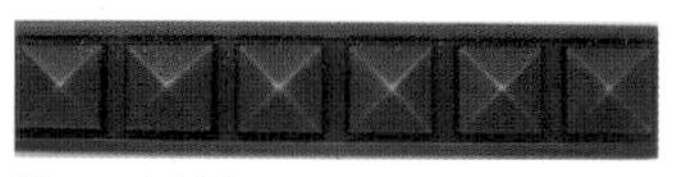
Pyramid Liner
St. Gaudens Metal Arts

Egg and Dart, polished bronze
Crossville

Tuscan Vineyard
Crossville

Pyramids on Checks Nickel Silver *Crossville*

Pyramid Liner, silver
St. Gaudens Metal Arts

Sausalito, bronze satin
Jeffrey Court

Tiburan, silver antique

Golden Gate, iron rustic

Quilted metal backsplash tiles from *Rigidized Metal.*

Medium Pyramid, bronze
St. Gaudens Metal Arts

Caramel bronze with tigereye

Odyssey, silver

Odyssey, verdigris

MT 120, rust nickel silver *Crossville*

MT 118, rust nickel silver

MT144, bronze field

MT 104, rust nickel field

Metal Crete, metal and concrete *Sonoma Cast Stone*

Acanthus jeweled liner, bronze with tigereye
St. Gaudens Metal Arts

Patine Metals rust nickel Silver, MT09 *Crossville*

Patina Metals rust nickel Silver, MT108 *Crossville*

Florentine liner, bronze
St. Gaudens Metal Arts

Segovia cornice liner, silvertone
St. Gaudens Metal Arts

Laurel liner, silver

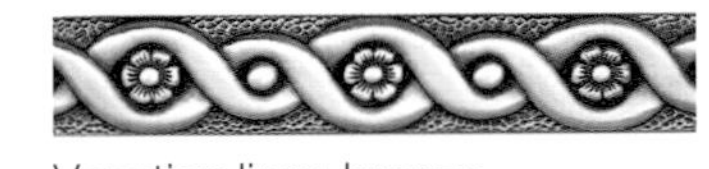

Venetian liner, bronze

Wave liner, silver

mosaic

Cypress border plain mosaic 12 x 12 by *Ann Sacks.*

Mediterranean border, plain mosaic *Ann Sacks*

Art Mosaic MA0005
Manet Tiles

Art Mosaic MA0009
Manet Tiles

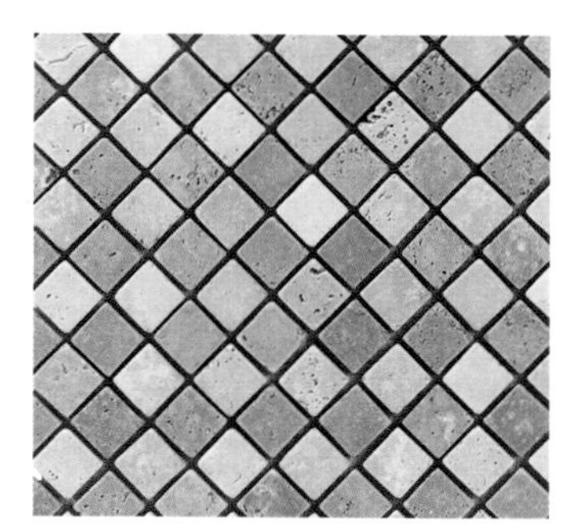

Knidos mosaic pattern
Jeffrey Court

Mosaic liner SL05 *Manet Tiles*

Mosaic liner CL01

Mosaic liner CL02

Braids B Micro *Jeffrey Court*

Coronado Micro

Monterey Micro

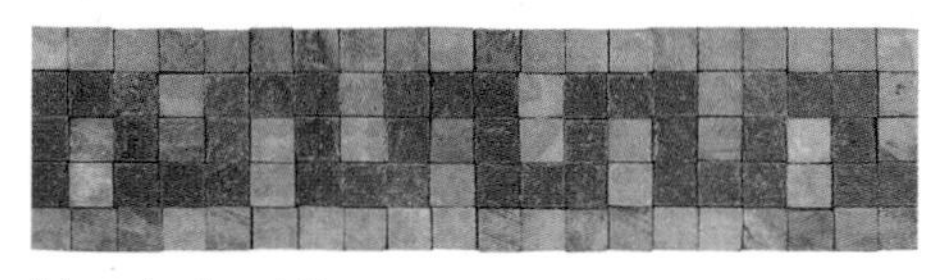

Mendocino Micro

Sierra Micro

Waves Micro

soapstone, gemstone & seashell

Fish Tail
Green Mountain Soapstone

Lily

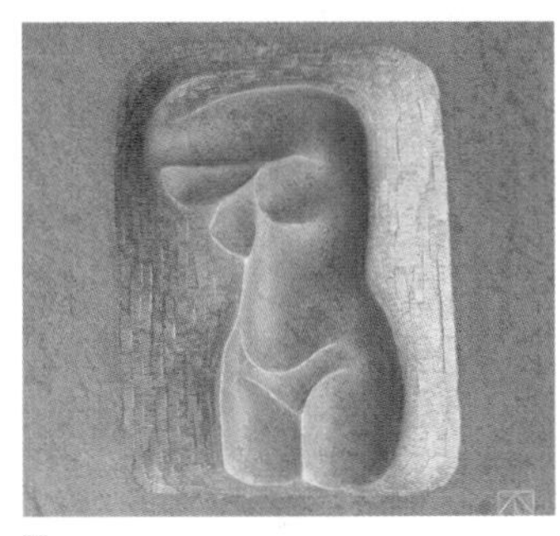

Torso

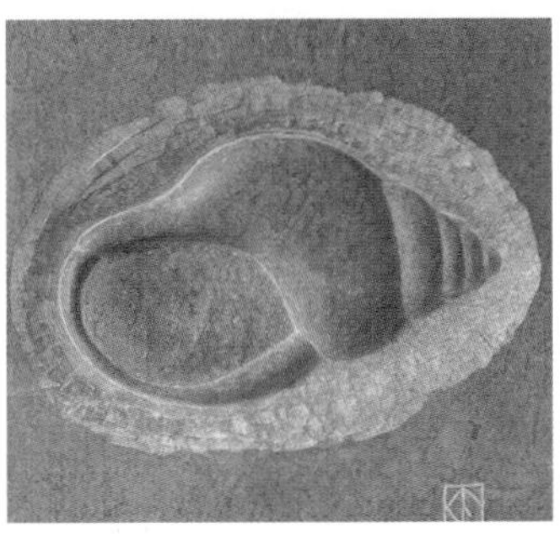

Nautilus

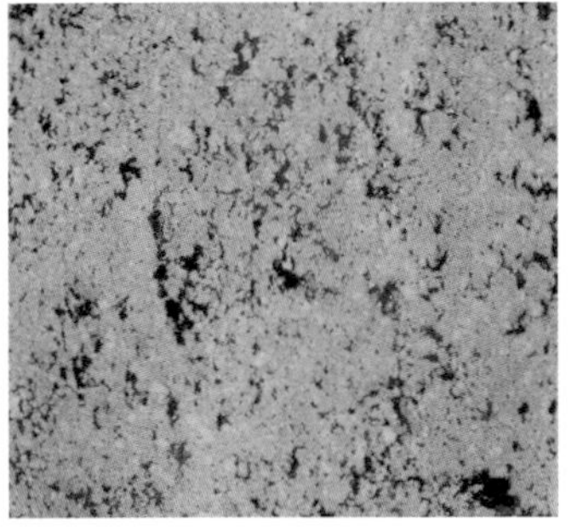

African Turquoise *Agape Tile*

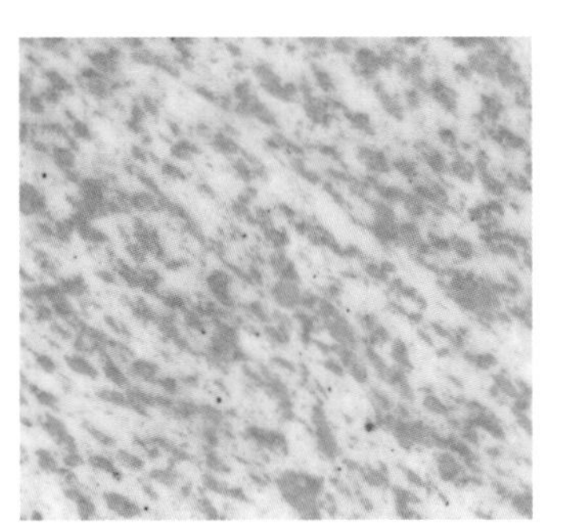

China Jade

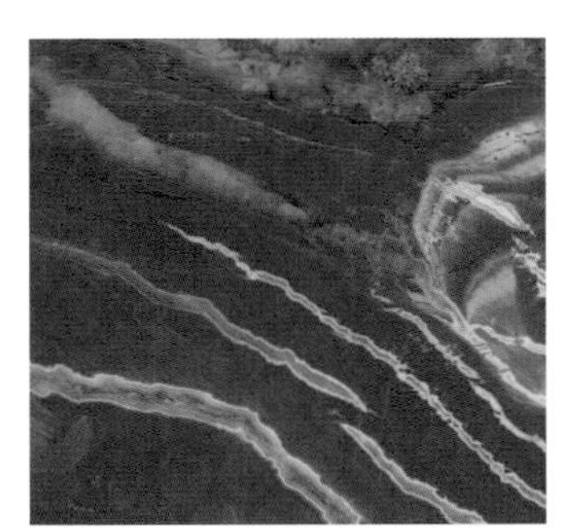

Red Jasper

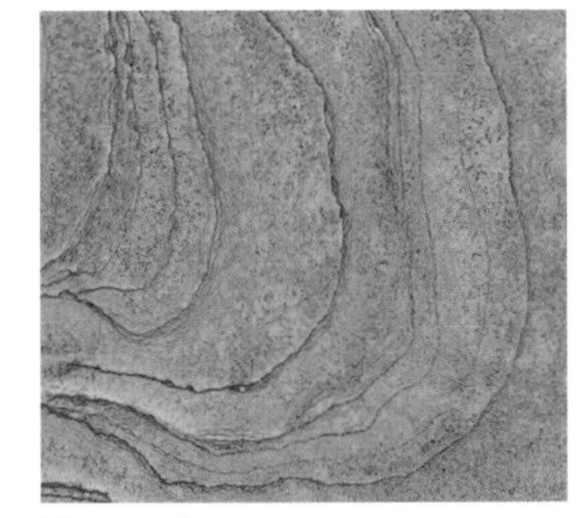

Picture Jasper

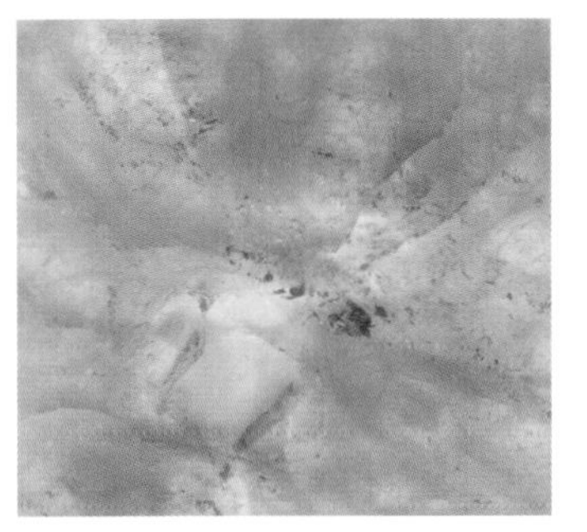

Amethyst *Agape Tile*

Yellow Aventurine

Rainbow Fluorite

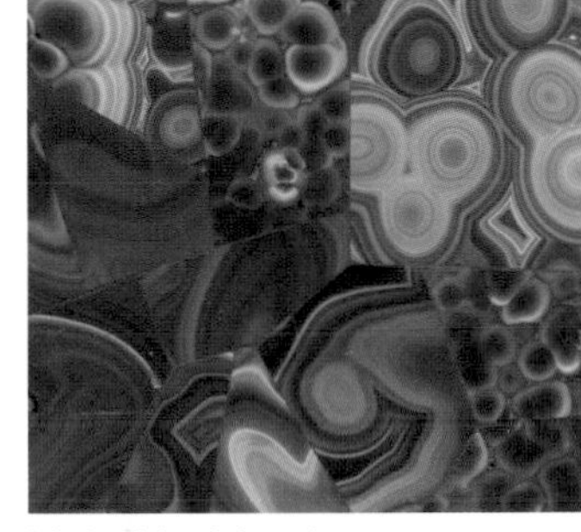

Malachite Mosaic

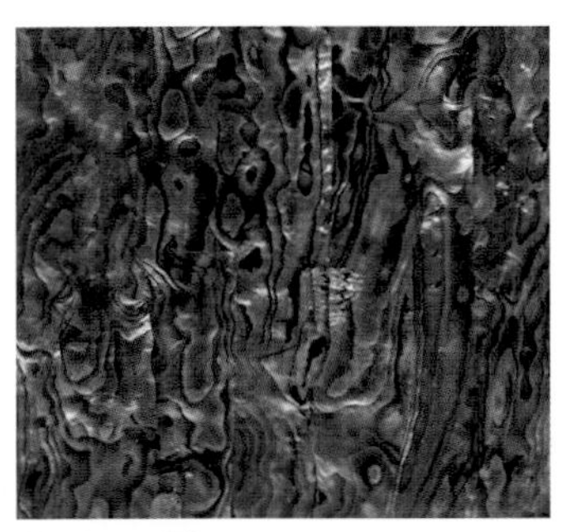

Abalone Tint

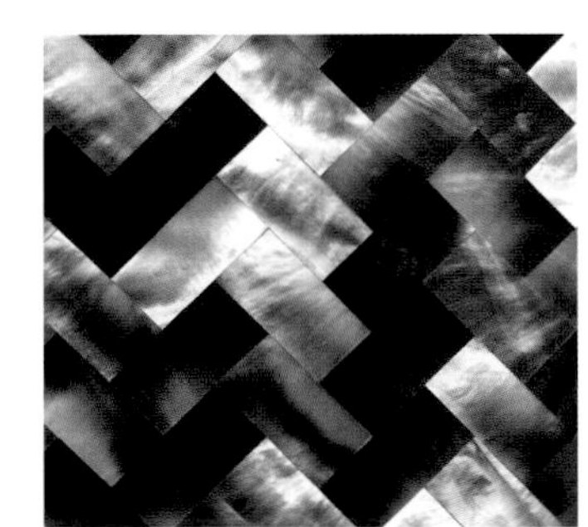

Backlit Mod Weave

Gold Tiger Eye

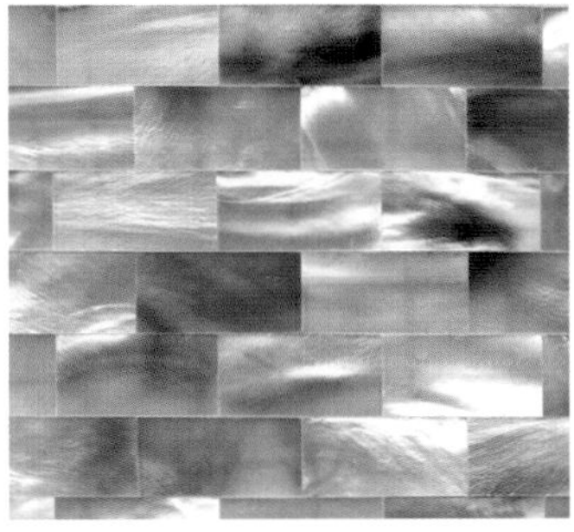

White Hammershell Brick

pressed tin

Pattern 24 in Copper, from *The American Tin Ceiling Co.*

Pattern 26, Bronze-Granite

Pattern 22, Custom Antique Copper

Pattern 7, Copper

Pattern 23, Creamy White Granite

Pattern 8, LADG

Pattern 14, Pale Rose Granite

Pattern 10, Mocha Granite Satin

Pattern 2, Pale Rose Granite

Pattern 13, Antique Rustic Copper

Pattern 1, Metallic Copper

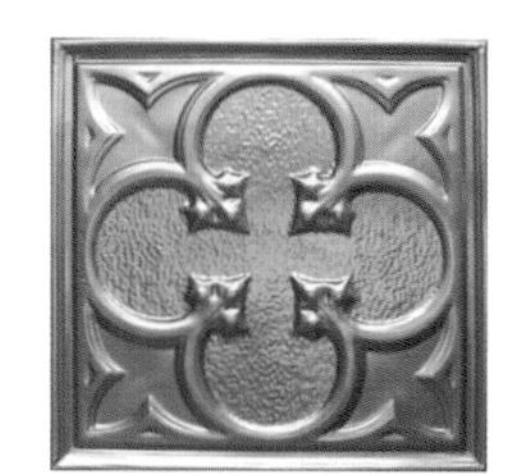

Pattern 12, Metallic Copper

Pattern 6, Royal Gold

Pattern 27, Silver

Pattern 14, Pewter

Pattern 3, Satin Copper

Pattern 31, Raw Tin

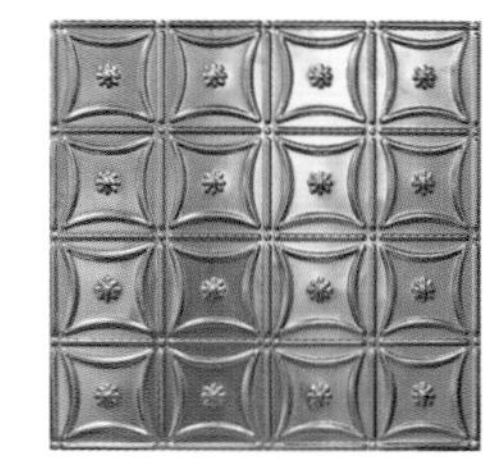

Pattern 30, Raw Tin

Pattern 25, Custom Antique Gold

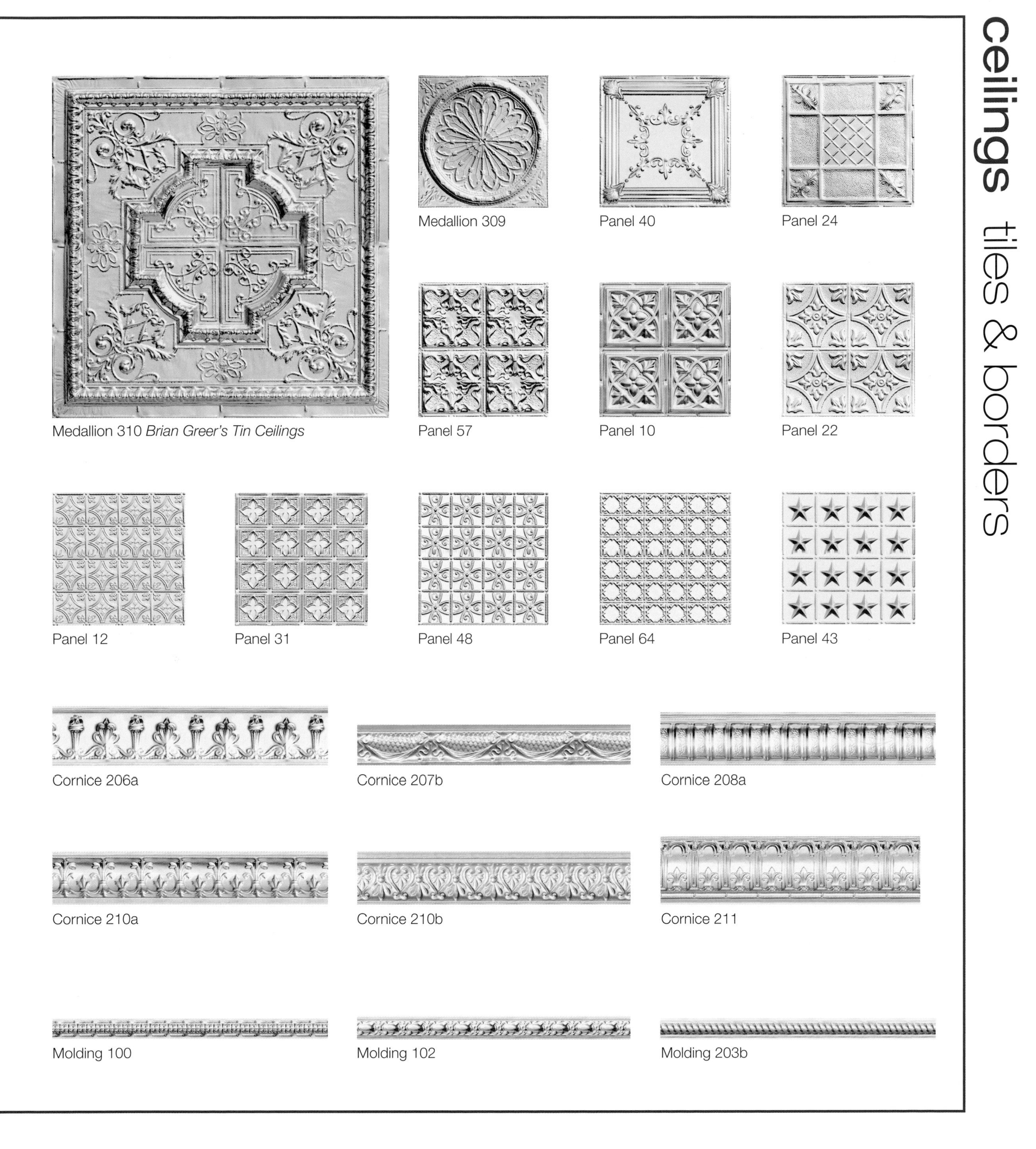

Medallion 310 *Brian Greer's Tin Ceilings*

Medallion 309

Panel 40

Panel 24

Panel 57

Panel 10

Panel 22

Panel 12

Panel 31

Panel 48

Panel 64

Panel 43

Cornice 206a

Cornice 207b

Cornice 208a

Cornice 210a

Cornice 210b

Cornice 211

Molding 100

Molding 102

Molding 203b

metal

Honest Aluminum 236
Gage Decorative Metal Ceilings

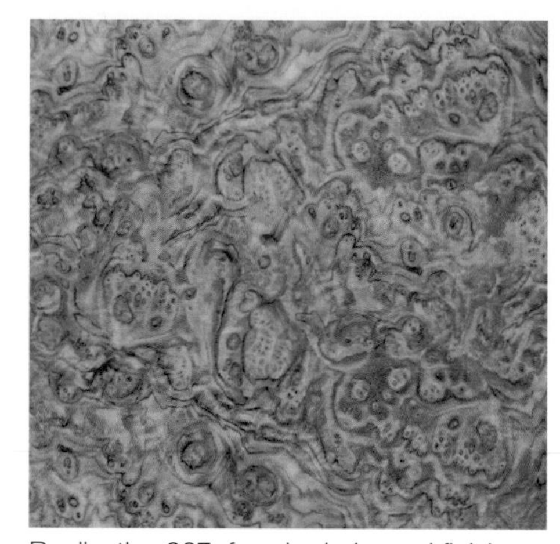
Replicative 267, faux burled wood finish, aluminum

Patterns, Textures & Colors tile 326, aluminum

Replicative tile 529, faux wood finish, aluminum

Honest Aluminum 601, silver finish

Custom Design 664, aluminum

Replicative tile 154, with a faux wood finish in aluminum from *Gage Decorative Metal Ceilings.*

plaster

Classic Panel *Above View*

Coffered Dentil

Contemporary Coffer

Gothic Coffer

Quad Wedge

Modulated Cube

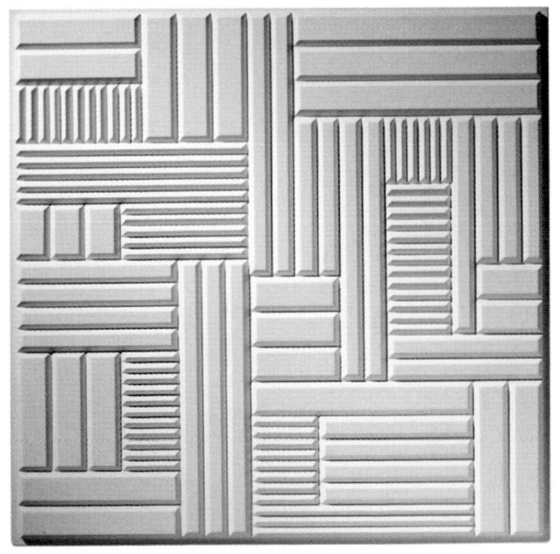

Emerald Cut

Bamboo 1

Palm Leaf

architectural details

Architecture really is like "frozen music" — it has themes that run through the building in order to create a sense of richness, with backgrounds that reappear and resolve themselves. You can see this clearly in things like moldings that step back three times on the first floor and twice on the second floor and once on the third floor.

JAMES CULLION, ARCHITECT

ONE OF THE BEST WAYS TO ADD PERSONAL EXPRESSION IN YOUR HOME is with architectural details such as columns, crown molding, carved millwork and other surface adornments. In traditional moldings and other architectural adornments, you will find elements of the historic styles that gave birth to them: Victorian, Baroque, Queen Anne, Georgian, Regency, and English and French Country, as well as the rebirths inherent in neoclassical and Romanesque, and more modern styles such as Craftsman. The more ornate the style, the more elaborate the embellishments, naturally. In Ranch and Cape Cod styles, moldings are characteristically restricted to simple friezes, plain baseboards and trim. Larger and heavier moldings can suit a rustic, country-style home. Beyond that, it's really only a matter of how elegant or ornate you want your home to appear.

Guilloche frieze crown molding (from the Design-Build Store), paired with subtle pilasters and a matching fireplace mantel sets a tone of casual elegance to this living room.

The main families of architectural adornments are door and window casings, crown moldings, plate and chair rails and friezes, wall moldings, ceiling medallions, domes and rosettes, corbels, and columns, including capitals and bases. Almost all of the architectural detail you see in a room (and to some extent on the exterior of a house) encompasses these elements, often used together to build up layers of decoration. Corbels, for instance, are normally used as decorative brackets to hold up a shelf or mantel. But they are also commonly incorporated into larger moldings to add detail, or built into a column or even a piece of furniture to add elegance. Corbels can also be elaborately carved, with foliage such as papyrus or acanthus leaf (a prickly Mediterranean herb), but also with cherubs or even faces. Decorative beams, coffered ceilings and ceiling medallions serve to draw the eye up, adding a touch of formality or rusticity, depending on the materials used. Soffits can hide wires, pipes and ducts or can be designed to contain concealed lighting that will wash ceiling, walls or both in a warm glow. Consider architectural salvage if you wish to make a one-of-a-kind statement.

However, if you have an ultra modern home, this kind of architectural detail won't be of much interest to you. The modern home is characterized by large spaces of unadorned material: stone, metal, shiny resins or exotic woods, without a corbel or a plate rail to break up the smooth lines or take away from the focus on color and texture. Contemporary homes have a little more leeway in that, employed judiciously, traditional architectural elements can be used — a couple of Doric columns flanking an arch in the living room, for example, or some crown molding in the dining room, can exist very comfortably beside your sleek silk couch and expanse of ipe wood walls.

Left: A simple round frame meeting at a square corner makes for a dramatic door because of its massive scale. Center: A rounded molding richly finishes off this bay window. Right: A nicely built-up crown molding. Notice the cove lighting hidden behind the molding at lower left.

You can use antique materials in contemporary homes — in fact we love the contrast. For example, we take modern spaces and put in wide plank floors and old beams to give the feeling of a loft. Or we use very elaborate, tall antique doors from France or England to enter a room full of very modern furniture and millwork.

BRIAN GLUCKSTEIN

Japanese-style beamwork in the skylight and sliding Chinese-style doors give this Rocky Mountain lodge an East-meets-West flavor.

moldings & trimwork

Crown moldings are probably the most important family of architectural adornments. Usually placed at or just below where there the wall meets the ceiling, they can be used in almost any style room; the more elaborate and built up, the more formal the look. The classic crown molding motifs are familiar to most of us: the Greek return, egg and dart and dentil, but there is almost no limit to the kinds of molding you can buy (see the Product section starting on page 313). Strictly speaking, the "crown" is the cap piece, installed at an angle at the top of a wall or cabinet, and a "cornice" is the trim installed along the top of a wall or above a window. The same motifs used as crown moldings can be used on the baseboards (sometimes known as "mopboards") that run up the wall from the floor.

Wall moldings add intricacy and richness to flat surfaces, and elements such as ceiling medallions and ceiling domes (a medallion that is concave and often recessed into the ceiling) are used to draw attention to fixtures. Size is also important, and oversize moldings are a relatively easy way to make a rather spectacular statement, as is painting the molding or trim a color that contrasts with color of the wall. One caveat here: a contrasting color will give your room a very dynamic, dramatic look that some people find too strong. Moldings are also used to define a space and to focus attention on some element within the space, such as a fireplace, door or window. A fireplace mantel is often made the focal point of a room and as such deserves special consideration. Pilasters or small columns can extend from the floor to a mantelshelf – the size of which is based on the size of the fireplace. The shelf usually looks best when it extends to or reaches beyond the width of the hearth; unless it is a raised hearth, in which case the mantel can be slightly narrower.

Left and center: Moldings add flourish and that finishing touch to a room. **Right:** A ceiling medallion is a way to focus attention on the light fixture while adding interest to an otherwise blank surface.

With no room for a corner molding, this one was painted on.

Traditionally, trimwork was made of carved wood (painted or varnished) and this is still the case. However, new materials have arrived that provide some interesting options. Plastic moldings are increasingly being used. Fiberglass and polyester blends or extruded polystyrene are the most common. Some are rigid but there are brands on the market that are so flexible they can be easily used to adorn circular ceilings and archways. Plastic moldings are easy to cut and install, take (latex) paint well and come in one piece rather than the combination of pieces some complex wooden moldings are made of. Medium density fiberboard moldings are much harder than plastic so it is harder to cut and needs to be attached with a nail gun. They often come already primed for easy painting.

Trim and moldings around doorways and windows hide mistakes and allow you to work with fairly low tolerances — that's what they are for. But the reveal is just the opposite. The reveal creates a shadow that distinguishes one element from the next and gives them definition. But they are hard to do. There is nowhere to hide with reveals because you are working with very high tolerances. It's like performing without a net.

LYNN APPLEBY, A TORONTO DESIGNER WITH A PENCHANT FOR REVEALS

A carved column set in a niche.

columns

Carved columns, traditionally reserved for neoclassical motifs, and adapted for Georgian, Federal and Greek Revival styles, can be mixed with just about any style of architecture you can name.

There are three orders of columns as organized in the classical architecture of ancient Greece: Doric, Ionic and Corinthian. The Doric is the oldest and the simplest. Doric columns have no base and only a simple capital (top) that is usually just a square sitting on a circle. The Parthenon in Athens has Doric columns. The Ionic order has fluted columns, a base normally made of stacked rings, and a scrolled capital, which all adds up to a more decorative column than the Doric. The Corinthian order has the same fluted columns and stacked base as the Ionic, but below the scrolled capital all kinds of decorative elements can be used, including leaves, flowers and lavish curlicues. Two other kinds of columns are the composite, which is a combination of the Doric and Ionian, and the Tuscan, which is a very simple column with a plain shaft and a simple capital and base. The Greek Revival style may have used columns literally, seeking to imitate a formal Greek temple, but don't be intimidated by all this classical history. Basic post-and-beam structures rely on columns as well. Although the Greeks constructed their columns of marble, modern builders can now turn to other, less costly materials such as wood, fiberglass, plastic, plaster and metal. Columns can be merely decorative or load-bearing — many a renovator has discovered the necessity for some sort of supportive columns after demolishing a wall to open up space only to discover that it was, in fact, a load-bearing wall.

Left: A column that is a combination of concrete and steel pipe frames the entrance to an artist's loft/studio. The materials, although modern, manages to look anciently monolithic. **Center:** A naturally occurring Y in a tree makes for an appropriate column in this log house. **Right:** These smooth (as opposed to fluted) columns with Doric bases and capitals are echoed by the pilasters on the left and right, all of which combine for a spectacularly neoclassical room.

architectural salvage

For anyone with an appreciation of the past, the best kinds of architectural adornments are those that already have a history of use. Beside the obvious high degree of craftsmanship found in most work done in centuries past, it is that ephemeral patina of use that is the most desirable aspect of antique architectural details. And it must be honest use; just compare the difference in charm contained in a worn brass lamp, dinged and scratched from the various storms it has had to endure, with a faux lantern that has been gently beat up a little in the shop. The same goes for everything that can be used in a home, from old doors and windows to light fixtures, columns, mantels, sculptures and iron railings.

Sometimes it's only a matter of a slight change in attitude. If you can see the beauty and potential in such blemishes as peeling paint or scuff marks and chips, a whole new world can open up. Perhaps the best example of this kind of "celebration of defect" is the log home. Since logs are by nature big, funky and imprecise, with dents and axe marks all part of the allure, almost anything else that happens to the wood is not only forgivable but embraced. We have had burn marks, birth dates and scratches made from a deer sharpening its antlers all proudly pointed out to us by log home owners. And we ourselves once owned a log cabin that had a particularly lopsided euchre score carved into a log wall by some happy player of long ago. You can't buy details like that.

Right: Antique corbels have potential as pieces of art, functional or purely decorative, in almost any home. **Opposite:** Ancient carved stone works awaiting new homes share company in one room of this architectural salvage warehouse. (Five O Seven Antiques)

Lots of windows, unfinished columns, wicker, a jumble of objects picked up on walks through nature, and a stone floor give this sunroom its casual charm.

Architectural salvage particularly suits certain period styles of residence, especially when the material is from the same era and style. But a good eye for detail can mate elements from any era or culture and make it work. Antique wooden columns from Pakistan can add an exoticism to a room done in a Southwestern desert style. And ancient stones can set off a wall or floor in a very contemporary setting.

Authentic salvage or antique details can be purchased through your architect, designer or a dealer who specializes in architectural salvage (see Sources chapter). A dealer generally has a yard or warehouse full of salvage that has been gathered from the four corners of the world. A day or two spent scavenging at such a place is recommended, if only for the ideas and possibilities it will suggest.

Or you could try to get your hands on architectural salvage by going to the source yourself. Old buildings are always being torn down to make way for new structures. If you are lucky enough to come across this kind of demolition, it is worth your while to try to buy some of the architectural details before they are sold off to a jobber or dealer.

products

architectural details

crown moldings

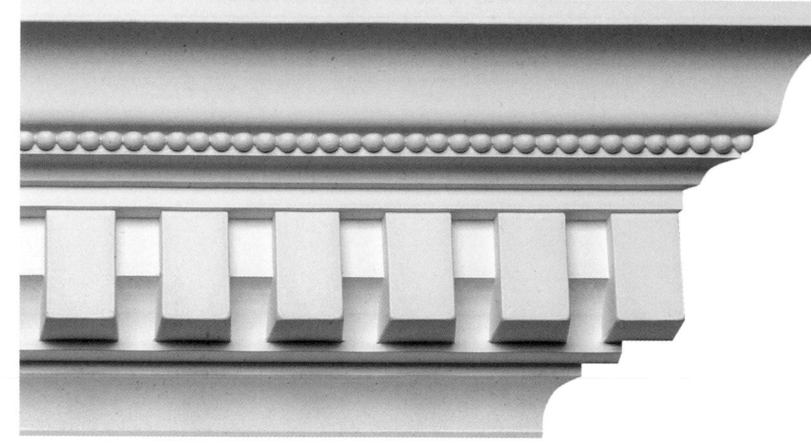

Dentil crown molding (T-165PL) in plaster from *Balmer Moldings.*

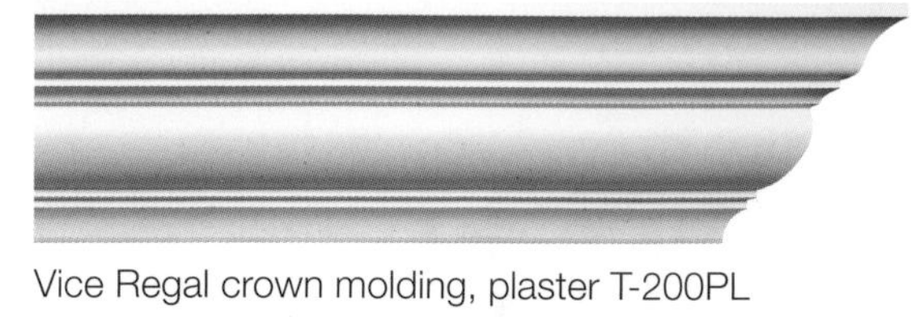

Vice Regal crown molding, plaster T-200PL

Gothic Arch crown molding, plaster T-400PL

cornices

Polyurethane cornice, J138110899
Richelieu Hardware

Polyurethane cornice, J138120899

Polyurethane cornice, J1455191299

Decorative Cornice, plaster 399
Design-Build Store

Bracket Cornice with Egg and Dart, plaster 403

Cove with Guilloche Cornice, plaster 421

Floral and Berry Cornice, plaster 370

Architectural cornice, plaster 371

Cameo cove cornice plaster 372

Plain Cornice, plaster 367

Architectural Cornice, plaster 368

Architectural Cornice, plaster 369

Grapevine Crown Molding, plaster T-600PL *Balmer Moldings*

Egg and Dart Crown Molding, plaster T-300PL

Duchess Leaf Crown Molding, plaster T-500PL

Classic Acanthus and Pearl Cornice, plaster 425 *Design Build Store*

Classically Proportioned, plaster 435

Ornate Acanthus, Bead, Barrel and Leaf Cornice, plaster 451

Acanthus Dentil and Leaf, plaster 453

Dentil, Egg and Dart, Lamb's Tongue Cornice, plaster 487

Egg and Dart with Double Ribbon Bundled Wreath Cornice, plaster 505

Dentil and Triglyph Cornice, plaster 481

Disks and Triglyph Cornice, plaster 513

Upright Leaf Cornice, plaster 600

decorative moldings

Polyurethane appliqué J40030399
Richelieu Hardware

Polyurethane appliqué J40050299

Polyurethane appliqué J1486851299

Reed and Ribbon, plaster M510
Balmer Moldings

Traditional Leaf, plaster M520

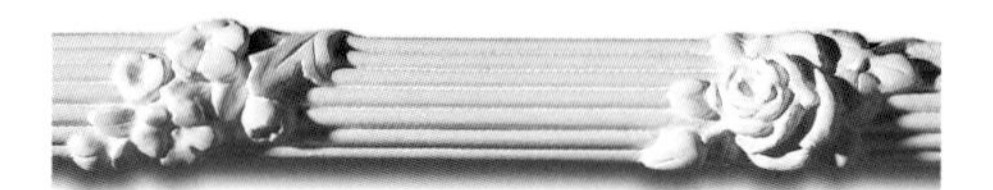

Royal Bouquet, plaster M530

Regal Pearls, plaster M540

Adams's Flute, plaster M550

Pearl Rail, plaster M560

Egg and Dart, plaster M570

Small Guilloche Frieze, plaster 643
Design-Build Store

Large Guilloche Frieze, plaster 643L

Guilloche Frieze, plaster 338

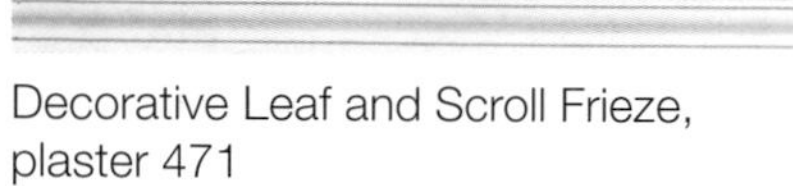

Decorative Leaf and Scroll Frieze, plaster 471

Greek Key Frieze, plaster 342

Rose Casing, hardwood MDCA DC-07
Architectural Adornments

Acanthus Crown Molding, hardwood MDCR DC-12

Floral Link Molding, hardwood MDLK DC-01

Shell Link Molding, hardwood MDLK DC-02

Egg and Dart Casing, hardwood MDCA DC-18

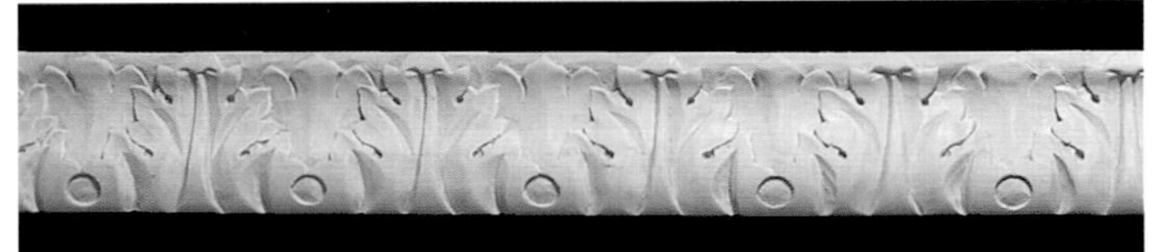
Acanthus Crown Molding, hardwood MDCR DC-17

Floral Vine Chair Rail MDCH DC-11

Acanthus Casing, hardwood MDCA DC-13

Dragon Casing, hardwood MDCA DC-14

Oval Floral Chair Rail, hardwood MDCH DC-16

Acanthus Chair Rail, hardwood MDCH DC-04

Acanthus Chair Rail, hardwood MDCH DC-05

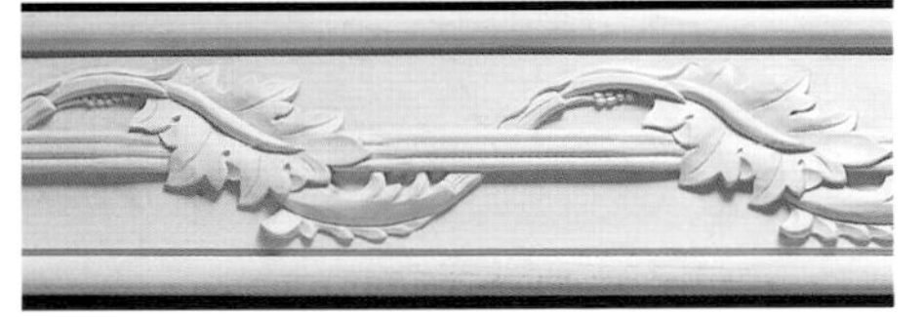
Acanthus Chair Rail MDCH DC-03

Rose Casing, hardwood MDCA DC-07

Acanthus Dentil Crown Molding, hardwood MDCR DC-09

decorative moldings

Seashell Appliqué, wood *Richelieu Hardware*

Wood appliqué

Carved wood block

Small Grape Bunch
hand-carved maple

Large Mallard hand-carved
lime

Swag Onlay, hardwood
Architectural Adornments

Center Balanced Onlay, wood

Wood appliqué *Richelieu*

Hand-carved Grape lime appliqué
Richelieu

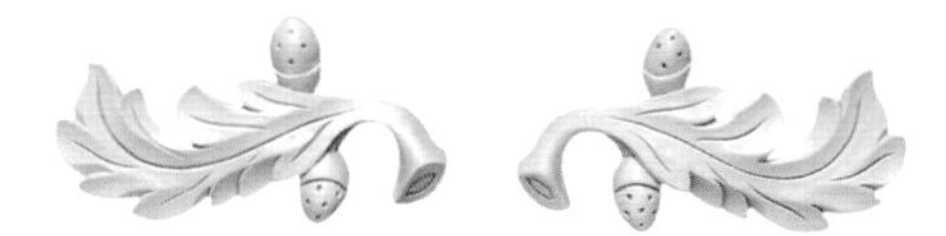
Hand-carved Flower lime appliqué

Hand-carved Leaf lime appliqué

Maple appliqué
Richelieu Hardware

Birch appliqué

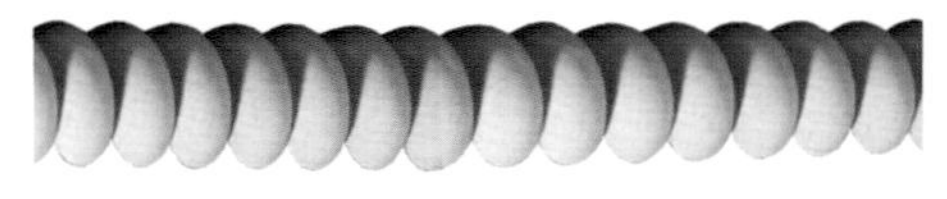
Cherry rope molding

Polyester molding
Richelieu Hardware

Polyester flat molding

Hand-carved wood and clay moldings
Ancient Excavations

Flexible polyester frieze 321600
Richelieu

Flexible polyester frieze 3203-00

Carved panel PADR DA-51
Architectural Adornments

Scroll Capital, hardwood CTAR DA-54
Architectural Adornments

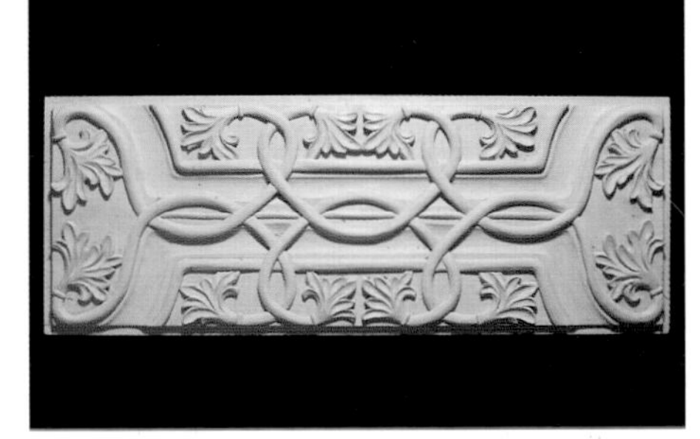

Rectangle Knotted Rope Panel, hardwood PAWA DA-48

Rectangle Knotted Rope Panel, hardwood PAWA DA-49

Square Acanthus Panel, hardwood PAWA DA-44

Square Phoenix Panel, hardwood PAWA DA-46

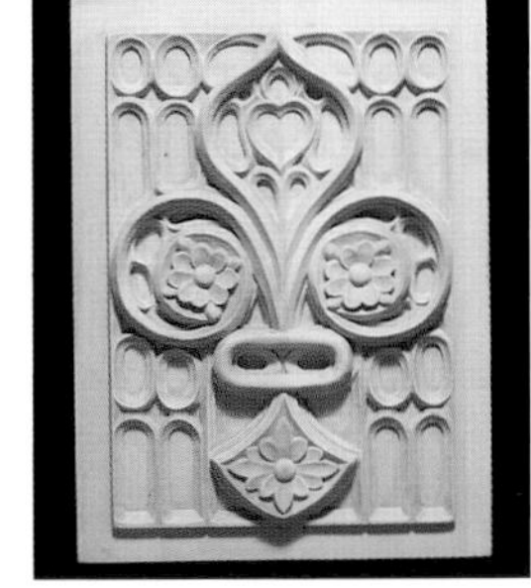

Rectangle Moorish Panel, hardwood PADR DA-50

Carved corbel COAR GW-42

decorative moldings

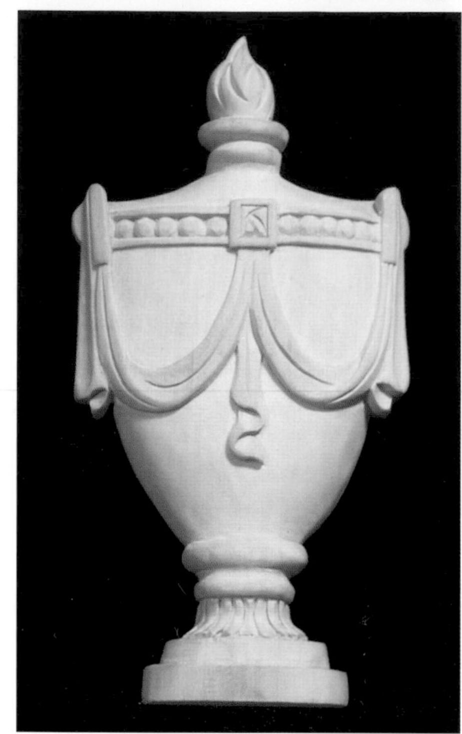

Compact Urn OLCP-GW-183, wood
Architectural Adornments

Lion Head, wood C-40
Michael Shea Woodcarving

Polyester keystone
Richelieu Hardware

Medallion Onlay, wood, OLMN-GW-13
Architectural Adornments

Corner Onlay, wood OLCP-GW-34

Center Balanced Onlay, wood
OLCB-GW-49

Center Balanced Dragon Onlay, wood
OLAS DA-57

Asymmetric Rose Onlay, wood OLAS
GW-186

Center Balanced Floral Onlay, wood OLCB
GW-27

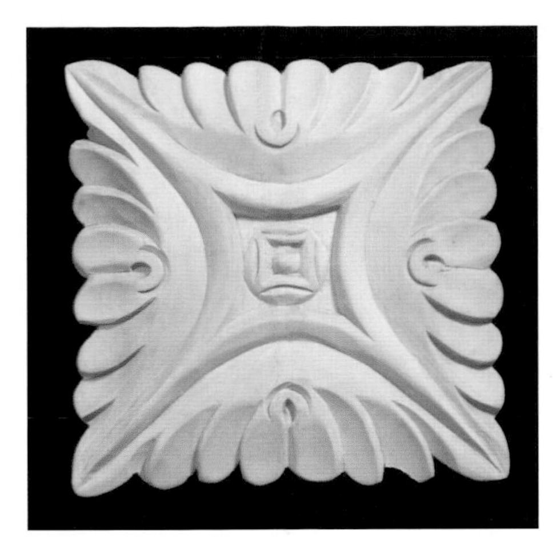

Square Acanthus Medallion, wood OLMN GW-07

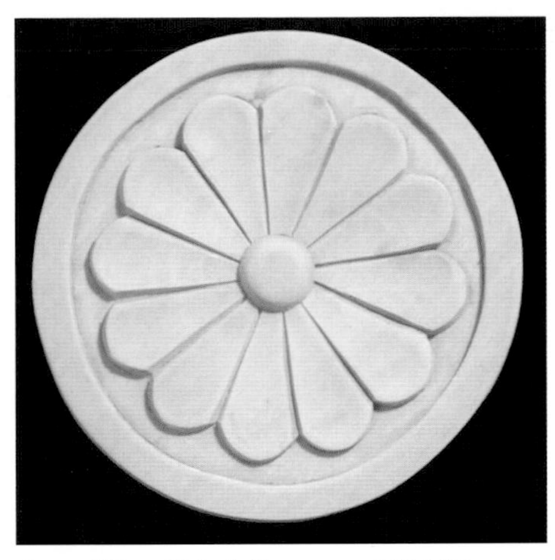

Round Floral Medallion, wood OLMN GW-47

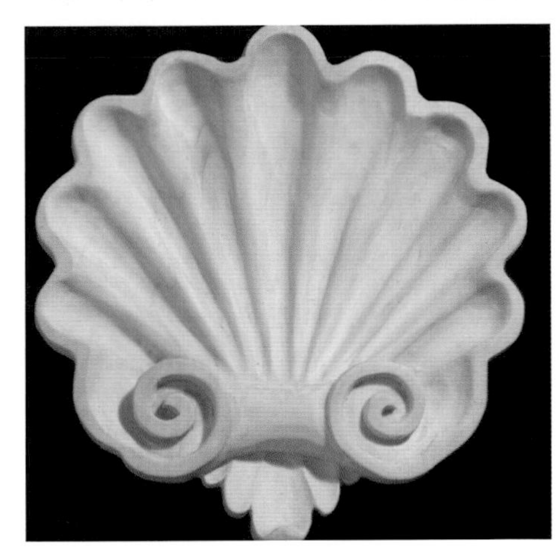

Compact Shell Onlay, wood OLCP GW-127

Large Round Acanthus Medallion, wood OLMN DA-41

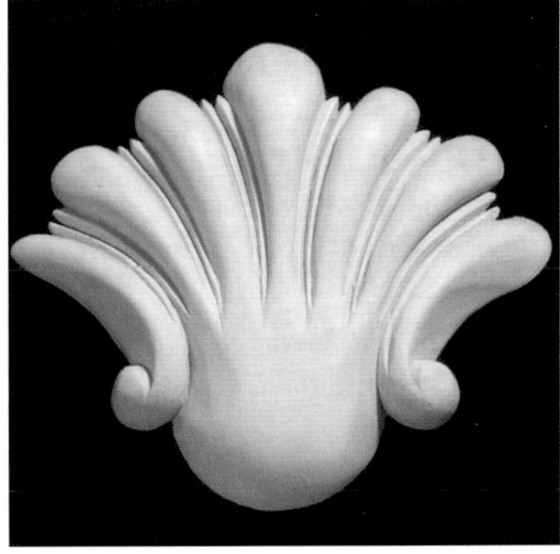

Compact Shell Onlay, wood OLCP GW-05

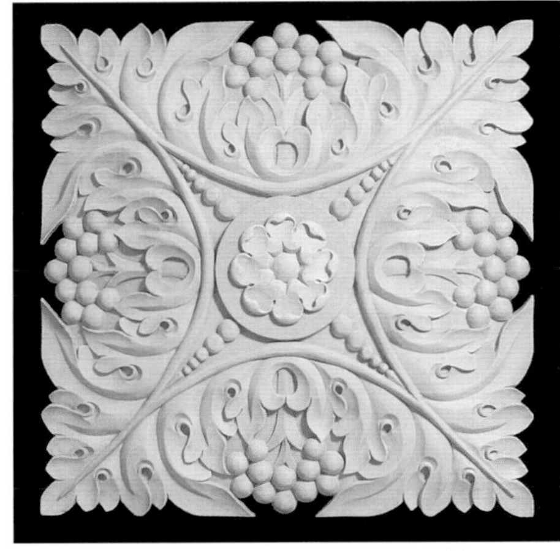

Large Square Berry Medallion, wood OLMN DA-45

Compact Basket Onlay, wood OLCP GW-182

Compact Acanthus Onlay, wood OLCP GW-06

Compact Floral Onlay, wood OLCP GW-69

fiberglass & plaster

Fiberglass corbels 0109 (side and front views) from *Richelieu Hardware.*

Acanthus Leaf with Pearl Bracket, plaster, 103S *The Design-Build Store*

Acanthus Leaf with Scroll Bracket, plaster, 128S

Contemporary Fluted Bracket, plaster, 129B

Acanthus Leaf with Scroll Bracket, plaster, 128

Contemporary Fluted Bracket, plaster, 138 *Design-Build Store*

Contemporary Fluted Bracket, plaster, 138S

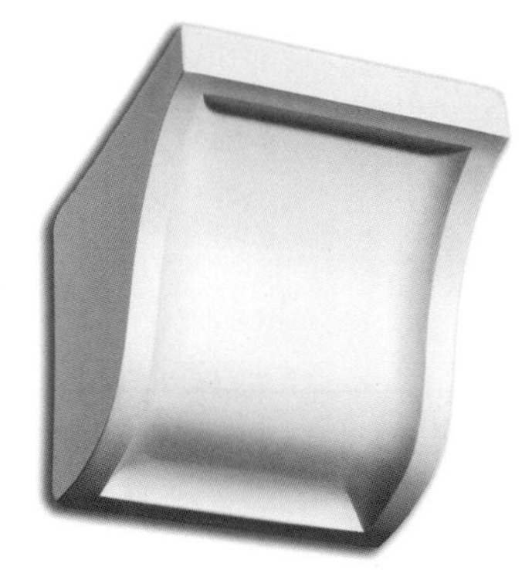

Crown Molding Divider Block, plaster B-104 *Balmer Moldings*

Crown Molding External Divider Block, plaster, B105

wood, lime & polyester

Acanthus Capital, wood
CTAR DA-11
Architectural Adornments

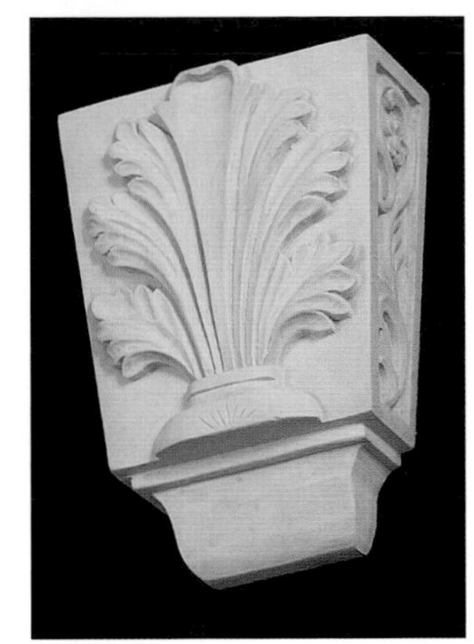
Acanthus Corbel, wood
COAR DA-81

Acanthus Capital, wood
CTAR DA-82

Leaf Corbel, wood
COAR DA-13

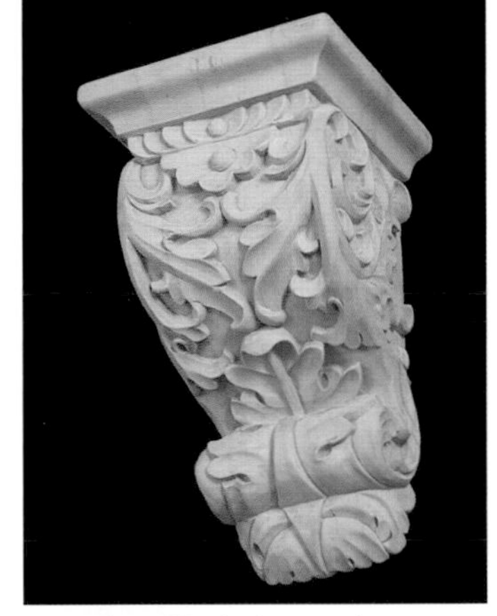
Acanthus Corbel, wood
COAR DA-17

Acanthus Corbel, wood
COAR DA-14

Acanthus Corbel, wood
COAR DA-80

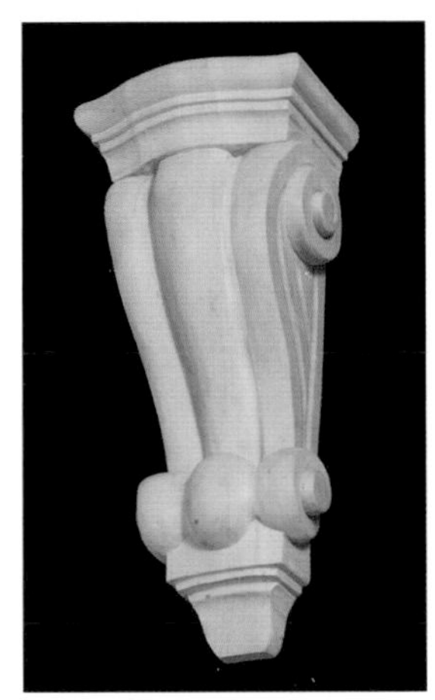
Scroll Corbel, wood
COAR DA-301

Lion's Head Corbel, lime 02300
Richelieu Hardware

Horse Corbel, C-59
Michael Shea Woodcarving

Corbel C-142, wood
Michael Shea Woodcarving

Polyester Corbel, 06700808
Richelieu

corbels & finials

Oak Corbel, 010500 *Richelieu Hardware*

Oak Corbel, 010800

Oak Corbel, 010801

Thin Polyester Corbels, 011500 *Richelie*

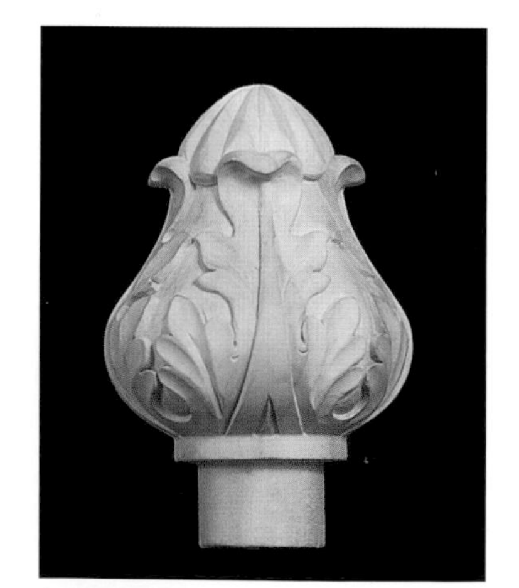

Acanthus Finial, wood FNAR DA-22 *Architectural Adornm*ents

Dogwood Finial, wood FNAR DA-23

An acanthus capital and column from *Richelieu Hardware* built into a bookshelf.

ceiling medallions

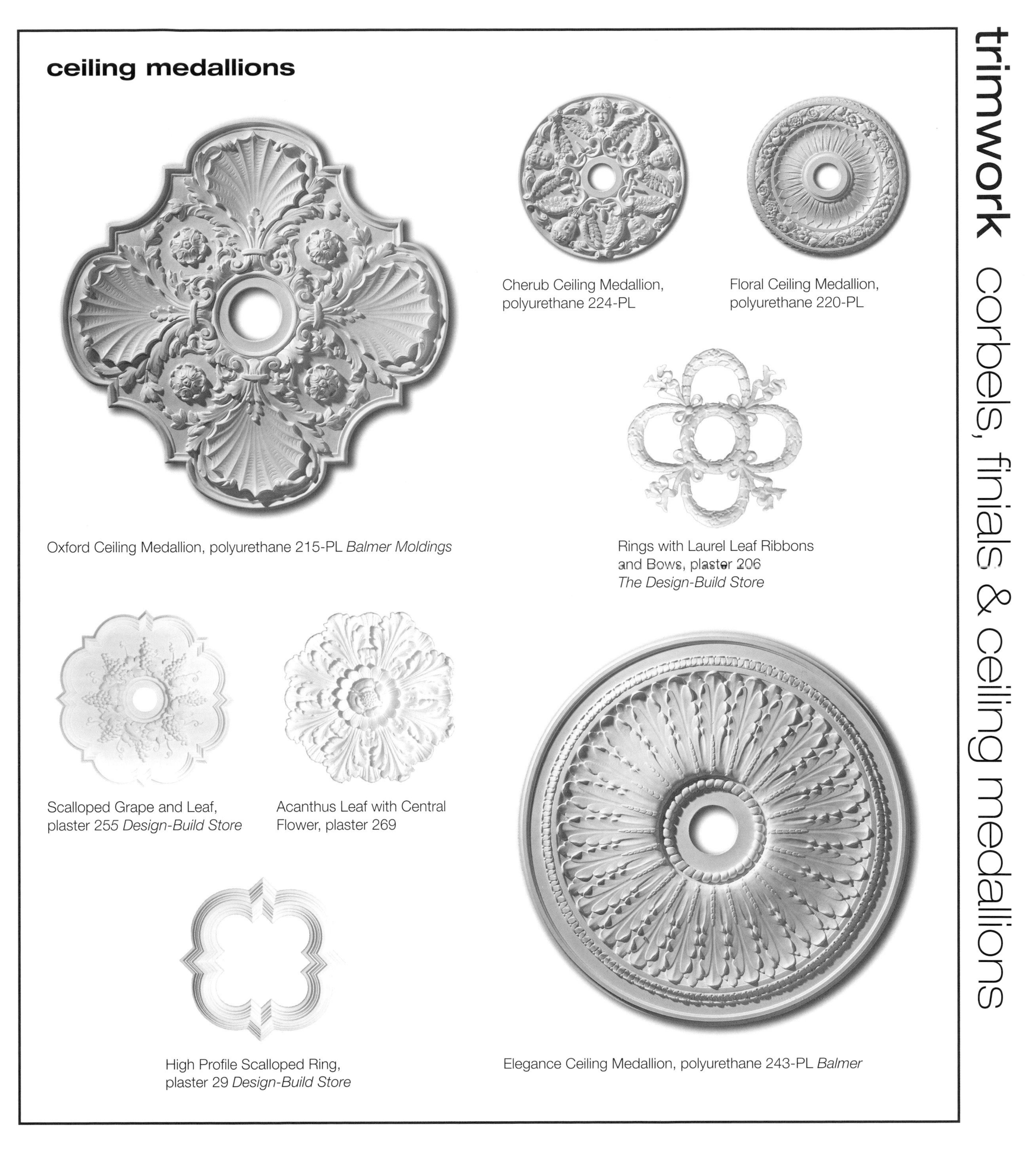

Oxford Ceiling Medallion, polyurethane 215-PL *Balmer Moldings*

Cherub Ceiling Medallion, polyurethane 224-PL

Floral Ceiling Medallion, polyurethane 220-PL

Rings with Laurel Leaf Ribbons and Bows, plaster 206 *The Design-Build Store*

Scalloped Grape and Leaf, plaster 255 *Design-Build Store*

Acanthus Leaf with Central Flower, plaster 269

High Profile Scalloped Ring, plaster 29 *Design-Build Store*

Elegance Ceiling Medallion, polyurethane 243-PL *Balmer*

capitals

A Corinthian capital (plaster 191) from *The Design-Build Store.*

Doric, plaster 181
Design-Build Store

Doric, plaster 180

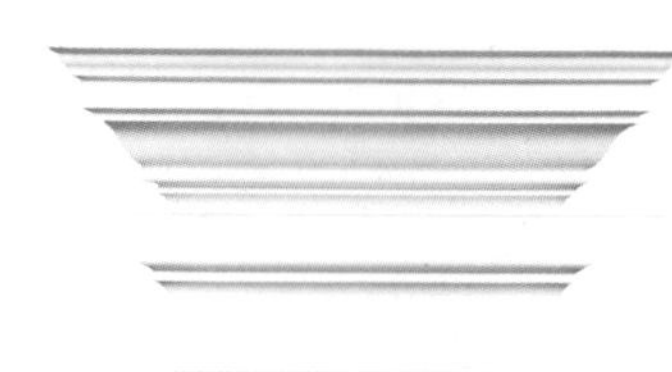

Doric, plaster 187-188

Doric, plaster 186

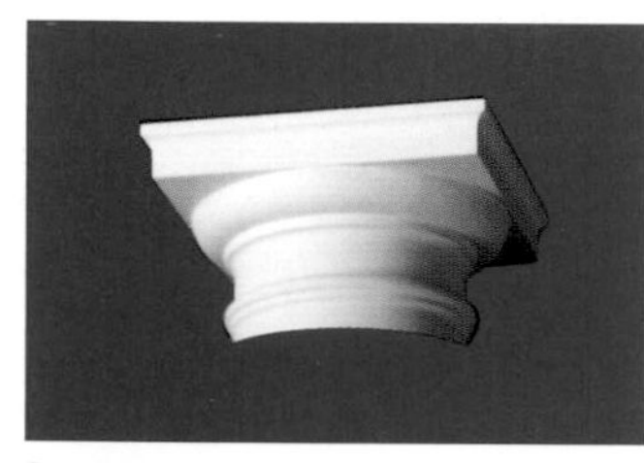

Doric, plaster/fiberglass, *Chadsworth's*

Roman Doric, plaster/fiberglass

Empire with Necking, plaster/fiberglass

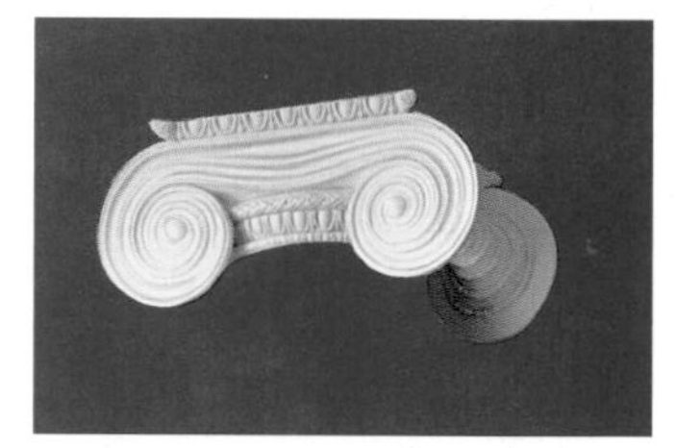

Greek Erechtheum, plaster/fiberglass

Gothic Renaissance, wood

Gothic, plaster/fiberglass

Roman Ionic, wood

Italian Renaissance Ionic Pellegrini, wood

Modern French Renaissance Orleans, cast marble

Modern French Ionic with Necking, cast marble

Roman Corinthian, plaster/fiberglass

Greek Corinthian, plaster/fiberglass

columns

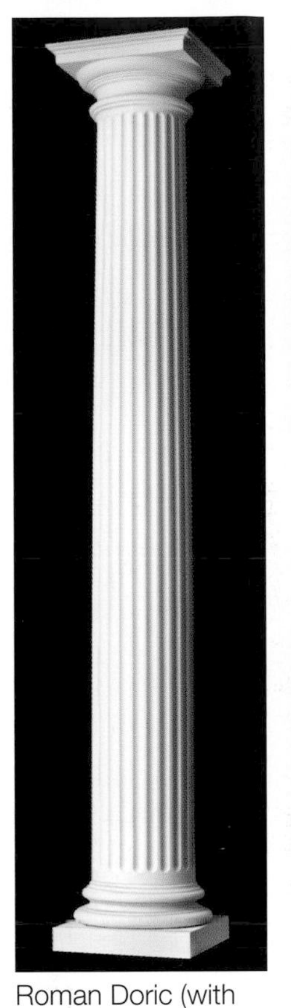
Roman Doric (with Attic base), wood
Chadsworth's

The Belley, wood

The Arlington, wood

Rope Twist, wood

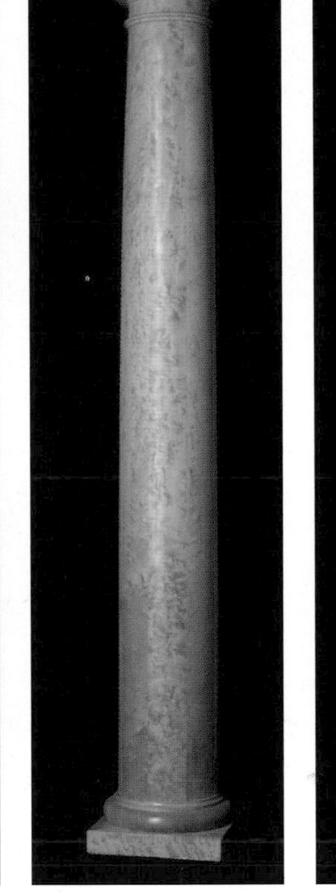
Tuscan Birdseye Maple

Tuscan Fluted, Cherry

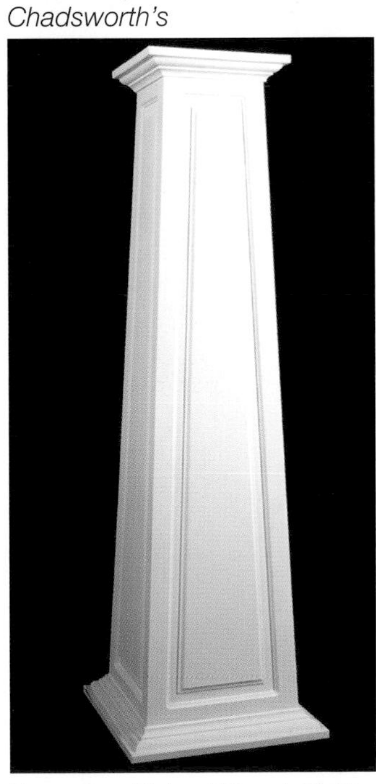
Bungalow Raised Panel, PVC

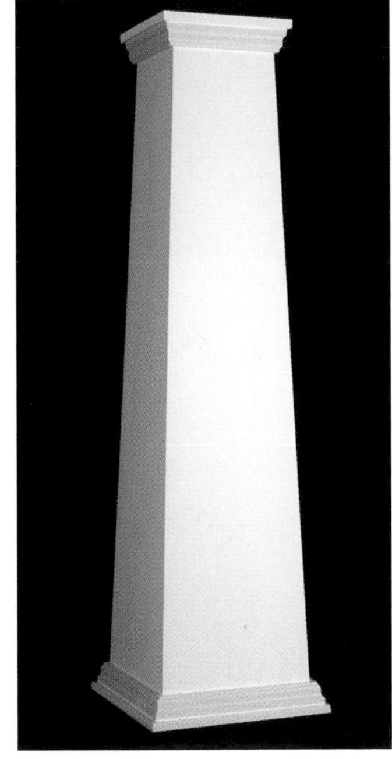
Bungalow Plain Panel, PVC

bases

1 Ring Column Base, plaster 149
Design-Build Store

Pilaster Base, plaster 159

Pilaster Base, plaster 158

carved stone

A king's ransom in carved stone salvage at Five O Seven Antiques.

Arrab, carved stone
Walter Arnold

Carved face
Ancient Excavations

Lion Keystone *Walter Arnold*

Mirth-Green Man

Prince *Ancient Excavations*

Laughing Face *Walter Arnold*

Flowers and Leaves
Ancient Excavations

Daffodils

Dragon *Ancient Excavations*

metal mesh

Multiple wire 008 *Universal Wire Cloth*

Alpha Stainless
TWP Inc.

Colorado Bronze

Colorado Stainless

Diana Brass

Fandango Brass

Durango Brass

Edison Brass *TWP Inc.*

Juliet Bronze

Antique Brass
88114BB
Richelieu Hardware

Brushed Nickel
881214195

Pewter 88286142

Slotted OPG Rectangular
Universal Wire Cloth

Slotted Opening Weave

Flat Top Weave

Multiple Metals
Universal

Triple Shoot Weave

Architectural Mesh

Antique Brass 88286BB
Richelieu Hardware

railings & balustrades

Railings available in bronze or aluminum by *Historical Arts & Castings.* (RL063)

Railing RL027

Railing RL014

Railing RL076

Railing RL076

Railing RL060

Wrought-iron Railing *Design-Build Store*

Bronze (or aluminum) railings RL027, RL0659
Historical Arts & Castings

Railing RL069: two views of the same staircase *Historical*

Railing RL0659 *Historical*

light & heat

We try to use what we call a “family of fixtures” that all have some relationship — maybe the same color glass or geometry with variations in size — so there is a harmony as you move through the house. We also like to keep it uncluttered so you don’t see a lot of cans on the ceiling or a lot of hardware, unless its something very beautiful that you want to show as an art object — a focal point.

BUZZ YUDELL

ONE OF THINGS THAT MAKES A HOUSE A HOME IS PROTECTION from the more frightening and uncomfortable aspects of life. We seek to ward off the kind of winter chill that seeps into the bones on a winter’s day, or the gloom and unknowns of a pitch black night. As in all sophisticated societies, we like to make a virtue of necessity and so we jump at the chance to take our need for a light source to fill our house with beautiful, multipurpose light fixtures that not only illuminate but make architectural statements on their own. Similarly the fireplace has become not just a place to cook and keep from freezing to death, but a comfort center employing the best materials and the latest technologies. Light and heat are two of the most creative ways we can turn a house into a home.

A spectacular Dale Chihuly chandelier dominates this contemporary dining room. Note the long, rectangular gas fireplace and the suspended, glassed-in office, top left.

lighting

Artificial light sources provide color, texture, drama and definition to a room and to the surfaces of the room. The fixtures themselves can be unobtrusive illumination sources, handsome design elements that complement their surroundings, or the spectacular focus of a room. To understand how light fixtures fit in to the overall architectural approach to home design, it helps to understand the nomenclature.

AMBIENT LIGHTING Sometimes known as "general lighting," this is the overall level of lighting in any particular room. In residential architecture you don't have to be so concerned with evenly lighting the entire room as you do in commercial applications because you don't use residential rooms in the same way as you use an office space. As a general rule, having a room that is lit from various

Anyone who has been to Chicago's Unity Temple, designed by Frank Lloyd Wright, will know where the inspiration for this majestic Craftsman-like chandelier comes from. Note the art glass in the coffered skylight. The door at the lower right leads to the vaulted brick passage seen on page 170.

What is really bad are recessed lights right above where somebody sits so that the light shines down on them, which is a very unattractive way to light a person. I put recessed light over a coffee table to highlight what is on the coffee table, or maybe a textured wall or picture, but never over a sofa.

LISA JAFFE

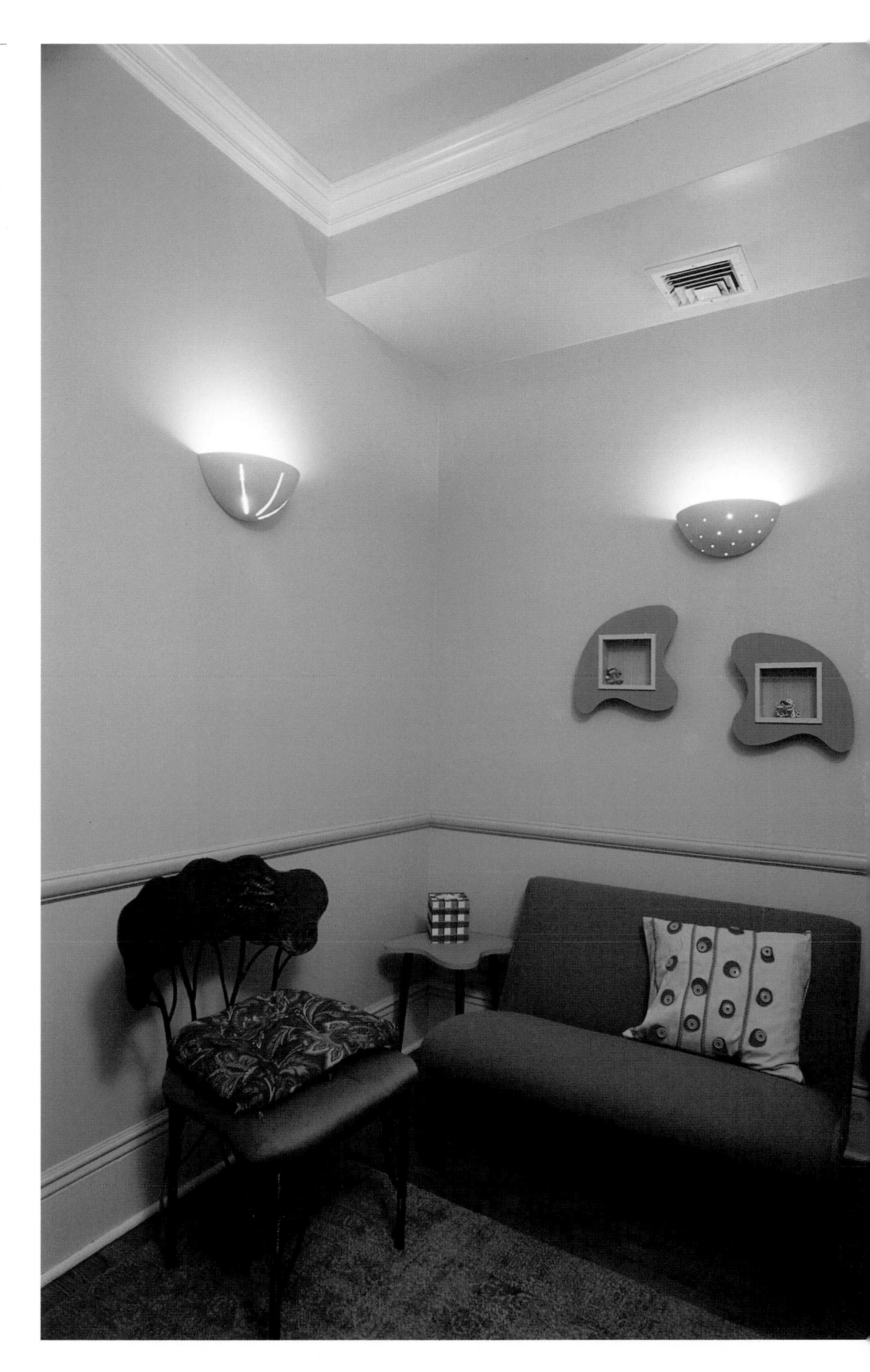

Wall sconces throw scalloped light onto the wall of this quirky landing. Using what could have been wasted space, architect Heather Faulding played with vibrant colors and moldings to come up with an almost cartoon version of the 1950s.

BOTILLERIA
CIELO

sources — floor and table lamps plus ceiling fixtures — is always more relaxing as well as versatile. A too-strong light will make most rooms uncomfortable to be in because of glare, so another general rule is that it's better to use more lights of lower wattage. A remote control that operates lights and fans can add to your sense of control and convenience because you can turn fixtures on and off or up and down from the lazy comfort of your sofa or bed.

ACCENT LIGHTING Normally used to create depth, drama or to highlight an object. To make it as dramatic as possible, it should be four to fives times brighter than the surrounding ambient light. Accent lighting is not flattering on the human form or face, so restrict it to objects — paintings, statues or vases.

TASK LIGHTING Under-cabinet lights make great task lighting in the kitchen. Positioning the light as close to the front of the cabinet as possible reduces glare. Overhead recessed lights can also function as task lights, but should be centered directly over working areas such as countertops to prevent having to work in someone else's shadow. For clothes closets, try using a fluorescent or halogen light that is as close to sunlight so you can see the true color of your clothes.

COVE LIGHTING Indirect lighting, usually hidden in valences, that emanates from the edges of a ceiling and washes the ceiling in a soft, diffused light. Cove lights use the ceiling as a light diffuser, creating a glow or ring of light around the perimeter of the room. This means of lighting can also be used to highlight the moldings or the shape of the ceiling, something that works particularly well in barrel-arched ceilings. Depending on the shape and color of the ceiling, its effects range from the dramatic to the cozy and warm. Cove lighting should never be too harsh, which is why having a dimmer switch installed is a good idea. The light source is usually incandescent, low-voltage xenon, or fluorescent. "Rope" lighting (see Fiber optic lighting) can also be used.

Opposite: The large scale of this oversized staircase makes a little house seem bigger. Los Angeles architect Barton Phelps also changed the perspective for everyone who enters by playing with the scale of the staircase, making it wide at the bottom and narrow at the top. The detailing on the stairs is over-scaled as well by having the bullnose larger than normal to make the stairs look bigger. The lights mounted on arms extending from the wall rise above the stairs create a cloud motif, one of the many references to the natural world in this house. Left: A beautiful 18th-century French light made of hand-blown glass and ormolu, a gilded brass. Center: An offbeat Art Deco ceiling globe from the 1930s. Right: A reproduction Colonial coach lamp fixture on a stone wall. Overleaf: The cove lighting in this living room works particularly well because the slight arch of the ceiling acts as a light diffuser.

MANAGERS
DRESS SMART
OUT of CONTROL
humanity

TRACK LIGHTING Although they only really look good in Contemporary homes, these individual lights that move along a conductive track are one of the most versatile lighting arrangements available. Each light is able to be positioned over 360 degrees, making it usable as either an ambient, accent or task lighting source. To highlight a picture on the wall but not shine in anyone's eyes, angle the track light at 30 degrees from the vertical. Halogen bulbs give the most true light and are therefore the best for lighting pictures.

RECESSED LIGHTING Unobtrusive and relatively inexpensive, recessed lighting (called "pots" or "cans" in the trade) can be used virtually anywhere, including the exterior of your home under the eaves. Nothing more than holes in the ceiling with the bulk of the light fixture hidden from view, they can be used as accent, ambient or wall-washing and wall-grazing lights (see below), depending on their size and positioning.

WALL GRAZING A form of accent lighting that positions lights in the ceiling not more than 12 inches from the wall. The light that is cast down the wall throws into relief the texture of interesting materials such as brick, stucco, stone and rough fabrics. It is also good for polished surfaces such as marble because it lights the surface without the usual accompanying glare.

WALL WASHING Down lights usually positioned closer together and further from the wall than lights used for wall grazing. Wall-washing lights eliminate the shadows that appear at the top of the wall when using wall-grazing down lights, as well as the more dramatic textural effects. These lights are best used when placed not more than 24 inches from the wall and between 24 and 36 inches apart.

When choosing lighting fixtures, you may want to consider not only the quality of light you desire but also the economic and environmental costs of the various kinds of bulbs.

Left: This Art Deco fixture from the 1930s had disappeared from the original home. When the new owners were renovating they went to an architectural salvage company to look for replacements and found it for sale. They recognized it from photos of the house and put it back in its place. **Center:** A formal dining room deserves an elegant centerpiece and gets one in this classic bell-shaped hurricane lantern based on an oil lamp of Georgian design. **Right:** Putting this recessed task lighting in slots over a kitchen counter was a way to organize light so it didn't spill all over the room but was more focused below or to the sides while still retaining a clean, streamlined ceiling.

types of bulbs

INCANDESCENT The ubiquitous light bulb consisting of a heated metal filament in an inert gas. The resulting light is a warm yellow that is very flattering. It can be used with a dimmer switch and is often the source for ambient light. Cheap and easy to install, but not energy efficient — most of the common incandescents have a life of only 750 hours and up to 85 percent of the energy it uses is turned into heat rather than light.

XENON A variation of the incandescent, it heats a carbon filament inside xenon gas for a brighter, more long-lasting light that burns with so little heat it doesn't need a glass cover. It does, however, require a transformer. Usually low voltage. Xenon lights work particularly well as under-cabinet lights in a kitchen, but your cabinet maker or supplier needs to know in advance to accommodate the transformer.

HALOGEN combines halogen gas with tungsten inside a quartz shell to yield a bright but warm white light. Although they require special fittings, halogen lights are good for task and accent lighting, particularly the low-voltage kind, and are often used for track lighting, floor lighting, and recessed cans. Slightly more expensive than incandescents, they also last around 3,000 hours and are 20 percent more efficient. Reflector lamps and lamps with infrared reflecting are available for added efficiency (40 and 60 percent more efficient than incandescent lamps, respectively).

FLUORESCENT Less heat and more light for your dollar (75 percent more efficient than incandescent bulbs and lasting up to 20,000 hours) but a relatively harsh, cold and unattractive white light unless you buy some of the new and more expensive color balanced varieties. Usually comes in long tubes or compact fluorescents (CFLs) in a variety of shapes and color temperatures. Can't be dimmed without special and costly equipment.

FIBER OPTICS In this high-tech alternative, beams of light from a single, centralized halogen or metal halide lamp travel optically through a flexible cable full of glass or acrylic fibers. This cable acts as a sort of light pipe that can actually pump the light to any location and emerge in any kind of bulb or diffuser you wish, including more than one. And, because the cables contain only light with no electricity, heat or UV rays involved, it is energy efficient and perfect for outdoor or even underwater applications, including swimming pools and ponds. Fiber optic cable is easy to install and can also be translucent, allowing for a glowing "rope" of light that can be used to highlight an architectural feature such as a staircases banister, cornice molding, or the eaves of a roofline. They are also sometimes displayed as tiny pinpoints of light to create a "starry sky" effect on the ceiling. One last benefit — because there is no bulb to burn out in the ceiling (the source bulb can be put anywhere convenient), fiber optics are the perfect choice for hard-to-get-at places. The only drawback is that it is relatively costly. Acrylic cables are cheaper and can be easily cut to length but can lose power over distance, while the glass cables cost more and have to be specially cut, but last longer and work better over distance.

A rattan ceiling fan's slow revolve conjures up sultry tropical night on this screened-in porch, exemplifying both their practical and design possibilities. The fan is a Belleria, by Fanimation.

ceiling fans

Historically, fans were the only way other than strategic placement of windows and shade to cool a house; the first large "motorized" fans, inspired by the popularity of hand fans, were originally powered by servants pulling on ropes. Workers in the factories of the Industrial Revolution are credited with inventing the first mechanical ceiling fan when they began attaching hand fans to the overhead steam-powered gears and shafts that drove their machines. Before the advent of central air conditioning, attic fans — a large fan installed in the attic to bring cooler outside night air in through screened windows and doors and create a continuous breeze throughout the house — were popular, especially in the southern United States. The first home ceiling fans came about with the introduction of reliable electricity in the late 1800s.

A ceiling fan can provide a pleasant breeze on hot days and add a year-round exotic, tropical flavor to a room. Beyond comfort and beauty, ceiling fans can reduce energy costs by up to 40 percent in the summer and 10 percent in the winter. They work by blowing breezes over exposed skin, which helps wick the heat off the body, but reversing the direction of the fan's spin in the evening will suck in the cooler night air outside and reduce the internal temperature of the house. For most efficient use, place the fan from between 8 and 9 feet from the floor, using a down rod to lower it if your ceiling is higher than 10 feet.

Today, ceiling fans come in a surprising array of styles and looks, from minimalist sci-fi modern to baroque Victorian, Chinese, Arts and Crafts and even the simulated palm leaf fans you used to see in old black-and-white jungle movies (see Products section). Fans that come with lights can be used to create a mood, facing up to hit the ceiling for a more general and subtle effect, or as accent or ambient lights when directed down.

fireplaces

The phrase "The hearth is the heart of the home" has meaning both figuratively and architecturally. Figuratively, in that there is nothing like the cheery glow of a fire to create a warm welcome in any home. Architecturally, because a fireplace is a great opportunity, an excuse for elaboration and design that can easily become the focal point of a room whether it contains a fire or not.

The style of your fireplace depends on the décor of the room; one will suit the egg and dart surrounds of a stately Georgian fireplace, another demands the sleek, angular line of a contemporary slab. But it is one of those architectural elements that calls for the same elaboration and expense that any centerpiece calls for. With a fireplace, you get to make a grand statement while at the same time creating one of the creature comforts that make a house a home.

One of the best ways to heighten the cozy aspect of a fireplace is to build it into an inglenook. An inglenook is a fireplace with a built-in seat, or more commonly a seat on either side of the fireplace, often set into a nook or recess so that the fireplace and the seat form a small and intimate seating area warmed by the fire.

This simple sandstone mantel and "dry stack" fireplace is pushed into the realm of the truly extraordinary by the metal work of Aspen artist Rick McClain. The leaves inside the firebox glow when the fire gets hot enough.

Opposite: The river-washed stones of this unique round fireplace were laboriously set by hand and seem to undulate and move as you look at them. **Left:** Architect Marilyn Lake (The Ideal Environment) knew there would be art hung over this fireplace and didn't want it to "roar too loud." So she had it made of a subdued material — a soft, golden sandstone-like limestone from Ontario's Owen Sound region. The shades of the flagstone floor are designed to go with the color of the stone.

From the point of view of décor, the mantel is one of the most important parts of a fireplace. Those convenient projections that hold everything from family pictures to Christmas stockings can be made from just about anything. The most popular materials are stone and wood, but we have seen everything from ceramic tile and concrete, to glass and metal. The only rule is that the fireplace harmonizes with the rest of the room. You can buy an antique, have your surround custom made, or buy one from one of the many reproduction companies that abound in North America.

Early American Colonial fireplaces were often simply made, with a timber slab as a mantel. Out west, early adobe fireplaces followed the minimalist, rounded shapes of the homes themselves. This approach is epitomized by the plaster "beehive" — or kiva — fireplaces found everywhere in the American Southwest. Because a kiva fireplace is traditionally made of adobe bricks, it has a lot of mass that stores and radiates the heat for hours, making it very efficient (see Masonry Heaters).

Fireplace surrounds during the 17th and 18th centuries were made of elaborately carved wood or stone and also of brick. Victorians liked their fireplaces made of cast iron, and this trend to metal continued during the Arts and Crafts and Art Nouveau periods, especially in the fireplace designs of Charles Rennie Mackintosh, who combined the artistic approach of William Morris with Scottish baronial design elements and stylized fauna and ceramic tile details set in the ironwork. Metal was used a lot in the early part of the 20th century because of the Art Deco movement's love of nickel and chrome, as well as modern materials such as bakelite plastic and Vitrolite (a pigmented plate glass, mechanically ground to a mirror finish). Today's fireplaces can be made of anything: stainless steel, concrete, ceramics, rubble stone and more.

Contemporary fireplaces often don't have a mantel or elaborate surrounds — or any surrounds for that matter. Instead they are flush to the wall and sometimes have an extended hearth that acts as a bench or a shelf. Some very effective fireplaces are simply a hole in the wall. Any shape of hole will do: a sensuous circle (see page 344) or a rectangular slit (see the photo on page 347). The key to success is how it works with the room.

Advances in technology have also made it possible to have at least some kind of fireplace anywhere in the house, even if you don't have a chimney. We are, as in most things these days, awash in choices and one of the questions facing homeowners when it comes to the fireplace is: wood burning, gas or electric? Taking one side or the other is like jumping into a discussion about religion or politics — opinions are strong and run deep.

To those who favor the traditional wood-burning fireplace, the benefits are as obvious as they are ancient. Wood fires crackle and snap. They are alive. The smoke-blackened hearth and ember-strewn firebox give character and honest

Left: An Italian marble aptly called Opera Fantastica forms the base of this fireplace. The ornamental work on the top and sides of the mantel is plaster that has been hand painted and antiqued to look like aged limestone. **Center:** A concrete mantel, black slate tile surround and stainless-steel mesh fire screen create a minimalist fireplace in bold colors and shapes. **Right:** Traditional "beehive" or Kiva fireplaces, popular in Southwestern style homes. The name comes from respectively: the shape of the hives bees were once kept in, and the circular stone rooms where certain Southwestern native peoples hold religious ceremonies. The white of the adobe goes perfectly with the natural colors of the wood and terra cotta tiles.

To be an effective heater, a fireplace must borrow some of the features perfected by woodstove designers over the last 20 years. These include gasketed, ceramic glass doors with an air-wash system to keep them clean, firebox insulation and internal baffling. An adjustable combustion air supply is also needed to control the burn rate and, therefore, the output of heat. The quick way to find them is to look for either factory-built fireplaces or fireplace inserts that are certified by the Environmental Protection Agency (EPA).

JOHN GULLAND, EXECUTIVE DIRECTOR OF THE WOOD HEAT ORGANIZATION INC.

The steel plates around this fireplace were intentionally rusted by misting on successive coats of salt water onto the panels, allowing them to rust and then brushing the surface with Scotch Brite scrubbing pads. Several coats of paste wax were then applied, and this wax is rubbed on again once a year. The rusted metal fits in with the history of this industrial building turned living space. Note the steel beam in the ceiling and the concrete floors. The fireplace is double sided so it can also warm the bedroom (behind the wall). The wood is white oak.

integrity to the fireplace. Even the manual labor involved in the hauling of wood, the building and poking of the fire and the cleaning out of the ashes is looked upon as part of an age-old ritual. Besides, modern fireplaces and wood stoves have become much more energy efficient and environmentally friendly. To this mindset, a gas or electric fireplace is more of an appliance, like a stove or a dishwasher — convenient no doubt, but an over-civilized affectation.

Gas enthusiasts take the view that the sheer, overwhelming convenience of flipping a switch to achieve a fire in the wall more than compensates for the slight phoniness of a gas flame pretending to be a fire, and ceramic lumps pretending to be logs. And that's not to mention the cleanliness aspect of not having to haul ashes and clean soot off of the limestone mantel, or the environmental horrors they imagine.

On the other hand, people who favor electric fireplaces have given up all pretense and admit that all they really want is a heat source that looks nice. They are more interested in the ideal of a fireplace and the aesthetics of the surround than in the actual fire.

All three camps have their points. To help you decide, there follows a more detailed description of the pros and cons of the three kinds of fireplaces.

WOOD-BURNING FIREPLACES This was how it all began. Throughout early history, the fireplace was the sole source of heat and the only way to cook. Essentially a hole in the wall connected to a chimney, it sucked air and heat out of the house, which the radiant heat it provided only partially made up for. Today's fireplaces can be just as inefficient if not built properly. For one thing, that rich woodsy smell you enjoy when your fire is burning may be pleasant, but it means your fireplace is back-venting smoke into your house.

There are many efficient, clean burning fireplaces on the market, however. And if you want to increases the efficiency of a beloved old fireplace, there are inserts that will fit into the hearth and make it work to EPA (Environmental Protection Agency) standards without changing the look.

One of the more famous fireplace designs is the Rumford, named after Count Rumford (Benjamin Thompson) who came up with the design in the late 1700s. Distinctive and aristocratic in design, it is recognized by its tall, shallow firebox designed to reflect heat back into the room, plus a streamlined throat to eliminate turbulence and carry away the smoke with little loss of heat.

MASONRY HEATERS If you really want to heat with wood, a masonry heater is the way to go. Used in Russia and the Scandinavian countries for centuries, masonry heaters are massive piles of block or stone with channels running through them to circulate and hold the heat, releasing it slowly and continuously day and night up to 24 hours after the fire has gone out. Their sheer mass provides a chance to display the material used to build them, resulting in walls of buttery soapstone, or concrete block with rough, pebbly aggregate showing through. They can also be used to collect solar energy if built near south-facing windows. Best of all, you can still have a visible fire while being extremely energy efficient. Drawbacks are they take time to heat up, weigh a lot, take up space and are relatively expensive.

Left: A spectacular rubblestone fireplace has one very intriguing aspect: the circular glass plate set in the left foreground of the hearth stone is a window that looks down onto a huge crystal that is penetrated by water-filled tubes that connect to the re-bar of the house. The crystal purportedly cleanses the energy of the home. The owner, who played a big part in the design, wanted a home that was very much "Southwest Zen." Center: The detailing of this oak and ceramic tile fireplace is a combination of Craftsman and Japanese influences. Right: This sleek black Art Deco fireplace is made of Vitrolite — a pigmented plate glass that is mechanically ground to a mirror finish with no distortions — that was very popular in the 1920s and 1930s. (See the sources chapter for a Vitrolite specialist.) The ornament of the mantel is an old radio antenna that "just seemed to fit." The andirons are brass skyscrapers.

GAS FIREPLACES The gas alternative has been around for decades but it has been gaining on wood rapidly, so much that most fireplaces in the big new housing developments are gas. They seem to fit in better with today's busy lifestyles that demand convenience over tradition. You can use timers that turn them on and off — for when you get up, when you get home from work. And if you like having a fireplace in the bathroom, you might want to consider a gas fireplace because you can turn it on with the flick of switch and have a thermostatically controlled warm room when you get out of the shower.

There are two kinds of gas fireplaces: ventless and vented. Manufacturers of ventless fireplaces claim that there are few if any byproducts of the burning gas vented into the room, but this has not assuaged all worries, so anyone installing a ventless gas appliance is urged to also install a carbon monoxide detector. Their biggest advantage is that, because they don't need a chimney, you can have them anywhere there is an available gas line, making some freestanding models essentially portable. A vented fireplace, on the other hand, expels combustion materials directly to the outside. This can be done through a thin metal vent that fits into any standard wall, but many people choose to vent their gas burner through an existing wood burning fireplace chimney, which is fine if it is installed by a trained technician. This kind of wood-to-gas conversion is called an "insert," when is installed in an existing masonry fireplace or an approved metal firebox. They come complete with accessories such as thermostats and realistic looking ceramic logs and embers.

Freestanding units, which are more like wood stoves in appearance, can be vented into existing chimneys, through the roof with new venting or out the wall directly behind the unit.

An intricately composed inlaid marble mantel is beautifully set off by ceramic tiles and a Victorian coal grate.

GAS LOGS This is the easiest way to take an old wood-burning fireplace and convert it to gas. It consists of a set of ceramic logs (or stones) that are connected to a gas line and set down in your existing fireplace by an experienced heating contractor or plumber. You have to keep the vent to the chimney at least partly open, and you don't get as much heat as you do with a conventional gas fireplace, but it's a quick and relatively cheap conversion that leaves you with the look of your old fireplace.
GEL BURNING FIREPLACES These operate through the burning of an odorless grain-alcohol gel. Each can of gel can burn for about two and a half to three hours. Since they are free-standing and don't have to be vented, they can be put anywhere in the house.
ELECTRIC FIREPLACES Relatively inexpensive to install but lacking in realism, electric fireplaces produce a glow rather than a flame. They are usually installed in an existing fireplace but also come as inserts, freestanding units, and zero clearance models. All you have to do is figure out how to hide the plug.

products

light & heat

chandeliers

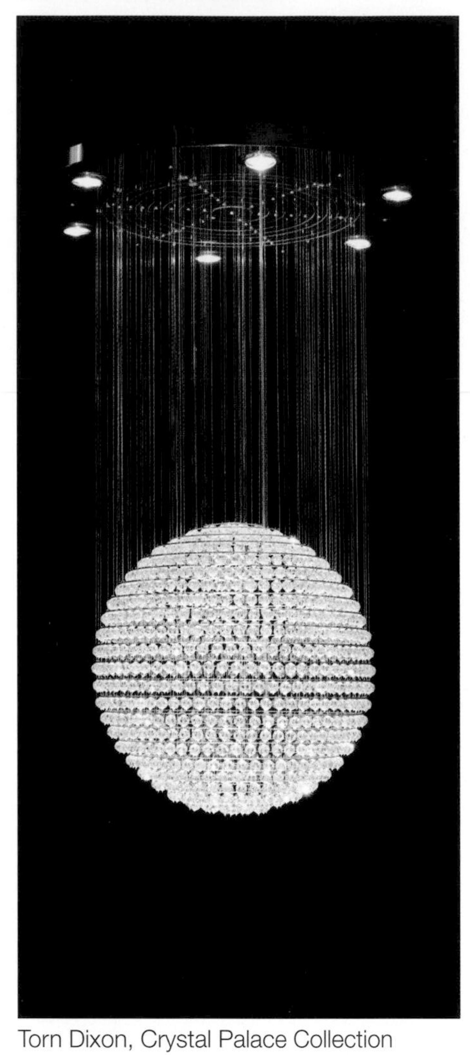

Torn Dixon, Crystal Palace Collection *Swarovski*

Jurgen Bey, Crystal Palace Collection *Swarovski*

Bella Iguazu *Elk Lighting*

Candles and Spirit, Van Egmond *Eurolite*

Bird, Foqus Collection *Eurolite*

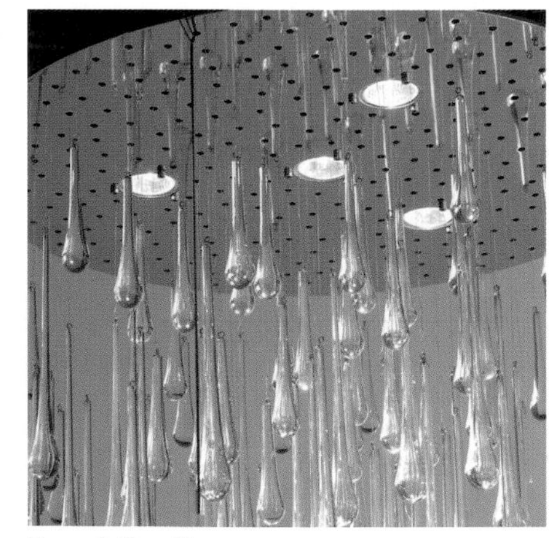

Egos II *Eurolite*

Universe, Foqus Collection *Eurolite*

Tromba, hand blown *Elk Lighting*

Quadralli Island Light *Fine Art Lamps*

Foyer Light *Hubbardton Forge*

Wrought Iron Chandelier *Hubbardton*

Wrought Iron Chandelier

Candelabra *Hubbardton*

Mansfield *Elk*

chandeliers

Valenciana Elk *Lighting*

Foyer Light *Elk*

Centennial Foyer Light *H.A. Framburg*

Vanezia *Elk*

Foyer Lamp

Chandelier *Elk*

Wrought Iron Chandelier

Candelabra Chandelier

Villa de Eleganza

Clock chandelier *Fine Art Lamps*

Bohemian Crystal Chandelier *Crystorama*

Candelabra Chandelier

Chrome Chandelier

French White Alabaster

Wrought Iron

Islander Chandelier *Fanimation*

Society Hill *Elk Lighting*

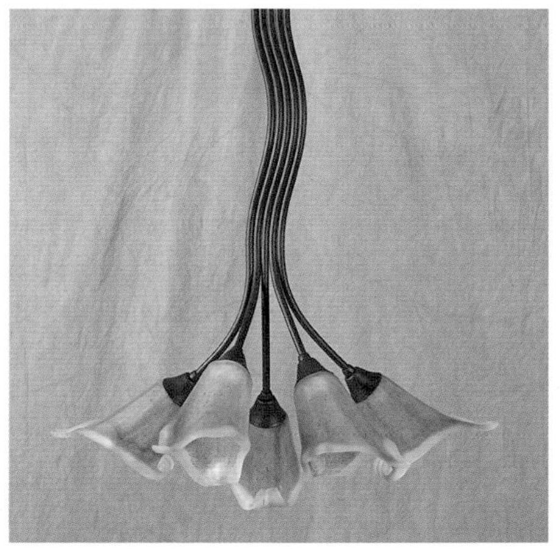
Vanizia with hand-formed Tulip glass

Classic Pull-down

ceiling fixtures

Mahogany Finish *Hubbardton Forge*

Semi Flush Steppe with Art Glass

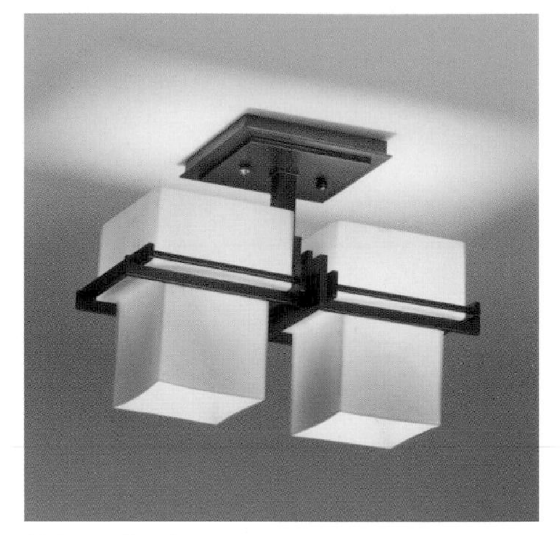

Kakomi Double

Valenciana *Elk Lighting*

Semi Flush with crystals *Elk*

Semi Flush with Hand-blown Glass *Fine Art Lamps*

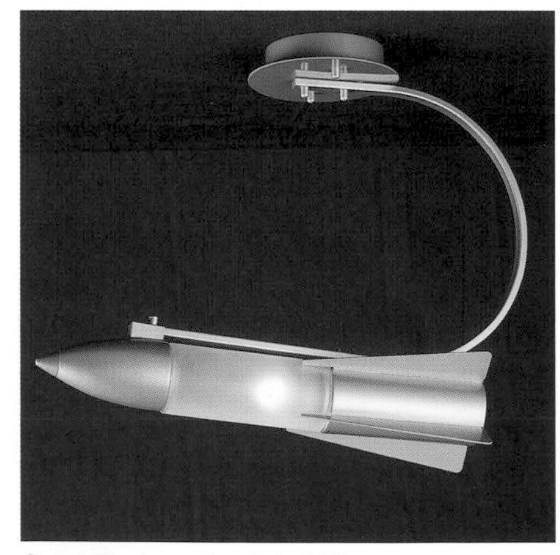

Semi Flush, satin nickel *Elk*

Semi flush 506-30 *Elk*

Semi Flush Mount Crystal *Crystorama*

pendants

Oval Macintosh Pendant *Hubbardton*

Adjustable Pendant

Ondrian Pendant

Prairie Pendant

Galileo Lumina *Eurolite*

Maritime Pendant
Elk Lighting

Slate Wafer Pendant
Hubbardton

Japanese Pendant, 233449
Fine Art Lamps

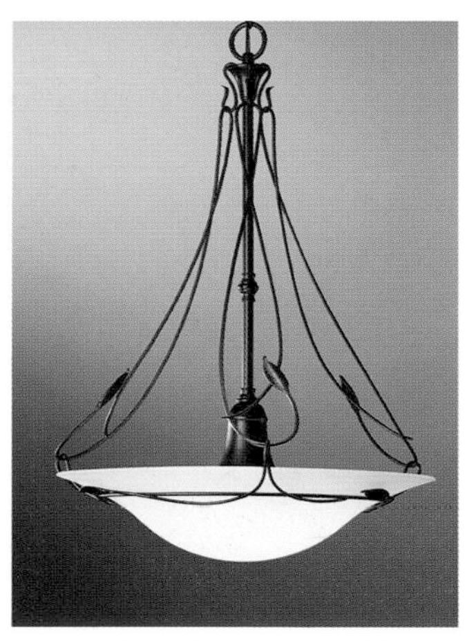
Pendant Leaf, 13-9501-07
Hubbardton

pendants

Tromba *Elk Lighting*

Black Finish Pendant
Fanimation

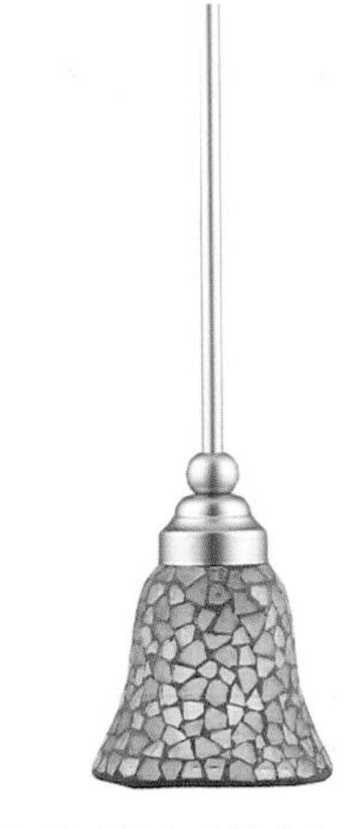

Satin Nickel Finish
Fanimation

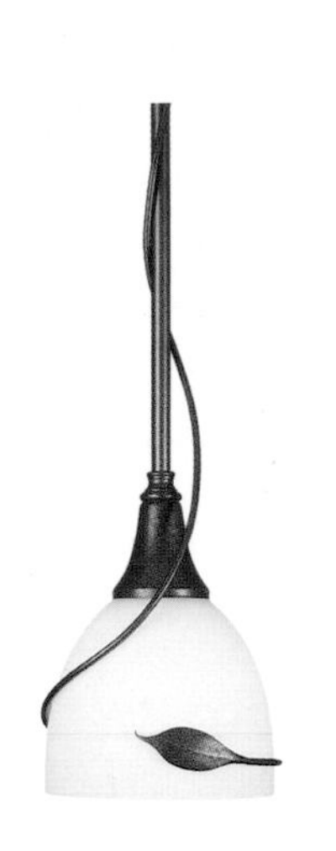

Leaf Pendant
Hubbardton Forge

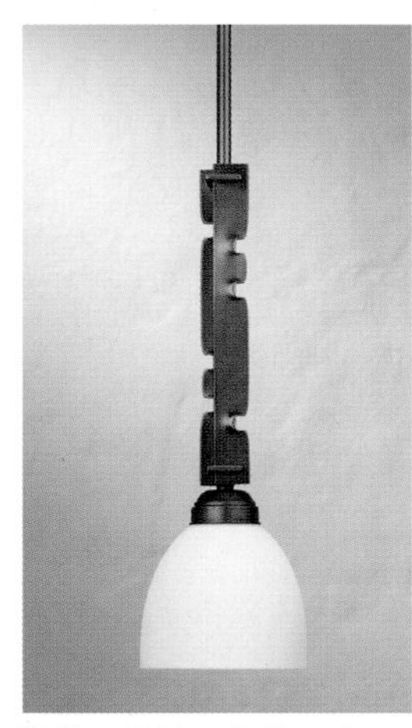

Fullered Notch Pendant
Hubbardton

Arts & Crafts style
pendant *Renaissance Antique Lighting*

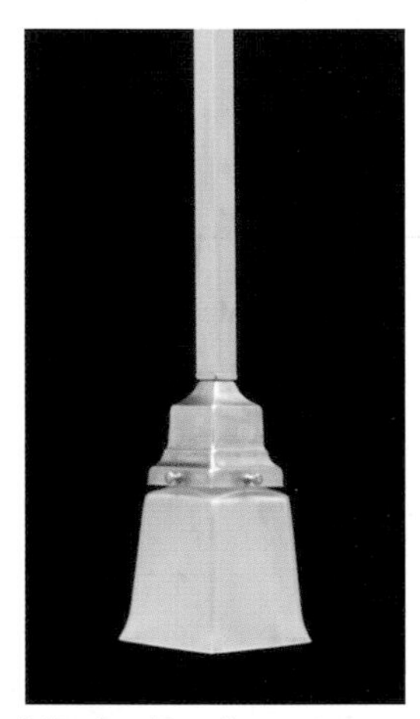

Mission Pendant
Renaissance

Hall Lantern, Bostonian
Renaissance

Antique Brass
Renaissance

Centennial Pendant,
French Brass
H.A. Framburg

Victorian pendant *Elk*

Arts & Crafts Pendant
Hubbardton

pot racks

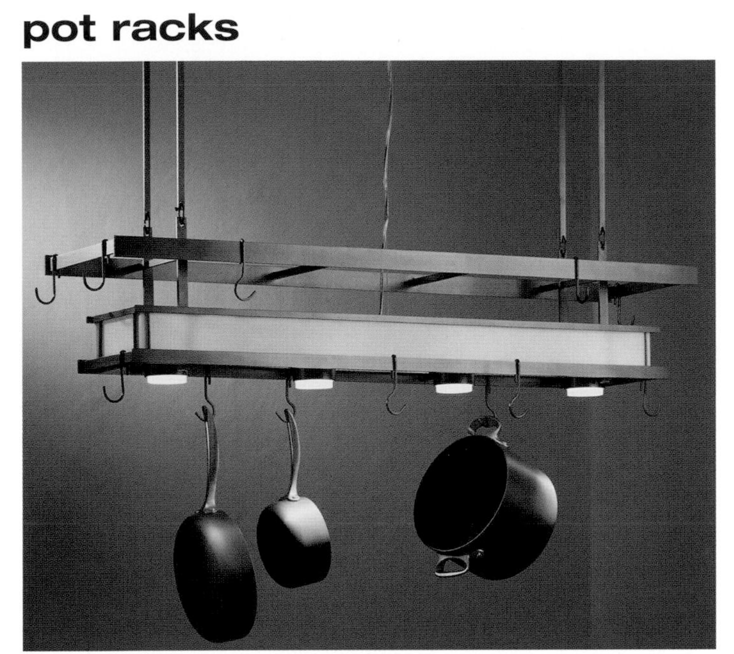

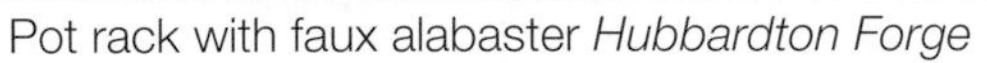

Pot rack with faux alabaster *Hubbardton Forge*

Island Light with pot hooks *Elk Lighting*

Pot Rack with 16 hooks *Hubbardton*

Pot Rack with smoked glass

Pot Rack with mahogany finish

wall mounts & sconces

Wrought iron lamps and a mahogany mirror from *Hubbardton Forge.*

Direct Wire wall sconce

Wrought Iron sconce

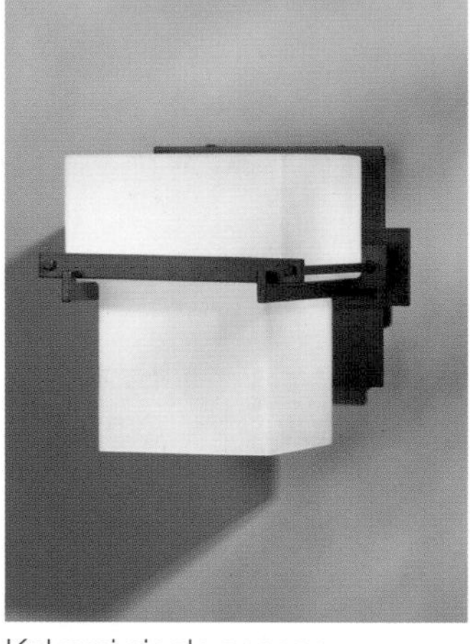

Kakomi single sconce

Halogen socket sconce

Plated Sconce *Hubbardton*

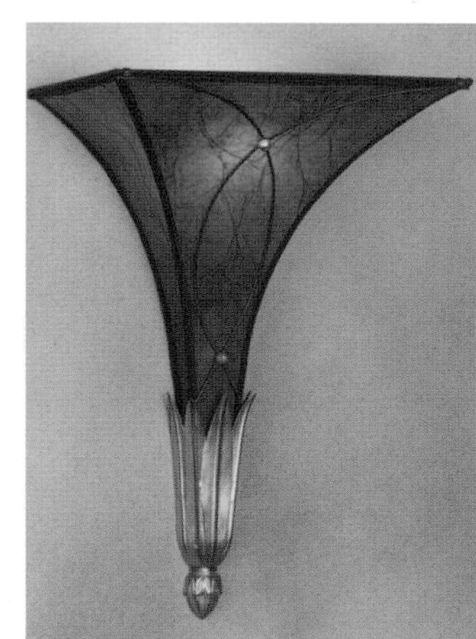

Valenciana *Elk Lighting*

Pearl Glass sconce *Elk*

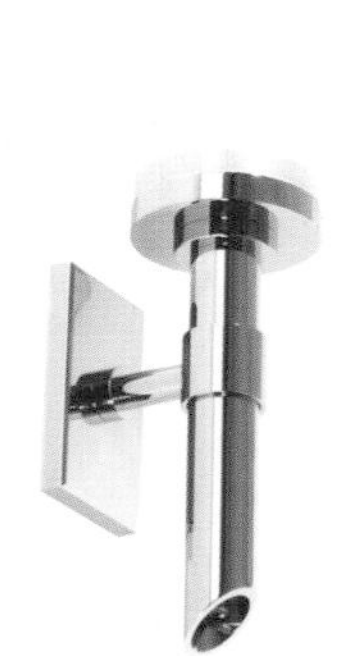

Wall sconce chrome and frosted glass *Waterdecor*

Wall mount with palms *Elk Lighting*

Moonband *Hubbardton Forge*

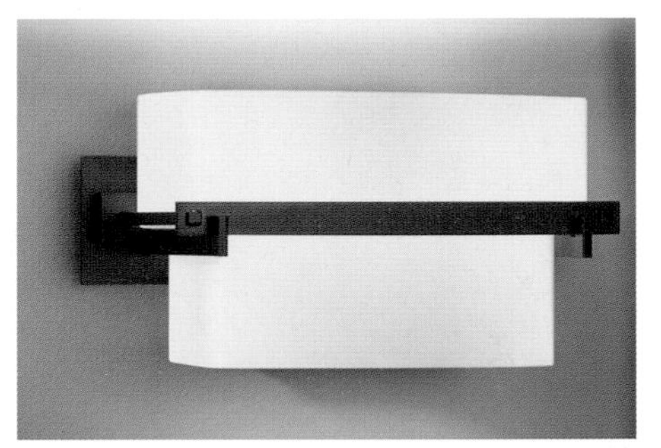

Kakomi Single

Orkje wall sconce, mat nickel *Eurolite*

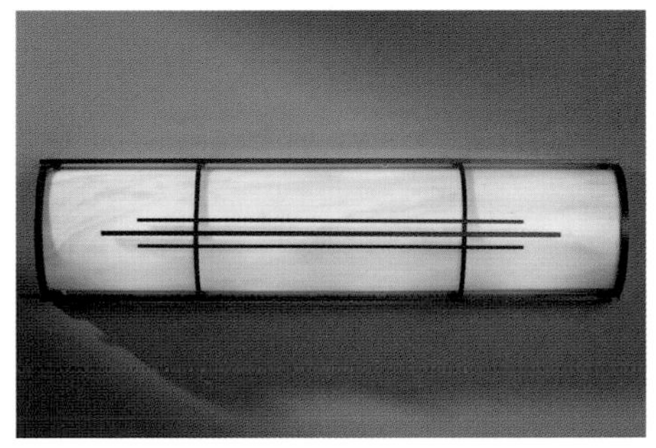

Bath Bracket *Hubbardton*

Forged Leaf and Stem

Metra 4-light *Hubbardton*

Hand-cut crystal on a wrought-iron frame *Crystorama*

wall mounts

Custom holder for flat glass shade *WPT Design*

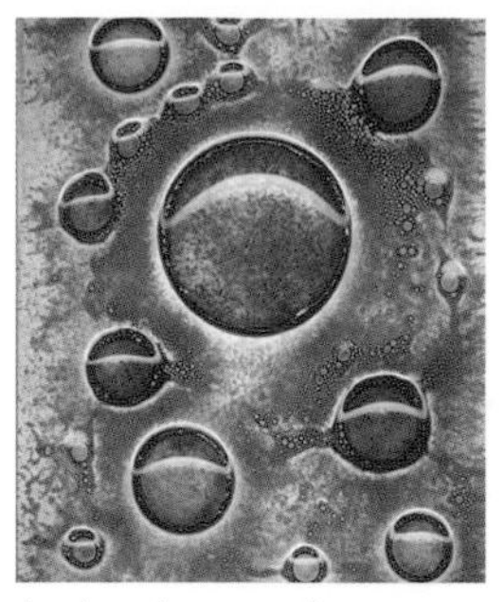
Amber Lemondrop

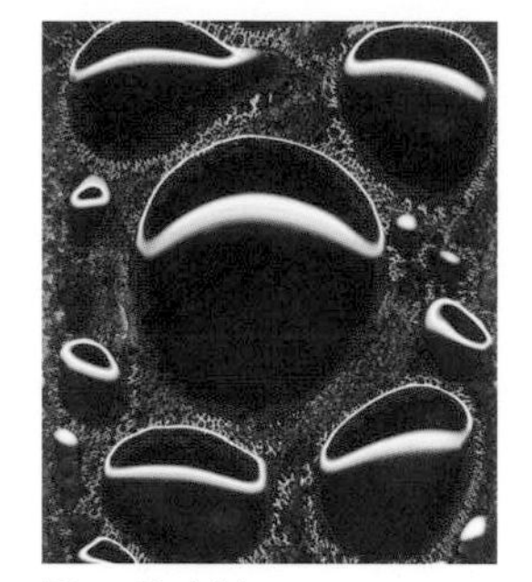
Blue Bubble

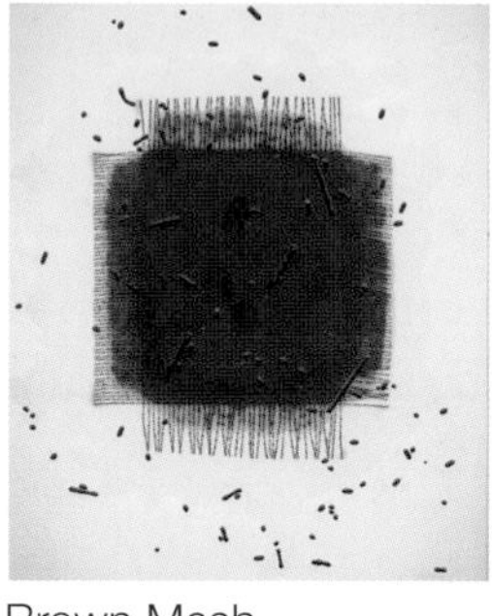
Brown Mesh

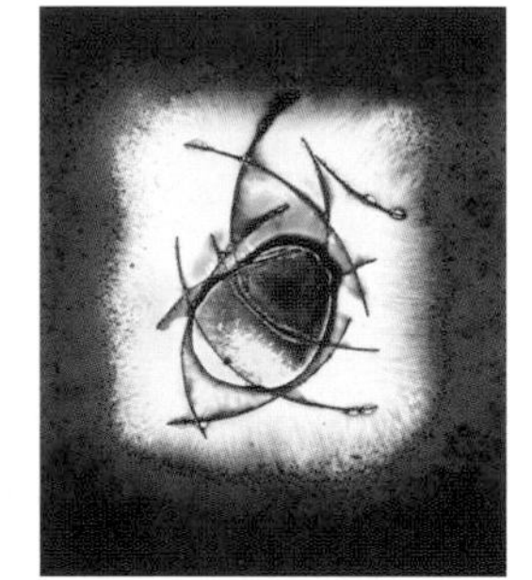
Barron

Honey Starfish

Gecko Frost

Jeweled single candle *Crystorama*

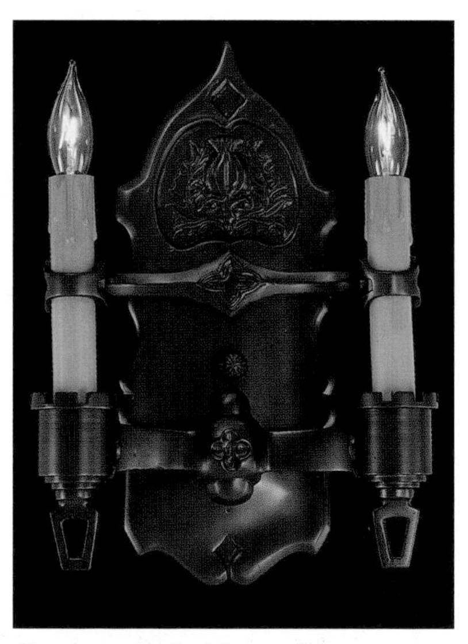
Pewter and double candle
H.A. Framburg

Double faux kerosene *Renaissance Antique Lighting*

wall plates

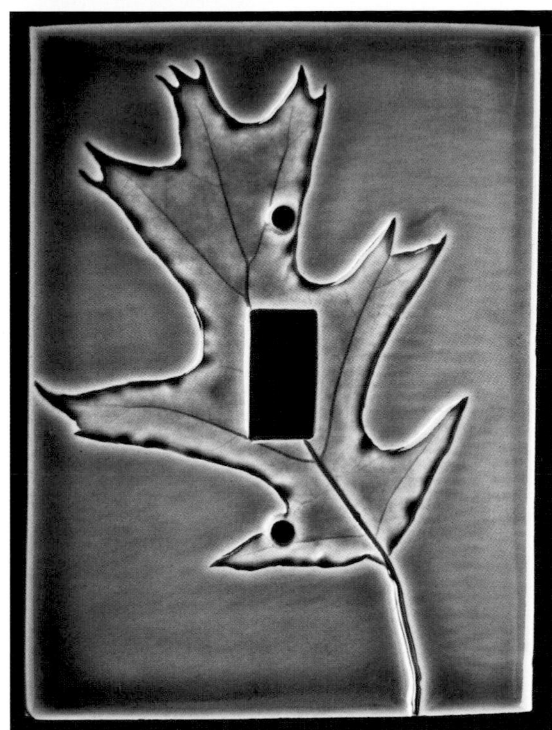
Oak Leaf ceramic light plate *Earth Marks*

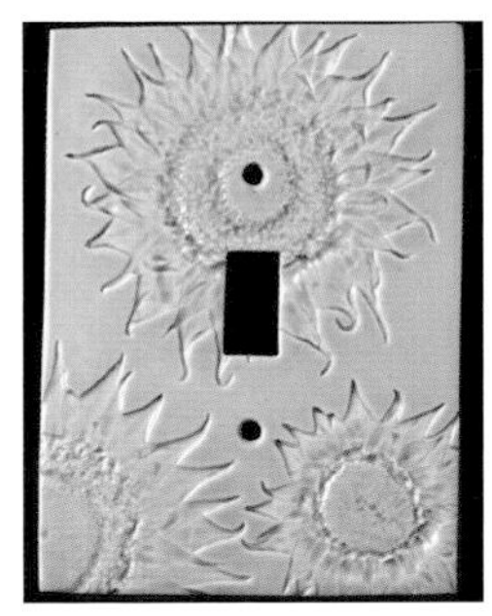
Sunflower *Earth Marks*

Daisy Double Light Switch

Rock star
Clay Squarod to Infinity

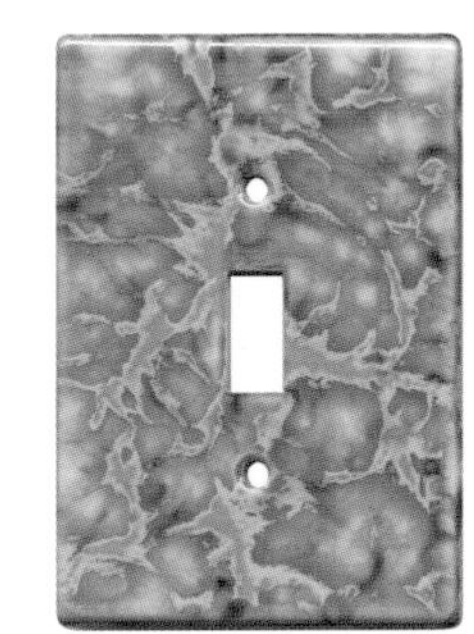
Spanish Moss

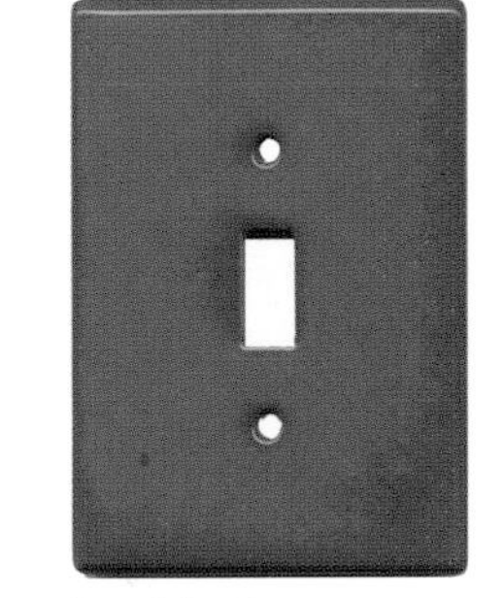
Steel Patina

An antique reproduction brass push switch from *House of Antiques.*

Metal switch plates and plug covers from *Acorn Manufacturing.*

wall mounts

Atlantis, extruded aluminum and tempered glass *Hinkley Lighting*

Atlantis, aluminum and satin opal glass

Aspen, solid brass and glass

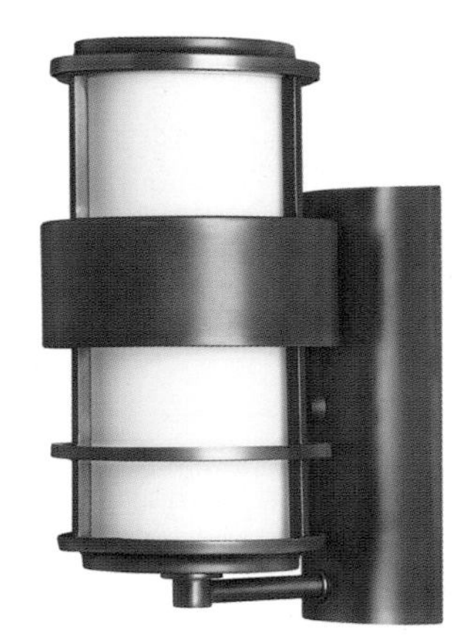

Saturn, solid brass

Dark Sky, solid brass

Outdoor sconce
Hubbardton Forge

Outdoor bracket, bronze finish
Kichler Lighting

Seaside

Modesto, bronze finish

hanging & post

Park Avenue, hematite and clear seedy glass *Hinkley Lighting*

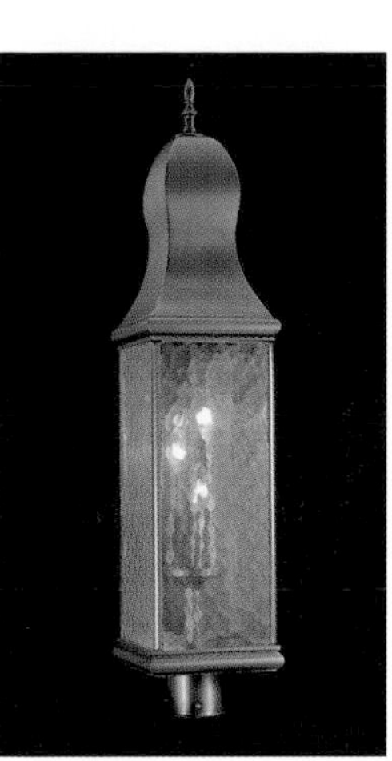

Marquis 9270 lantern, Harvest Bronze *H.A. Framburg*

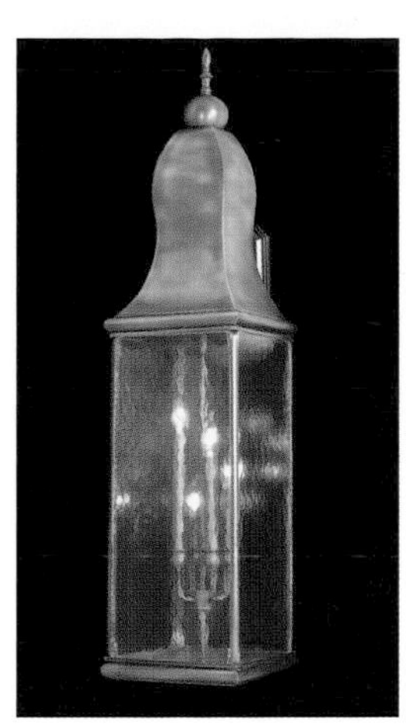

Marquis 9275 lantern Harvest Bronze

Marquis 9265 lantern, Raw Copper

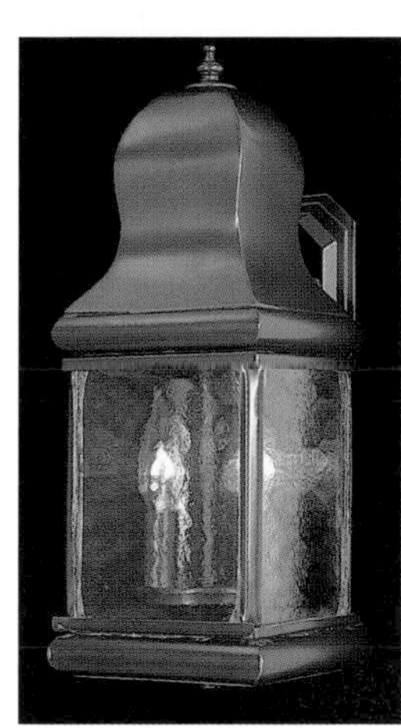

Marquis 9260 lantern, Mahogany Bronze

Kimora Japanese lantern *Fanimation*

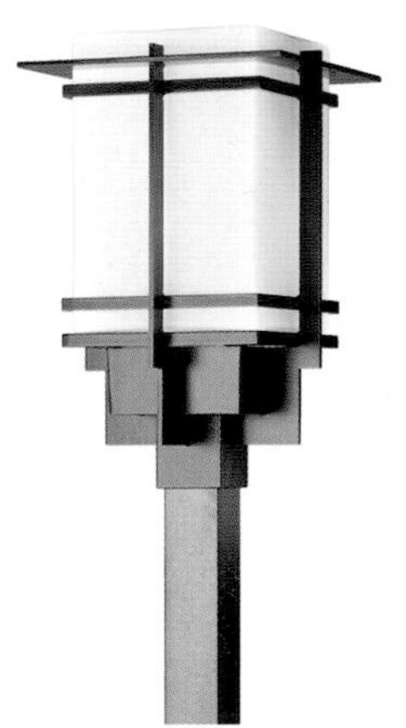

Tourou Post Light *Hubbardton Forge*

Tourour Bollard and Post *Hubbardton*

Aspen, solid brass and glass *Hinkley Lighting*

Plantation, metal and glass *Hinkley*

Alameda Pole, solid brass *Kichler Lighting*

ceiling fans

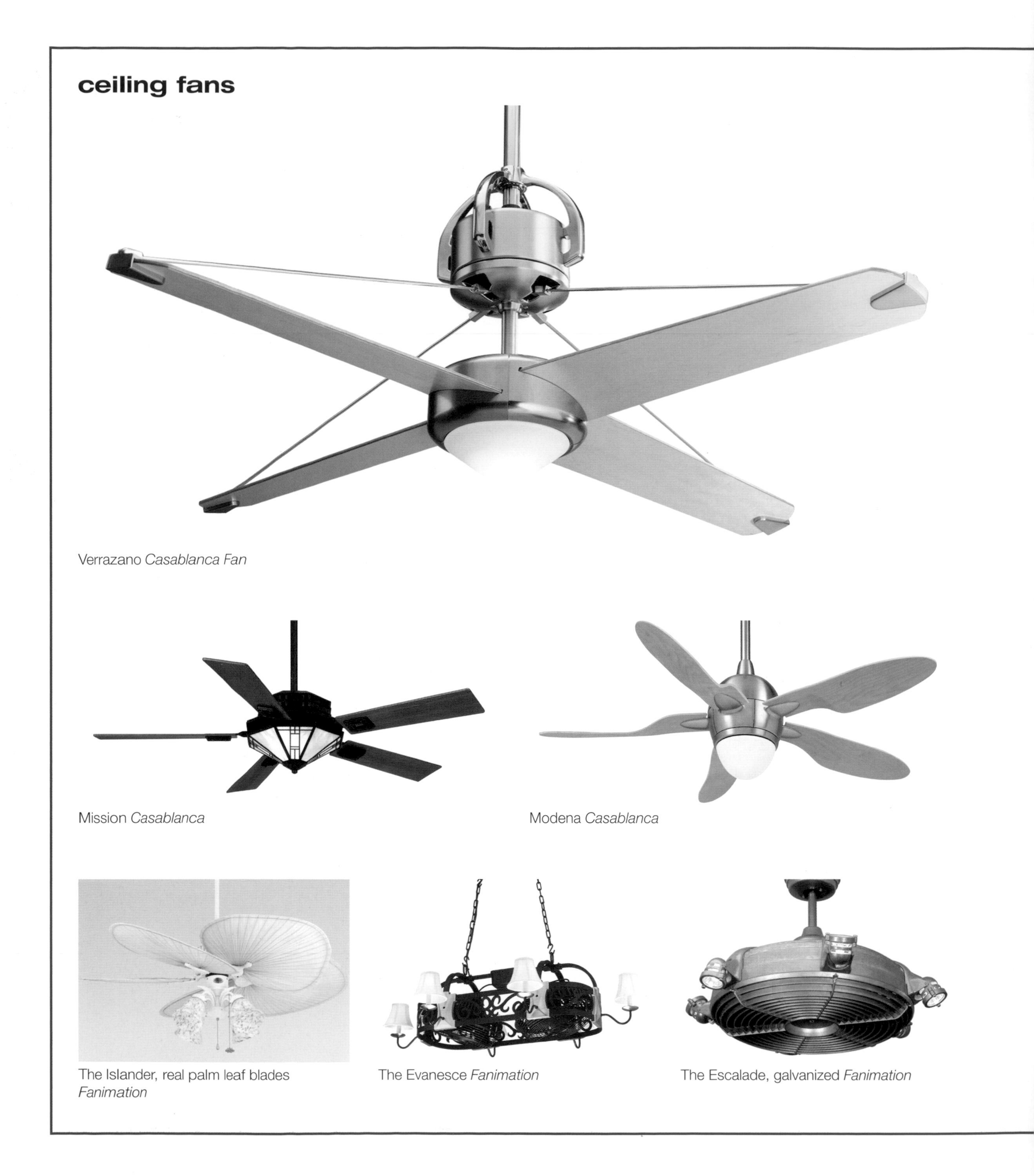

Verrazano *Casablanca Fan*

Mission *Casablanca*

Modena *Casablanca*

The Islander, real palm leaf blades *Fanimation*

The Evanesce *Fanimation*

The Escalade, galvanized *Fanimation*

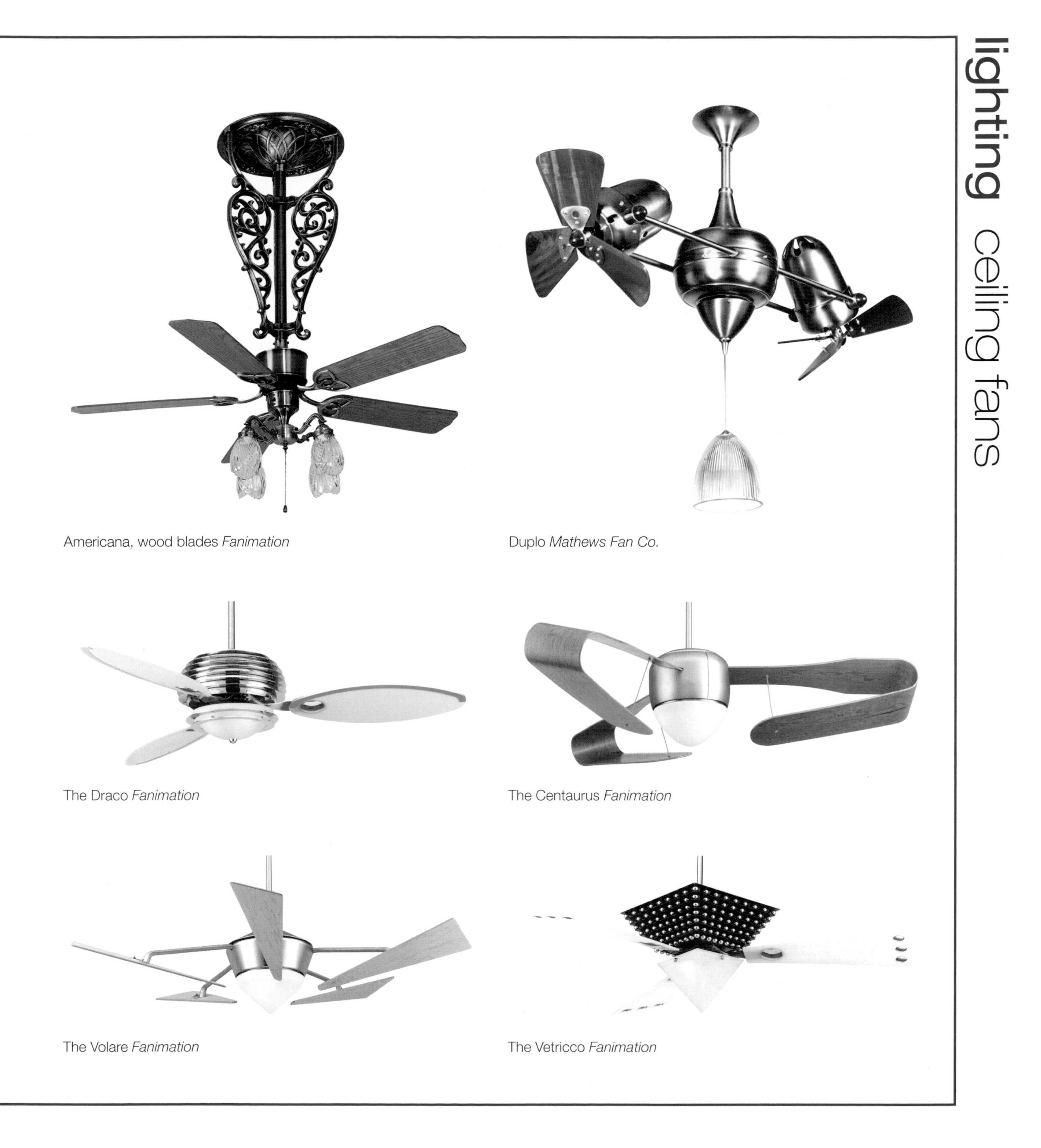

Americana, wood blades *Fanimation*

Duplo *Mathews Fan Co.*

The Draco *Fanimation*

The Centaurus *Fanimation*

The Volare *Fanimation*

The Vetricco *Fanimation*

gas

DV38 direct vent *European Home*

Vision gas with brushed surround

Natural gas and liquid propane log set

Pegasus 9603 direct vent gas insert
Blaze King

Contemporary direct rear vent with gold-plated doors *Blaze King*

A ventless gas fireplace means you can install it virtually anywhere.
Off the Wall Fireplaces

Burning Bush ventless
Off the Wall Fireplaces

Clarity ventless

Quote ventless

Flow Titanium ventless

electric

Taiko *Vermont Castings*

Regal Cove

Chateau Vista

Roman Rustic

Chateau

Sandridge 24 gas logs

wood-burning

Parlor with ceramic glass *Blaze King*

Princess catalytic insert with gold-plated doors *Blaze King*

Masonry heater
Green Mountain Soapstone

stone

ThinCast stone mantel *Balmer Moldings*

Louis XV gypsum cement mantel 4101 *Balmer*

Alexis, Estate Collection, cast limestone *Siteworks*

Elizabeth, Chateau Collection, cast limestone

Rothschild, Estate Collection, cast limestone

wood

Red Oak Leaf mantel *Richard Cornelius*

Aspen Bough Mantel *Richard Cornelius*

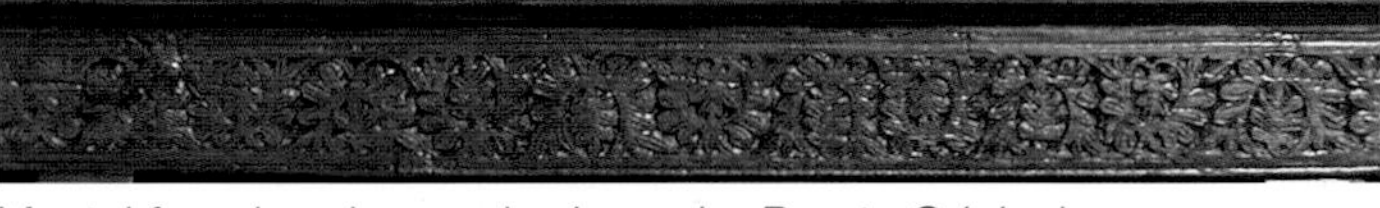

Mantel from hand-carved salvage *La Puerta Originals*

Mantel from hand-carved salvage *La Puerta Originals*

A mantel including surround made of concrete and recycled glass from *Vetrazzo.*

The Kensington, wood *Pearl Mantels*

The Olivia, wood

The Princeton

firebrick

Stone surround *Stone Forest*

Herringbone firebrick
Rhodes Architectural Stone

Firebrick close up
Rhodes Architectural Stone

kitchen fittings

The kitchen is the new heart of the home, the social center. People are now inviting their guests into their kitchens rather than the kitchen being hidden behind two doors that flop open. The preparation is now part of the experience – it's performance art!

LYNN APPLEBY, TORONTO DESIGNER

Designed by Lynn Appleby and implemented by Poggenpohl, this contemporary stainless-steel kitchen has floating shelves that keep upper cabinets from interrupting the wall plane — a divergence from traditional kitchen design.

THE EARLY EVOLUTION OF THE NORTH AMERICAN KITCHEN progressed something like this: from firepit to fireplace to wood stove and table, and then to the traditional stove, fridge and sink triangle with a tiny bit of counter space. Even if you were wealthy, the space was very basic and utilitarian because, after all, it was only the cook who had to work there. The 1960s saw eat-in kitchens as the first step toward a more open plan kind of home. This was often as simple as including a "breakfast nook," but also meant a larger kitchen overall with the added importance that increased square footage brings. This opening up of the kitchen and its connection to the rest of the house continued for the next decade or so, as did the introduction of time-saving devices such as microwaves and Cuisinarts.

By the 1980s and 1990s, some fancy touches had begun to show up in North American kitchens — work islands with granite countertops, cappuccino makers and custom cabinetry. The kitchen continues to evolve during first decade of the 21st century, with the most striking evolution being the one to the "professional kitchen," the kind with a six-burner gas stove, hanging pot racks and stainless steel surfaces of a sort previously associated only with the business end of a restaurant.

The modern kitchen has evolved into a showplace, not only for our increasingly sophisticated gourmet techniques, but for the glittering technology of that most sensuous of arts. For that you need some flash and style — you might choose elements such as luminous stainless steel appliances, rough slate floors that pick out the colors of warm imported woods in your cabinetry, composite plastics for your freeform sink, concrete this and cultured stone that. New and old varieties of tiles were combined with metals — copper, zinc and bronze. People have begun haunting the antique shops and stoneyards of Europe to find nontraditional — for kitchens, that is — materials to further personalize their kitchen spaces. We have discovered how much fun it is to juxtapose different materials: a limestone-topped island along with a tinted concrete counter; or a slate backsplash with hard maple counters — virtually anything you think works.

A state-of-the-art Christopher Peacock kitchen has an entry arch that is echoed in the stove nook.

The false ceiling in this kitchen is called a dropped soffit. It was used because architect Richard Landry needed a place to put the recessed task lights and the dropped soffit provided the solution, with indirect lighting installed above. The pot rack is hung from the joists above the ceiling, which is lit at night by the same indirect lighting. The task lighting can be toned down at night, with the overheads shining on the beams above.

The modern kitchen has a lot more going on than just food preparation. It has become a focal point of home life. It can be the after-school, after-work debriefing center, the homework hub, the Sunday morning reading-the-paper oasis, the Saturday night watch-the-game haven. Besides all variety of built-in appliances, from drawer dishwashers to warming drawers to wine fridges and countertop woks and grills, kitchen cabinetry can also house TVs and stereo systems, facilitate internet, satellite radio and cable connections, books and video games. You may not have the space for a full-fledged great room, but even just a corner with a cabinet for the entertainment electronics, a couch and a fireplace — or a pair of sofas and a fireplace to make an inglenook, or a window seat that looks out into the garden — can make it a room that you want to hang out in, not just cook and eat in.

Work islands, often with a second sink for making prep work easier while limiting trips to a main counter sink, are almost a necessity in most modern kitchens. They not only provide more workspace but act as a seating area for family or guests who want to be near the action. A raised section of counter on the island is also good for screening off some of the kitchen mess, especially if the work area is within sight of the dining room.

Islands, extended countertop peninsulas and even free-standing cupboards can also be used to replace walls as a way of delineating the kitchen space. This is particularly effective if you are creating an "open-faced" kitchen that is meant to be part of the family room, dining room complex. What you need is the mere suggestion of an area rather than solid, obtrusive walls that interrupt the feeling of inclusiveness. Other means to achieve a sense of openness while marking space are dividers such as shoji screens and glass walls, which can be set on rollers to screen off or open up the kitchen at will.

Opposite: An East meets West kitchen with rice paper and poplar shoji screens closable during dinner parties. The light fixture over the island is hung from stainless-steel rods. The counter is granite, the cupboards are anigre (a kind of maple only more mottled) and the floors are maple. Left: These luminescent walls are handmade paper from Thailand laminated on Plexiglas. Center: A clean, uncluttered look was achieved here in part by minimizing the number of upper cabinets (upper cabinets cast shadows and can make it feel less like an open room). Translucent glass in the cabinets also contributes reflectivity and makes the cabinets less obtrusive. The floors and cabinets are maple with a clean finish. Right: Putting the oven in the wall instead of under the stove burners allows for under-the-stove pull-outs for the orderly and convenient storage of pots and pans. This model is stainless steel. Overleaf: This kitchen countertop, light fixture and more — are all Corian. The floor is mahogany that has that has been stained chocolate. The mirror on the right was installed in order to have the room reflect back at you as you enter, instead of looking into the column.

sinks

Another kitchen necessity about which your thinking can be expanded beyond the traditional is the sink. Your old double sink can be replaced by two or even three purpose-oriented sinks, with perhaps an articulated spigot over the stove to fill the pasta pot. If you do want a double sink, the two sinks can be of different shapes and sizes. Your sink can be much bigger and more elaborate than you might be used to. Sinks fabricated out of a material such as stone or metal can be extended with a matching drainboard built into the countertop all of a piece. Old-fashioned apron-front sinks, some with elaborately carved fronts, are also good for that European country kitchen look.

There are three basic ways to mount a sink: overmount, undermount and integral.

OVERMOUNT (SELF-RIMMING) The simplest way to install a sink: an overmount sink is designed to just drop into a hole in the countertop. Because the sink rim protrudes above the countertop, it has to be sealed at the edges and is a little harder to clean.

UNDERMOUNT Installed from below: this approach gives a clean look to the countertop and allows you to wipe debris from the countertop right into the sink.

INTEGRAL Most often seen with manufactured surface sinks but also with natural materials such as granite, the integral combines the sink and the countertop in one continuous piece, sometimes with a built-in drainboard.

You certainly have more choice in materials for kitchen sinks today than your great grandmother did, and if you look around, you may find even more creative vessels than the options described here.

CAST IRON Big, heavy and usually old-fashioned looking, these sinks are made of molten iron cast into a shape and coated with a shiny protective layer of porcelain (white or colored). These are best for holding water temperature but murder on glasses or crockery if dropped.

CERAMIC (SOMETIMES KNOWN AS FIRECLAY) Looks the same as cast iron but is not quite as strong. Because of the manufacturing process, ceramic sinks can be sculpted into elaborate shapes. Small ceramic sinks are often used in places such as the butler's pantry or the bathroom, or as a secondary sink on a kitchen island. For a delicate-looking alternative, try a hand-painted and fired ceramic sink.

STAINLESS STEEL The most modern look of all, easy to maintain and the best for hygiene and resistance to corrosion. It will scratch, however, but that will lesson its tendency to show up fingerprints. Pick a heavier gauge (16 to 18) to decrease noise.

STONE Looks good with anything and is durable. Stone sinks can be part of an integral sink and drainboard. Apron sinks are often made of stone, sometimes carved. Granite, marble, soapstone and onyx are the most common materials for stone sinks.

MANUFACTURED SURFACE Often made of acrylic polymers, but these sinks can be made of almost anything the manufacturer thinks will work. They have a continuous, homogenous look to them and can be an integral part of the countertop.

GLASS As in other kitchen and bath fittings, glass is the hot newcomer, although you won't find many kitchen sinks of this material because even tempered glass will break if a cast iron soup pot is dropped into it. Until recently most often used in vessel sinks, glass is now used in almost every kind of sink.

A concrete countertop complements a rustic French country sink and terra cotta floor tiles.

The elegant lines of an old-fashioned high-neck bridge faucet are reminiscent of a stately tea service, suitable for a French Country kitchen or a more formal Georgian setting. The honed marble makes a perfect backdrop.

kitchen faucets

Faucets are either "wall" or "deck-mounted," meaning they either sprout out of the wall above the sink, or are attached to the countertop or sink itself. A faucet can either be a "single hole" (with the spout and hot and cold controls all housed in the one unit), "centerset" (a hot and cold control with a spout in the middle, all sitting on a single base), or "widespread" (a hot and cold control separated by a spout in the middle, all of which are separate). A spout that is connected to the taps above the level of the counter is called a "bridge." With farmhouse or apron sinks, the faucets sometimes have to be wall mounted, usually in the backsplash. All of these configurations may require different hole arrangements in the sink, so make sure your fixtures match your sink. The main thing to remember is that single-hole faucets are more stylish, but widespread, especially those with levers, are the most practical, especially for children and seniors, because they are easier to manipulate.

High-arch or gooseneck spouts are terrific for filling up large pots and big double and triple sinks, but can be too large for the scale of a smaller kitchen. The higher the arc of the spout, the more elegant these faucets look, which makes them a welcome accompaniment for traditional-looking kitchens but not so much for contemporary ones. Some faucets have a built-in spray that detaches from the faucet; matching separate spray and liquid soap dispensers are available for most faucets.

Brushed or matte finishes are easier to maintain because they don't show scratches or water spots as much as polished finishes.

Pot fillers: Cold-water-only spigots that come out of the wall over the stove burners are designed to fill large pots of water for boiling. The newest versions have valves at the side of the stove rather than on the handle so you don't have to put your arm over the hot flame or element to turn it on or off. Another innovation is an articulated joint in the middle of the spout to increase flexibility or, even better, a spout that is not fixed but swings out to fill and back to rest against the back wall, out of the way when not in use.

cabinets

Kitchen cabinets, even more than countertops, flooring and appliances, define the style of your kitchen. Think of them as furniture hung on the wall. Cabinets will probably account for up to half of what you spend on your kitchen, and they are also the investment that will receive the most use. You therefore want them to be a happy combination of quality, function and style.

As can be seen from the products featured in this book, which include only a fraction of what is available, cabinet styles and finishes are almost limitless, encompassing everything from wood (plain and molded) and glass to metal and even woven mesh. Wood or laminate cupboard doors are of three basic kinds: raised panel, recessed panel and the smooth front, slab panels that are the most

Gothic cabinetry highlights 19th-century kitchen with beadboard paneling. The arch for the clock was specially made.

appropriate for Contemporary kitchens. In terms of construction, however, there are essentially two kinds: face-frame and frameless. In face-frame cabinets, the open front of the cabinet is faced with a frame, and the doors and drawer fronts sit on that frame. The doors are smaller than the frame of the cabinet and the hinges, visible at the side of the door, are mounted on the front, face frame. This means of construction has a more complex, traditional look. Frameless cabinets usually have drawers and doors that cover the entire box of the cabinet for a cleaner, more European look. The hinges used in frameless construction are most often mounted inside the cabinet and invisible from the outside.

As elsewhere in your house, in your kitchen you can mix and match cabinet fronts and finishes to good effect, contrasting solid below-counter cupboards with glass-fronted top shelves, different kinds of wood top and bottom, and even totally different materials such as wood and metal.

The height of the cabinets is also variable. Above-counter cabinets can stop a few inches or even a few feet from the ceiling. This not only lets light in but creates a feeling of openness, something that is particularly important in small kitchens. This top space can also be used for decorative objects such as plates or baskets. Of course, cabinets can also continue all the way to the ceiling, to be finished off by crown molding (the traditional approach) or reveals (a more contemporary look) where they meet the ceiling. Some kitchen designers feel that wall cabinets clutter up the walls and will reduce or even eliminate them entirely, relying on below-counter cupboards or an adjacent pantry for storage. The bottoms of most cabinets have either a recessed "toe space" or a wide base (the wider, the more European the look) called a "plinth."

A plain alternative to cabinets and cupboards with doors is the open shelf. Simple and inexpensive, open shelves can give a kitchen a decidedly more

Left: Cupboard fronts of metal mesh. **Center:** Semicircular notches instead of hardware contribute to the clean, streamlined look of these kitchen drawers. **Right:** Ebony inlay and slender steel hardware give depth to these light satinwood drawers.

This guest house kitchen is made of economical materials but does not scrimp on style. The kitchen cabinets are maple plywood. The floor is concrete that has been scored to prevent cracking. The ceiling is made of Tetun, a brand name for aspen wood that has been shredded and bonded with Portland cement. This material was used because it absorbs sound, unlike hard surfaces such as drywall, concrete and glass.

informal flavor and are good for displaying china, glassware and decorative objects. However, the downside is that all that open space can be less than streamlined when it reveals the clutter that cabinet doors are meant to conceal. A general reluctance to have a less-than-beautiful collection of half-empty cereal boxes on display is one reason people shy away from glass-fronted cabinets. Frosted glass is a good compromise here, or you can have opaque fronts that are relieved by one or two glass fronts for anything you would like to display, such as special glassware or dishes.

When deep, "pot" drawers were introduced in kitchens to allow easy access to bulky items, including pots and pans, they were welcomed with open arms. Today pull-out shelves, concealed by a regular cabinet door and available in a variety of configurations and sizes and for many different purposes, give the same access to the backs of cabinets, although they do involve an extra motion — opening the cabinet doors and then pulling out the drawer. Any pull-out or pot drawer that will see heavy duty should have the hardware that the drawer slides on (called a "glide") mounted on the side rather than near the bottom for greater strength. The more the glides extend, the more storage space you have. Other innovations in hardware for cabinet doors include spring-loaded hinges that pop open with a push — perfect if your hands are full — and even step-on openers, for slide-out garbage and recycle centers.

The key to a good kitchen is the placement of the classic triangle of work stations: cooking area, prep area, clean up area. The danger of a large kitchen is that you can spread out these functions too much.

MICHAEL FULLER, ARCHITECT

Toronto designer Dee Chenier went to great lengths to make her narrow Victorian home work. The cabinet that separates the kitchen from the hallway is set away from the wall by three or four inches so it appears as a free-standing cabinet while letting more light in. To save on space, the smallest possible appliances were used, and the cupboards to the right have been recessed six inches into the wall, with "layback" hinges, so they open flat against the wall. To give maximum space around the fridge and ovens, the island does not extend in front of them.

A lot of people think that stainless steel is cold, but it's not. It's only after you get over the first use of it — because it does scratch — that it gets this beautiful kind of patina to it that reflects the warmth and color of the furnishings and other materials around it. But of course people are getting over the idea that everything has to be new and shiny and not show wear — they are beginning to understand that wear is a good thing

BARTON MYERS, ARCHITECT WITH AN APPRECIATION FOR OFF-THE-SHELF METAL

countertops

Since, architecturally speaking, the kitchen is all about work surfaces and practical considerations, the finishing materials you use and how you use them are the key to its success. With so many different materials available to create your countertop, you may find some difficulty in choosing just one. As it turns out, it's sometimes a good idea to mix materials for different work stations in the kitchen: matching stone with stainless steel, for instance, or butcher block with concrete and slate.

Natural stone is still one of the best materials for countertops, as well as floors and even sinks. It is the choice that gives the greatest sense of pure luxury as well as a timeless style that never goes out of fashion. Although they look indestructible, stone countertops are actually not. In fact, for all its hardness, stone is somewhat porous and must be sealed to protect it from stains, especially from acids. Stone comes in different finishes: polished for that shiny, two-feet deep look; honed, for a smooth, soft matte finish; and flamed, or blow-torched for more textured look. Following are the most usual materials used for those most important of kitchen surfaces.

GRANITE A perennial choice because of its dazzling look (due to the crystals of quartz, mica and feldspar trapped in its matrix) and durability. The polished look can be spectacular, but don't overlook the matte finish appeal of honed granite, which also shows wear less than polished. It should be sealed.

MARBLE Sensitive when it comes to stains, marble has to be sealed even more often and more carefully than granite. You can't beat its smooth appearance and rich hues, though. It is everyone's favorite for rolling out pastry dough, which is why some people have a special marble counter section for just that purpose, no matter what material the rest of the counter is made of.

SLATE If it's good enough to spend centuries on your roof, it should be tough enough for your kitchen. Extremely durable and stain resistant, slate is softer than granite but less porous. It comes in dark colors such as green, gray, black, dark purple and red and is often used right out of the quarry with rough edges and even drill holes showcasing the honest nature of the stone. Getting a long, thick slab or two and using it in continuous pieces for the counter is the key.

SOAPSTONE Not as soft and porous as you might think. Acids don't bother it and stains and nicks can be sanded out, which is a good thing, because soapstone chips easily, especially around the edges. It comes in various dark shades and strata, but can be darkened even more by rubbing oil into it. You can even buy soapstone sinks and drainboards.

LIMESTONE Looks weathered and much warmer than marble or granite, but stains very easily and so is high maintenance. Try Jerusalem limestone for a hardier variety. Limestone's suitability depends on its quality. Here is an easy way to test the porosity of unsealed limestone: Put a few drops of water on the limestone and wait about ten minutes. Use a clean towel to wipe dry. Superior quality will be completely dry in less than one minute. Poor-quality limestone absorbs water and remains wet for an hour or more.

LAVA STONE Lava rock glazed with enamel in any color you want and fired in a kiln. What emerges is an extremely hard, crackled surface sold under the French brand name Pyrolave.

NATURAL QUARTZ As durable as granite and not as susceptible to staining, but without the irregularities that cause some people to feel granite is busy.

ENGINEERED QUARTZ Science never sleeps and one of its latest marvels is engineered quartz — essentially, real quartz that has been ground and mixed with resin to provide uniformity of color and pattern. It is heat, stain and scratch resistant, doesn't have to be sealed and comes in polished, honed and sandblasted finishes in either solid colors or a particulate mix. It is very low maintenance. Most engineered quartz is 90 percent stone and is uniform in color. Some brand names are Zodiaq Stone, Silestone and Cambria.

STAINLESS STEEL The workaday look of stainless steel used to be reserved for the no-nonsense world of the professional kitchen. Recently, however, this most durable and forgiving of materials made the leap to residential kitchens. The benefits are heat, water and stain resistance that is second to none, plus it is one of the more hygienic materials you can choose, which is why it is now used on everything from countertops and backsplashes to cabinets and appliances. (A study by the Hospitality Institute of Technology and Management determined that stainless steel was the easiest of all materials to make microbe free.) You can warm up the colder aspects of the material by contrasting it with wood or tile in proximity. Stainless steel can also be patterned or ground into different swirls and textures. Avoid the highly polished version and opt for the easier-to-maintain matte finish that will minimize fingerprints. Stainless steel is often attached to plywood to give it strength and deaden its sound, and is fairly expensive to fabricate. Backsplashes come in various patterns.

Left: Honed white Carrara marble and cupboards and an island of quarter-sawn white oak stained to match the antique limestone floor provide an Old World warmth to this contemporary kitchen. **Center:** This two-piece custom countertop is a French product called Pyrolave and is made of orange-glazed volcanic stone. **Right:** A stone country sink with matching drainboard works well with the more contemporary oak cupboards and stainless steel pulls.

ZINC Reputed to have at least as good antimicrobial properties as stainless steel, zinc has been used to make oyster bars for years, as well as all of those cafe counters in Paris. Originally shiny, it quickly takes on a dull patina much like pewter. Although it will cut, stain and tarnish, this is considered part of its patina. Repolishing to remove the tarnishing, and resanding to remove stains is always a possibility. Zinc backsplashes are often quilted.

COPPER Copper surfaces take on the same kind of patina that zinc does — only sooner because copper is even softer. When it is used inside, you won't get the green verdigris you do outside.

LAMINATES While lacking the luxury image of stone or the hip allure of metal, laminate countertops (two or more materials bound together and manufactured) have many pluses, including their relative low cost, their ease of installation and maintenance, and the wide variety of color and pattern options. You can even get images — clouds or bricks or flowers — imprinted onto the surface. Another advantage is that the color and pattern runs all the way through, lessening the damage of scratches. Formica is the brand name of one of the first and most popular laminates.

MANUFACTURED SURFACES Manufactured — or solid-surface — counters are made by blending acrylic polymers and stone-derived materials. One of the most popular brands is Corian. Manufactured by DuPont in over 100 colors, Corian is durable, easy to clean and doesn't need to be sealed. It does cut, however, and needs to be protected from high heat.

CONCRETE Concrete can be tinted to just about any color, although there will be a certain amount of variation in any one particular piece, which proponents see as a plus. It is as durable as you would imagine, although it is very porous and does have to be sealed, while excessive heat could discolor the sealant. (For that reason, some manufactures advise having raised metal ridges installed in the concrete as a place to put hot pots and pans.) Cracking can be a problem, but small hairline or surface crazing is considered a character feature, just like veins in stone. Although it has a natural matte finish, concrete can come with both a gloss and high-gloss seal, which can be augmented with periodic rubbings of butcher's wax. The color often deepens into a burnished glow over time. One of its best attributes is that it can be formed into virtually any shape, making it a fun material, especially when it comes to edges. You can also embed objects such as tiles or shells in the surface. There is even a product called Vetrazzo on the market that embeds glass in concrete countertops to dazzling effect.

CERAMIC TILE Ceramic tile is beautiful and offers an almost endless palette to choose from, but is hard to clean and to roll out dough on it. Use glazed tiles that won't stain (but will scratch over time), but seal the grout or the stains will show there. (For more information, see the Tiles section in the Floors, Walls & Ceilings chapter.) Tiles often lose out as a countertop choice to materials such as stone and manufactured solids that provide a cleaner look because they have no grout lines. Tiles are still a common choice for backsplashes, however.

The island in this kitchen has three distinct parts to it, made of honed black granite and butcher block maple. The lighting consists of two pendant lights as task lighting plus recessed lights in ceiling. The floor is stained santos mahogany.

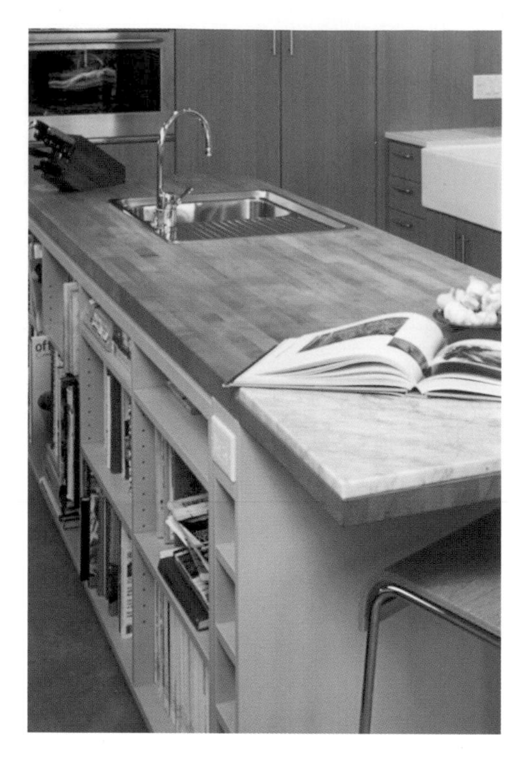

Left: This maple countertop highlights a simple and inexpensive kitchen of wood and steel. The floor is polished concrete. **Center:** Ceramic tiles embedded in a concrete countertop exemplify the accommodating nature of concrete. Other material can also be sunk into this material, such as shells, wood, metal and glass. **Right:** Different shades of amber tile on a checkerboard countertop.

GLASS Glass is finding a new life as a gleaming and easy-to-clean countertop surface, but it does scratch. You can purchase translucent colors, which are produced by tinting the back of the glass. Opaque colors are produced by adding a special opaque coating to the back of the glass.

WOOD Always a solid choice, wood is also good for its softening effect on the colder materials such as stainless steel. Although the preferred cutting-board material is hard maple, you can also make countertops out of furniture-grade woods such as cherry, teak and mahogany — but you will either have to treat it with kid gloves or get to like the stained and nicked patina. The wear marks in wood can be more easily fixed than in most countertop materials, and because there are so many different varieties, you can be very creative in coming up with custom designs.

PAPER-BASED One of the lesser-known products on the market is a countertop made of paper that has been treated with resin, then pressed and baked to create solid sheets. Reportedly more durable than hardwood and coming in a variety of muted earth tones, it is designed to provide a soft look and ambience to the kitchen. It is heat resistant and sanitary and, like most materials, is resistant to stains, but prolonged contact with acids and other strong solutions will stain it eventually. Stains can usually be cleaned up with normal commercial cleaners, while refinishing can deal with particularly bad stains and nicks.

exhaust hoods & fans

Exhaust hoods and fans perform the practical functions of getting rid of the smoke, odors and excess moisture that are the byproducts of cooking, while providing task and heat lighting over the stove. But today's architectural exhaust hoods are also echoing the evolution of the kitchen itself in becoming design elements in their own right. One way to focuses attention on the hood is to build the stove and hood into an alcove. Hoods can be made of sculptured steel or copper, or as inserts in blocky wooden structures with molding that complements the style of cupboards and walls, or in spectacular versions fashioned from cement and cast stone.

There are several basic kinds of exhaust systems, including the under-the-cabinet wall-mount hood, wall-mount chimney style hood, island hood, and pop-up or surface-mounted downdraft vents.

The under-the-cabinet hood is the one you see in most kitchens. Usually fitted between two wall cupboards and jutting out over the stove, it is generally made of sheet metal or aluminum. Small and utilitarian, it is unobtrusive and often the least costly hood choice, but there are upscale models of better quality available. The chimney style of hood is usually built right into the wall over the

This clean, colorful Italian-designed kitchen by Valcucine is dominated by cabinets made of high-gloss lacquer (Artematica Red Gloss) on an aluminum frame. The exotic looking exhaust fan is a Valcucine design manufactured by MaxFire; above it is a curved glass shield that contains and conveys kitchen vapors into the hood.

The stove hood for this Arizona home was custom made from metal lath frame that was then plastered. The countertop is concrete, as is the floor.

stove. Sometimes it is stand-alone but it can be built up to form an alcove around the stove, which is excellent for ventilation purposes because it has a better draft to pull up the cooking vapors. Island hoods, which are suspended over an island cooktop in the middle of the room, are subject to cross drafts that can interfere with the draft. A good rule of thumb is the wider the hood, the higher up you can place it over the cooktop. Be sure to get a hood and fan that are suitable for the kind and size of range or cooktop you have. Those with fans under 600 CFM (removing up to 600 cubic feet of air per minute) are not recommended for professional-style ranges. (The average under-the-counter hood operates in the 175 to 250 CFM range.)

The less common pop-up variety is for people who don't like the look of big hoods but want some kind of exhaust system. The pop-up vent is discreetly built into the stove back, usually protruding only six to twelve inches above the cooking surface. (Some they even retract into the stove back to disappear entirely when not in use.) The surface-mounted kind is flat and built right into the cook surface, often by a grill. The drawback to this type of exhaust fan is that they are generally not as effective as the hood type.

Although primarily functional, a hood, especially a chimney-type hood, can be faced in material (everything from tile, steel and wood to adobe and brick) to make it blend in with the rest of the kitchen (see above). Some alcove hoods are so built up they form an eye-catching architectural element all on their own.

products

kitchen fittings

granite

Blue Polar Volga *Genesee Cut Stone & Marble*

Academy Black

Almond Mauve

Azul Platino

Bordeaux Dark

Blue Marinace

Brown Cocoa

Brown Santa Fe

Café Bahia

Crema Azul

Costa Esmeralda

Delicatus

Giallo

Ghibli

Green Marinace

Giallo Vincenza

Juparana Colombo

Juparana Juliet

Kashmir White *Genesee Cut Stone & Marble*

Labrador Antique

Lilas Gerais

Lac Du Bonnet

Mountain Green

Nevada Gold

Uoro Black

Pacific Blue

New Imperial Red

Piracema White

Pokarna Green

Rainbow

Red Marinace

Rockville Beige

Tan Brown

Verde Fontaine

Verniz Tropical

Violetta

Waterfall Green

Yellow Juparana *Genesee Cut Stone & Marble*

marble

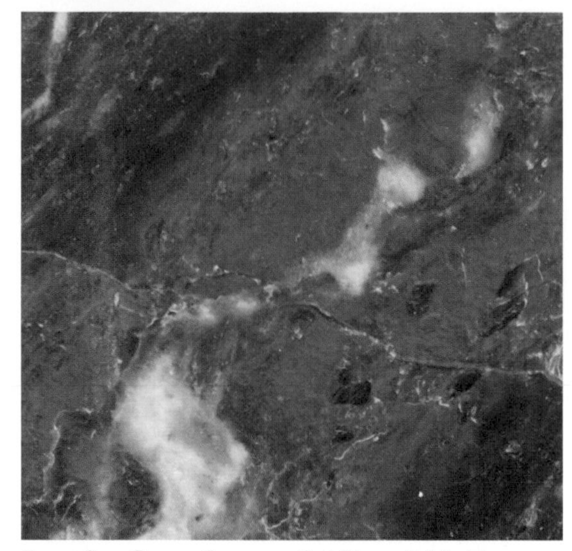

Deep Sea Green *Genesee Cut Stone & Marble*

Afyon White *The Architectural Stone Co.*

Aqua Marina

Arco Iris

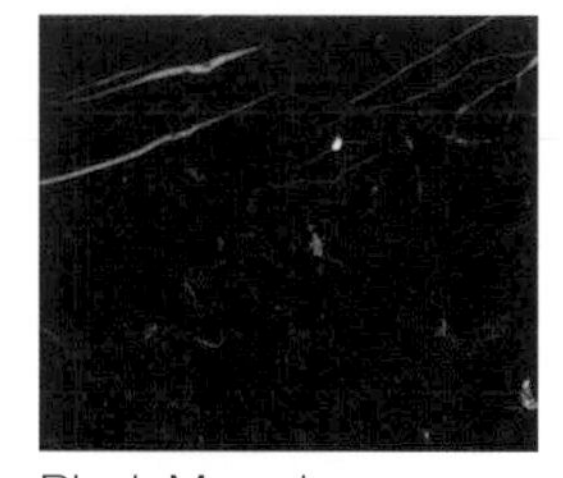

Black Marquina *Genesee*

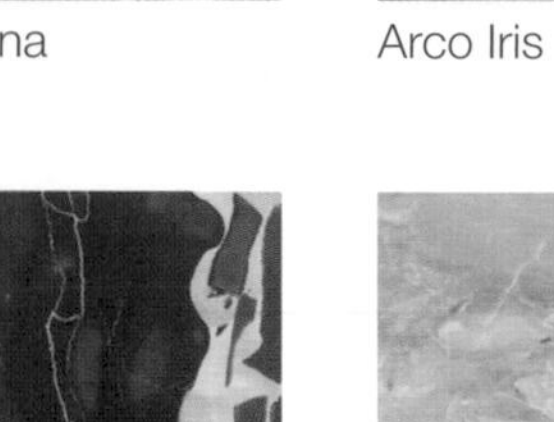

Black & Gold Italian

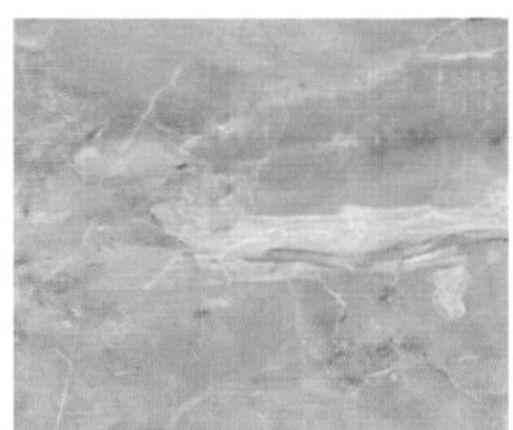

Breccia Oniciata

Dakota Mahogany

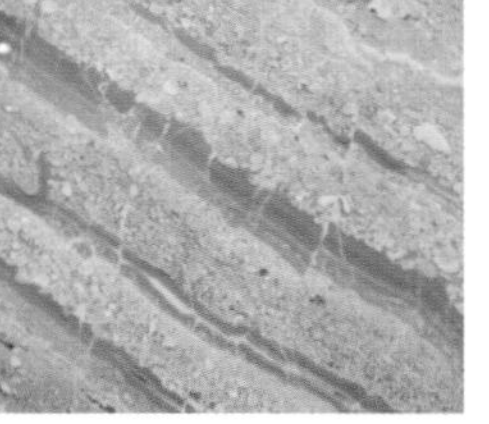

Diana Reale

Emerald Green

Fior Di Pesco

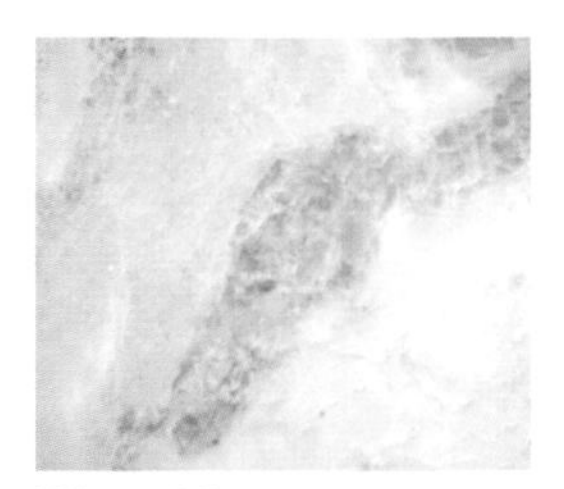

Wispy Mint

travertine

Navona Travertine

Classic Roman

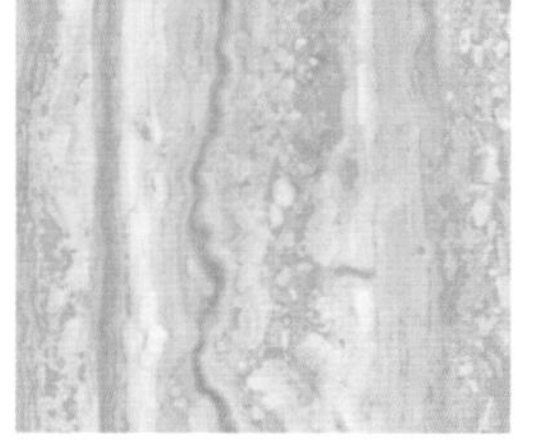

Silver Travertine

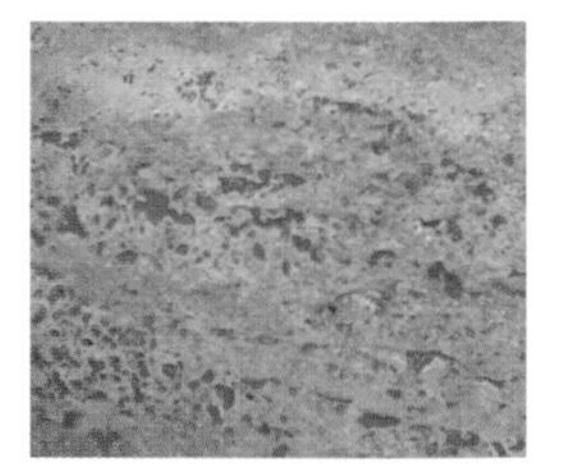

Noce Travertine *Architectural Stone*

onyx

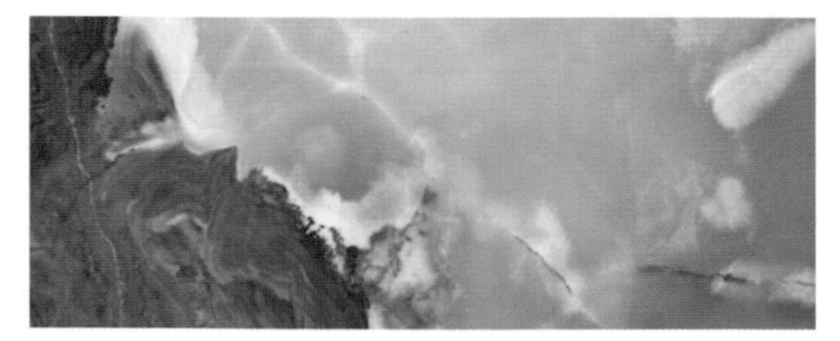

Classic Green Onyx *Architectural Stone*

Multi Brown Onyx

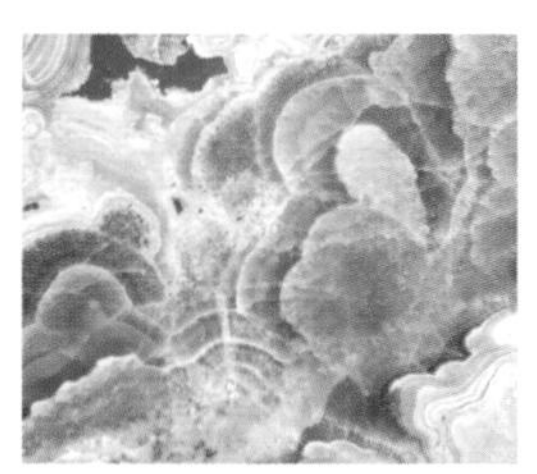

Verde Esmeraldo

Fantasy Onyx

limestone

Buff Indiana Limestone Bush Hammer Finish
Genessee Cut Stone & Marble

Buff Indiana Limestone
Machine Tooled Finish

Buff Indiana Limestone
Polished Finish

Buff Indiana Limestone
Diamond Sawn Finish

Buff Indiana Limestone
Honed Finish

Buff Indiana Limestone
Shot Sawn Finish

Gray Indiana Limestone
Machine Smooth Finish

Honed Jurastone Beige

Rosa Tea

Thassos

Varigated Indiana Machine
Smooth Finish

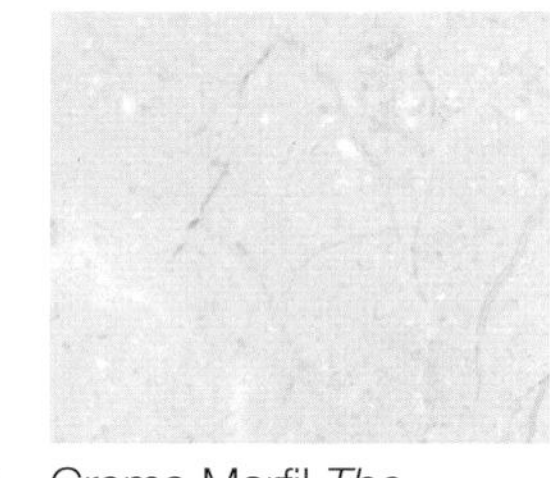

Crema Marfil *The Architectural Stone Co.*

slate

Green Cleft Slate
Genesee Cut Stone & Marble

Black Cleft Slate

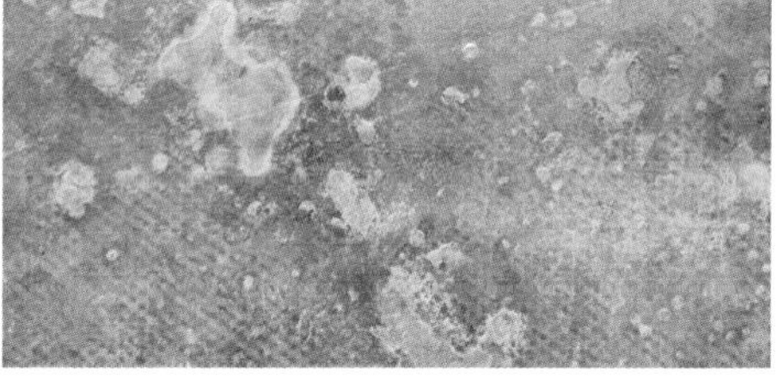

Multicolor Slate

soapstone

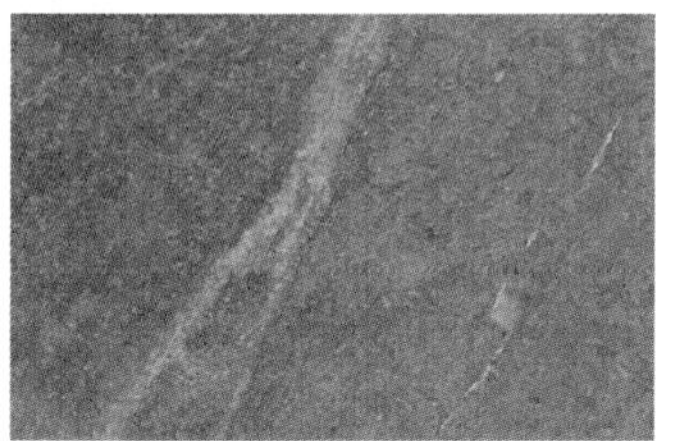

Original, Natural
Green Mountain Soapstone

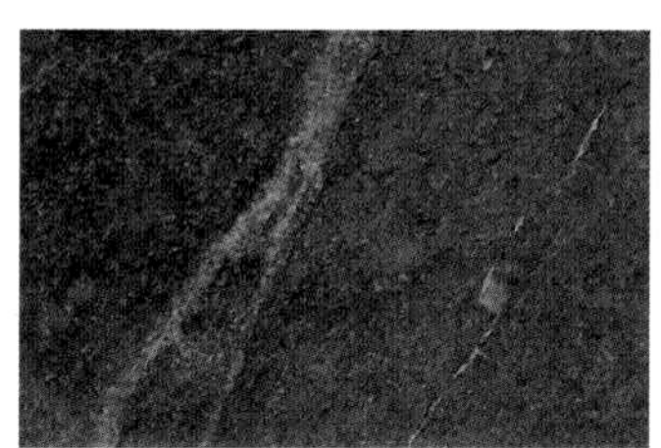

Original, Oiled

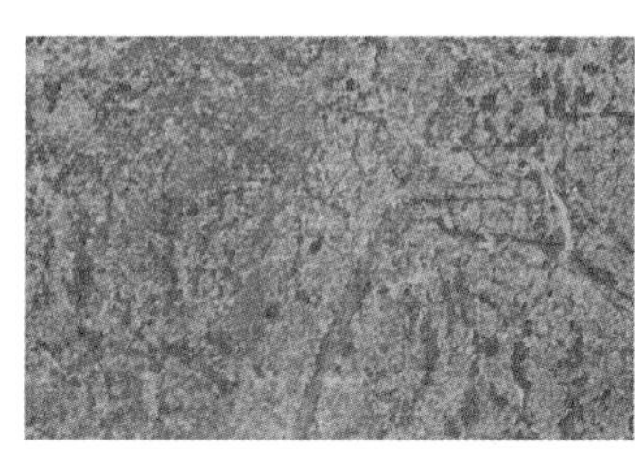

Ice Flower, Natural

Ice Flower, Oiled

engineered marble

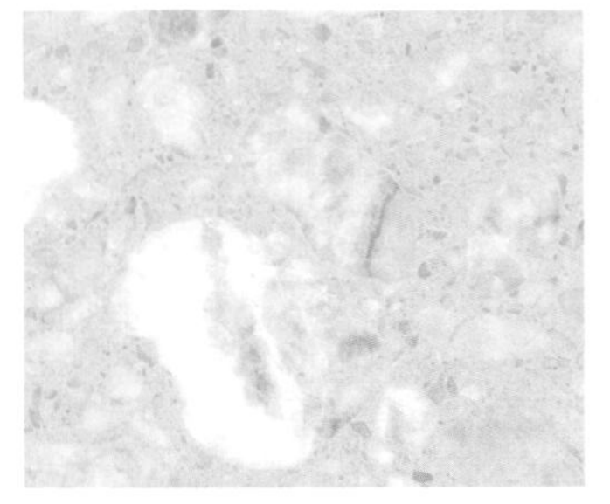
Rover Fior DiPesco
Genessee Cut Stone & Marble

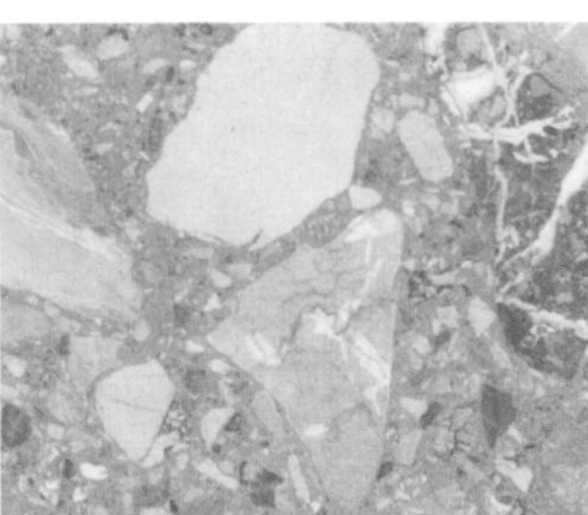
Rover Breccia Pernice

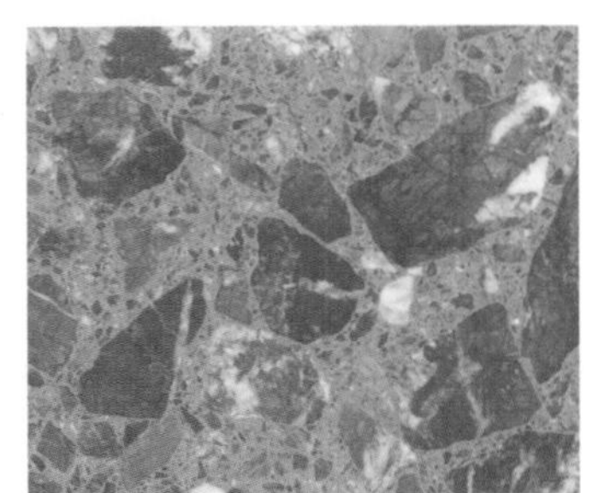
Rover Grigio Carnico

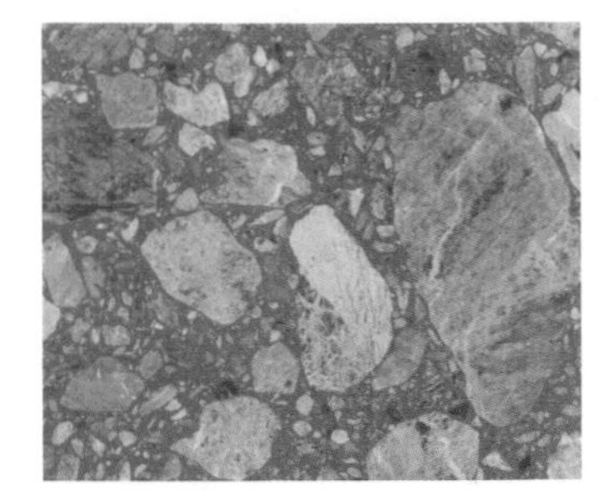
Rover Verde Alpi

engineered gemstone

Concetto Indigo *CaesarStone*

Concetto Intaglio

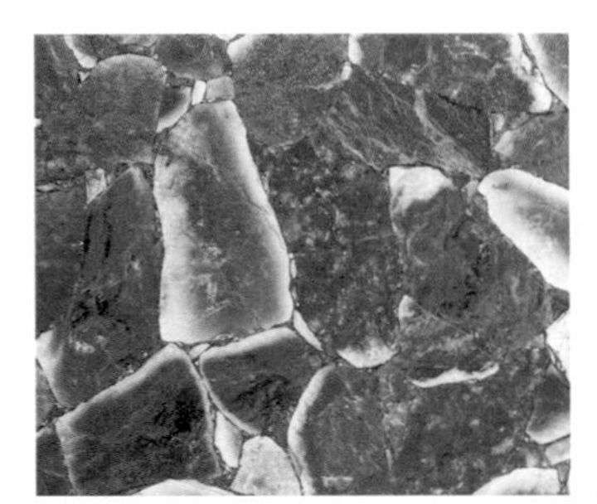
Concetto Patina

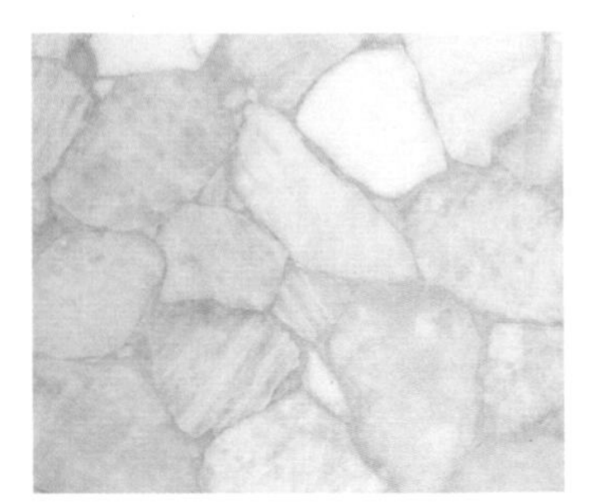
Concetto Puro

engineered quartz

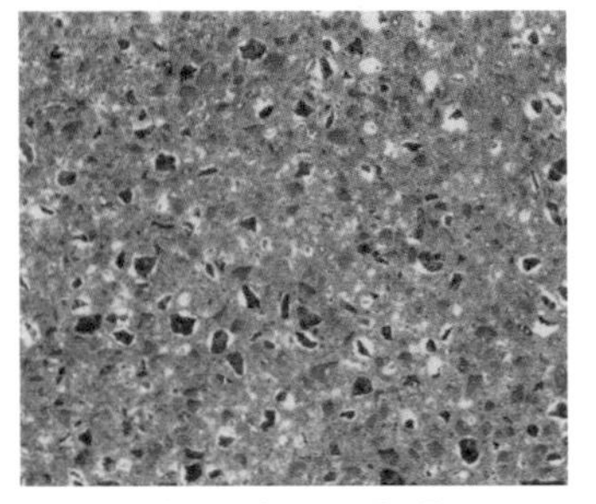
Zodiaq Argo Green *DuPont*

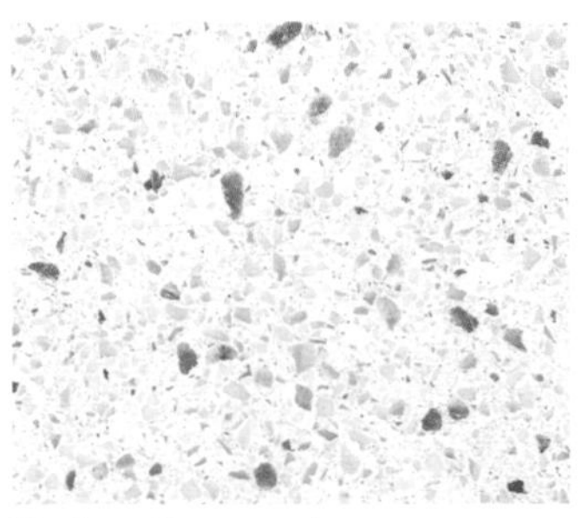
Zodiaq Cappuccino

Zodiaq Celestial Blue

Zodiaq Antique Pearl

Zodiaq Cinnamon

Zodiaq Indus Red

Zodiaq Meteor Gray

Zodiaq Mosaic Gold

metal

Stainless steel suits a contemporary kitchen in dark wood tones. *Brooks Custom*

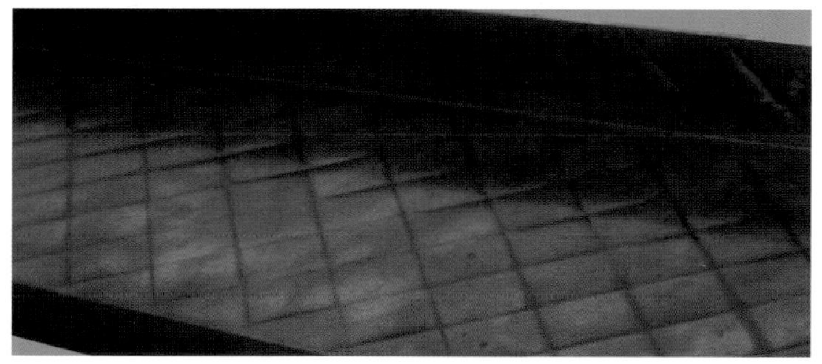

Quilted copper countertop *Brooks Custom*

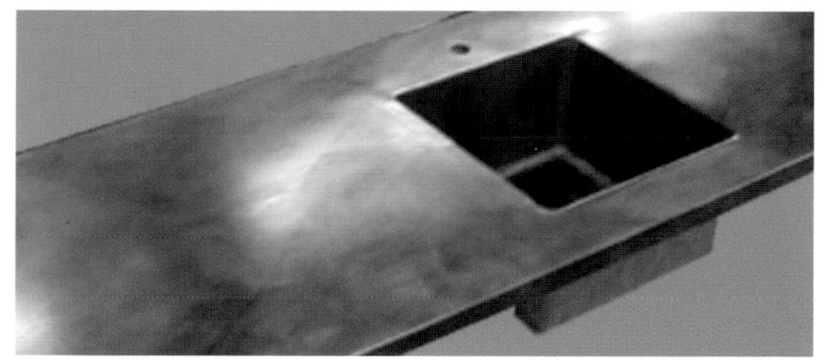

Copper counter with aged finish

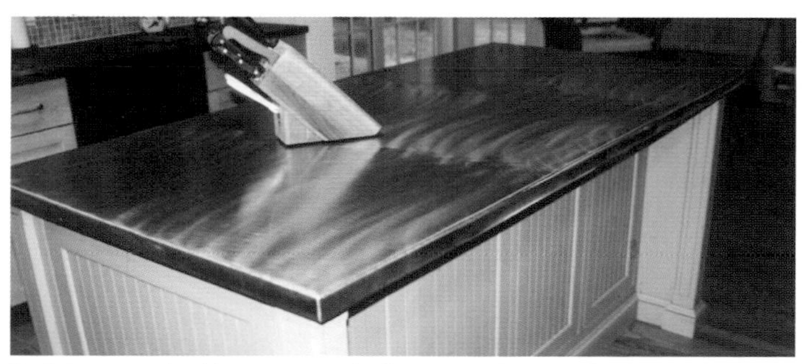

Stainless steel with custom grind pattern *Brooks Custom*

Textured stainless steel counter *Brooks*

solid surface

Solid surfaces can look like almost any kind of stone or solid.
Meganite

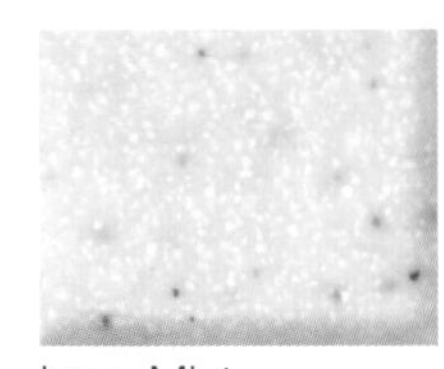

Ivory Mist
Meganite

Pesto Mist

Silver Mist

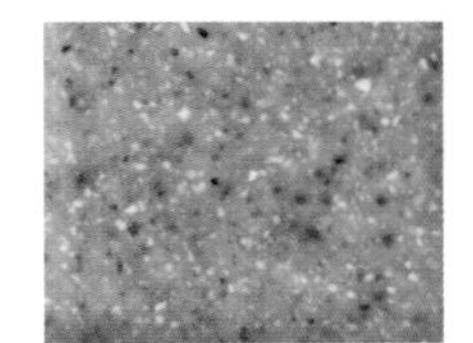

Mauve Mist

Galaxy Mist

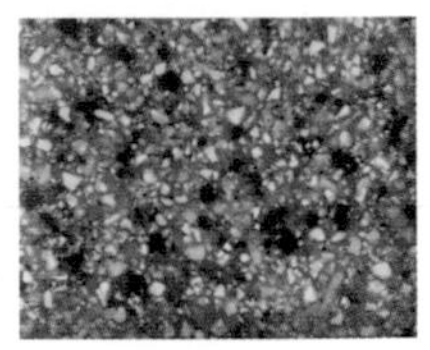

Madeira Mist

Camel Mist

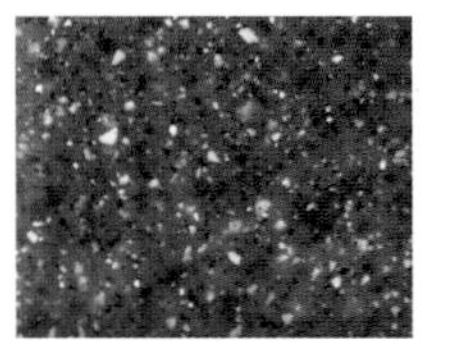

Amazon Mist

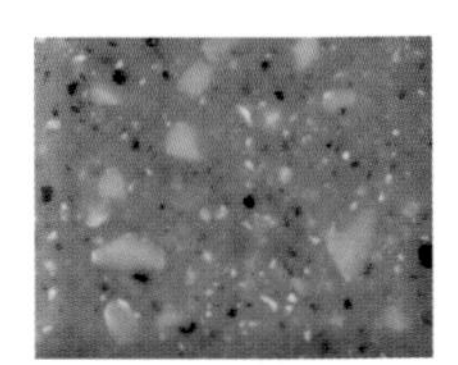

Platinum Granite

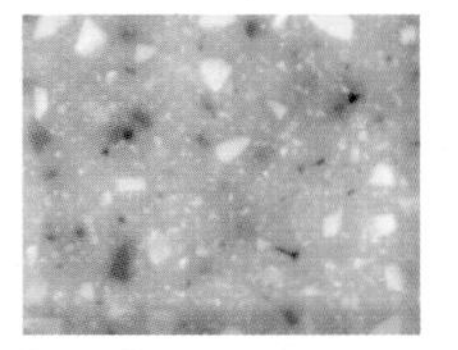

Rye Granite

Copper Granite

Sonora Granite

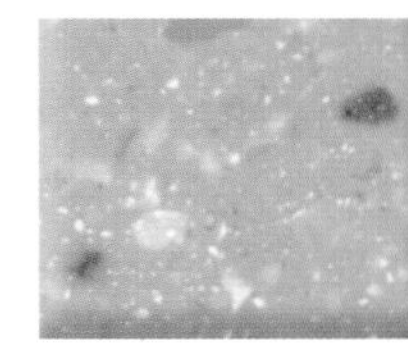

Green Tea Granite

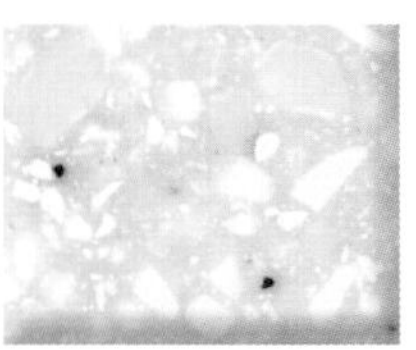

Sanibel Granite

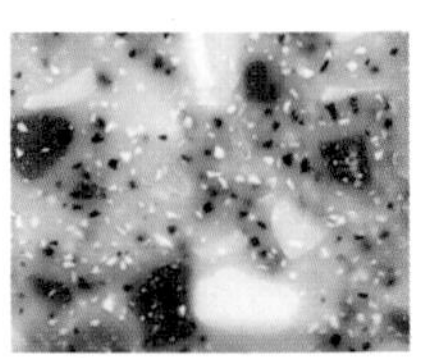

Pinnacle Bolder

Corian Mohave
DuPont

Corian Storm Blue

Corian Mandarin

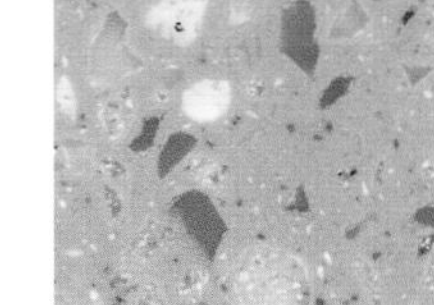

Corian Acorn

Corian Adobe

Corian Anthracite

Corian Suede

Corian Fossil

Corian Sun

Corian Hot

Corian Graphic Blue

Corian Glacier White

wood

Teak with mitered corners *Brooks Custom*

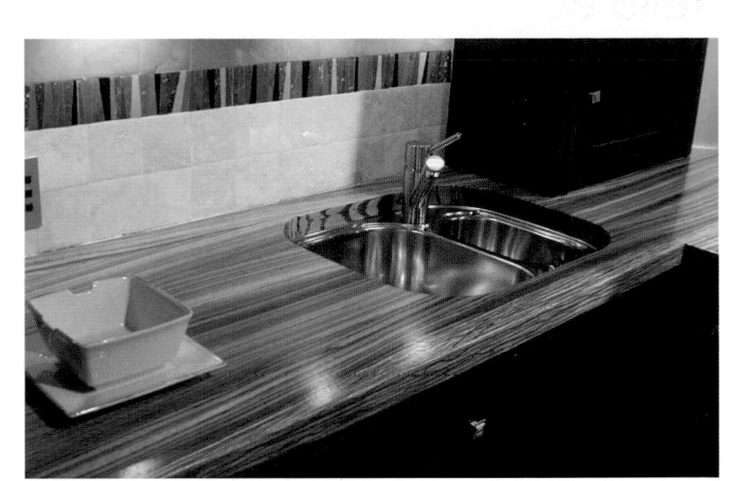

Zebra wood

Maple end grain butcher block

Interlocking joint butcher block

End grain butcher block

Mesquite wood countertop
La Puerta Originals

paper

The subtle tones and baked-on strength of paper countertops by *Richlite.*

vetrazzo

Millefiori, recycled glass and concrete *Vetrazzo*

Firehouse Red countertop

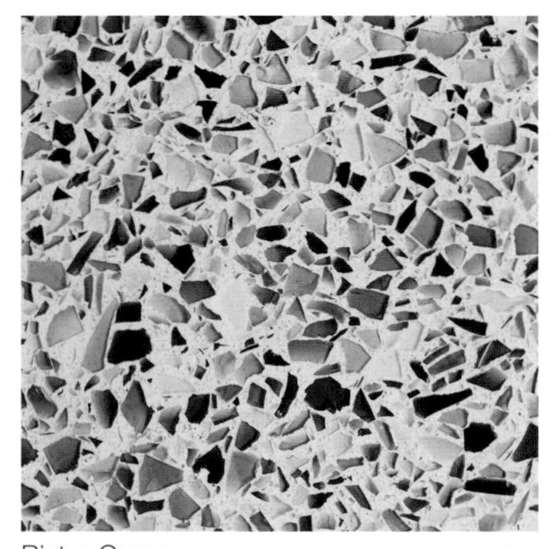
Bistro Green

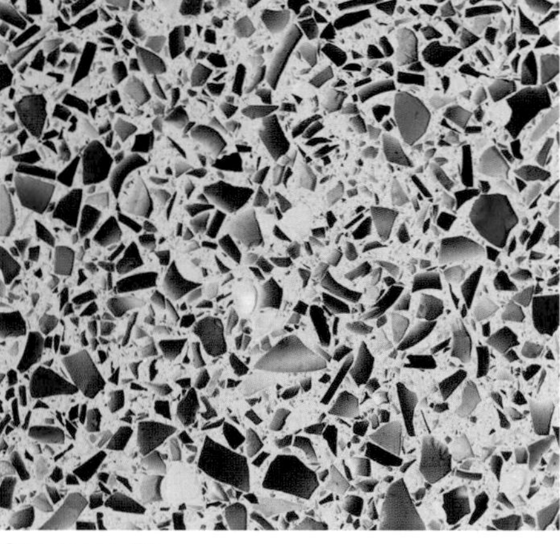
Charisma Blue

Palladian Gray

Firehouse Red Close up

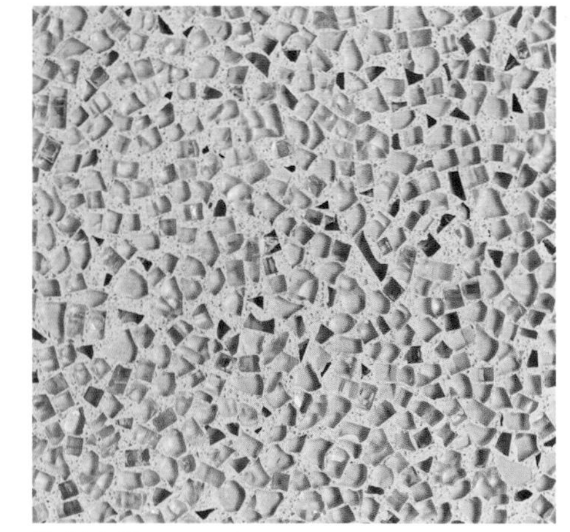
Cubist Clear

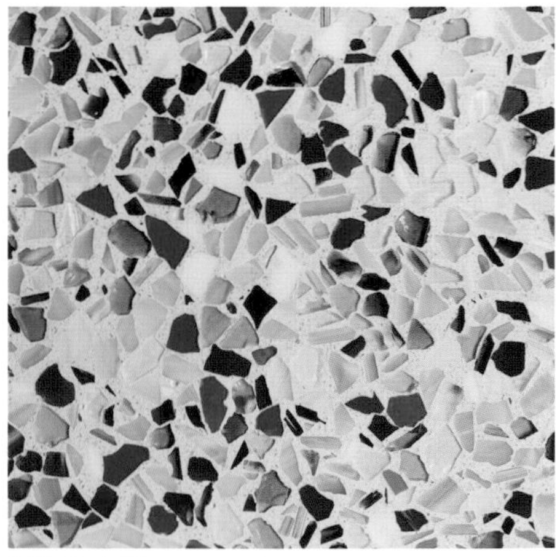
Millefiori Close up

concrete

Scored concrete countertop in a muted earth tone from *Sonoma Cast Stone.*

Concrete countertop with simple sink and faucet
Sonoma Cast Stone

Concrete slab island *Brooks Custom*

glass

Concrete and glass island *Brooks Custom*

flat

Tarsia II *Wood-Mode*

Vanguard Plus

Vanguard Stainless

Vanguard Plus Birdseye

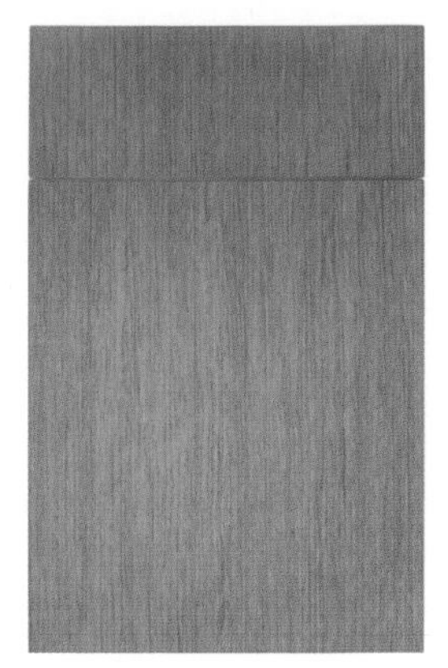
Vanguard Plus Rift Cut

raised

Barrington Vertical
Crown Point Cabinetry

Camden Vertical

Cameo Vertical

Cathedral Vertical

Lexington Raised Square
Wood-Mode

Embassy Raised Curved
Wood-Mode

Hallmark II Raised Square *Wood-Mode*

Classic Vertical
Crown Point

Galleria Raised
Wood-Mode

Square Edge Barcelona
Wood-Mode

Provence, French Country line, wood

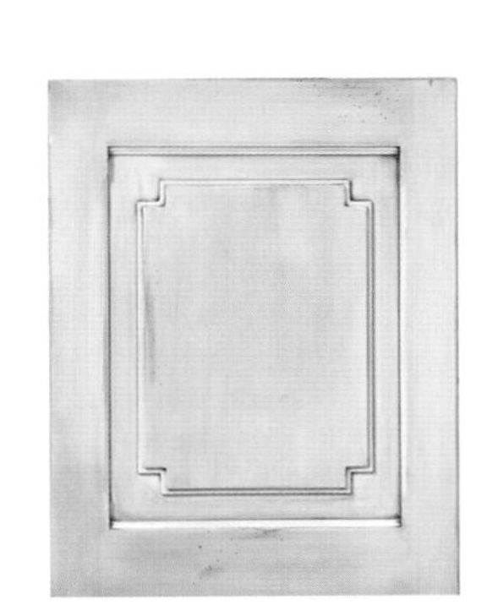

Rochelle, French Country line, wood

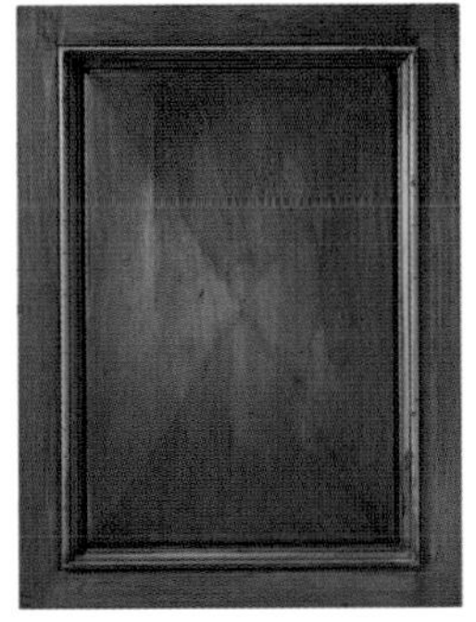

Toulouse, French Country line, wood

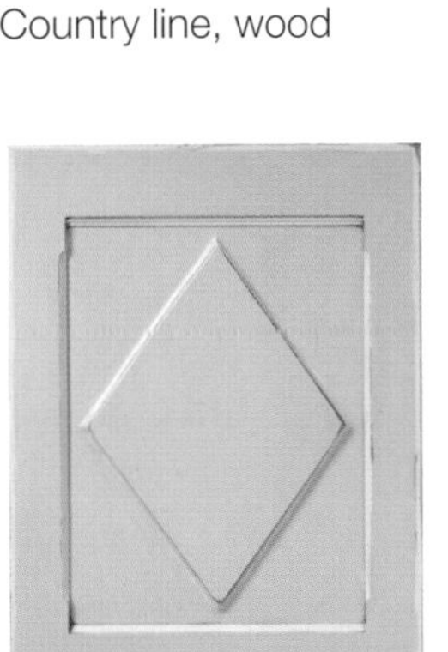

Avignon, French Country line, wood

Woven cabinets with stone, granite and dark wood combine in this kitchen by *Wood-Mode.*

Kensington Raised *Wood-Mode*

Ridgefield Vertical *Crown Point*

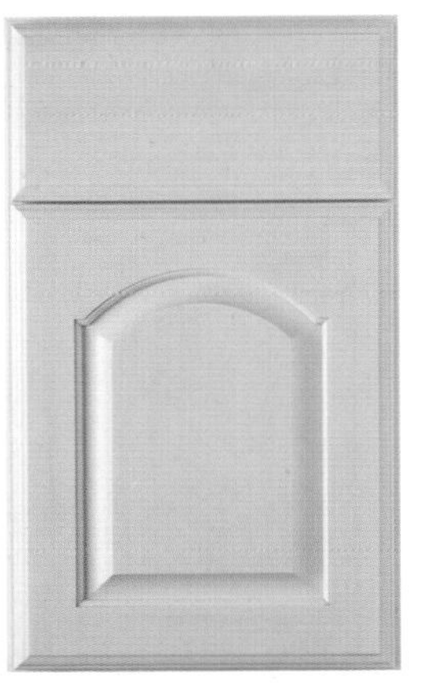

Hallmark II Raised *Wood-Mode*

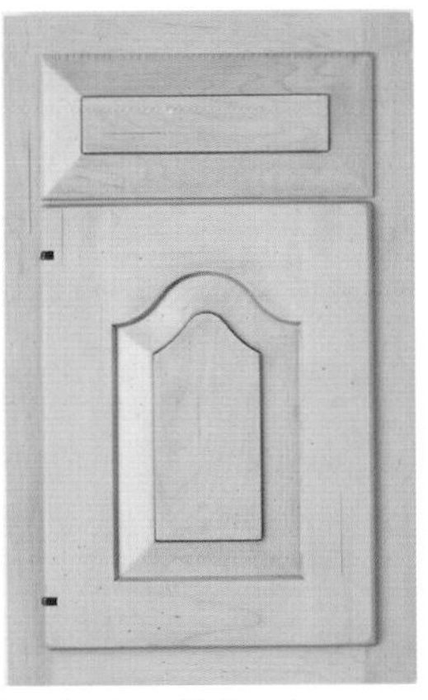

Lexington Raised Curved *Wood-Mode*

Lancaster Raised *Wood-Mode*

recessed

Sturbrdige Recessed
Wood-Mode

Newport Vertical
Crown Point

Portsmouth Vertical

Old Cupboard Vertical

Canterbury Vertical

Craftsman Vertical
Crown Point

Bradford Vertical
Crown Point

Sonoma Recessed
Wood-Mode

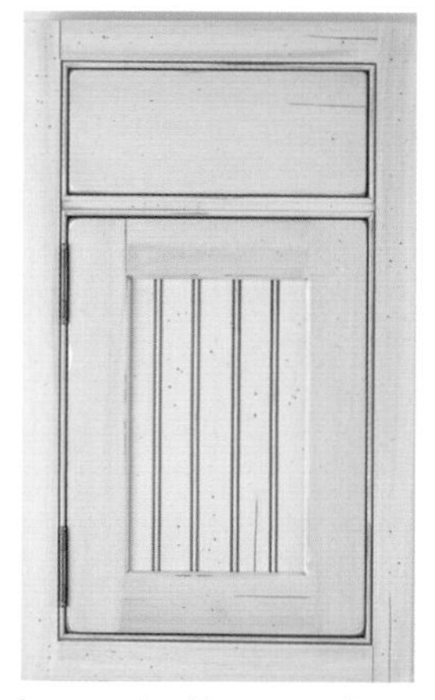
Lancaster Recessed
Wood-Mode

Essex Recessed
Wood-Mode

Barcelona Recessed
Wood-Mode

Concord Recessed
Curved *Wood-Mode*

Westbrook Vertical
Crown Point

Olive Ash
Wood-Mode

Herringbone
Wood-Mode

insert

Brushed Metal Laminate
Wood-Mode

Silver Dot Laminate

Vertical Lattice

Arts & Crafts Insert

Birdseye Maple

Ground Glass

Ribbed Glass

Astral Glass

Mist Glass

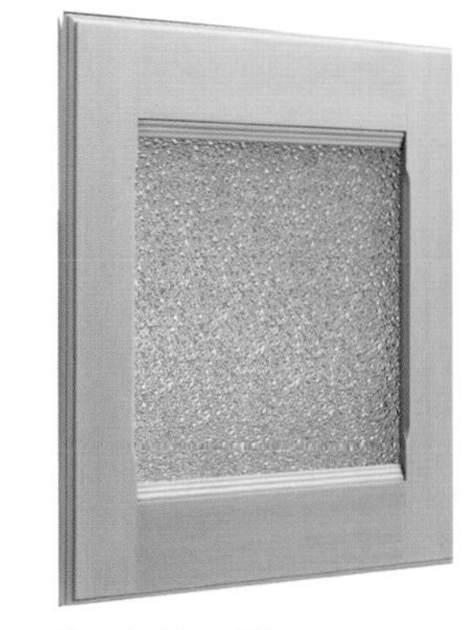
Crystal Ice Glass

Beveled Glass
Wood-Mode

Gothic Mullion

Optional Leaded Glass

Type B Contemporary

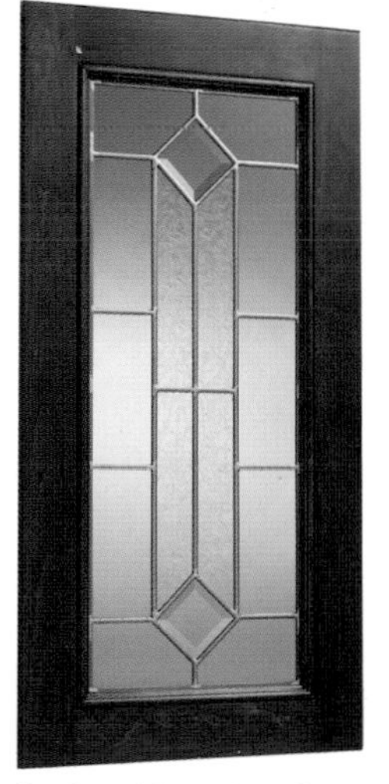
Optional Leaded Glass

metal

Copper sink
Brooks Custom

Distressed copper sink
Brooks

Circular copper undermount bar sink
Diamond Spa

Stainless steel bar sink in a concrete counter top
Brooks

Double stainless steel sinks
Brooks

Stainless steel vintage
Julien

Stainless steel, pewter
Bates & Bates

Cocina weathered brass
Bates & Bates

Avalon, white bronze
Rocky Mountain Hardware

stone, concrete & solid surface

Granite, large bowl-small bowl
Swanstone

KSDB solid surface double bowl
Swanstone

Soapstone double sink
Green Mountain Soapstone

European sink carved from solid block of Macael marble *Allante*

Concrete with built-in drainboard
Brooks Custom

Rough stone apron sink
Stone Forest

Granite Ascend Bowl
Swanstone

Integrated concrete chef's sink
Sonoma Cast Stone

Hand-painted bar sink
Purple Sage

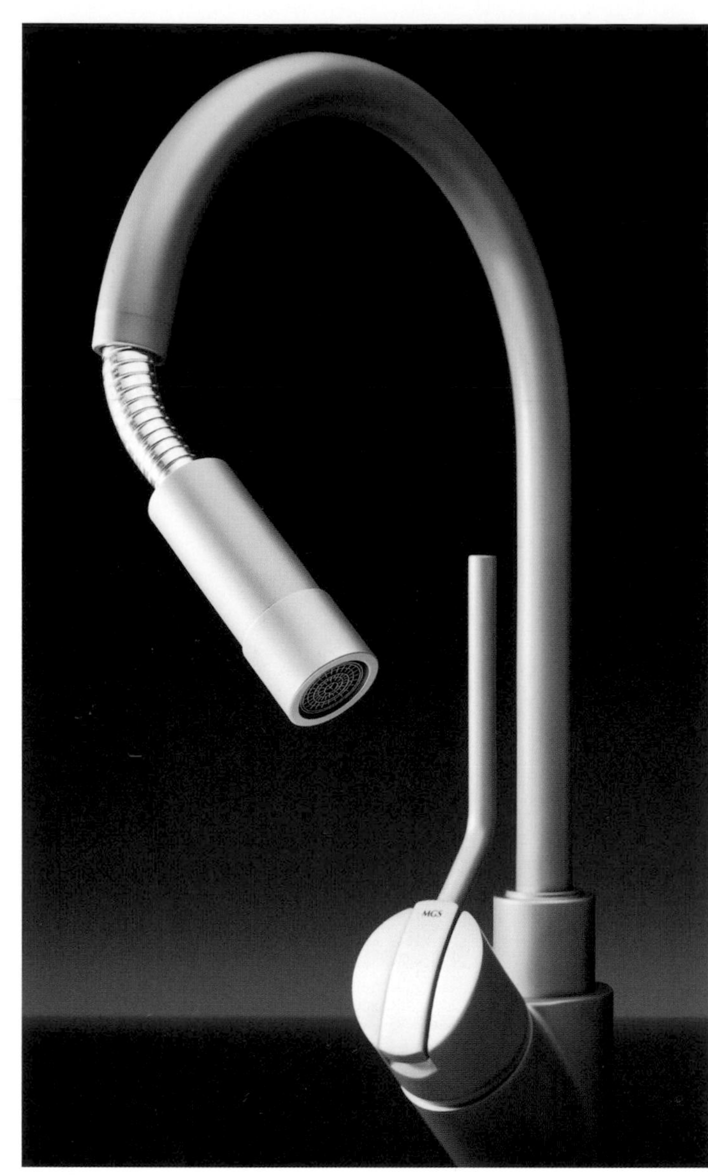

Unico pull-out spout hose, stainless steel *MGS*

Hamilton faucet with spray *American Standard*

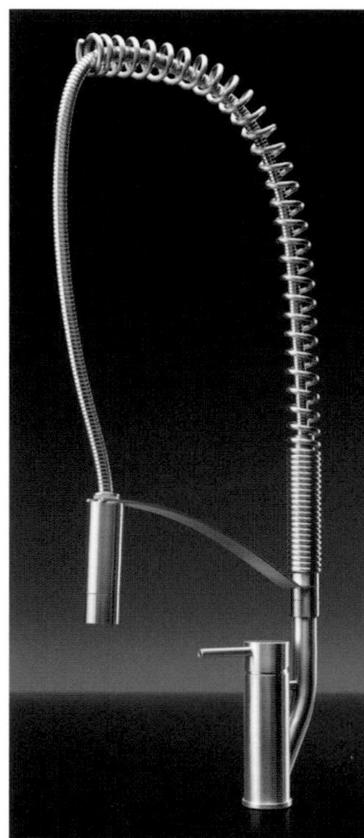
Randa, KL flexible faucet *MGS*

Randa, glazed polish

Boma, glazed polish

Trevi chrome *Delta*

Trevi higharc bronze *Brizo*

Ladylux Cafe 35A25F *Grohe*

Deck mounted LB60 with Olumpus levers in silicon bronze *Rocky Mountain Hardware*

Single hole faucet, bronze *American Standard*

Alliance bridge single spout *THG*

Sonoma single-hole faucet with hand spray, oiled bronze finish *Signature Hardware*

Single handle faucet *THG*

Culinaire bridge faucet with hand spray, solid brass *American Standard*

Bridge faucet with swivel spout and cross handles, solid brass *Signature Hardware*

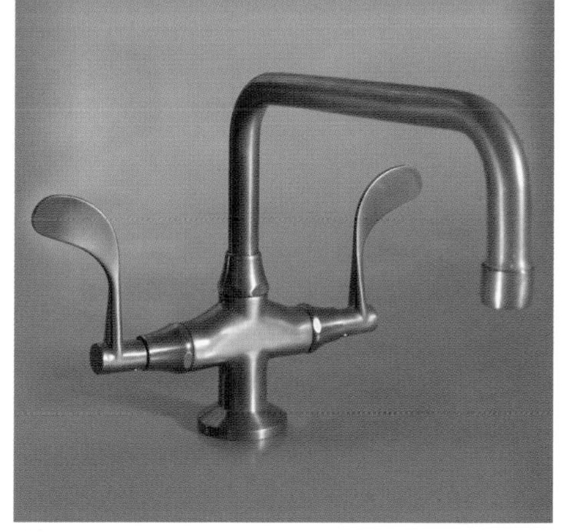

Wingnut levers, copper faucet *Sonoma Cast Stone*

wood & stone

Brentwood, cast stone kitchen hood
Old World Stoneworks

Wood and stainless *Stanisci Design*

Appliqué and Beadboard G Series

Extra Tall T Series

All Wood W Series, includes spice rack

Wood vent hood with vertical appliqués
Richelieu Hardware

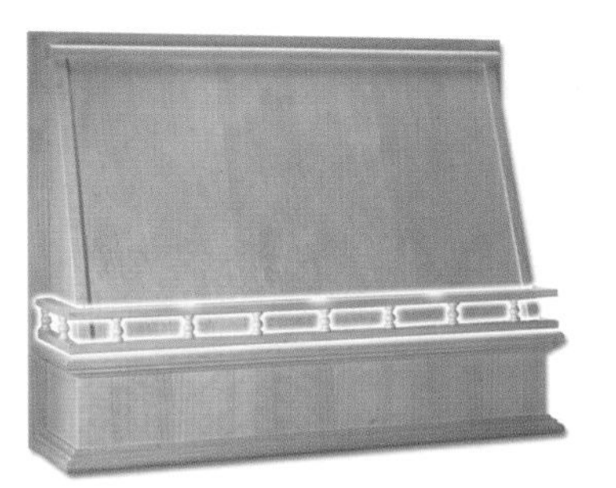

Wood vent with rail appliqué

Plain wood vent

Hand-carved antique hood *La Puerta Originals*

All Wood Series *Stanisci Design*

metal

Shade wall vent, Cheng Collection, stainless steel with folding hood *Zephyr Ventilation*

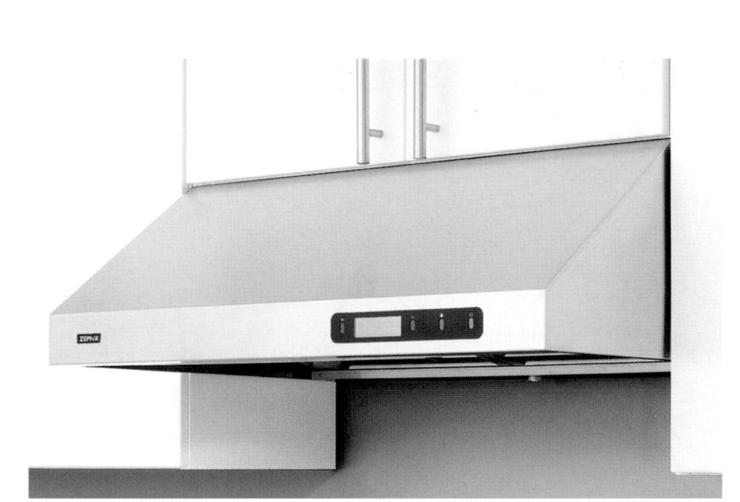

Stainless steel Tempest undercounter

Menhir island vent, Elica Collection, stainless steel

Concave wall vent, Elica Collection, stainless steel with curved aluminum wing

Trapeze island vent, Cheng Collection, stainless steel

Futura wall vent, Elica Collection, stainless steel and glass

Rangemaster, stainless steel and glass *Broan-Nutone*

Eclipse, stainless steel, retractable to stovetop

Silhouette, steel and glass

bathroom fixtures

The bathroom has now come into its own as a room, and you enjoy it as you would any room, like the living room or dining room. Today it is much more like a spa, a place where you spend time relaxing. Using materials like stone and glass and falling water adds to that sense of serenity.

LYNN APPLEBY, A DESIGNER WITH A CREATIVE FLAIR FOR THE USE OF WATER

Original fixtures and black and amber tile set off what was once actress Gloria Swanson's bathroom. It is smaller than today's spa-like retreats but no less charming.

THE BATHROOM, ONCE THOUGHT OF AS MERELY A DISCREET PLACE TO HIDE the plumbing, has now been given the option of becoming the in-house equivalent of the luxury spa — a convenient getaway from a hectic and intrusive world. It can be a bracing way to prepare for the day ahead, or a reward at the end of a long day. To free up your imagination, think of the bathroom as an integral part of an open master bedroom suite that sees little or no separation between bedroom, changing room and bath. This metamorphosis can be best seen in the way water is introduced into the room — not just with the usual workmanlike faucet and showerhead, but with "rainforest" showers and "rain towers," "spray surrounds," "infinity tubs" and "fountain" faucets, along with an ever-increasing array of taps, controls, basins and surrounds that are more works of art than mere practical toggles and pulls.

Left: A faucet that is more fountainhead or mini waterfall. Center: The reflective quality of stainless steel makes it cool but not cold, giving this vanity an almost sculptural quality to it. Because this apartment is all about being able to see through, past and in between spaces, there is a one-way mirror behind the sink. When a motion detector activates a light fixture above the mirror, it turns the glass into a mirror. When you move away, the light goes off and you can see through it into the bathtub area behind it. The beautifully textured column was original to this converted industrial building. The floors are slate; note the indirect lighting under the steps. Right: Understanding that anyone sitting on a toilet has inevitably to look at the business under the sink, architects at Toronto's The Ideal Environment improved the view by lavishing attention and money on the plumbing hardware. Opposite: From the waterfall drain into the infinity tub (by Kohler) and the corner windows, this master bathroom is all about water and light. Note the stainless-steel sink, tub surround and ribbed shower floor. The shower faucets and rainshower are from Grohe.

Generally speaking, bathroom trends follow what is happening in kitchens, but there are a few timeless ways to treat a bathroom. One is to think of it is as a Roman bath, with natural materials such as stone (marble, granite, limestone, slate and travertine — a hard limestone that can be polished to resemble marble), lots of matte metal finishes (the soft look of nickel, gleaming copper, brushed bronze, wrought iron), and flowing water that conjure up a soothing, relaxing "hardscape" of rounded shapes and liquid sound. A bathroom can also go in the opposite direction — to a modernity that is all about straight lines and hard surfaces such as acrylic, stainless steel and glass, all put together with a minimalist sensibility.

When it comes to glass in the bathroom, sandblasted and acid-etched glass can be used to great effect to let in light while still providing privacy. But with all of the attention paid to creating these elaborate "hardscapes"— spectacular tubs, walk-in showers made with imported marble, and more — you may choose to use clear glass instead, not only to give the space a little dazzle, but to show off all that gleaming tableau as well.

The concept of the bathroom as a room that one lives in and not just a utility room has also opened the door to the introduction of furniture in the bathroom. Think armoires and built-in cabinetry to soften the hard edges of the traditional bathroom. Woods such as mahogany, teak, wenge and even cherry can be very appealing. (It should be noted, however, that although they add a beautiful element of warmth, most woods should be coated with some kind of water repellent to keep the moisture out.)

One good idea is to put the toilet in its own compartment so the rest of the space can be wide open. Another nice design trick is to use pieces of furniture — antique or contemporary — to serve as storage, and to house sinks and even

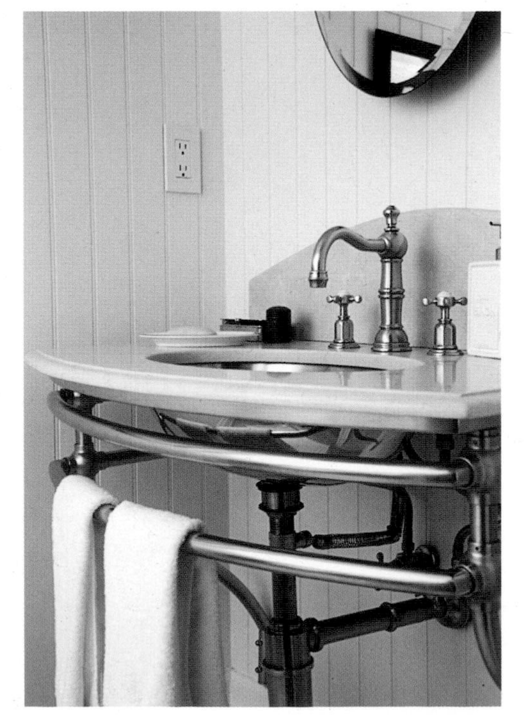

I like to build closets in the master bath so you can access your clothes. This is a merging of the traditional dressing room with the master bath. To make this work, you have to build the closets and doors like fine cabinetry.

JOHN DOUGLAS, ARCHITECT

The Asian motif is strong in this bathroom. The floor consists of pebbles that have been set in cement and ground smooth — a kind of oversized terrazzo. The architects built a platform around the Japanese soaking tub mainly because they wanted to have a perch with which to look out the window. The steps up to the tub are made of granite, hand-hammered in China. The top is honed granite.

A seven-foot-long Kohler tub dominates this bathroom. Note how the beadboard used in the ceiling, walls and tub surround ties the room together. Light from the large eyebrow window and the amount of glass in the shower stall give the room a sunny, open feel. **Overleaf:** Toronto designer Brian Gluckstein wanted his bathroom to look like an extension of the other rooms in his house, not like a utility room. The cast iron pedestal tub by Kohler and a curved bookcase (filled with old leather-bound Reader's Digests he got at auction) do the trick. Add the terrycloth armchairs and the fireplace (out of frame to the right), and this is a luxurious room to relax in.

bathtubs. To create interest with what might be otherwise ordinary objects such as sinks, buy a vessel sink made by a ceramic artist or one of jewel-like stones or glass, which can then be displayed above the counter as art objects. You can even install TVs and sound systems in the bathroom.

Why not? We certainly spend enough time there, and this way you can be soothed and entertained at the same time. Other ways of introducing luxury and pampering into the bathroom are heated towel warmers and heated floors — either electronic radiant heat or a heated water-coil system, called "hydronic," which is the most energy-efficient type of radiant heat. Why not install a fireplace to extend the feeling of warmth and coziness? Or you might choose exercise equipment and a massage table, or a stainless steel or cedar sauna and free-standing hot tub. Seating could include a chaise lounge or a pair of terry-cloth armchairs for relaxing after a long soak. You are limited only by your available space and budget, and even within those restraints, with the right choices and materials, you can create your own restorative spa retreat.

sinks

If you are tired of seeing the traditional counter-installed, overmount sink, consider from many choices of above-counter vessel basins, made of everything from onyx and marble to fused or hand-blown glass, carved granite and ceramics. Depending on the artistry involved, a vessel basin can elevate the sink from a place to wash your hands to a work of art that can be the centerpiece of the bathroom. Wall-mounted sinks are also an option, as are custom countertops made from purposefully workaday materials such as cement board (a drywall-like material impervious to water) marine plywood, and textured concrete. (For more on countertops, see the Kitchen Fittings chapter.)

In a house that is rustic yet refined, this rough-cast concrete sink has a textured stainless steel bowl that perfectly sets it off.

Left: A Corian sink. Note how the long, thin light fixtures fit the proportions of the mirror. **Center:** A stunning glass sink set off by a copper bowl is further embellished by a polished granite countertop. **Right:** A metal basin set flush in an antique table.

As in the kitchen, stainless steel sinks are an effective choice for those wanting a cool, tough, sterile effect that works well in a loft or ultra-modern home. Copper sinks are also an option, making a strong design statement that can be the centerpiece of your bathroom if you like that metal's color and buttery luminance. And the durability and color selection of solid-surface sinks in materials such as Corian make them viable options, often in sink-countertop combinations, as are cultured stone sinks made from a mixture of crushed stone and polyester resins. (One caveat: sinks with a gel-coat finish may blister and peel with age.)

Ceramic sinks, usually in some kind of mosaic, can also look good if you like the texture and ambiance of a more ancient style. There are basically two types of ceramic sinks: the "tile in" type, which is set below or in line with the tile surface and, although trickier to install, has a much cleaner, continuous and easier-to-clean style. The other is the "self rimming" type, which is easy to install but harder to clean because it tends to collect dirt and water at its elevated rim when the counter is wiped.

And the good old pedestal sink, last popular in the 1920s and 30s, is a great choice for a period look, mainly because of its stately, vertical lines.

Beyond what is readily available, you can also have your own sinks and tubs designed and built from scratch in any kind of material (how about a polished mahogany sink?) or combination of materials. This is the ultimate in bathroom self expression.

One more good bathroom idea to consider is the installation of electronic water taps, which turn on automatically when you hold your hands underneath them. Long used in public washrooms, they are showing up in residences, especially guest powder rooms, because of their hygienic benefits.

showers

One of the best places to let technology run wild is in the shower. Larger, more sculpted and full of futuristic knobs and sprays, the shower has gone beyond the traditional stall with a plastic curtain, and has become more than just an adjunct to the bathtub.

First of all, a separate shower, not the old practical shower-and-tub arrangement or the glassed-in model with the goldfish etched into the glass, makes a lot of sense, especially if you want to highlight the bathtub. At the very least, think about the elegant, simple lines of frameless glass doors. For true luxury, nothing beats the walk-in shower. Bigger, more an area of the bathroom than a separate enclosure, the walk-in shower opens up the space for more functions than just showering. You can, with the installation of a built-in sink, taps and perhaps even a fogless mirror, shave in the beard-softening atmosphere. You can also install a steam apparatus that turns it into a steam room, just the place to warm up after a day on the ski slopes. You can then cool off with a spray from the shower. Stone or tile seats built into the shower walls allow you to steam or shave in comfort, sometimes with two benches so you can stretch out your legs to shave them. Walk-in showers also have a practical aspect for our aging population, as they are much easier to get in and out of.

Showers can be as complex as you want. High-tech, multiple spray units that surround you with soothing mist or invigorating massage, complete with "his and hers" controls, are now made by just about every manufacturer. You can even get a chromatherapy device for your bath or shower that allows you be bathed in the light of different hues that purportedly affect your mood and feeling of well-being, and aromatherapy devices that achieve the same thing.

Left: Dark wood and marble give this bathroom a very masculine touch, as do the teak drainboards in the shower. Note the opaque windows inside the shower stall. **Center:** A modern take on a Santa Fe style bathroom uses simply white tiles. Note the open beam ceiling and rough boards — another Southwestern touch. **Right:** The rich material of this artfully hung shower curtain partially hides a tiny triangular shower that makes the most of a small space.

I like to incorporate daylight into the bathrooms to make the mornings pleasant — so we try to open the master bathroom up to an outside area so you can have air and light — often with a little spa outside the bathroom, which would include a whirlpool and waterfalls.

JOHN DOUGLAS, AN ARIZONA ARCHITECT WHO DESIGNS FOR THE DESERT

A small walk-in shower wrapped in glistening mosaic tiles, lit from above by a skylight.

You can do everything but work on your car in this walk-in shower. Note the two shower heads, one that can be used sitting down on the marble-topped bench, and the stand-up sink for shaving.

Yet despite the availability of high-tech gadgets galore, you may be one of those people who yearn for more simple, if luxuriant, pleasures. For example, you may be happy to settle for a handheld shower wand combined with a rainforest shower. There are other relaxing ways to deliver water, such as showers that spill the water out of the wall from a metal or stone shelf, to simulate the sound and feel of a waterfall. This idea can be taken one step further in an out-of-doors shower. Generally an outdoor shower has much to do with rinsing the salt or chlorine we get from swimming off the skin, but it can also bring this notion of a spa-style luxury fantasyland out the door and into the backyard. Sensuality is taken to even greater heights with the sensation of being naked outside, where warm breezes waft, as water sluices down from a rock or bronze fountainhead and washes over the body.

bathtubs

The old shallow, enameled metal tub with its rounded rectangular shape is looking a little old hat. Just about everything else is not. In fact there is a plethora of shapes and sizes that go backward and forward in time. The biggest change in tubs is in the size. Longer, deeper and with more choice in shape, this change mirrors our desire to use the tub as a place to soak and steam away our jangled nerves while the fast, efficient shower is our choice for actually getting clean.

With that in mind, an interesting option if you are bored with the standard North American tub is the Japanese soaking tub. Often enclosed in wood or tile and sometimes freestanding, these ancient, vertical soakers come complete with a short stairway and often a wooden top that encloses the tub when not in use. The fact that these tubs rise up and are enclosed can give the bathroom an extra special dimension without taking up much floor space.

Variations on the old cast-iron clawfoot tub abound, but most are bigger, with elaborate, antique-looking and French-inspired-faucet arrangements. The traditional cast-iron models with gleaming porcelain enameled finishes are still available and there are countless variations on the theme. Try to find an old one that can be re-enameled.

A tub sunk into a platform of polished black granite. Huge windows just above the tub look out over a spectacular view.

Tubs aren't used in a practical sense anymore — to get clean, you take a shower. Tubs are more to enjoy yourself. We often put reading lights and horizontal window slits near the tub, so you can look out into a controlled environment, perhaps with lights illuminating it at night.

JOHN DOUGLAS, AN ARCHITECT WITH A FEEL FOR CONNECTING ROOMS WITH THE OUTSIDE

Encasing a free-standing tub in wood so that it resembles a piece of furniture is also a possibility, using everything from salvaged panels to new moldings or distressed barnboard and bamboo stalks. And a bathtub ground out of a large boulder, polished on the inside and left rough and unfinished on the outside, can make an awesome centerpiece for your bathroom. Old-fashioned metal tubs but made of zinc or copper can be equally impressive.

To counter the move toward the "vertical bathroom" driven by the popularity of the shower, bathtub manufacturers are customizing their bathtub products to include specialized aspects such as an indented foot-washing area at one end of the tub, or mini tubs that can turn a small, under-the-stairs two-piece into a full-use bathroom. Concrete tubs are also a possibility. They look good, come in any color you wish, and can be custom made to your body's contours. They tend to suck heat out of the water, however, so are sometimes constructed with built-in heating coils.

One edgy bathtub that can look great in a loft or ultra-modern home is the one-piece, institutional stainless-steel tub, sink and toilet ensemble. The more it looks like prison issue, the better.

Right: As dramatic as any film set, this round, free-standing tub with its tumbled Roman-style tile is backlit by a marble perfume cabinet and cove lighting. The curtains not only add the feeling of luxury but deaden the sound that normally rings off the hardscape of any tiled room. **Opposite:** Panels from an 18th-century French apartment went into the construction of this bath surround in a New England farmhouse.

toilets

The most notable recent innovation in toilets has been the creation of energy-efficient models that use less water per flush. But there are other advances worth a look: styles that have built-in heating coils in the seat, or a flush-with-no-hands foot pedal, or lids that automatically open and close, or that have a soft closing apparatus that won't let them slam. Then there is Toto's Neorest 600 Toilet, which senses your approach and lifts the lid, and then closes the lid and flushes when you walk away.

And, to complete that institutional look for hip bathrooms, wall-mounted urinals (they are, if nothing else, environmentally correct) are making an appearance. Put in a stainless steel, suicide-proof toilet and a terrazzo floor and you're done.

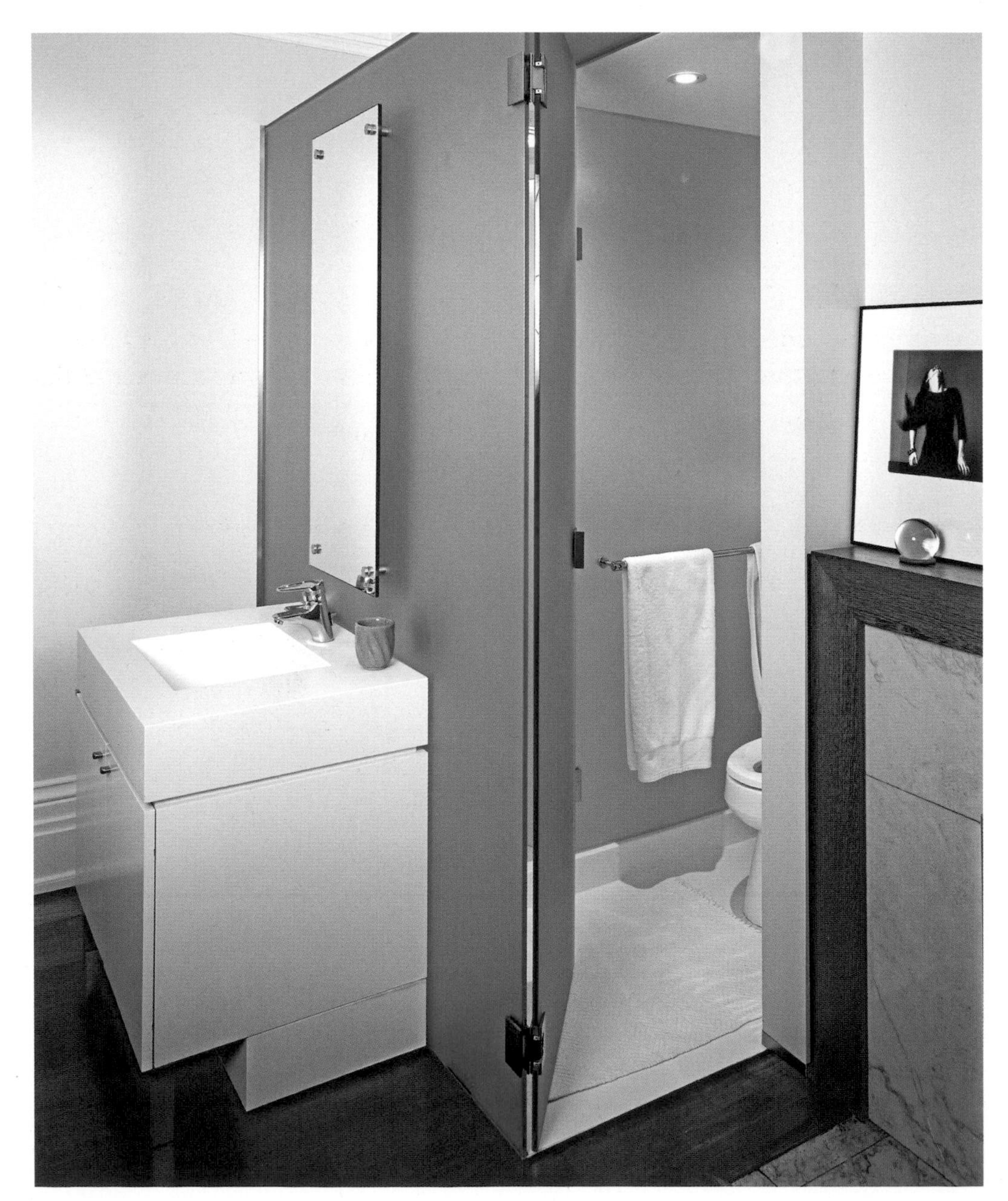

Cleverly designed master baths keep the toilet area private by compartmentalizing it with walls — high, low, or sometimes glass. Acid-etched glass and a Corian sink combine for an elegantly simple bathroom. The fireplace on the right adds a cozy touch to this very open master suite.

bidets

In Europe you often see another ceramic bowl beside the toilet, but without a seat. This is a bidet, a fixture that has been popular in Europe and Latin America for years but has only recently caught on in North America. Simply put, a bidet is used for personal hygiene immediately after using the toilet. Originated in France in the early 1800s (the name stems from the French word for pony, because you usually sit astride the bidet as you would sit on a horse). The first bidets were merely a shallow tub with a low stool in the middle that allowed a quick wash of the nether regions without having to completely undress. Today bidets come in various configurations, some of which consist of a bowl that can be filled with water from a tap but the more popular ones with jets of water of controlled temperature that spray up at the body. Some even have heat units built in that dry the body as well. Contrary to popular myth (at least in North America), bidets are designed to be used by men as well as women.

taps, faucets & showerheads

These are the functional details that adorn the larger elements of a bathroom. Made from materials as diverse as metal, glass, bamboo and stone, they are an opportunity to splurge as well as get creative. Bathroom faucets come in three basic categories: lavatory sets (for sinks); tub sets (for bathtubs); and shower sets (for showers, including showerheads and handheld sprays). Each product line a faucet manufacturer offers will have lavatory, tub and shower set examples of the line so you can match them in your bathroom. Just keep in mind that the more elaborate the design, the harder your faucet set will be to keep clean.

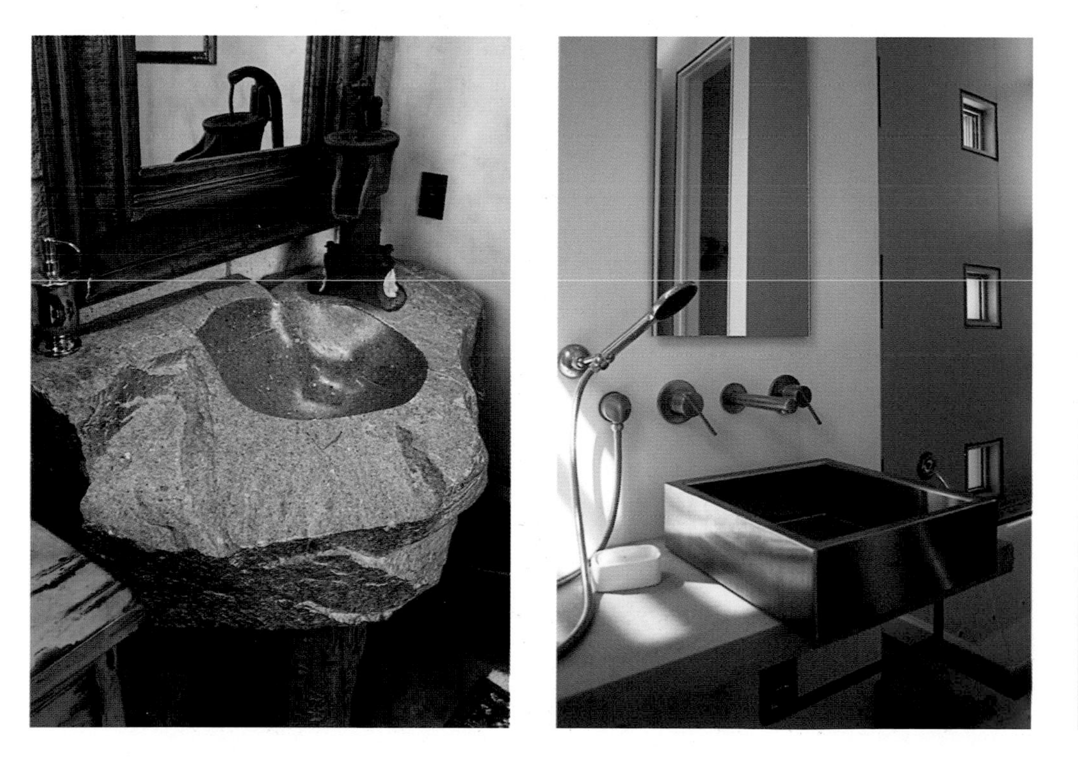

Left: Architect Richard Landry found this granite boulder in the rubble pile of a stonemason. A carved and polished basin and some hand-hewn beams turned it into a sink that is also a conversation piece. **Center:** A square stainless-steel sink seemingly balanced on the edge of the counter helps define this contemporary bathroom. **Right:** Bamboo-like copper faucets and a stone vessel sink set in a concrete vanity.

Opposite: This walk-in shower has it all: marble slab counters, fogless shaving mirror, piped-in music, and lots and lots of space. Small ceramic tiles and a glass sink add sparkle, while the caged light fixture adds an industrial touch. There are two rainforest shower heads: one for sitting and one for standing. **Left:** An old-fashioned clawfoot tub set against a dramatic barnboard and round-beamed frame.

Chrome faucets and fixtures are always a practical choice because chrome is so durable and easy to clean. Brass and bronze are sometimes desired from a design point of view, but these materials tend to tarnish, scratch and discolor over time. Happily, there is a new technique called physical vapor deposition (PVD), which deposits a clear protective coating on hardware. Be careful: it doesn't stand up well to abrasive cleaners. For more information on faucets, see the Kitchen Fittings chapter.

Especially when designing a master ensuite bath, you can be a bit more adventurous when it comes to choosing your faucets. The ultimate in anti-high-tech, a rainforest shower, which can be as large as 12 inches across, lets water gently fall from holes directly above in an experience that is more like warm rain than bracing blast. Or you can install an inset shower grid in the entire ceiling, programmed to provide every kind of water experience from a misty day to a full-blown deluge complete with lightning and perhaps a little Wagner piped in for full effect. One kind of tub fixture that combines function with beauty also has a body sprayer on it — the antique style is sometimes called a "French telephone" (see page 431).

When choosing from the dizzying array of finishes available, you may want to decide to use one finish throughout your home for a homogenous look, or select contrasting finishes for a more complex décor. In terms of levers and lever finishes, most manufacturers offer several options. Brushed or matte finishes are easier to maintain because they don't show scratches or water spots as much as polished finishes. A roundup of the most common materials and finishes follows:

CHROME Since the 1930s, chrome has been the winner in terms of durability, low maintenance and cost. Its popularity is also one its drawbacks, as it is seen everywhere. Also, the unrelieved brightness of chrome may be shunned by those who long for more subtle or sensuous materials that are brushed or otherwise matte. (And the more highly polished the finish, the more it will show water spots and fingerprints.)

BRASS Popular a hundred years before chrome, brass boasts a certain historic authenticity to accompany its charming patina. Brass needs plenty of polishing to maintain the shine, but many manufacturers apply a coating to keep it bright.

NICKEL Nickel has gone in and out of style over the past century and is currently back in, mainly due to its versatility. Nickel can be made to look like almost anything, from the matte subtlety of brushed or satin finishes, to the high shine of the polished variety. Brushed metal can look good with other metal contrasting detailing such as shiny brass or copper.

STAINLESS STEEL Stainless steel is usually chosen to go with stainless-steel sinks and tubs. It can be chemically aged or antiqued.

ENAMELED PLASTIC As expected, enameled plastic is inexpensive and available in a wide variety of non-metallic colors. It can come with metal detailing.

Left: A shallow steel sink floats like a glob of mercury in a glass countertop. **Center:** Put a bookcase lined with leather-bound books, an elaborately mullioned round window, a chandelier and a marble corner sink and you have a powder room that looks charmingly like a small men's club. **Right:** Sometimes simplicity of design and materials works best, as in this metal background and shelf supporting a polished stone sink with the outside left rough.

products

bathroom fixtures

acrylic

Atlantis *Neptune.* Metal is available in chrome or brushed nickel.

Neptune R2

Ruby

Zen 3466R

Victoria

Zen 3466R

Victoria

solid surface & polyethylene

Mini Bath 26 *Durat*

Mini Bath 27

Soikko Bath 30

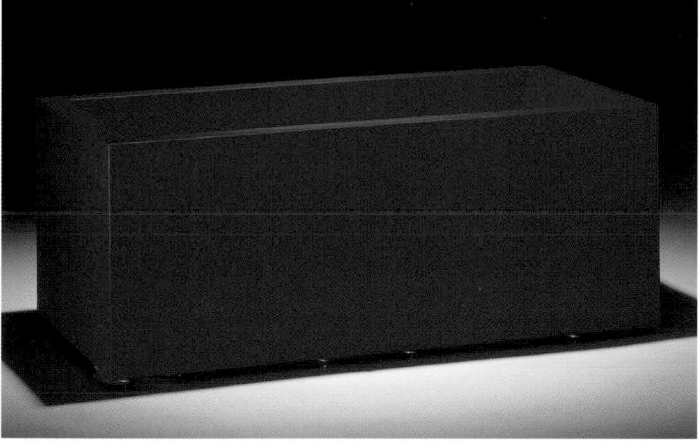

Mini Bath 44

Soikko Bath 41

Two Italian glow-in-the-dark polyethylene tubs with either an internal multicolored or white LED lighting *Lavabo*

Rapsel Kea *Lavabo*

metal

Stainless Steel Dish tub *Diamond Spa*

Shenandoah, hickory and copper

Waterfall stainless steel

Cast iron double slipper tub with gothic feet *Signature Hardware*

French Bateau (boat bath) with riveted skirt, cast iron *Signature Hardware*

Stainless Steel Japanese soaking tub with front skirt *Diamond Spa* (all)

Copper Japanese soaking tub

Stainless Steel Bamboo tub

stone

Carved stone boulder, *Stone Forest*

Carrera marble Roman tub *Stone Forest*

Stone tub with rolled rim *Stone Forest*

Carrera marble tub

Carved granite boulder tub

Double Wave soaking tub, concrete *Sonoma Cast Stone*

wood

Agape Woodline wooden tub *Lavabo*

manufactured material

Wash Basin, manufactured stone sink and steel base *Durat*

Vati Basin, manufactured stone *Durat*

Basin Kehto

Kippo Shelf basin, manufactured stone *Durat*

Italian polyethylene sinks with embedded LED lights *Lavabo*

metal

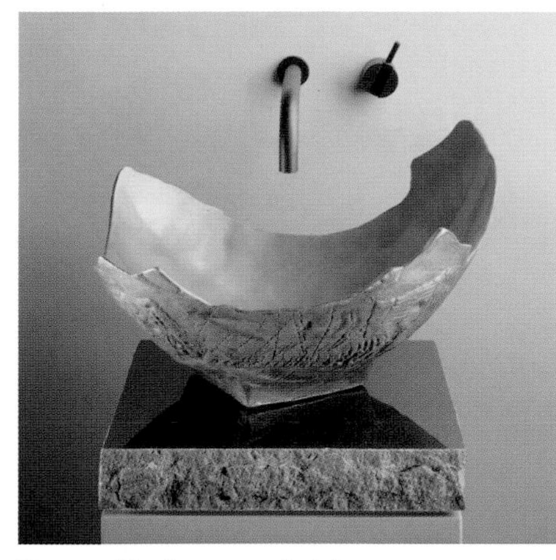

Bronze Chalice vessel sink
Stone Forest

Bronze Verona vessel sink

Hand-crafted copper vessel sink

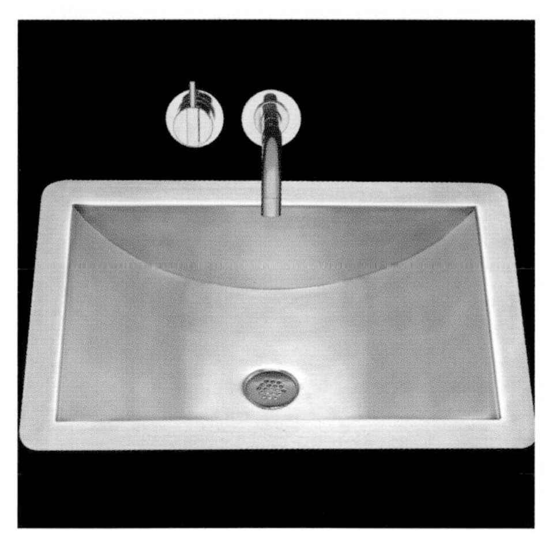

Stainless Steel Wave sink in satin finish, S1218 *Bates & Bates*

Sand-cast alloy Vallarta, A1517V

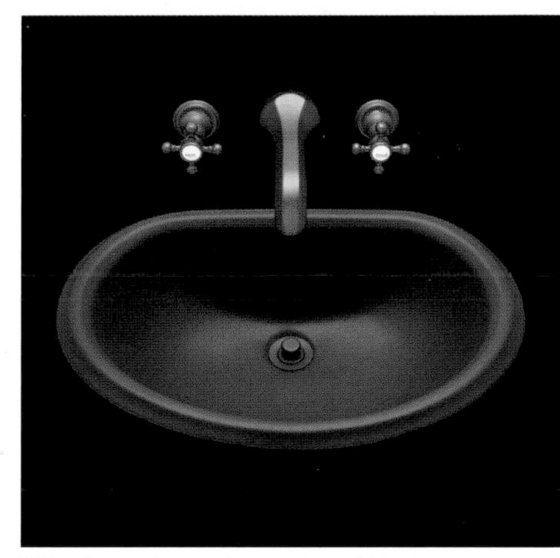

Weathered cast iron, Jane P1521

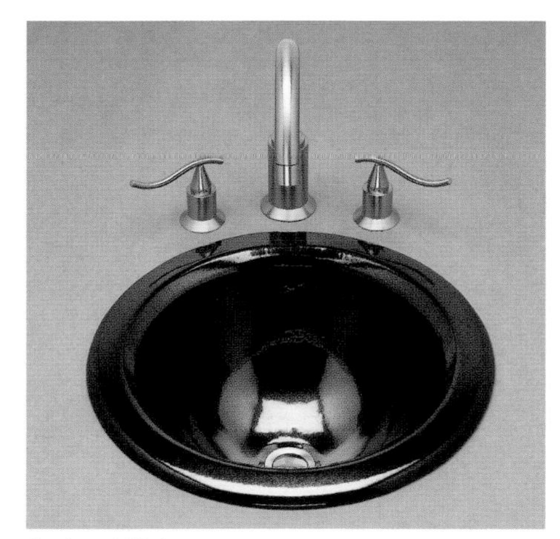

Amber Wildpewter
Bates & Bates

Luna Lisa bronze vessel
Firebowl

Concept II polished nickel, CP-16
Stone Forest

stone

Antonio lupi barrel, floor-mounted pietra juta stone sink *Lavabo*

Farmhouse sink in beige granite, C04-10 *Stone Forest*

Zen vessel blaxk granite C36-BL *Stone Forest*

Concrete Zin vessel *Sonoma Cast Stone*

Milano Vessel honed carrara mable, C-52 *Stone Forest*

ceramic

Hand-painted counter sink *Purple Sage*

Wall mounted fire clay basin *Sonia*

Hartley square countertop basin, porcelain *Signature Hardware*

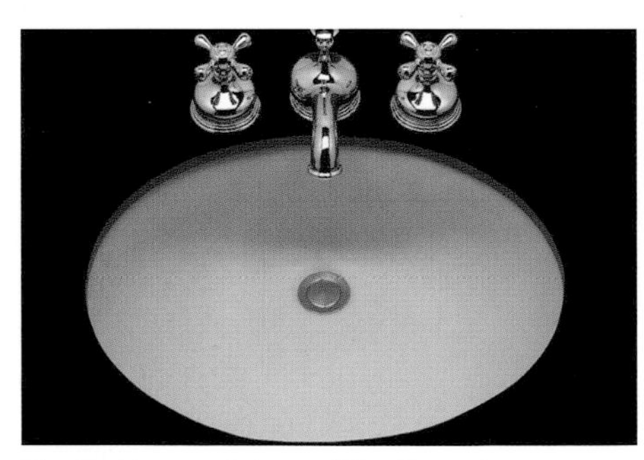

Doreen in sandstone, Weathered Collection *Bates & Bates*

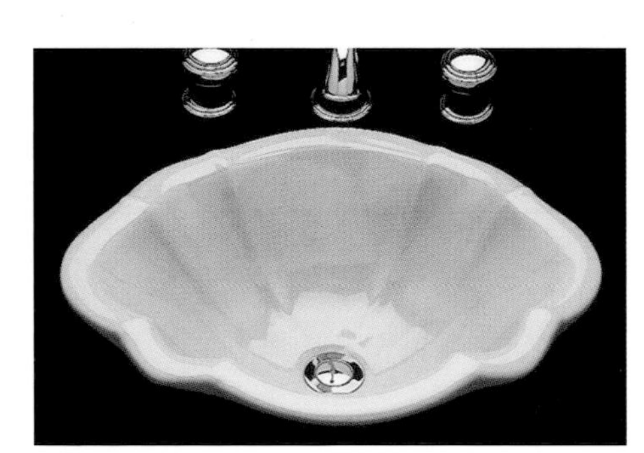

Erin in Ice Gray, ceramic *Bates & Bates*

Heritage countertop sink, vitreous china (porcelain) *American Standard*

glass

Glass sink GS-102 *Cantrio Koncepts*

Solid glass GS-107

Glass sink GS-112

Delta 16 glass sink *Sonia* (all)

Frosted basin

Turin frosted glass vanity

wood

Rectangular wooden sink WS-007 *Contrio Koncepts* (all)

Round wooden sink

toilets

Ideal Standard (small). This compact wall unit integrates a wc and a bidet with a toilet roll holder and chrome faucet for the bidet *Lavabo*

Pozzi-ginori easy.02 ceramic toilet from Italy. Wall or surface mounted

Simas Frozen wall mounted toilet and bidet

Parma *Neptune*

Murano *Neptune*

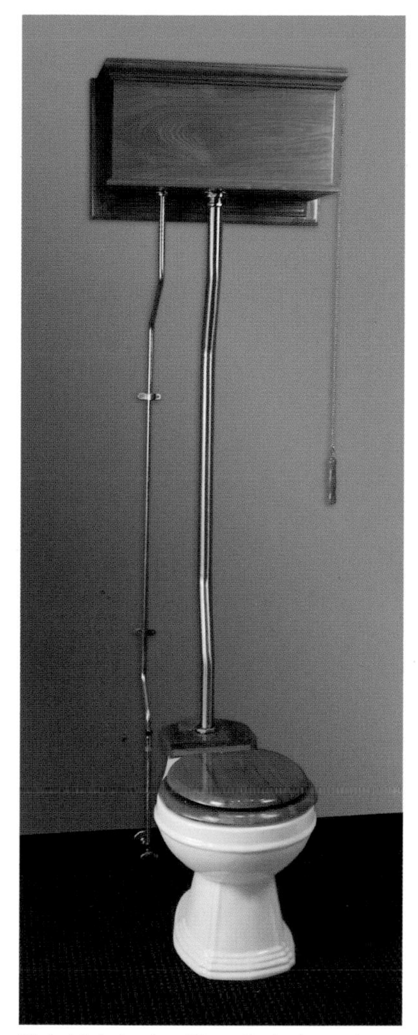

Oak high tank with pull chain and Victorian bowl *Signature Hardware*

showers

Aquatower 3000, 12-24 water jets *Grohe*

Antonio Lupi Sella, wall mounted ceramic *Lavabo*

Hand-painted urinal *Purple Sage*

Agape Chiicciola standalone shower in translucent Parapan *Lavabo*

sink faucets

Ardsley vessel bath faucet
American Standard

Art Deco 2, chrome *THG*

Tenso, single-handle center set *Grohe*

Gavroche, chrome *THG*

Shao Cross, chrome *THG*

Oceania 2, Cristal de Lalique Collection
THG (all)

Faubourg Metal a Manettes

Shao Lever

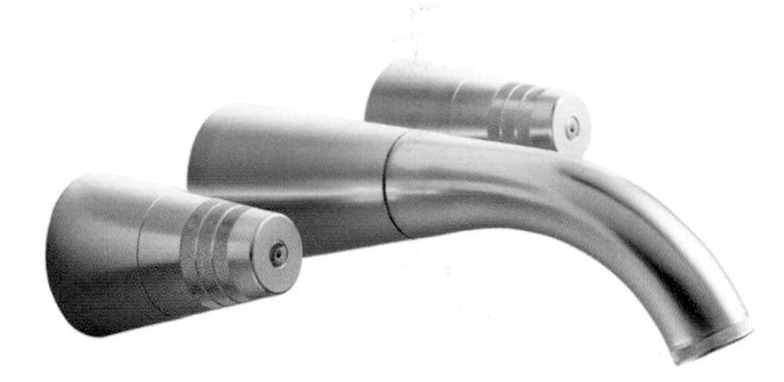

Wall-mounted vessel faucet F1 *Grohe*

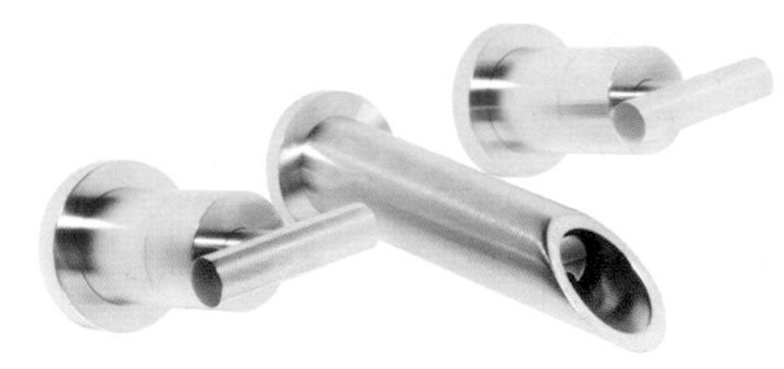

Wall-mounted lav set, polished chrome *Waterdecor*

Profile, Jamie Drake Collection *THG*

Bamboo, Crystal and Chrome, Cristal de Lalique *THG* (all)

Beluga, lever handles

Cubica

Profile, Jamie Drake Collection *THG*

Fl single-hole center set, dual lever design, matte aluminum finish *Grohe*

sink faucets

Charleston, cross handles *THG*

Versailles lav set, antique nickel *THG*

Amour de Trianon, Classical Romance Collection *THG*

CiXX hand-forged brass *Sonoma Forge*

Tall Lav Faucet, 03302-393, polished copper *Waterdecor*

1900 lav set *THG*

WherEver deck mount, rustic copper finish *Sonoma Forge*

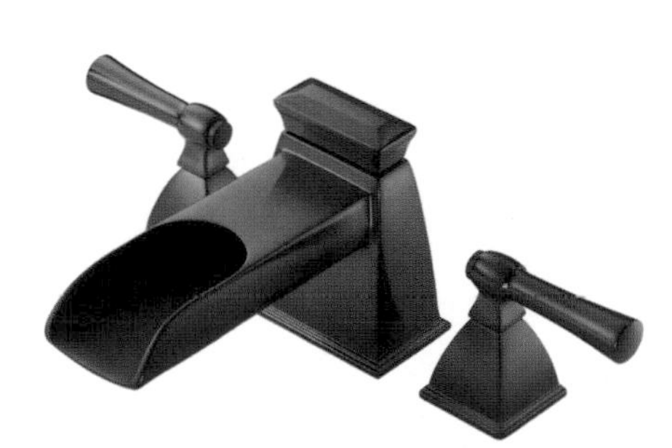

Vesi Channel, brushed bronze *Brizo*

Frou Frou with levers *THG*

tub fills

Tangara spillway tub filler *THG*

Copper tub fill Diamond Spa

Round escutcheons, Roswell knobs and Arched spout in white bronze *Rocky Mountain Hardware*

Tete Lion spout *THG*

Art Deco tub filler set *THG*

Fantinala tub spout *Lavabo*

bidet taps

Amarilis two-handle bidet faucet *American Standard*

Hampton cross bidet faucets *American Standard*

Townhouse bidet set *American Standard*

showerheads

Ceiling-mounted rain shower, polished copper lacquer *Waterdecor*

Metropolis shower head, Cristal de Lalique *THG*

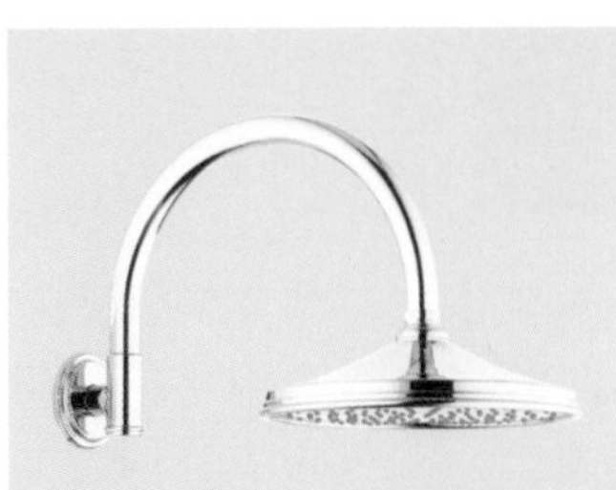

Rainshower head *Grohe*

Modern rainfall shower head *Signature Hardware*

Art Deco Shower unit *THG*

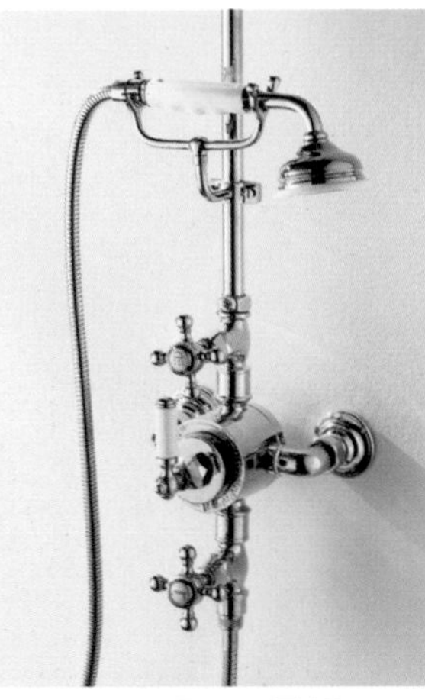

Bain Douche Retro *THG*

Custom shower *Grohe*

Freehander with rotating shower heads *Grohe*

In-wall shower set *MGS*

Thermostatic shower column with bath spout *MGS*

Shower and tub fill *Sonoma Forge*

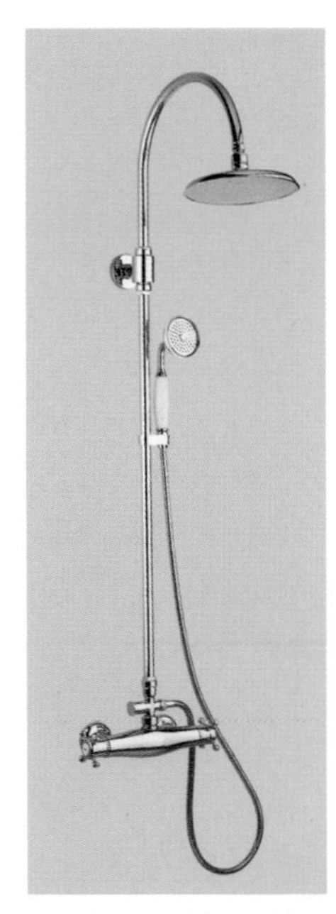

Tradition a Manettes shower head *THG*

modular fixtures

Teak and stainless-steel modular bathroom suite system by Troy Adams *Julien (all)*

Teak cover slides back to reveal toilet

Bathtub cover

Sliding teak cover over stainless steel sink can be removed

A modern take on the traditional Japanese soaking tub in stainless steel and teak

hardware

We're using a lot more metals, and emphasizing metals that change such as antique brass. Brass was almost completely out, but now it has come back but in an unlacquered style so that it dulls and turns almost a bronze. Antique nickel is very nice but also polished nickel for door and cabinet hardware and for details on fireplaces and cabinetry, lighting and drapery hardware.

BRIAN GLUCKSTEIN, DESIGNER

YOUR CHOICES OF HARDWARE — DOOR AND CUPBOARD PULLS AND handles, knobs and hasps and grillwork, hinges and fittings — those most inherently useful objects, can elevate the very nice to the absolutely spectacular. Or they can make your kitchen cabinets, for instance, look like your modest great-grandmother flashing assorted piercings and tattoos when she takes off her coat at church — not quite right! For that reason alone, you can't spend too much time deciding on and selecting your hardware.

Fortunately there is an almost overwhelming variety of hardware, decorative and utilitarian, to choose from. Two broad categories dominate the industry: old and new. The "old" consists of either authentic antiques or reproductions. "Antique" can, of course, include nominally modern styles such as Art Deco and Art Moderne.

The ageless beauty of handcrafted door hardware.

A selection of antique reproduction hardware from Lee Valley Tools. Clockwise from top left: Cast-steel shelf bracket, three sand-cast solid brass Victorian bin pulls, a cast-steel Victorian screen door bracket, cast-steel shelf bracket, cast-steel handrail bracket; mid left: Victorian screen door hardware; center: three cast door hinges.

This latter path has been made easier by the plethora of historical reproduction companies (see Sources) that specialize in taking a sample or a photograph of a hasp, a piece of trim or a doorknob and reproducing it exactly. Many of these companies are small or sole-proprietor forges that use the age-old traditions of the blacksmith to pound out intricate contrivances from hot metal, or use old-time casting technologies to make reproductions out of iron or aluminum. And of course the "new," or contemporary, category comes from the thousands of manufacturers big and small that design and make hardware to suit every residential style and work with every functional piece.

For most of us, the hardware choices that will demand the most attention are in the devices designed to open and close doors, drawers and cabinets. These fall into several main categories.

KNOBS Knobs are good for kitchen cabinets and drawers because their generally round edges don't accumulate food and are easy to clean. Knobs also tend to be more decorative than pulls and so can offer greater options where you are more interested in making a design statement. And, because they are usually attached with a central screw and don't have to span the space that the more elongated pulls do, material that is relatively fragile, such as glass, can be used (in high-use areas, a more durable acrylic can be substituted for glass). Requiring only one hole makes it easier to replace knobs, whereas pulls can vary in length and have to be matched with the existing drill holes (knobs are usually from between ¾ inch to 2 inches wide). Knobs attached to drawers should be the same size or bigger than those on the doors.

PULLS Because they are anchored with two screws at either end, pulls can be made from flexible material such as rope and leather as well as rigid materials like metal or wood. However, because they are exposed to more pressure in the middle, they have to be made of fairly durable material. A drop pull is a regular pull that swivels at each end and usually hangs down a little for a more elegant appearance. A good rule of thumb is that the longer the pull, the more contemporary it will look.

CUP, OR BIN PULLS These hollow, half-moon pulls instantly say "antique" and are therefore perfect for a more traditional look, although their downscale, hardware store appearance makes them suitable in any kind of minimalist, industrial or loft setting. Cup pulls are usually made of metal.

If you want the clean, simplistic look of no hardware at all, consider dispensing with hardware altogether and choosing cutouts, holes or spring hinges that pop open with a push.

Left: An abundance of texture in this iron doorknob pounded by hand, set in a weathered barnboard door. Center: A simple rope pull as a door knob. Right: This brass pull, which resembles a half shell, works best where simplicity and sensuous curves are called for.

Left: These door pulls are made from Elk "knuckles," or knee bones. **Center:** The clean lines of a stainless-steel doorknocker with a pewter finish and a basket weave accent, from Bouvet Hardware. **Right:** Stainless steel against a blue-stained wood background.

Traditionally, metal hardware (the "hard" aspect of hardware) was hand-forged using heat to make the metal malleable and then beating it into the desired shape. Today, this "traditional" look is underlined with a roughness that incorporates small beat marks or intentional imperfections. Ironically, the hallmark of the blacksmith's art used to consist of a smooth, unblemished surface that the craft persons of old were remarkably adept at supplying.

The trick with hardware finish is to make sure it goes with all of the other hardware and décor aspects of the room it is in. Will those long, stainless-steel pulls go with your Shaker cabinets? Will the glass knobs complement the brushed nickel faucets? Another detail to consider is the screw heads — do they match the finish of the hardware?

Like food (stilton cheese and port is a good example), there are some classic hardware matches. Porcelain used with soapstone or granite, for instance, always looks good together. Forged iron or stainless steel and dark oak is another good match. And stainless-steel or satin nickel hardware always looks good with stainless steel appliances. Wooden knobs are a natural choice with Colonial architecture. Using natural materials — stone, unlacquered brass, oil-rubbed bronze, iron, unleaded copper — and not protecting them from the ravages of time, weather and wear gives you what is known as a "living finish" that is becoming more and more popular.

One caveat to this old and worn look: not everything called an "antique" actually is. An "antique original" is a piece that was actually made in the period it comes from, while an "antique reproduction" is a copy of the original (often made with extremely high skill levels), and an "antique interpretation" is a hardware element that is based on a period style but with the modern home in mind.

These issues are just as important with door hardware: doorknobs, plates, hinges and knockers. Door hardware not only has to look good, it has to — in the case of exterior doors — also provide an element of security as well. To be safe, it's best to use hardware that has the highest security rating, which is ANSI Grade 1. And hinges that are extruded are better than those that have been stamped.

Door hardware comes in the usual period styles in North America. The hand-worked, battered and dimpled Colonial style goes with any period style up to Victorian and is often used on interior doors to complement hand-scraped floors and beams. The Victorian style itself is characterized by elegant curves, but also by the white porcelain and glass doorknobs that were introduced in the Victorian era. Art Deco is known for its hard angles and hard surfaces, often of

A collection of doorknobs that have photographic images baked onto them.

chrome, ebony, plastic or nickel. The strength and cool, post-industrial good looks of stainless steel makes it very compatible with Contemporary door styles, often using knobs or levers that are very simple in style and used without a plate. Stainless steel comes with a satin (matte) polished finish.

Another modern material to think of for door hardware (usually pulls) is colored nylon. Normally seen on doors leading to public buildings such as hospitals or libraries, colored nylon (either solid or as a covering for metal) comes in a wide range of colors and, with its déclassé charm, is perfect for any post-modern home. (In the near future, look for streamlined hardware that is keyless and electronic, perhaps with a card swipe made of wrought iron, bronze or some other material normally used for keyholes.)

It is important that the door hardware you choose fits the overall style of the house. Having said that, there is a certain amount of leeway with hardware, in that you can use it to accent one particular area or room despite (and in some cases, *because* of) being a different style.

Entry hardware especially can make a strong impression. The lockset — deadbolt plus door handle combination — should be of a scale and material and ornateness (or lack of decoration) that suits the door. Thus a simple, straight-lined chrome lockset would work well on a plain mahogany slab entrance door on a modern Ranch home. A Tudor-style home, on the other hand, with dark wood trim and a heavy oak door, might be better served by heavy, burnished bronze hardware, perhaps including an ornate door knocker that in its design echoes the carvings in the door. The following pages and the Sources section at the back of this book offer ideas and possibilities to finish off your home with unique and fabulous hardware.

Left: This custom designed door handle is called a break-away lever, and is meant to look like two big handles that break away in the middle. The handles are brushed stainless steel and the door is mahogany. **Center:** The white bronze of this rectangular thumb latch entry set by Rocky Mountain Hardware gives it a contemporary flair. **Right:** Polished bronze Heroic door pulls from the Hedgerow Collection, Martin Pierce Hardware.

products

hardware

knobs & plates

Acanthus Bead with Rosettes
Charleston Hardware

Parasol Rice Dummy with Square Rosettes

Butterfly with Rosettes

Edwards: Eastlake knob set
Rejuvenation Hardware

Colonial Revival Putman
Rejuvenation Hardware

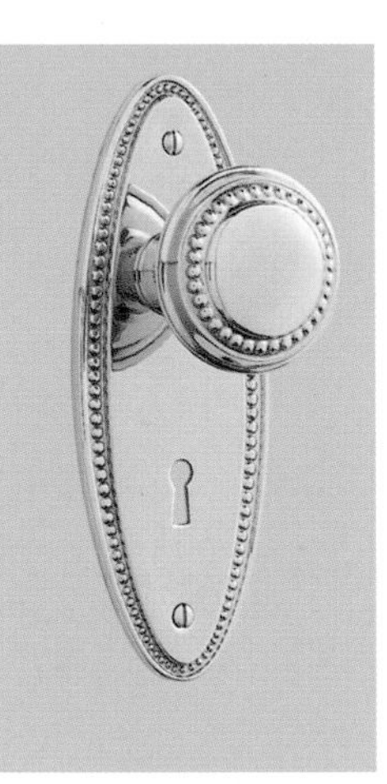

Durham beaded oval style *Rejuvenation*

Blue Birds Doorknob
Charleston Hardware

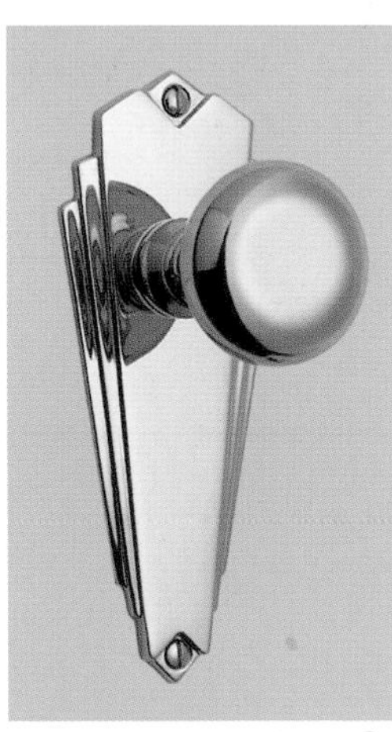

Art Deco Jenkins
Rejuvenation

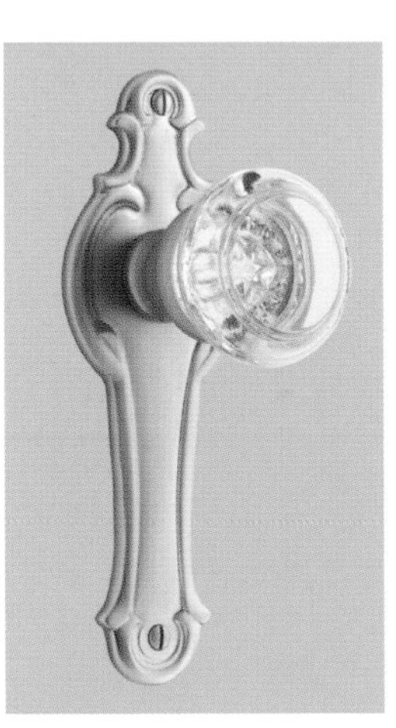

Wentworth Colonial Revival*Rejuvenation*

Oriental Doorknob with Rosettes
Charleston Hardware

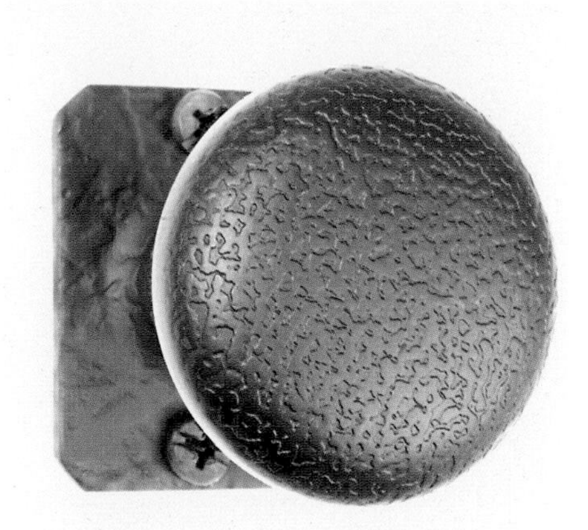

Rugbi, rough iron
Acorn Manufacturing

Rosette Trim
La Porcelaine

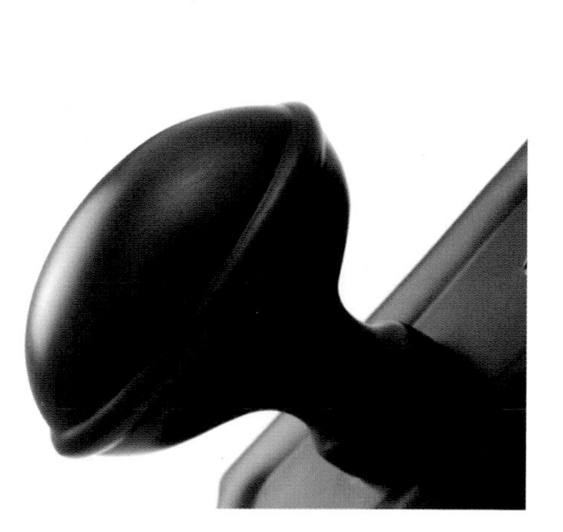

Oil-rubbed bronze
Belvedere

Ornate set, antiqued brass
Whitechapel

Brass, satin nickel

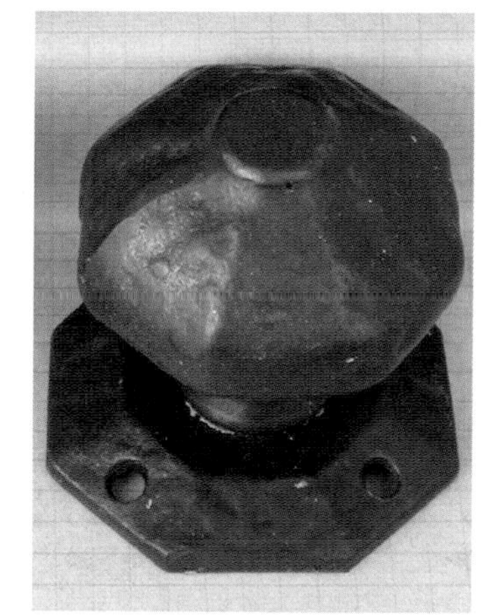

Black Iron Octagonal Rustic

Black Iron Flower Rustic

Rosette set, oil-rubbed bronze 2228 *Belvedere*

Rosette set, European pewter 2232

Satin Nickel 2226

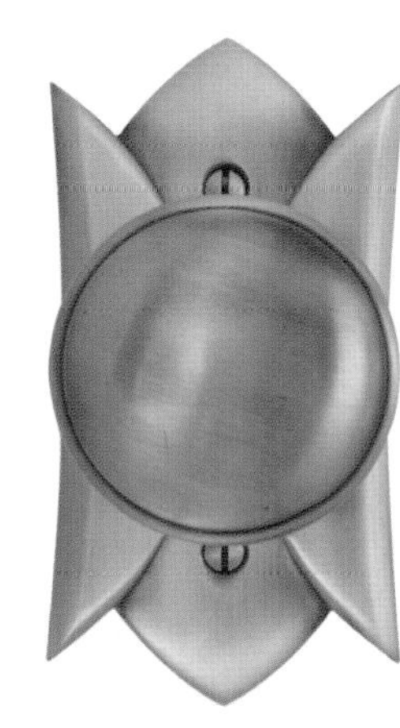

Satin Antique Brass 2236

Bronze patina *Bouvet*

handles & pulls

Porcelain handle with rosette trim 2391
La Porcelaine

Porcelain handle with rosette trim 2393

Porcelain handle with rosette trim 2396

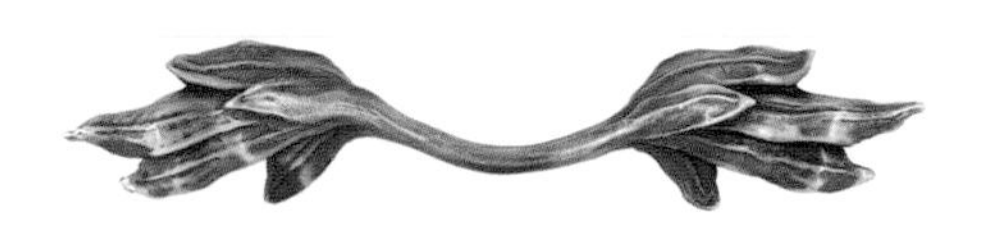

Willow passageway lever
Martin Pierce Hardware

Leaf cabinet pull
Martin Pierce Hardware

Rough iron lever
Acorn Manufacturing

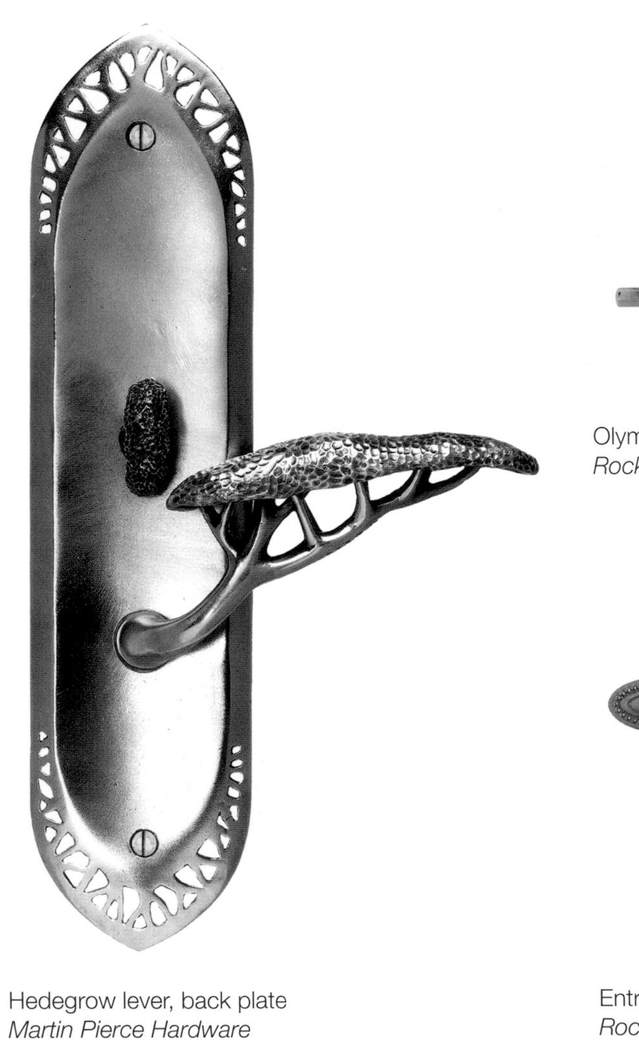

Hedegrow lever, back plate
Martin Pierce Hardware

Olympus lever in white bronze, dark patina
Rocky Mountain Hardware

Bronze Patina entry set 2601
Bouvet Hardware

Entry set E591 and E597
Rocky Mountain Hardware

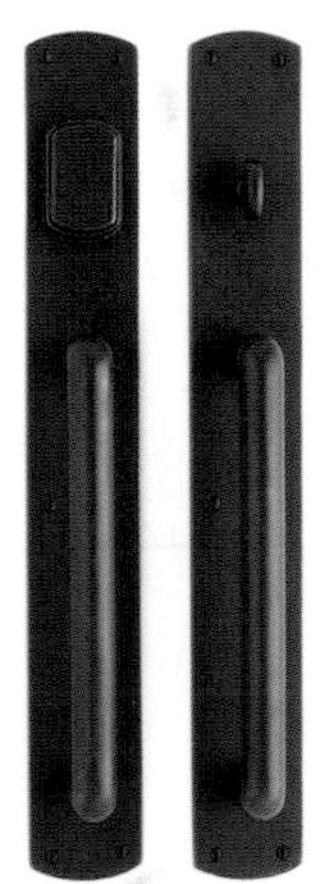

Entry set G501 and G502

Black Iron Rattail Rustic lever set *Whitechapel*

Adobe Style *Acorn*

Adobe Style

Lever set *Bouvet*

Iron entry handles
Rocky Mountain Hardware

Basket weave handle with rosette trim *La Forge*

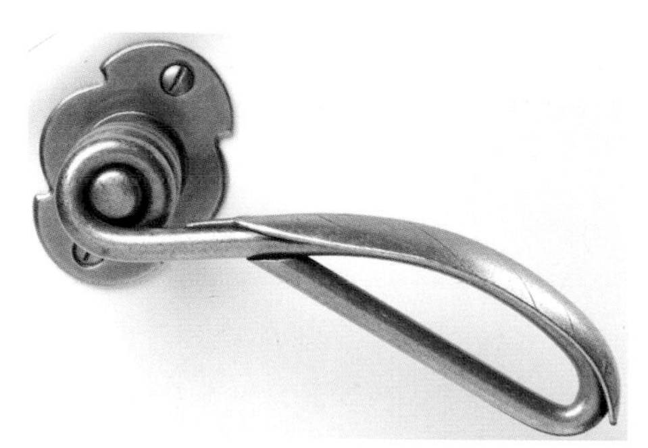

Lever Hematites Collection *Bouvet*

Pull, satin nickel *Belvedere*

Rosette lever, oil-rubbed bronze *Belvedere*

hinges, hasps & grilles

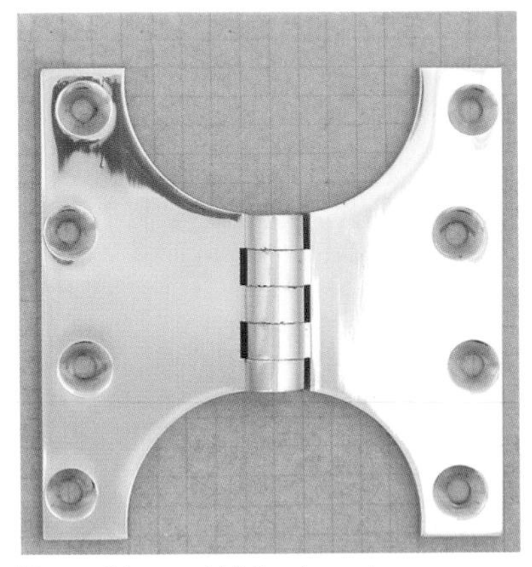
Brass hinges *Whitechapel*

Forged hinge
House of Antiques

Forged-iron hinges *Acorn Manufacturing*

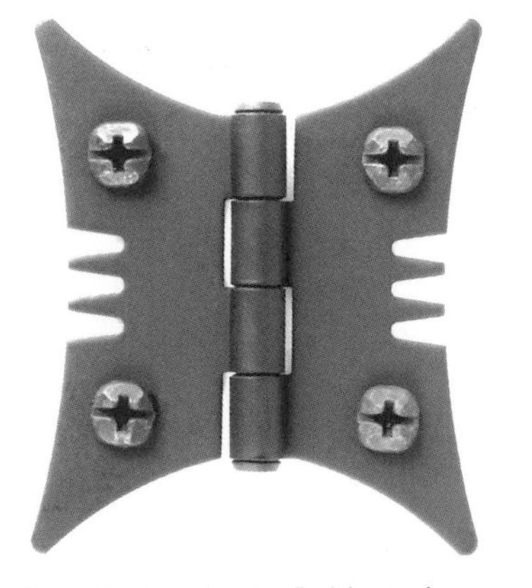
Smooth iron butterfly hinge *Acorn*

Smooth iron cabinet latch *Acorn*

Back iron hasp *Whitechapel*

Forged speaker grilles *Kayne & Son*

knockers & bells

Ring door knocker *Whitechapel*

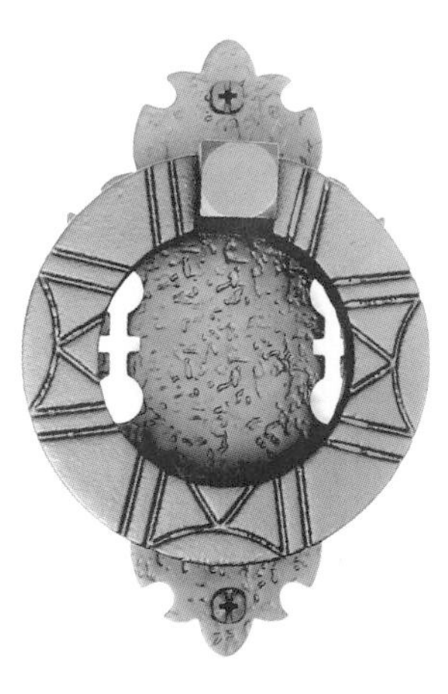
Forged doorknocker *Acorn*

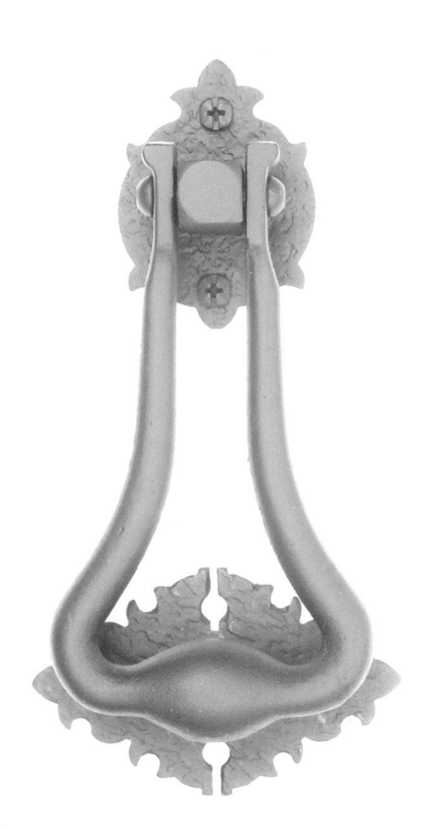
Warwick Iron *Acorn*

Antiqued doorknocker *Bouvet*

Forged doorknocker *La Forge*

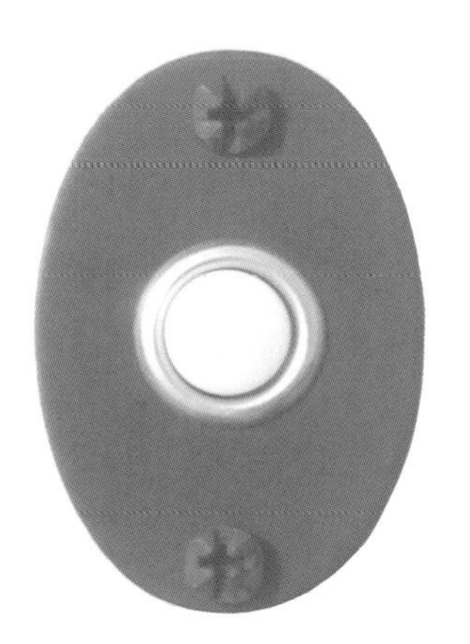
Smooth iron bell push *Acorn*

Adobe Style bell push *Acorn*

Ornate antique bell push
House of Antiques

knobs

Small Isabella knob *Susan Goldstick*

Cabinet knobs *La Porcelaine*

Sand cast knob *Du Verre*

Stainless steel knob *Acorn*

Pyramid knob *Talisman Foundry*

Steel knob *Acorn*

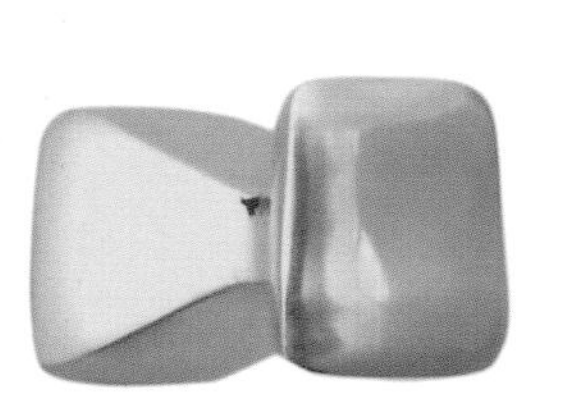

Polished aluminum knob *Du Verre*

Polished aluminum knob

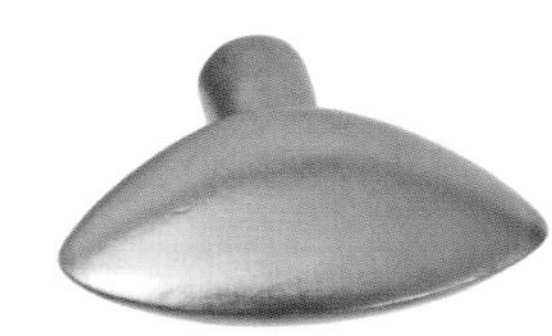

Matte brass knob

Black aluminum knob

Cabinet knob on cabinet escutcheon *La Porcelaine*

Grid Talisman *Foundry*

Mission knob *Rejuvenation*

Laurel Dot, bronze *St. Gaurdens Metal*

Palm knob, silver

Palm Knob, bronze

Illuminated knob Leo colored *Freego Design*

Leo White illuminated knob *Freego Design*

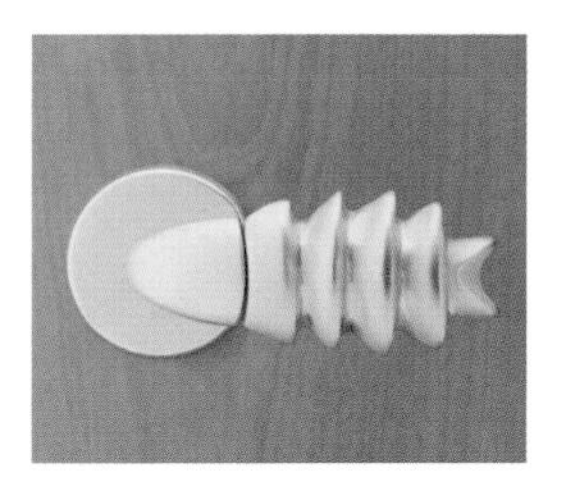

Cult knob *Freego*

Contemporaine Cubica *THG*

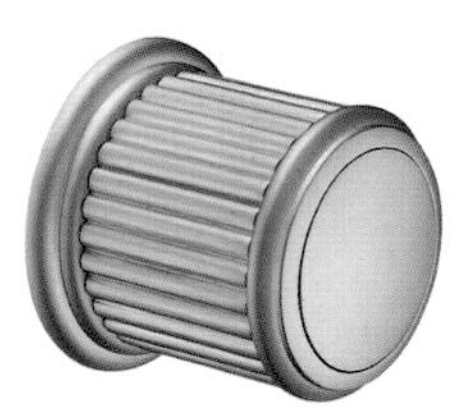

Contemporaine Acropole *THG*

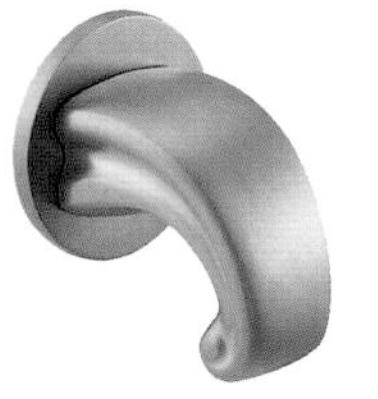

Baroque Capitole

Cristal Chantilly Sapphire

Baroque Chantilly

Pierre Cheverny

Cristal de Lalique Bellagio

Soapstone knobs *Green Mountain Soapstone*

Cristal de Lalique Cactus *THG*

Metroplis

Cristal de Lalique Thais

Baroque Elysee *THG* (all)

Cristal Luna

Pierre Medicis

Polished aluminum knobs *Du Verre*

handles & pulls

Jeweled handles *Susan Goldstick*

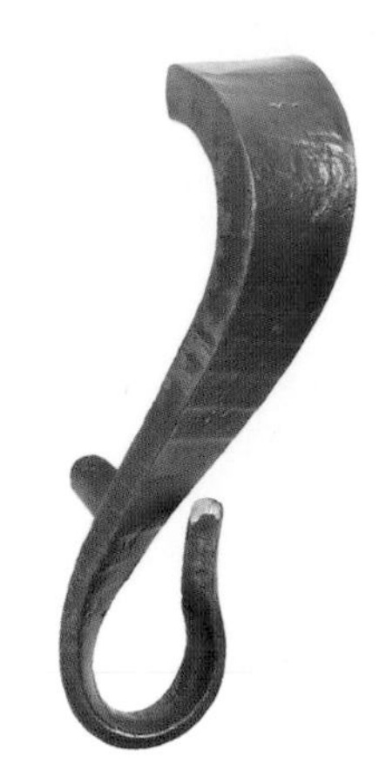

Heinz Pfleger forged *Du Verre*

Heinz Pfleger forged

Heinz Pfleger forged

Porcelain door pull *La Porcelaine*

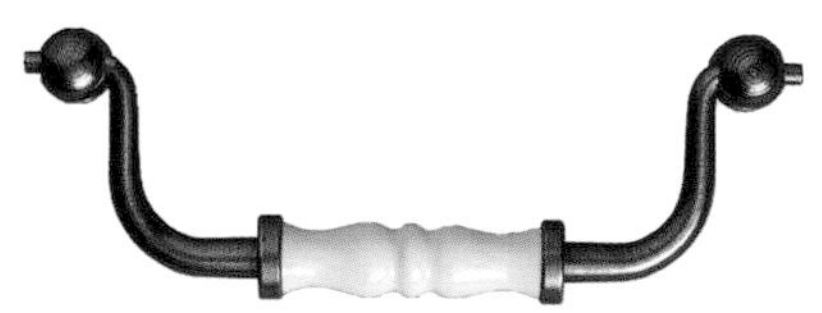
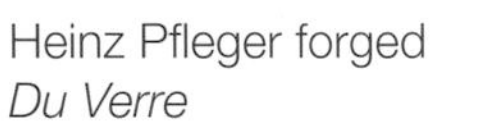

Porcelain cabinet drop pull *La Porcelaine*

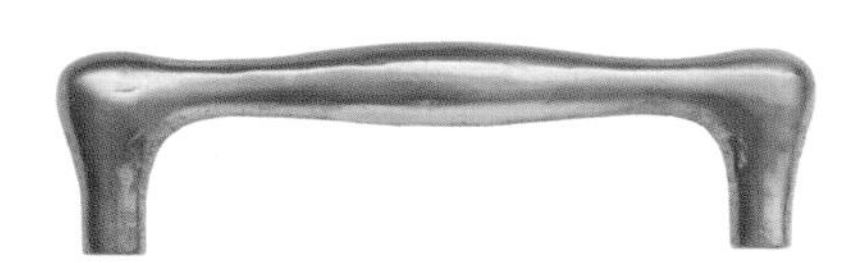

Brass Foundry Art pull *Talisman*

Iron Art pull 1R8JP *Acorn Manufacturing*

Iron Art pull 1PHJP *Acorn*

Black Matte iron pull FC17-B *Du Verre*

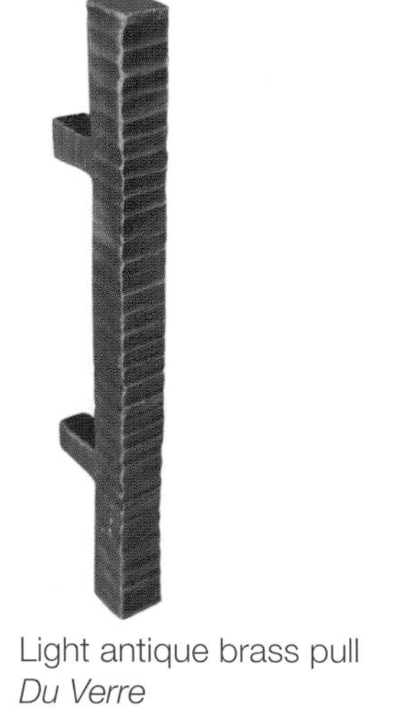

Light antique brass pull *Du Verre*

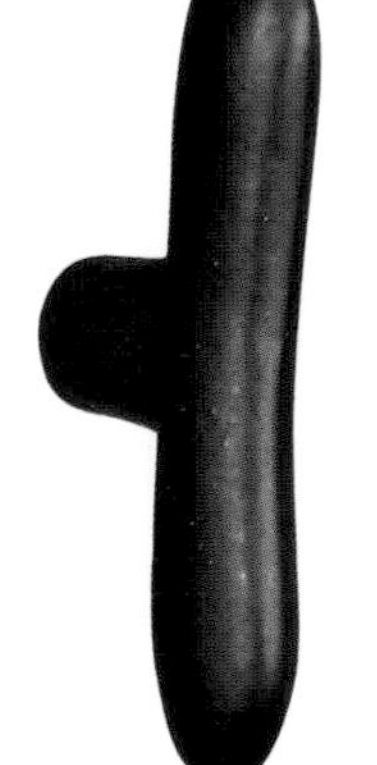

Black aluminum pull

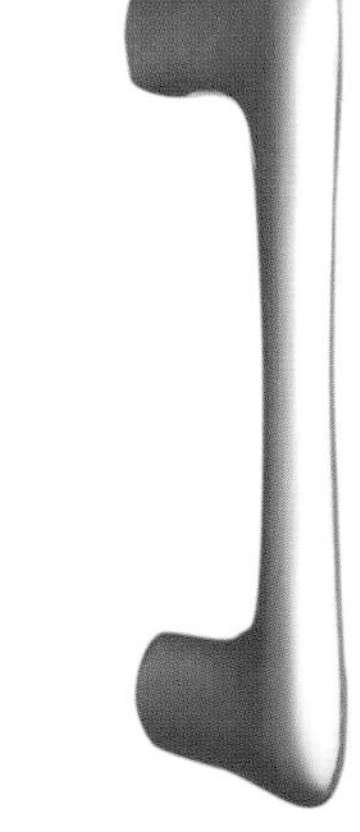

Matte brass pull

Black aluminum pull

Bamboo pull in cast aluminum

Bin pull, brass *Acorn Manufacturing*

Bin pull, black *Acorn*

Black aluminum pull *Du Verre*

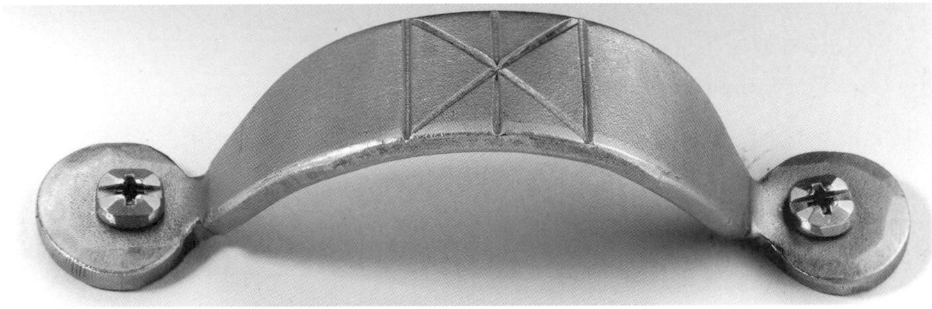

Forged-iron pull *Acorn*

Stainless-steel Colonial pull *Acorn*

Black aluminum pull *Du Verre*

Textured rectangular pull stainless steel *Acorn*

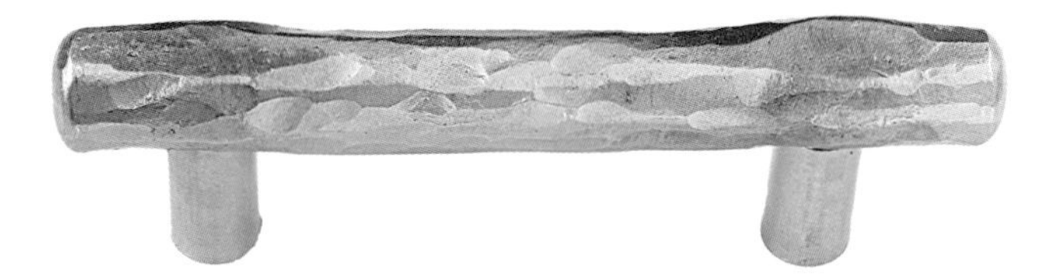
Stainless-steel branch pull *Acorn*

Light antique brass pull *Du Verre*

sources

architects & designers

Aleks Istanbullu Architects
1659 11 Street, Ste 200
Santa Monica, CA 90404
Tel: 310-450-8346
Fax: 310-399-1888
aasken@ai-architects.com
www.ai-architects.com

Andre Kikoski, Architect
180 Varick Street, Ste 1316
New York, NY 10014
Tel: 212-627-0240
Fax: 212-627-0242
akikoski@akarchitect.com
www.akarchitect.com

Barton Myers Associates Inc.
1025 Westwood Boulevard
Los Angeles, CA 90024–2902
Tel: 310-208-2227
Fax: 310-208-2207
mail@bartonmyers.com
www.bartonmyers.com

Barton Phelps & Associates
5514 Wilshire Boulevard, 10th Floor
Los Angeles, CA 90036-3829
Tel: 323-934-8615
Fax: 323-934-3289
bpala@aol.com
www.bpala.com

Breath Architects: Martin Liefhebber, Marco Vandermaas
177 First Avenue
Toronto, ON M4M 1X3
Tel: 416 469-0018
Fax: 416 469-0987
info@breathebyassociation.com
www.breathebyassociation.com

Burns and Beyerl Architects, Inc.
300-601 South LaSalle Street
Chicago, IL 60605
Tel: 312-663-0222
Fax: 312-663-9552
bbachicago@aol.com
www.bbaworld.com

Cecconi Simone
1335 Dundas Street W
Toronto, ON M6J 1Y3
Tel: 416-588-5900
Fax: 416-588-7424
info@cecconisimone.com
www.cecconisimone.com

DCA Design Inc.
220 Avenue Road
Toronto, ON M5R 2J4
Tel: 416-531-8155
Fax: 416-531-6558
www.dcadesign.ca

Dean/Wolf Architects
40 Hudson Street,
Sixth Floor
New York, NY 10013
Tel: 212-385-1170
Fax: 212-385-1174
office@dean-wolf.com
www.dean-wolf.com

The Design-Build Store
1046 Yonge Street
Toronto, ON M4W 2L1
255 Sunrise Avenue, Ste 200
Palm Beach, FL 33480
Toll-free tel: 866-889-9919
www.designbuildstore.com

Douglas Architects and Planning
7522 E McDonald Drive
Scottsdale, AZ 85250
Tel: 602-951-2242
Fax: 602-951-0165
johndouglas@douglasarchitects.com
www.douglasarchitects.com

Eric Brandt Architect
PO Box 1014
2940 Southwest Drive
Sedona, AZ 86339
Tel: 928-203-9918
Tel (cell): 928-821-3617
eba@sedona.net
www.brandtarchitect.com

Faulding Architecture and F2Inc.
2-106 E 19th Street
New York, NY 10003
Tel: 212-253-1513
Fax: 212-358-7053
infor@f2inc.com
www.f2inc.com

Garret Eakin Architect
1000 Woodbine
Oak Park, IL 60302
Tel: 708-660-0315
Fax: 708-660-0316
garreteakin@hotmail.com
garreteakin@comcast.net

Gluckstein Design Planning
234 Davenport Road
Toronto, ON M5R 1J6
Tel: 416-928-2067
Fax: 416-928-2114
www.glucksteindesign.com

Gordon Ridgely Architects & Associates Inc.
281 Avenue Road
Toronto, ON M4V 2G8
Tel: 416-921-5130
Fax: 416-921-5209
Info@gordonridgelyarchitects.com
www.gordonridgelyarchitects.com

Harry Teague Architects
412 N Mill Street
Aspen, CO 81611
Tel: 970-925-2556
Fax: 970-925-7981
info@teaguearch.com
www.harryteaguearchitects.com

Horticultural Design
1610 Bayview Avenue
Toronto, ON M4G 3B7
Toll-free tel: 800-771-2558
Tel: 416-488-7716
Fax: 416-488-7191
info@horticulturaldesign.com
www.horticulturaldesign.com

The Ideal Environment (Marilyn Lake and Brian Lee, Architects)
398 King Street E
Toronto, ON M5A 1K9
Tel: 416-363-7199
Fax: 416-363-5834
carola@theidealenvironment.com.
www.theidealenvironment.com

Jaffe Architectural Group Ltd.
920 N Franklin Street, Loft 401
Chicago, IL 60610
Tel: 312-475-1800
Fax: 312-475-1900
admin@jaffearchgroup.com
www.jaffearchgroup.com

James Cullion Architects
4 Claremont Park
Boston, MA 02118
Tel: 617-266-9356
Fax: 617-267-0648
jc@jamescullionarchitects.com
www.jamescullionarchitects.com

Johnson Chou Design
56 Berkely Street
Toronto, ON M5A 2W6
Tel: 416-703-6777
Fax: 416-703-7009
mail@johnsonchou.com
www.johnsonchou.com

Kuwabara Payne McKenna Blumberg Architects
322 King Street W, Third Floor
Toronto, ON M5V 1J2
Tel: 416-977-5104
Fax: 416-598-9840
kpmb@kpmbarchitects.com
www.kpmbarchitects.com

Landry Design Group Inc.
11333 Iowa Avenue
Los Angeles, CA 90025
Tel: 310-444-1404
Fax: 310-444-1405
contactus@landrydesign.net
www.landrydesigngroup.com

Levitt Goodman Architects
572 King Street W, Ste 300
Toronto, ON M5V 1M3
Tel: 416-203-7600
Fax: 416-203-3342
dean@levittgoodmanarchitects.com
www.levittgoodmanarchitects.com

Lipkin Warner Design and Planning
23400 Two Rivers Road, Ste 44
PO Box 2239
Basalt, CO 81621
Tel: 970-927-8473
Fax: 970-925-8487
dwarner@lipkinwarner.com
www.lipkinwarner.com

LOT-EK Architects
55 Little West
New York, NY 1001
Tel: 212-255-9326
info@thesnowshow.net
www.lot-ek.com

Lynn Appleby Design
12 Camden Street, Ste 201
Toronto, ON M5V 1V1
Tel: 416-361-0218
Fax: 416-361-1428
mail@lynnappleby.com
www.lynnappleby.com

Marpillero Pollak Architects
1-132 Duane Street
New York, NY 10013
Tel: 212-619 5560
Fax: 212-619 5561
mpstudio@aol.com
www.newyorkarchitects.com

Martynus-Tripp, Inc.
658 N Crescent Heights Boulevard
Los Angeles, CA 90048-2210
Tel: 323-651-4445
Fax: 323-651-4442
folks@martynus-tripp.com
www.martynus-tripp.com

Menders, Torrey and Spencer
123 N Washington Street
Boston, MA 02114
Tel: 617-227-1477
Fax: 617-227-2654
dtorrey@mendersarchitects.com
www.mendersarchitects.com

Michael Fuller Architects
Aspen, Basalt, Telluride, CO
Tel: 970-925-3021
Fax: 970-927-5366
contact@mfullerarchitects.com
www.mfullerarchitects.com

Michael Trapp Antiques
7 River Road
West Cornwall, CT 06796
Tel: 860-672-6098
Fax: 860-672-3489
www.michaeltrapp.com

Moore Ruble Yudell Architects
933 Pico Boulevard
Santa Monica, CA 90405
Tel: 310-450-1400
Fax: 310-450-1403
info@mryarchitects.com
www.moorerubleyudell.com

Paul Froncek, Architect
110 Corson Place
Ithaca, NY 14850
Tel: 413-204-0690
paulfroncek@yahoo.com

Paul Gordon Architect
24 Glenvale Boulevard
Toronto, ON M4G 2V1
Tel: 416-489-3728
Fax: 416-489-8831
pgordonarch@vif.com
www.paulgordonarchitect.com

Richard Tremaglio
5 Story Street
Cambridge, MA 02138
Tel: 617-492-4469
rctarchitect@earthlink.net
www.rctarchitect.com

Ronald Holbrook & Associates Landscape Architects
390 Dupont Street
Toronto, ON M5R 1V9
Tel: 416-924-8165
Fax: 416-924-7164
holbrook@bellnet.ca

Ruhl Walker Architects
60 K Street
Boston, MA 02127
Tel: 617-268-5479
Fax: 617-268-5482
info@ruhlwalker.com
www.ruhlwalker.com

Sanba Inc.
Arizona Office
2675 West Highway 89A, Ste 449
Sedona, AZ 86336
Tel: 928-282-3755
Fax: 928-282-4083
New York Office
341 Lafayette Street, Ste 767
New York, NY 10012
Tel: 212-431-6121
Fax: 212-431-6140
info@sanba.com
www.sanba.com

StellarGray Fine Homes
3661 N. Campbell Avenue
PMB 208
Tucson, AZ 85719
Tel (cell): 520-603-3128
Fax: 520-319-1526
info@stellargray.com
www.stellargray.com

Steven Ehrlich Architects
10865 Washington Boulevard
Culver City, CA 90232
Tel: 310-838-9700
Fax: 310-838-9737
info@s-ehrlich.com
www.s-ehrlich.com

Strasman Architects
One Atlantic Avenue, Ste 214
Toronto ON M6K 3E7
Tel: 416-588-1800
Fax: 416-588-1009
info@strasmanarch.com
www.strasmanarch.com

Timbersmith Log Construction Ltd.
Head Office
10 Mistaken Island
Nanoose Bay, BC V9P 9B6
Tel: 250-248-8048
Fax: 250-248-8046
Info@TheTimbersmiths.com
Eastern Branch
Gen. Del. Hillsdale, ON L0L 1V0
Tel: 705-725-2585
Fax: 705-725-2590
Info@TheTimbersmiths.com
www.thetimbersmiths.com

Timmerhus
Main Office
Tel: 303-449-1336
Fax 303-449-9170
ed@timmerhusinc.com
East Coast Office
Tel: 603-643-2002
Fax: 603- 643-5651
elevin@valley.net
www.timmerhusinc.com

T. Jenkins (Tony Jenkins)
2935 Latimer Road, RR 1
Elginburg, ON K0H 1M0
Tel: 613-353-7262
tony@kingston.net
www.historiclogandtimber.com

products & services

adobe

Adobe Builder
www.adobebuilder.com

Adobe Building Supply
5609 Alameda Pl. NE
Albuquerque, NM 87113
Tel: 505-828-9800
www.abslumber.com

Adobelite
10031 Southern SE
Albuquerque, NM 87123
Tel: 505-291-0500
Fax: 505-291-9317
adobelite@adobelite.com
www.adobelite.com

Mark Chalom, architect
52 Calimo Circle
Santa Fe, NM 87505
tel: 505-983-1885
www.markchalom.com

Clay Mine Adobe
6401 West Old Ajo Highway
Tucson, AZ 85735
Tel: 520-578-2222
Fax: 520-578-1721
info@claymineadobe.com
www.claymineadobe.com

Compadre Custom Construction, Inc
Santa Fe, NM (Kevin Pino)
Tel: 505-438-2933

R.S. Davin
Bayfield, CO
Tel: 970-884-9465
rrdavins@frontier.net

Earth Block Inc.
PO Box 3605
Pagosa Springs, CO 81147
Tel: 970-883-2456
earthblock@juno.com
www.earthblockinc.com

Homes By Marie
PO Box 2777
Corrales, NM 87048
Tel: 505-342-1532
Fax: 505-342-1579
marie@homesbymarie.com
www.homesbymarie.com

C.E. Laird
627 Paseo del Bosque,
Albuquerque, NM 87114
celaird2@hotmail.com
www.lairdadobe.com

Mule Creek Adobe
547 Highway 78
PO Box 33
Mule Creek, NM 88051
Tel: 505-535-2973
gefjon@gilanet.com
www.mulecreekadobe.com

New Mexico Earth
310 Paseo Del Norte NE
Albuquerque, NM 87113
Tel: 505-898-1271

Old Pueblo Adobe Co.
9353 N Casa Grande Highway
Tucson, AZ 85743
Toll-free tel: 800-327-4705
Tel: 520-744-9268
Fax: 520-744-8057
oldpuebloadobe@qwest.net
www.oldpuebloadobe.com

Poling Custom Homes
PO Box 747
Placitas, NM 87043
Tel: 505-867-3200
Tel (cell): 505-980-5280
rhpoling@polingcustomhomes.com
www.polingcustomhomes.com

Rainbow Adobe
Alpine, TX
Tel: 915-837-7191 (Steve Belardo)
www.adobebuilder.com

William Stoddard
Albuquerque, NM
Tel: 505-898-6733
wstodd6733@aol.com

Hans Sumpf Adobe
40101 Avenue, 10
Madera, CA
Tel: 559-439-3214
hanssumpf@aol.com

art deco societies

The Art Deco Society of Los Angeles
PO Box 972
Hollywood, CA 90078
Tel: 310-659-3326
artdecola@sbcglobal.net
www.adsla.org

The Art Deco Society of New York
385 Fifth Avenue
New York, NY 10016
Tel: 212-679-DECO
info@artdeco.org
www.artdeco.org

Art Deco Society of Toronto
Tel: 416-467-0742
timdeco@sympatico.ca
www.artdecotoronto.com

Art Deco Society of Washington
PO Box 42722
Washington, DC 20015
Tel: 202-298-1100
info@adsw.org
www.adsw.org

Canadian Art Deco Society
800–626 Pender Street
Vancouver, BC V6B 1V9
Tel: 604-688-1216
www.luxton@portal.ca

Chicago Art Deco Society
PO Box 1116
Evanston, IL 60204–1116
Tel: 847-869-5059
decoben@comcast.net
www.chicagoartdecosociety.com

Detroit Area Art Deco Society
PO Box 1393
Royal Oak, MI 48068-1393
Tel: 248-582-DECO (3326)
membership@daads.org
www.daads.org

architectural adornments

Ancient Excavations
697 Queen Street E
Toronto, ON M4M 1G6
Tel: 416-406-3005
Fax: 416-406-5982
sculptor@ancientexcavation.com
www.ancientexcavation.com

Agrell Architectural Carving
319 Dolan Avenue
Mill Valley, CA 94941
Tel: 415-381-9474
Fax: 415-381-9475
www.agrellcarving.com

Antique Hardware & Home
39771 Highway 34
PO Box 278
Woonsocket, SD 57385
Toll-free tel: 800-237-8833
Tel: 605-796-4425
Fax: 605-796-4085
antique.hardware@cabelas.com
www.antiquehardware.com

Architectural Stone Company
1900 Preston Road, Ste 267 PMB 100
Plano, TX 75093
Tel: 972-769-8379
Fax: 972-599-7638
contact@architecturalstone.net
www.architecturalstone.net

Atelier Jouvence Stonecutting and Tile
329 West 18th Street, Ste 611
Chicago, IL 60016–1120
Tel: 312-492-7922
Tel (cell): 773-251-1581
Fax: 312-492-7923
contact@atelierjouvence.com
www.atelierjouvence.com

B&H Art-In-Architecture
341 Lafayette Street
PO Box 76, New York, NY 10012
Tel: 718-858-6613
Fax: 718-522-0342
bandchen@yahoo.com
www.traditional-building.com

Boston Valley Terra Cotta
6860 South Abbott Road
Orchard Park, NY 14127
Toll-free tel: 888-214-3655
Tel: 716-649-7490
Fax: 716-649-7688
info@bostonvalley.com
www.bostonvalley.com

Chelsea Decorative Metal Company
8212 Braewick Drive
Houston, TX 77074
Tel: 713-721-9200
Fax: 713-776-8661
www.metals.about.com

Cleveland Quarries
230 West Main Street
S. Amherst, OH 44001
Toll-free tel: 800-248-0250
Tel: 440-986-4501
Toll-free fax: 800-649-8669
Fax: 440-986-4531
amst@amst.com
www.clevelandquarries.com

Da Vinci & Co. Ltd.
616 Merrick Road
Lynbrook, NY 11563
Tel: 516-596-6666
Fax: 516-596-6094
sales@davincimoldings.com
www.davincimoldings.com

DeSantana Handcarved Stone
PO Box 907
Higley, AZ 85236
Tel: 866-932-2783
Fax: 480- 664-8682
info@desantanastone.com
www.desantanastone.com

DMS Studios
5–50 51st Avenue
Long Island City, NY 11101
Tel: 718-937-5648
Fax: 718-937-2609
dmsstudios@mindspring.com
www.dms-studios.com

Eye of the Day
Garden Design Center
4620 Carpenteria Avenue
Carpenteria, CA 93013
Tel: 805-566-0778
Fax: 805-566-0478
www.eyeofthedaygdc.com

Fairplay Stonecutters
PO Box 0355
Oberlin, OH 44074–0355
Tel: 440-775-7878
Fax: 440-775-7979
meadows@fairplaystonecarvers.com
www.fairplaystonecarvers.com

Fine's Gallery
11400 South Cleveland Avenue
Fort Myers, FL 33907
Tel: 239-277-0009
Fax: 239-418-0321
www.finesgallery.com

Five O Seven Home and Garden Antiques
50 Carroll Street
Toronto, ON M4M 3G3
Tel: 416-462-0046
kari@507homeandgarden.com
www.507antiques.com

Homes By Marie
PO Box 2777
Corrales, NM 87048
Tel: 505-342-1532
Fax: 505-342-1579
marie@homesbymarie.com
www.homesbymarie.com

Hyde Park Architectural Decorating
130 Salt Point Turnpike
Poughkeepsie, NY 12603
Tel: 845-483-1340
Fax: 845-483-1343
info@hydeparksurrounds.com
www.hydeparksurrounds.com

Jean Pierre Jacquet Stone Carving
5710 Hawkes Terrace
Edina, MN 55436
Tel: 952-925-5543
Fax: 952-929-2301
Jean@handcarvedstone.com
www.handcarvedstone.com

La Puerta Originals
4523 State Road
Santa Fe, NM 87505
Tel: 505-984-8164
Fax: 505-986-5838
info@lapuertaoriginals.com
www.lapuertaoriginals.com

Maine Millstones
Pratt's Island
Southport, ME 04576–0228
Tel: 207-633-6091
emery@mainemillstones.com
www.mainemillstones.com

Melton Classics Inc.
PO Box 465020
Lawrenceville, GA 30042
Toll-free tel: 800-963-3060
Fax: 770-962-6988
mclassics@aol.com
www.meltonclassics.com

New England Cast In Stone Manufacturing, Inc.
257 Clinton Street
PO Box 158, 132 Main Street
Springfield, VT 05156
Tel: 802-885-7866
Fax: 802-885-7870
info@newenglandcastinstone.com
www.newenglandcastinstone.com

Nichols Bros. Stoneworks, Ltd
20209 Broadway
Snohomish, WA 98296
Toll-free tel: 800-483-5720
Fax: 425-483-5721
sales@nicholsbros.com
www.nicholsbros.com

NIKO Contracting Co., Inc. (metal)
3434 Parkview Avenue
Pittsburgh, PA 15213
Tel: 412-687-1517
Fax: 412-687-7969
info@nikocontracting.com
www.nikocontracting.com

W.F. Norman Corp.
214 N. Cedar
PO Box 323
Nevada, MO 64772
Toll-free tel: 800-641-4038
Tel: 417-667-5552
Fax: 417-667-2708
ceilings@wfnorman.com
www.wfnorman.com

Pineapple Grove Designs
PO Box 1121
Boynton, FL 33425
Toll-free tel: 800-771-4595
Fax: 561-586-0845
info@pineapplegrove.com
www.pinapplegrove.com

Five O Seven Home and Garden Antiques

Precision Development
5711 Clarewood
Houston, TX 77081
Tel: 713-667-1310
Fax: 713-667-0515
www.pdcaststone.com

Profile Moldings
PMB 222, 250 "H" Street
Blaine, WA 98230
Toll-free tel: 888-882-0641
info8@profilemouldings.com
www.crown-molding.com

Rhodes Architectural Stone
2011 E Olive Street
Seattle, WA 98122
Tel: 206-709-3000
Fax: 206-709-3003
johnt@rhodes.org (West)
trevorb@rhodes.org (North, Midwest)
andrewg@rhodes.org (East)
www.rhodes.org

Rutland Gutter Supply
10895 Rocket Boulevard
Orlando, FL 32824
Tel: 407-859-1119
Fax: 407-859-1123
rutlandguttersupply@yahoo.com
www.rutlandguttersupply.com

Sierra Concrete Design, Inc.
2110 S. Anne Street
Santa Ana, CA 92704–4409
Tel: 714-557-8100
Fax: 714-557-8109
sales@sierraconcrete.com
www.sierraconcrete.com

Stone Age Designs
Atlanta
351 Peachtree Hills Avenue, Ste 501-A
Atlanta, GA 30305
Tel: 404-350-3333
Fax: 404-842-0936
Florida
935 Orange Avenue, Ste A
Winter Park, FL 32789
Tel: 407-628-5577
Fax: 407-628-5565
info@stoneagedesigns.net
www.stoneagedesigns.net

StoneDecora
442 - 14622 Ventura Boulevard
Sherman Oaks, CA 91403
Tel: 818-986-1171
Fax: 818-907-0343
Info@StoneDecora.com
www.stonedecora.com

Stone Legends
301 Pleasant Drive
Dallas, TX 75217
Toll-free tel: 800-398-1199
Fax: 214-398-1293
sales@stonelegends.com
www.stonelegends.com

Stonex Cast Products Inc.
127 Squankum – Yellowbrook Road
Farmingdale, NJ 07727
Tel: 732-938-2334
Fax: 732-919-0918
info@stonexonline.com
www.stonexonline.com

Stonesculpt
Palo Alto, CA
Tel: 650-575-9683
Fax: 650-322-5002
Lobykin@bigfoot.com
www.customstonecarving.com

Sun Precast Co. Inc.
PO Box 423, Ridge Road
McClure, PA 17841
Tel: 570-658-8000
Fax: 570-658-8008
sales@sunprecast.com
www.sunprecast.com

Barbara Tattersfield Design Inc.
Palm Beach, FL Showroom
Tel: 561-833-3443
Fax: 561-833-3414
Corona del Mar, CA Showroom
Tel: 949-675-0600
Fax: 949-675-0601
info@btattersfielddesign.com
www.btattersfielddesign.com

Thunderstone LLC
3300 South 6th Street
Lincoln, NE 68502
Tel: 402-420-2322
Fax: 402-420-2542
cconiglio@castone.com
www.castone.com

Towne House Restorations Inc.
592 Johnson Avenue
Brooklyn, NY 11237
Tel: 718-497-9200
Fax: 718-497-3556
info@THRcaststone.com
www.thrcaststone.com

Traditional Cut Stone
1860 Gage Court
Mississauga, ON L5S 1S1
Tel: 416 652-8434
Fax: 905 673-8434
info@traditionalcutstone.com
www.traditionalcutstone.com

Robert Young & Sons
25 Grafton Avenue
Newark, NJ 07104
Tel: 973-483-0451
Fax: 973-483-0185
ryoungandsons@aol.com
www.traditional-building.com

historic renovation & reconstruction

Adams Architectural Wood Products
300 Trails Road
Eldridge, IA 52748
Toll-free tel: 888-285-8120
Tel: 563-285-8000
Fax: 563-285-8003
www.adamsarch.com

Allegheny Restoration
PO Box 18032
Morgantown, WV 26508
Tel: 304-594-2570 / 304-255-0074
Fax: 304-594-2810
www.alleghenyrestoration.com

American Dream Post & Beam
151 Laten Knight Road
Cranston, RI 02823
Tel: 401-822-3122
dplsk@aol.com
www.netimbers.com

Architectural Reproductions Inc.
525 N. Tillamook Street
Portland, OR 97227
Toll-free tel: 888-440-8007
Tel: 503-284-8007
Fax: 503-281-6926
admin@archrepro.com
www.archrepro.com

The Barn People
2218 US Route 5 N
Windsor, VT 05089
Tel: 802-674-5898
Fax: 802-674-6310
barnman@sover.net
www.thebarnpeople.com

B&H Art-In-Architecture
341 Lafayette Street
PO Box 76
New York, NY 10012
Tel: 718-858-6613
Fax: 718-522-0342
bandchen@yahoo.com
www.traditional-building.com

Cathedral Stone Products, Jahn Mortars
7266 Park Circle Drive
Hanover, MD 21076
Toll-free tel: 800-684-0901
Tel: 410-782-9150
Fax: 410-782-9155
info@cathedralstone.com
www.jahnmortars.com

Cleveland Quarries
230 West Main Street
South Amherst, OH 44001
Toll-free tel: 800-248-0250
Tel: 440-986-4501
Toll-free fax: 800-649-8669
Fax: 440-986-4531
amst@amst.com
www.clevelandquarries.com

Country Road Associates
PO Box 885
63 Front Street
Millbrook, NY 12545
Tel: 845-677-6041
Fax: 845-677-6532
info@countryroadassociates.com
www.countryroadassociates.com

Timothy J. Dunn
2402 Bredell Street
Louis, MO 63143
Tel: 314-645-4317
Fax: 314-647-6276.
vitrolite@earthlink.com.
www.vitrolitespecialist.com.

European Reclamation
4520 Brazil Street
Los Angeles, CA 90039
Tel: 818-241-2152
Fax: 818-547-2734
htc@wgn.net
www.historictile.com

First Period Colonial Preservation/Restoration
PO Box 31
Kingston, NH 03848
Tel: 603-642-8613
Fax: 603-642-6727
pothier@mindspring.com
www.firstperiodcolonial.com

William Gould Architectural Preservation, LLC
102 Angel Road
Pomfret Center, CT 06259
Tel: 860-974-3448
Fax: 860-974-3448
historic@historic-architecture.com
www.historic-architecture.com

Heartwood Building and Restoration
26 Forget Road
Hawley, MA 01339
Tel: 413-339-4298

Heather & Little Limited
3205 14th Avenue
Markham ON L3R 0H1
Toll-free tel: 800-450-0659
Tel: 905-475-9763
Fax: 905-475-9764
info@heatherandlittle.com
www.heatherandlittle.com

Historic Structures
3711 Cumberland Street NW
Washington, DC 20016
Tel: 202-686-0135
Fax: 202-686-0135
ortado@starpower.net
www.historicstructuresdc.com

T. Jenkins (Tony Jenkins)
2935 Latimer Road, RR 1
Elginburg, ON K0H 1M0
Tel: 613-353-7262
tony@kingston.net
www.historiclogandtimber.com

Joseph Millworks
37123 Hansen Lane
Baker City, OR 97814
Tel/Fax: 541-894-2347
randy@josephmillworks.com
www.josephmillworks.com/about.htm

Kronenberger and Sons Restoration, Inc.
80 E Main Street
Middletown, CT 06457
Tel: 860-347-4600
Fax: 860-343-0309
kronenberger-sons@snet.net
www.kronenbergersons.com

Dan Lepore & Sons Co.
501 Washington Street
Conshohocken, PA 19428
Tel: 610-940-9888
Fax: 610-834-9226
greglep@danlepore.com.
www.danlepore.com

Marlowe Restorations
PO Box 402
Northford, CT 06472
Tel: 203-484-9643
Fax: 203-484-2398
www.marlowerestorations.com

National Trust for Historic Preservation
1785 Massachusetts Avenue, NW
Washington, DC 20036–2117
Tel: 800-944-6847
www.nationaltrust.org

New England Cast In Stone Manufacturing, Inc.
257 Clinton Street
PO Box 158
Springfield, VT 05156
Tel: 802-885-7866
Fax: 802-885-7870
info@newenglandcastinstone.com
www.newenglandcastinstone.com

Nicholson & Galloway, Inc.
261 Glen Head Road
Glen Head, NY 11545
Tel: 516-671-3900
Fax: 516-759-3569
TomC@nicholsonandgalloway.com
www.nicholsonandgalloway.com

Quality Restoration Works, LLC.
20 Jay Street, Room 201
Brooklyn, NY 11201
Tel: 917-575-8545
Fax: On request

Restoration Works
1345 Stanford Drive
Kankakee, IL 60901
Tel: 815-937-0556
Fax: 815-937-4072
gwallace@restorationworksinc.com
www.restorationworksinc.com

Re-View
1235 Saline
North Kansas City, MO 64116
Tel: 816-746-9339
Fax: 816-746-9331
www.re-view.biz

S and H Construction, Inc.
26 New Street
Cambridge, MA 02138
Tel: 617-876-8286
Fax: 617-864-1850
mail@shconstruction.com
www.shconstruction.com

Smith Restoration Sash
122 Manton Avenue, Unit 714
Providence, RI 02909
Tel: 401-351-1222
Fax: 401-351-1245
windows@smithrestorationsash.com
www.smithrestorationsash.com

The Tile Man Inc.
PO Box 329
Louisburg, NC 27549–6923
Toll-free tel: 888-263-0077
Fax: 919-853-6634
Fax Vintage Tile: 919-556-2072
info@thetileman.com
E-mail Vintage Tile:
loriak7@earthlink.net
www.thetileman.com

Towne House Restorations Inc.
592 Johnson Avenue
Brooklyn, NY 11237
Tel: 718-497-9200
Fax: 718-497-3556
info@THRcaststone.com
www.thrcaststone.com

Traditional Cut Stone
1860 Gage Court
Mississauga, ON L5S 1S1
Tel: 416-652-8434
Fax: 905-673-8434
info@traditionalcutstone.com
www.traditionalcutstone.com

U.S. Heritage Group Inc.
3516 N Kostner Avenue
Chicago, IL 60641
Tel: 773-286-2100
Fax: 773-286-1852
www.usheritage.com

Vulcan Supply Corporation
PO Box 100
Westford, VT 05494
Toll-free tel: 800-659-4732
Fax: 802-893-0534
info@vulcansupply.com
www.vulcansupply.com

Watertrol Inc.
PO Box 162
Cranford, NJ 07016
Tel: 908-389-1690
Fax: 908-389-9480
info@watertrolinc.com
www.watertrolinc.com

Wood Natural Restorations
3038 Woodlane Avenue
Orefield, PA 18069
Tel: 610-395-6451
Fax: 610-395-4662
ken@woodnatural.com
www.woodnatural.com

The WoodSource
6178 Mitch Owens Road
PO Box 700
Manotick, ON K4M 1A6
Tel: 613-822-6800
Fax: 613-822-6803
layton@wood-source.com
www.wood-source.com

Robert Young & Sons
25 Grafton Avenue
Newark, NJ 07104
Tel: 973-483-0451
Fax: 973-483-0185
ryoungandsons@aol.com

architectural salvage/antiques/materials

Adkins Architectural Antiques
3515 Fannin Street
Houston, TX 77004
Toll-free tel: 800-522-6547
Tel: 713-522-6547
adkins@adkinsantiques.com
www.adkinsantiques.com

Albion Doors & Windows
PO Box 220
Albion, CA 95410
Tel: 707-937-0078
bysawyer@mcn.org
www.knobsession.com

Antique Hardware & Home
19 Buckingham Plantation Drive
Bluffton, SC 29910
Toll-free tel: 800-422-9982
Tel: 843-837-9789
treasure@hargray.com
www.antiquehardware.com

Antique Market
4280 Main Street
Vancouver, BC V5V 3P9
Tel: 604-875-1434
info@antiquemarketvancouver.com
www.anantiquemarket.com

Architectural Emporium
207 Adams Avenue
Canonsburg, PA 15317
Toll-free tel: 724-746-4301
sales@architectural-emporium.com
www.architectural-emporium.com

Artefacts
46 Isabella Street
PO Box 513
St Jacobs, ON N0B 2N0
Tel: 519-664-3760
Fax: 519-664-1303
chris@artefacts.ca

Sylvan Brandt
651 E Main Street
Lititz, PA 17543
Tel: 717-626-4520
Fax: 717-626-5867
dean@sylvanbrandt.com
www.sylvanbrandt.com

Carlson's Barnwood Company
8066 N 1200 Avenue
Cambridge, IL 61238
Tel: 309-522-5550
Fax: 309-522-5123
info@carlsonsbarnwood.com
www.carlsonsbarnwood.com

Chen & Ragen
2100 E Union Street
Seattle, WA 98122
Tel: 206-325-2456
Fax: 206-325-7779
info@chenragen.com
www.chenragen.com

Crescent City Architectuals
3033 Tchoupitoulas
New Orleans, LA 70115
Tel: 504-891-0500
Fax: 504-891-1895
cca@architectural-salvage.com
www.architectural-salvage.com

Five O Seven Home and Garden Antiques
50 Carroll Street
Toronto, ON M4M 3G3
Tel: 416-462-0046
kari@507homeandgarden.com
www.507antiques.com

Governor's Antiques & Architectural Materials
8000 Antique Lane
Mechanicsville, VA 23116
Tel: 804-746-1030
governorsantiques@earthlink.net
www.governorsantiques.net

Heritage Co. II Architectural Artifacts and Design
116 E Seventh Street
Royal Oak, MI 48067
Tel: 248-547-0670
marisa@heritagecoii.com
www.heritagecoii.com

Historic Houseparts
540 South Avenue
Rochester, NY 14620
Tel: 585-325-2329
Toll-free tel: 888-558-2329
houseparts@msn.com
www.historichouseparts.com

D. Litchfield & Co. Ltd.
3046 Westwood Street
Fort Coquitlam, BC
Tel: 604-464-7525
Fax: 604-944-1674
demo@dlitchfield.com
www.dlitchfield.com

Materials Unlimited
2 West Michigan Avenue
Ypsilanti, MI 48197
Toll-free tel: 800-299-9462
Fax: 734-482-3636
materials@materialsunlimited.com
www.materialsunlimited.com

New York Salvage
35 Otsego Street
Oneonta, NY 13820
Tel: 607-432-9890 & 607-433-9890
Fax: 607-432-4119
elm@stny.rr.com
www.architiques.net

Off the Wall Architectural Antiques
Third S.E. Lincoln
PO Box 4561
Carmel, CA 93921
Tel: 831-624-6165
www.imperialearth.com/OTW/

Old Barn Wood Company
1911 Draper Street
Baraboo, WI 53913
Tel: 608-356-8849
david@old-barn-wood.com
www.old-barn-wood.com

Old Home Supply
1801 College Avenue
Fort Worth, TX 76110
Tel: 817-927-8004
oldhome@swbell.net
www.theoldhomesupply.com

Olde Good Things
Manhattan
124 West 24th Street
New York, NY 10011
Toll-free tel: 888-551-7333
Tel: 212-989-8401
Los Angeles
1800 South Grand Avenue
Los Angeles, CA 90015
Tel: 213-746-8600 or 8611
Tel (cell): 213-210-7675
Scranton, Pennsylvania
400 Gilligan Street
Scranton, PA 18508
Toll-free tel: 888-233-9678
Tel: 570-341-7668

Olde Wood Limited
Tel: 866-208-WOOD
Fax: 240-536-6571
info@oldewoodltd.com
www.oldewoodltd.com

Orlando Liquidators, Inc.
1016 Savage Court
Longwood, FL 32750
Tel: 407-332-6206
Fax: 407-332-6208
info@OrlandoLiquidators.com
www.orlandoliquidators.com

Phoenix Antiques
PO Box 2426
Peterborough, ON K9J 7Y8
Tel: 705-872-8029
mccrea.ant@sympatico.ca
www.phoenixant.com

Portland Architectural Salvage
919 Congress Street
Portland, ME 04101
Tel: 207-80-0634
preserve@portlandsalvage.com
www.portlandsalvage.com

Resource Recovery
PO Box 13644
Pensacola, FL 32591
Tel: 866-653-4769
sales@heartpine.org
www.heartpine.org

The Renovator's Supply
Renovator's Old Mill Dept. 2467
Millers Falls, MA 01349
Toll-free tel: 800-659-2211
Tel: 413-423-3556
www.rensup.com

Renovators Resource Inc.
6040 Almon Street
Halifax, NS B3K 1T8
Toll-free tel: 877-230 7700
Tel: 902-429 3889
Fax: 902-425 6795
frontdesk@renovators-resource.com
www.renovators-resource.com

The Tile Man Inc.
PO Box 329
Louisburg, NC 27549–6923
Toll-free tel: 888-263-0077
Fax: 919-853-6634
Fax Vintage Tile: 919-556-2072
info@thetileman.com
E-mail Vintage Tile:
loriak7@earthlink.net
www.thetileman.com

Timeless Timbers
2200 E Lake Shore Drive
Ashland, WI 54806
Tel: 715-685-9663
Fax: 715-685-9620
Info@TimelessTimber.com
www.timelesstimber.com

Uniquities Architectural Antiques
5240–1a Street SE
Calgary, AB T2H 1J1
Tel: 403- 228-9221
Fax: 403-283-9226
info@uniquities-archant.com
www.uniquities.ca

Vintage Log & Lumber, Inc.
RR 1 Box 2F, Glen Ray Road
Alderson, WV 24910
Toll-free tel: 877-653-5647
Tel: 304-445-2300
Fax: 304-445 2249
sales@vintagelog.com
www.vintagelog.com

White River Salvage and Antiques Co.
1325 West 30th Street
Indianapolis, IN 46208
Toll-free tel: 800-262-3389
Tel: 317-924-4000
Fax: 317-920-1343
whiteriversalvage@whiteriversalvage.com
www.whiteriversalvage.com

Zaborski Emporium
27 Hoffman Street
Kingston, NY 12401
Tel: 845-338-6465
zaborskiemporium@aol.com
www.stanthejunkman.com

bathrooms

Bathroom Machineries
495 Main Street
Murphys, CA 95247
Toll-free tel: 800-255-4426 (orders only)
Tel: 209-728-2031
tom@deabath.com
www.deabath.com

Crown Point Cabinetry
462 River Road
Claremont, NH 03743
Toll-free tel: 800-999-4994
Fax: 800-370-1218
info@crown-point.com
www.crown-point.com

Christopher Peacock Cabinetry
148 The Merchandise Mart
Chicago, IL 60654
Tel: 312-321-9500
Fax: 312-321-9510
Chicago@peacockcabinetry.com
Greenwich Showroom
2 Deerfield Drive
Greenwich, CT 06830
Tel: 203-862-9333
Fax: 203-302-3411
Greenwich@peacockcabinetry.com
Boston Showroom
One Design Center Place
Boston, MA 02210
Tel: 617-204-9292
Fax: 617-204-9293
boston@peacockcabinetry.com
Naples Showroom
6320 Shirley Street, Ste 102
Naples, FL 34109
Tel: 239-596-8858
Fax: 239-596-0358
naples@peacockcabinetry.com
www.peacockcabinetry.com

Poggenpohl U.S. Inc.
350 Passaic Avenue
Fairfield, NJ 07004
Tel: 973-812-8900
Fax: 973-812-9320
info@poggenpohl-usa.com
www.poggenpohl-usa.com
Poggenpohl Canada
8611 Weston Road
Unit 30 & 31
Woodbridge, ON L4L 9P1
Tel: 905-264-2137
Fax: 905-264-0786
info@poggenpohlcanada.com
www.poggenpohlcanada.com

THG USA
6601 Lyons Road
Coconut Creek, FL 33073
Tel: 954-425-8225
Fax: 954-425-8301
info@thgusa.com
www.thgusa.com

bathtubs

Antique Hardware & Home
39771 SD Highway 34
PO Box 278
Woonsocket, SD 57385
Toll-free tel: 800-237-8833
Tel: 605-796-4425
Fax: 605-796-4085
antique.hardware@cabelas.com
www.antiquehardware.com

BainUltra
956, chemin Olivier
Saint-Nicolas, QC G7A 2N1
Toll-free tel: 800-463-2187
Tel: 418-831-7701
Toll-free fax: 800-382-8587
Fax: 418-831-6623
info@BainUltra.com
www.bainultra.com

Diamond Spas
760 S. 104th Street
Broomfield, CO 80020
Toll-free tel: 800-951-7727
Tel: 303-665-8303
Fax: 303-664-1293
www.diamondspas.com

Lavabo
lavabo@giant.co.uk
www.giant.co.uk

Marble Concepts
6017 – 86 Street
Edmonton, AB T6E 2X4
Toll-free tel: 866-241-2735
Tel: 780-466-9342
Fax: 780-465-1087
ken@marbleconcepts.com
www.marbleconcepts.com

Mirolin
60 Shorncliffe Road
Toronto, ON M8Z 5K1
Toll-free tel: 800-647-654
Toll-free tel (Canada only):
800-463-2236
Tel: 416-231-9030
Fax: 416-231-0929
www.mirolin.com

Rhodes Architectural Stone
2011 E Olive Street
Seattle, WA 98122
Tel: 206-709-3000
Fax: 206-709-3003
johnt@rhodes.org (West)
trevorb@rhodes.org (North, Midwest)
andrewg@rhodes.org (East)
www.rhodes.org

Signature Hardware
Toll-free tel: 866-855-2284
sales@signaturehardware.com
www.signaturehardware.com

Stone Forest
PO Box 2840
Santa Fe, NM 87504
Toll-free tel: 888-682 2987
Tel: 505-986-8883
Fax: 505-982-2712
info@stoneforest.com
www.stoneforest.com

bathtubs: reproduction/ antique etc.

Antique Hardware & Home
39771 SD Highway 34
PO Box 278
Woonsocket, SD 57385
Toll-free tel: 800-237-8833
Tel: 605-796-4425
Fax: 605-796-4085
antique.hardware@cabelas.com
www.antiquehardware.com

Baths From the Past/ Besco Plumbing
83 E Water Street
Rockland, MA 02370
Toll-free tel: 800-697-3871
Tel: 781-871-8530
www.bathsfromthepast.com

Country Plumbing
PO Box 408
Exeter, CA 93221
Tel: 559-592-9115

Clawfoot Supply
2700 Crescent Springs Pike,
Erlanger, KY 41017
Tel: 859-581-1373
Fax: 877-682-4192
sales@clawfootsupply.com
www.clawfootsupply.com

Off the Wall Architectural Antiques
PO Box 4561
Third S.E. Lincoln and Fifth
Carmel, CA 93921
Tel: 831-624-6165
www.imperialearth.com

Ole Fashion Things
402 SW Evangeline Thruway
Lafayette, LA 70501–7139
Toll-free tel: 888-595-2284
Tel: 337-234-4800
ohjrd@classicplumbing.com
www.classicplumbing.com

Shop 4 Classics
5726 Lowell Avenue
Merriam, KS 66202
info@shop4classics.com
www.shop4classics.com

Vintage Tub and Bath
534 West Green Street
Hazleton, PA 18201
Toll-free tel: 877-868-1369
Tel: 570-450-7925
supply@vintagetub.com
www.vintagetub.com

bathrooms: sinks, faucets

Allante International
3801 Market Street
Wilmington, NC 28403.
Toll-free tel: 800-695-0805
Toll-free fax: 910-763-4213
info@allante.net
www.allante.net

American Standard
PO Box 6820
1 Centennial Plaza
Piscataway, NJ 08855–6820
Toll-free tel: 800-442-1902
www.americanstandard-us.com

Antique Hardware & Home
39771 SD Highway 34
PO Box 278
Woonsocket, SD 57385
Toll-free tel: 800-237-8833
Tel: 605-796-4425
Fax: 605-796-4085
antique.hardware@cabelas.com
www.antiquehardware.com

B & C Custom Hardware and Bath
23461 A Ridge Route Drive
Laguna Hills, CA 92653
Toll-free tel: 800-333-1111
Tel: 949-859-6073
www.customhardware.net

Bamboo Accents
(bamboo water spouts)
40 Longwood Drive
San Rafael, CA 94901
Tel: 415-454-6260
Fax: 415-454-6391
bambooaccents@hotmail.com
www.bambooaccents.com

Bates & Bates
7310 Alondra Boulevard
Paramount, CA 90723
Toll-free tel: 800-726-7680
Tel: 562-808-2290
Fax: 562-808-2289
website@batesandbates.com
www.batesandbates.com

Brizo
www.brizo.com

Cantrio Koncepts
16-3400 14th Avenue
Markham, ON L3R 0H7
Tel: 905-474-4888
info@cantrio.ca
www.cantrio.ca

Cheviot Products
200–1594 Kebet Way
Port Coquitlam, BC V3C 5M5
Tel: 604-464-8966
Fax: 800-211-2555
www.cheviotproducts.com

Duravit
1750 Breckinridge Parkway, Ste 500
Duluth, GA 30096
Toll-free tel: 888-387-2848
Tel: 770 –931-3575
Fax: 770-931-8454
Toll-free fax: 888 387 2843
info@us.duravit.com
www.duravit.us

Firebowl Artisan Sinks
2310 E 75th Street
Chicago, IL 60649
Tel: 312- 933-6656
Fax: 773-768-8003
info@studioelab.com
www.evanglassman.com

Grohe America Inc.
241 Covington Drive
Bloomingdale, IL 60108
Tel: 630-582-7711
Fax: 630-582-7722
Grohe Canada Inc.
1226 Lakeshore Road E
Mississauga, ON L5E 1E9
Tel: 905-271-2929
Fax: 905-271-9494
www.groheamerica.com

Hastings Tile & Bath Inc.
230 Park Avenue S
New York, NY 10003
Tel: 212-674-9700
Fax: 212-674-8083
nycsales@hastings30.com
Chicago
120 Merchandise Mart
Chicago, IL 60654
Tel: 312-527-0565
chicagosales@hastings30.com
www.hastingstilebath.com

Julian
Toll-free tel: 800-461-3377
Tel: 418-687-3630
Toll-free fax: 866-397-9090
Fax: 418-687-9129
www.julian.ca

Kohler Co.
Kohler, WI 53044
Toll-free tel: 800-456-4537
www.us.kohler.com

MGS Designs USA
20423 State Road 7, F6–291
Boca Raton, FL 33498
Tel: 561-218-8798
Fax: 561-218-8799
eric@mgsdesigns.com
www.mgsdesigns.com

Neo-Metro Collection
A Division of
Acorn Engineering Company
PO Box 3527
City of Industry, CA 91744
Toll-free tel: 800-591-9050
Tel: 626-855-4854
Fax: 626-937-4725
info@neo-metro.com
www.neo-metro.com

Neptune
6835 rue Picard
St-Hyacinthe, QC J2S 1H3
Tel: 450-773-7058
Fax: 450-773-5063
www.bainsneptune.com

Purple Sage Collections
1526B Howell Mill Road
Atlanta, GA 30318
Toll-free tel: 866-357-4657
Tel: 404-351-4445
Fax: 404-351-8250
info@purplesagecollections.com
www.purplesagecollections.com

Rhodes Architectural Stone
2011 E Olive Street
Seattle, WA 98122
Tel: 206-709-3000
Fax: 206-709-3003
johnt@rhodes.org (West)
trevorb@rhodes.org (North, Midwest)
andrewg@rhodes.org (East)
www.rhodes.org

RMG Stone Products Inc.
PO Box 807
E Hubbardton Road
Castleton, VT 05735
Tel: 802-468-5636
Fax: 802-468-8968
sales@rmgstone.com
www.rmgstone.com

The Rubinet Faucet Company
141 Caster Avenue
Woodbridge, ON L4L 5Y8
Toll-free tel: 800-461-5901
Tel: 905-851-6781
Fax: 905-851-8031
customer_service@rubinet.com
www.rubinet.com

Saint Tropez Stone Boutique
25 Evelyn Way
San Francisco, CA 94127
Tel: 415 2594820
Fax: 832-565-1100
sales@sainttropezstone.com
www.sainttropezstone.com

Sheldon Slate
General Information
38 Farm Quarry Road
Monson, ME 04464
Tel: 207-997-3615
Fax: 207-997-2966
john@sheldonslate.com
Inquiries
Fox Road
Middle Granville, NY 12849
Tel: 518-642-1280
Fax: 518-642-9085
www.sheldonslate.com

Sonia
Toll-free tel: 888-766-4287
Tel: 954-572-5454
www.sonia-sa.com

Sonoma Forge
133 Copeland Street
Petaluma, CA 94952
Toll-free tel: 800-330-5553
Tel: 707-789-9130
Fax: 707-789-.9201
info@sonomaforge.com
www.sonomaforge.com

Stone Forest
PO Box 2840
Santa Fe, NM 87504
Toll-free tel: 888-682 2987
Tel: 505-986-8883
Fax: 505-982-2712
info@stoneforest.com
www.stoneforest.com

THG USA
6601 Lyons Road
Coconut Creek, FL 33073
Tel: 954-425-8225
Fax: 954-425-8301
info@thgusa.com
www.thgusa.com

Waterdecor
3579 E Foothill Boulevard, Box 418
Pasadena, CA 91107
Toll-free tel: 877-222-9644
Tel: 626-568-0900
Fax: 626-744-0377
www.waterdecor.com

bricks, blocks & pavers

Acme Brick Co.
PO Box 425
Fort Worth, TX
Tel: 817-332-4101
Fax: 817-390-2409
contact@brick.com
www.brick.com

Atelier Jouvence Stonecutting and Tile
329 West 18th Street, Ste 611
Chicago, IL 60016–1120
Tel: 312-492-7922
Tel (cell): 773-251-1581
Fax: 312-492-7923
contact@atelierjouvence.com
www.atelierjouvence.com

Belden Brick Sales & Service, Inc.
386 Park Avenue South
New York, NY 10016
Tel: 212-686-3939
Fax: 212-686-4387
david@NYNJBrick.com
www.nynjbrick.com

Carolina Ceramics Brick Company
9931 Two Notch Road
Columbia, SC 29223
Tel: 803-788-1916
Fax: 803-736-5218
carolinaceramics@carolinaceramics.com
www.carolinaceramics.com

Church Brick
118 Burlington Road
Bordentown , NJ 08505
Tel: 609-298-0090
Fax: 609-298-4278
www.churchbrick.com

Clay Mine Adobe
6401 West Old Ajo Highway
Tucson, AZ 85735
Tel: 520-578-2222
Fax: 520-578-1721
info@claymineadobe.com
www.claymineadobe.com

Cleveland Quarries
230 West Main Street
S Amherst, OH 44001
Toll-free tel: 800-248-0250
Tel: 440-986-4501
Toll-free fax: 800-649-8669
Fax: 440-986-4531
amst@amst.com
www.clevelandquarries.com

Coronado Stone Products
11191 Calabash Ave
Fontana, CA 92337
Toll-free tel: 800-847-8663
sales@coronado.com
www.coronado.com
Coronado Canada
6061 90th Ave SE
Calgary, AB T2C 4Z6
Tel: 403-203-0881
Fax: 403-250-2929

Gavin Historical Bricks
2050 Glendale Road
Iowa City, IA 52245
Tel: 319-354-5251
info@historicalbricks.com
www.historicalbricks.com

General Shale Brick
PO Box 3547
Johnson City, TN 37602
Toll-free tel: 800-414-4661
www.generalshale.com

Glen Gery Brick
PO Box 7001 1166 Spring Street
Wyomissing, PA 19610–6001
Tel: 610-374-4011
www.glengerybrick.com

Endicott Clay Products Co. & Endicott Tile Ltd.
PO Box 17
Fairbury, NE 68352
Tel: 402-729-3323
Fax 402-729-5804
endicott@endicott.com
www.endicott.com

The Henry Brick Company
PO Box 850
Selma, AL 3670
Tel: 334-875-2600
Fax: 334-875-7842
cpalmer@henrybrick.com
ww.henrybrick.com

Kuhlman Corporation
1845 Indian Wood Circle
Maumee, OH 43537
Toll-Free: 800-669-3309
Tel: 419-897-6000
Fax: 419-897-6061
info@kuhlman-corp.com
www.kuhlman-corp.com

Monarch Stone
PO Box 2684
Carefree, AZ 85377
Tel: 480-220-8096
Fax 480-595-0545
info@monarchstone.com
www.monarchstone.com

Old Carolina Brick Company
475 Majolica Road
Salisbury, NC 28147
Toll-free tel: 800-536-8850
Tel: 704-636-8850
Fax: 704-636-0000
www.handmadebrick.com

Orco Block
11100 Beach Boulevard
Stanton, CA 90680
Toll-free tel: 800-473-6726
Tel: 714-527-2239
Fax: 714-895-4021
sales1@orco.com
www.orco.com

Pine Hall Brick
PO Box 11044
2701 Shorefair Drive
Winston-Salem, NC 27116–1044
Toll-free tel: 800-334-8689
Fax: 336-725-3940
paverinfo@pinehallbrick.com
www.americaspremierpaver.com

Robinson Brick Co.
Toll Free: 800-477-9002
Tel: 303-783-3000
Fax: 303-781-1818
www.robinsonbrick.com

Rhodes Architectural Stone
2011 E Olive Street
Seattle, WA 98122
Tel: 206-709-3000
Fax: 206-709-3003
johnt@rhodes.org (West)
trevorb@rhodes.org (North, Midwest)
andrewg@rhodes.org (East)
www.rhodes.org

ceiling fans

Casablanca Fan Company
761 Corporate Center Drive
Pomona, CA 91768
Toll-free tel: 888-227-2178
Toll-free tel (in Canada): 800-388-3382
Tel: (Toronto) 416-247-9221
www.casablancafanco.com

Craftmade International, Inc.
PO Box 1037
Coppell, TX 75019–1037
Toll-free tel: 800-486-4892
Tel: 972-393-3800
Fax: 877-304-1728
customerservice@craftmade.com
www.craftmade.com

Emerson Fan Company
USA
Emerson
PO Box 4100
8000 West Florissant Avenue
St. Louis, MO 63136–8506
Tel: 314-553-2000
Canada
Emerson Electric Canada Ltd.
9999 Highway 48
Markham, ON L3P 3J3
Tel: 905-294-9340
www.emersonfans.com

Casablanca Fan Co.

Fanimation
10983 Bennett Parkway
Zionsville, IN 46077
Toll-free tel: 888-567-2055
Toll-free fax: 866-482-5215
www.fanimation.com

Hunter Fan Company
2500 Frisco Avenue
Memphis, TN 38114
Toll-free tel: 888-830-1326
www.hunterfan.com

Mathews Fan Co.
1881 Industrial Drive
Libertyville, IL 60048
Tel: 847-680-9043
Fax: 847-680-8140
www.mathewsfanco.com

Minka Group
1151 W. Bradford Court
Corona, CA 92882
Tel: 951-735-9220
Fax: 951-735-9758
sales@minkagroup.net
www.georgekovacs.com

Monte Carlo Fan Company
740 SW Loop 820, Ste 110
Fort Worth, TX 76115
Toll-free tel: 800-519-4092
Tel: 817-927-5100
Fax: 817-927-2124
info@montecarlofans.com
www.montecarlofans.com

ceilings

Above View Tiles, Inc.
4750 S Tenth Street
Milwaukee, WI 53221
Tel: 414-744-7118
Fax: 414-744-7119
Headquarters@aboveview.com
www.aboveview.com

Armstrong Ceilings
2500 Columbia Avenue
PO Box 3001
Lancaster, PA 17604
Tel: 717-397-0611
www.armstrong.com

Global Specialty Products, Ltd.
976 Highway 212 E
Chaska, MN 55318
Toll-free tel: 800-964-1186
Tel: 952-448-6566
Fax: 952-448-6550
Toll-free fax: 800-964-0630
info@surfacingsolution.com
www.surfacingsolution.com

Homes By Marie
PO Box 2777
Corrales, NM 87048
Tel: 505-342-1532
Fax: 505-342-1579
marie@homesbymarie.com
www.homesbymarie.com

ceilings, metal

AA-Abbingdon Affiliates Inc.
2149 Utica Avenue
Brooklyn, NY 11234
Tel: 718-258-8333
Fax: 718-338-2379
info@abbingdon.com
www.abbingdon.com

The American Tin Ceiling Co.
1825 60th Place E
Bradenton, FL 34203
Toll-free tel: 888-231-7500
www.americantinceilings.com

Brian Greer's Tin Ceilings
1572 Mannheim Road
RR 2 Petersburg, ON N0B 2H0
Toll-free tel: 866-846-9710
Tel: 519-743-9710
Fax: 519-570-1443
bg@tinceiling.com
www.tinceiling.com

Chelsea Decorative Metal Company
8212 Braewick Drive
Houston, TX 77074
Tel: 713-721-9200
Fax: 713-776-8661
www.metals.about.com

Classic Ceilings
902 E. Commonwealth Avenue
Fullerton, CA 92831
Toll-free tel: 800-992-8700
Fax: 714-870-5972
ceilings@classicceilings.com
www.classicceilings.com

Gage Decorative Metal Ceilings
803 S Black River Street
Sparta, WI 54656
Tel: 608-269-7447
Fax: 608-269-7622
www.gageceilings.com

Global Specialty Products, Ltd.
976 Highway 212 E
Chaska, MN 55318
Toll-free tel: 800-964-1186
Tel: 952-448-6566
Fax: 952-448 6550
Toll-free fax: 800-964 0630
info@surfacingsolution.com
www.surfacingsolution.com

M-Boss
4400 Willow Parkway
Cleveland, OH 44125
Toll-free tel: 866-886-2677
Tel: 216-441-6080
sales@mbossinc.com
www.mbossinc.com

NIKO Contracting Co., Inc.
3434 Parkview Avenue
Pittsburgh, PA 15213
Tel: 412-687-1517
Fax: 412-687-7969
info@nikocontracting.com
www.nikocontracting.com

W.F. Norman Corp.
214 N Cedar
PO Box 323
Nevada, MO 64772
Toll-free tel: 800-641-4038
Tel: 417-667-5552
Fax: 417-667-2708
ceilings@wfnorman.com
www.wfnorman.com

Valley Tin Works
59 Berlin Street
Spring Grove, PA 17362
Tel: 717-229-9834
info@valleytinworks.com
www.valleytinworks.com

ceilings, glass

Dale Chihuly
Chihuly Studio
1111 NW 50th Street
Seattle, WA 98107-5120
Tel: 206-781-8707
Fax: 206-781-1906
www.chihuly.com

chimney pots

Jack Arnold
7310 South Yale
Tulsa, OK 74136
Toll-free tel: 800-824-3565
Pat@jackarnold.com
www.jackarnold.com

Buckley Rumford Co.
1035 Monroe Street
Port Townsend, WA 98368
Tel: 360-385-9974
Tel (cell): 360-531-1081
Fax: 360-385-1624
buckley@rumford.com
www.rumford.com

The Chimney Pot Shoppe/Chimneypot.com
1915 Brush Run Road
Avellia, PA 15312
Tel: 724-345-3601
info@chimneypot.com
www.chimneypot.com

The Copper Shop
115 N First Avenue
Haubstadt, IN 47639
Tel: 812-768-5008
info@coppershop.net
www.coppershop.net

Rutland Gutter Supply
10895 Rocket Boulevard
Orlando, FL 32824
Toll-free tel: 877-859-1119
Tel: 407-859-1119
Fax: 407-859-1123
www.rutlandguttersupply.com

Superior Clay Corporation
PO Box 352
Uhrichsville, OH 44683
Toll-free tel: 800-848-6166
Tel: 740-922-4122
Fax: 740-922-6626
catalog@superiorclay.com
www.superiorclay.com

countertops

3-form
2300 South 2300 W, Ste B
Salt Lake City, UT 84119
Toll-free tel: 800-726-0126
Tel: 801-649-2500
Fax: 801-649-2699
info@3-form.com
www.3-form.com
New York Showroom
48 West 21st, Ste 1002
New York, NY 10010
Tel: 212-627-0883
Fax: 212-627-0898
mlyle@3-form.com

All Granite & Marble
1A Mount Vernon Street
Ridgefield Park, NJ 07660
Tel: 201-440-6779
Fax: 201-440-6855
www.marble.com

Avonite Surfaces
7350 Empire Drive
Florence, KY 41042
Toll-free tel: 800-354-9858
Tel: 859-283-1501
Fax: 859-283-7378
sales@avonitesurfaces.com

Bedrosians Ceramic Tile and Stone
4285 N Golden State Boulevard
Fresno, CA 93722
Tel: 559-275-5000
Fax: 559-275-1753
bedrosians@aol.com
www.bedrosians.com

Brooks Custom
15 Kensico Drive
Mount Kisco, NY 10549
Tel/Fax: 800-244-5432
rb@brookswood.com
www.brookswood.com

Cambria
Le Sueur, MN
Toll-free tel: 866-226-2742
www.cambriausa.com

Canadian Soapstone
3423 Torbolton Ridge Road,
Woodlawn, ON K0A 3M0
Tel: 613-832-4256
Fax: 613-832-0539
soapstonecounters@operamail.com
www.soapstonecounters.com

Caruso Marble & Stone Corporation
169 Lodi Street
Hackensack, NJ 07601
Tel: 201-343-2840 or 201-343-2843
Fax: 201-343-2845
ccmarble59@yahoo.com
www.carusomarble.com

Cheng Concrete Exchange
Ste 220,
11000 W 78th Street
Eden Prairie, MN 55344
Tel: 510-849-3272
Fax: 510-549-2821
products@chengdesign.com
www.concreteexchange.com

Concrete Countertops Canada
RR 8
Newcastle, ON L1B 1L9
Tel: 416-450-7716
info@concretecountertopscanada.com
www.concretecountertopscanada.com

Cornerstone Masonry
PO Box 83
Pray, MT 59065
Tel/Fax: 406-333-4383
info@warmstone.com
www.warmstone.com

Countercast Designs Inc.
118, 6875 King George Highway
Surrey, BC
Toll-free tel: 888-787-2278
Tel: 604-542-1322
info@countercast.com
www.countercast.com

Dauter Stone Inc.
5230 - 1st Street SW
Calgary, AB T2H 0C8
Tel: 403-253-3738
Fax: 403-252-3672
dauter.stone@telus.net
www.dauterstone.com

Dupont (Corian & Zodiaq)
USA
Toll-free tel: 800-441-7515
Tel: 302-774-1000
Canada
E. I. du Pont Canada Company
PO Box 2200, Streetsville
Mississauga, ON L5M 2H3
Toll-free tel: 800-387-2122 (North American callers only)
Tel: 905-821-5193
Fax: 905-821-5057
Information@can.dupont.com
www.dupont.com

Exotic Metal Interiors
Tel: 877-715-5483
Fax: 905 507-1044
exotic@primeliteusa.com
www.exoticmetalinteriors.com

Floor Tile and Slate Co.
1209 Carroll Avenue
Carrollton, TX 75006
Toll-free tel: 800-446-0220
Tel: 972-242-6647
Fax: 972-242-7253
sales@floortileandslate.com
www.floortileandslate.com

Genesee Cut Stone & Marble
5276 South Saginaw Road
Flint, MI 48507
Tel: 810-743-1800
Fax: 810-694-7901
stone@gcsm.com
www.gcsm.com

Green Mountain Soapstone
680 E Hubbardton Road
PO Box 807
Castleton, VT 05735
Tel: 802-468-5636
vance@greenmountainsoapstone.com
www.greenmountainsoapstone.com

Grotto Designs
530 - 4 Street
Canmore, AB T1W 2H2
Toll-free tel: 866-262-3966
Tel: 403-678-5779
Cell: 403-512-3116
Fax: 403-678-1863
grottod@telusplanet.net
www.grottodesigns.com

KlipTech Composites
2999 John Stevens Way
Hoquiam, WA 98550
Tel: 360-538-9815
Fax: 360-538-1510
joel@kliptech.com
www.kliptech.com

La Puerta Originals
4523 State Road
Santa Fe, NM 87505
Tel: 505-984-8164
Fax: 505-986-5838
info@lapuertaoriginals.com
www.lapuertaoriginals.com

Meganite Inc.
1254 E Lexington Avenue
Pomona, CA 91766
Toll-free tel: 800-836-1118
Tel: 909-464-2908
Fax: 909-590-1628
www.meganite.com

Pacific Crest Industries, Inc. (Spectra Metals)
2411 Pomona Road
Corona, CA 92880
Toll-free tel: 877-674-3026
Tel: 951-520-0517
Fax: 951-520-0618
www.pacificcrestind.com

RMG Stone Products Inc.
PO Box 807
E Hubbardton Road
Castleton, VT 05735
Tel: 802-468-5636
Fax: 802-468-8968
sales@rmgstone.com
www.rmgstone.com

Robin-Reigi
48 W 21st Street, Ste 1002
New York, NY 10010
Tel: 212-924-5558
Fax: 212-924-2753
robin@robin-reigi.com
www.robin-reigi.com

U.S. Quartz Products
CaesarStone USA
11830 Sheldon Street
Sun Valley, CA 91352
Tel: 818-394-6000
Fax: 818-394-6006
info@caesarstone.com
CaesarStone USA NY
36-16 19th Ave
Astoria, NY 11105
Tel: 718-777-9780
Fax: 718-777-9784
ny@caesarstoneus.com
www.caesarstoneus.com

Richlite Corp.
624 E 15th Street
Tacoma, WA 98421
Toll-free tel: 888-383-5533
Fax: 253-383-5536
info@richlite.com
www.richlite.com

Ridalco
1551 Michael Street
Ottawa, ON K1B 3T4
Toll-free tel: 800-445-1097
Tel: 613-745-9161
Toll-free fax: 800-268-6526
Fax: 613-745-6452
info@ridalco.com
Website:www.ridalco.com

Rigidized Metals Corporation
658 Ohio Street
Buffalo, NY 14203
Toll-free tel: 800-836-2580
Tel: 716-849-4760
Fax: 716-849-0401
masonbowen@rigidized.com
www.rigidized.com

Soapstone Canada
H.A. Ness & Co. Inc.
PO. Box 11, Station U
Toronto, ON M8Z 5M4
Toll-free tel: 800-668-6377
Tel: 416-231-1645
Fax: 416-231-0231
glen.ness@soapstonecanada.com
www.soapstonecanada.com

Specialty Stainless.com
Toll-free tel: 800-836-8015
Tel: 716-893-3100
Fax: 716-893-0443
info@specialtystainless.com
www.specialtystainless.com

Stone Source
New York
215 Park Avenue South
New York, NY 10003
Tel: 212-979-6400
Fax: 212-979-6989
Boston
691A Somerville Avenue
Somerville, MA 02143
Tel: 617-666-7900
Fax: 617-666-7901
Washington
1400 16th Street NW
Washington, DC 20036
Tel: 202-265-5900
Fax: 202-265-5902
Chicago
414 N Orleans, Ste 403
Chicago, IL 60610
Tel: 312-335-9900
Fax: 312-970-5660
Philadelphia
Tel: 215-482-3000
Fax: 610-783-0156
info@stonesource.com
www.stonesource.com

Swan Stone
One City Centre, Ste 2300
St. Louis, MO 63101
Toll-free tel: 800-325-7008
Fax: 314-231-8165
infomail@swanstone.com
Non-U.S. E-mail: info@swanstone.com
www.theswancorp.com

Topstone
3601 Range Road
Temple, TX 76504
Toll-free tel: 866-774-9197
Tel: 254-774-9197
Fax: 254-774-7631
Sales@PSITopstone.com
www.psitopstone.com

Valley Countertops
30781 Simpson Road
Abbotsford, BC V2T 6X4
Toll-free tel: 800-506-9997
Fax: 604-852-9066
info@valleycountertops.com
www.valleycountertops.com

Vermont Marble, Granite & Soapstone Co.
1565 Main Street
Castleton, VT 05735
Tel: 802-468-8800
Fax: 802-217-1044
info@soapstone-co.com
www.soapstone-co.com

Vetrazzo
1414 Harbour Way S, Ste 1400
Richmond, CA 94804
Tel: 510-234-5550
Fax: 510-234-7890
info@vetrazzo.com
www.counterproduction.com

Wilsonart International, Inc.
2400 Wilson Place
PO Box 6110
Temple, TX 76503-6110
Toll-free tel: 800-433-3222
www.countertop.com

doors

Abraxis Art Glass, Inc.
212 West Laporte Avenue
Fort Collins, CO 80521
Tel: 970-493-7604
Fax: 970-493-7644
Info@Abraxisartglass.com
www.abraxisartglass.com

Adams Architectural Wood Products
300 Trails Road
Eldridge, IA 52748
Toll-free tel: 888-285-8120
Tel: 563-285-8000
Fax: 563-285-8003
www.adamsarch.com

Alamo Designs, Inc.
317-11765 West Avenue
San Antonio, TX 78216
Toll-free tel: 877-442-5266
Tel: 972-762-5472
Fax: 800-859-4716
mail@alamodesigns.com
www.alamodesigns.com

Albion Doors & Windows
PO Box 220
Albion, CA 95410
Tel: 707-937-0078
bysawyer@mcn.org
www.knobsession.com

Amberwood Doors Inc.
80 Galaxy Boulevard, Unit 16
Toronto, ON M9W 4Y8
Toll-free tel: 800-861-3591
Tel: 416-213-8007
Fax: 416-213-8009
info@amberwooddoors.com
www.amberwooddoors.com

Architectural Millwork
3522 Lucy Road
Millington, TN 38053
Tel: 901-327-1384
Tel (cell): 901-218-0616
Fax: 901-353-1645
samtickle@aol.com
www.a-millwork.com

Architectural Traditions
9280 E. Old Vail Road
Tucson, AZ 85747
Tel: 520-574-7374
www.architecturaltraditions.com

Arimoto Design & Woodworking, Inc.
1800 Columbus Avenue
Pittsburgh, PA 15233-2247
Tel: 412-231-2310
Fax: 412-231-5445
tadao@arimotowood.com
www.arimotowood.com

Artglass & Metal
4033 Skyline Road
Carlsbad, CA 92008
Tel: 760-390-2994
jay@artglassandmetal.com
www.artglassandmetal.com

Artistic Windows & Doors
10 South Inman Avenue
Avenel, NJ 07001
Toll-free tel: 800-278-3667
Tel: 732-726-9400
Fax: 732-726-9494
info@artisticdoorsandwindows.com
www.artisticdoorsandwindows.com

Arya Enterprises
2692 Dellinger Drive
Marietta, GA 30062
Tel: 678-770-1007
Fax: 678-560-3244
sales@aryaenterprises.com
www.aryaenterprises.com

Asselin U.S.A.
714-2870 Peachtree Road NW
Atlanta, GA 30305–2918
Tel: 404-419-6114
Fax: 404-419-6116
contact@asselinusa.com
www.asselinusa.com

The Bay Area Door Exchange, Inc
PO Box 8234
St. Petersburg, FL 33738
Toll-free tel: 800-588-4291
Tel: 727-724-3667
Doorexch@aol.com
www.bayareadoorexchange.com

Beautifully Carved Doors
PO Box 905
Ucluelet, BC V0R 3A0
Tel: 250-726-7755
Fax: 250-726-7555
info@carveddoors.com
www.carveddoors.com

Chautauqua Woods
PO Box 130
134 Franklin Avenue
Dunkirk, NY 14048
Tel: 716-366-3808
Fax: 716-366-3814
customerservice@cwdoors.us
www.cwdoors.us

Cienega Creek Mesquite
HC 1, Box 945
Sonoita, AZ 85637
danny@mesquitedoor.com
www.mesquitedoor.com

Richard Cornelius
Cornelius Enterprises
4155 Jay Street
Wheatridge, CO 80033
Tel: 303-463-0236
Fax: 303-463-0237
rich@richdoor.com
www.richdoor.com

Craftsmen In Wood
5441 West Hadley Street
Phoenix, AZ 85043
Tel: 602-296-1050
Fax: 602-296-1052
www.craftsmeninwood.com

Crisp Door and Window
1004 FM 1960 Bypass Road E
Humble, TX 77338
Tel: 281-540-5551
Fax: 281-540-5552
www.crispdoor.com

Cutting Bros. Inc.
3680 Muskoka Road 118 West
Port Carling, ON P0B 1J0
Tel: 705-765-1615
Fax: 705-765-1613
cuttingbrosinc@bellnet.ca
www.cuttingbrosinc.com

Chesney's, Jasper Conran Collection

Da Vinci & Co. Ltd.
616 Merrick Road
Lynbrook, NY 11563
Tel: 516-596-6666
Fax: 516-596-6094
sales@davincimoldings.com
www.davincimoldings.com

Dazzle Glazz
6605 Jefferson Highway
Baton Rouge, LA 70806–8104
Toll-free tel: 888-717-9484
Tel: 225-216-9484
duncan@dazzleglazz.com
www.dazzleglazz.com

Doors By Decora
3332 Atlanta Highway
Montgomery, AL 36109
Toll-free tel: 800-359-7557
Tel: 334-277-7910
Fax: 334-279-5470
www.doorsbydecora.com

Doorways to the West
3250 County Road 255
Westcliffe, CO 81252
Tel: 719-783-9271
inquiries@doorwayswest.com
www.doorwayswest.com

Ellenburg & Shaffer Glass Art Studio
344 S. Elm Street
Greensboro, NC 27401
Tel: 336-271-2811
Fax: 336-271-8000
info@stainedglassdomes.com
www.ellenburgandshaffer.com

Emerald Stained Glass & Door
2357 Harper Road
Beckley, WV 25801
Tel: 304-255-3072
Fax: 304-255-3073
information@emeraldsg.com
www.emeraldsg.com

Enjo Architectural Millwork
16 Park Avenue
Staten Island, NY 10302
Toll-free tel: 800-437-3656
sales@enjo.com
www.enjo.com

Extraordinary Doors
Santa Cruz, CA
Tel: 831-465-1470
Fax: 831-465-1471
robin@extraordinarydoors.com
www.extraordinarydoors.com

Galaxy Mailboxes/European Home
10 Corey Street
Melrose, MA 02176
Tel: 781-662-1110
Fax: 413-832-4879
info@europeanhome.com
www.europeanhome.com

Grabill Windows & Doors Incorporated
7463 Research Drive
Almont, MI 48003
Tel: 810-798-2817
Fax: 810-798-2809
tgrabill@bignet.net
www.grabillwindow.com

GrandRiverSupply.com
1421 Camino del Pueblo
Bernalillo, NM 87004
Toll-free tel: 877-477-8775
Tel: 505-867-4110
Fax: 505-867-9711
www.grandriverdoor.com

Hide A Door
8331 F.M. 1960 West, Ste K
Humble, TX 77338
Toll-free tel: 888-771-3667
www.hideadoor.com

House of Doors
3466 West Broadway
Vancouver, BC V6R 3B3
Tel: 604-737-1748
Fax: 604-737-1748
houseofdoors@hotmail.com
www.houseofdoorsandwindows.com

Hyde Park Architectural Decorating
130 Salt Point Turnpike
Poughkeepsie, NY 12603
Tel: 845-483-1340
Fax: 845-483-1343
info@hydeparksurrounds.com
www.hydeparksurrounds.com

JELD-WEN Inc.
PO Box 1329
Klamath Falls, OR 97601
Toll-free tel: 800-535-3936
www.jeld-wen.com

Joseph Millworks
37123 Hansen Lane
Baker City, OR 97814
Tel/Fax: 541-894-2347
randy@josephmillworks.com
www.josephmillworks.com

Knock on Wood
159 Argyle Street N
Caledonia, ON
Tel: 905-765-0442
Fax: 905-765-0442
doors@mountaincable.net
www.customscreendoors.ca

La Puerta Originals
4523 State Road
Santa Fe, NM 87505
Tel: 505-984-8164
Fax: 505-986-5838
info@lapuertaoriginals.com
www.lapuertaoriginals.com

Loewen Windows and Doors
77 Highway 52 W
Box 2260
Steinbach, MA R5G 1B2
Toll-free tel: 800-563-9367
Tel: 204-326-6446
www.loewen.com

Marvin Windows and Doors
PO. Box 100
Warroad, MN 56763
Toll-free tel US: 888-537-7828
Toll-free-tel Canada: 800-263-6161
www.marvin.com

Michaels Artistic Iron Doors
9102 Unit K
Firestone Boulevard
Downey, CA 90241
Tel: 562-861-3900
Fax: 562-861-7088
sales@michaelsirondoors.com
www.michaelsirondoors.com

Molyneux Designs
PO Box 1637
Weaverville, CA 96093
Tel: 530-623-2268
Fax: 530-623-1671
md@CustomMade.com
www.molyneuxdesigns.com

Native Wood Art.com
Gibson, BC
Tel: 604-886-2465
davey@dccnet.com
www.nativewoodart.com

Next Door Company
1840 N Commerce Parkway, Ste 1
Weston, FL 33326
Toll-free: 888-791-4450
Fax: 954-772-8466
info@nextdoorco.com
www.nextdoorco.com

Old Iron Doors
2192F Parkway Lake Drive
Birmingham, AL 35244
Toll-free tel: 877-261-4800
Tel: 205-985-4800
Fax: 205-985-4100
sales@oldirondoors.com
www.oldirondoors.com

Pella Windows and Doors
102 Main Street
Pella, IA 50219
Tel: 641-621-1000
www.pella.com

Redwood Burl .Com
34-1834 Allard
Eureka, CA 95503
Tel: 707-441-1658
gbuckjr@redwoodburl.com
www.redwoodburl.com

Restoration Works
1345 Stanford Drive
Kankakee, IL 60901
Tel: 815-937-0556
Fax: 815-937-4072
gwallace@restorationworksinc.com
www.restorationworksinc.com

Simpson Door Company
400 Simpson Avenue
McCleary, WA 98557
Toll-free tel: 800-952-4057
simpsondoor@brandner.com
www.simpsondoor.com

Sperlich Lighting Art Glass and Doors
7005 N Waterway Drive
Miami, FL 33155
Tel: 786-388-7522
Fax: 786-388-7523
sales@sperlich.com
www.sperlich.com

Spirit Elements, Inc.
200-6672 Gunpark Drive,
Boulder, CO 80301
Toll-free tel: 800-511-1440
Tel: 303-998-1440
Fax: 800-511-1421
cs@spiritelements.com
www.spiritelements.com

Therma-Tru Doors
1750 Indian Wood Circle
Maumee, OH 43537
Toll-free tel: 800-843-7628
www.therma-tru.com

Touchstone Woodworks
PO Box 112
Ravenna OH 44266
Tel: 330-297-1313
tinawalters@touchstonewoodworks.com
www.touchstonewoodworks.com

Upstate Door, Inc.
26 Industrial Street
Warsaw, NY 14569
Toll-free tel: 800-570-8283
Tel: 585-786-3880
Fax: 585-786-3888
www.upstatedoor.com

Jack Wallis Doors & Wallis Stained Glass
2985 Butterworth Road
Murray, KY 42071
Tel: 270-489-2613
Fax: 270-489-2187
wallis@wk.net
www.wallisstainedglass.com

Weston Millwork Company
722 Washington Street
Weston, MO 64098
Tel: 816-640-5555
Fax: 816-386-5555
info@westonmillwork.com
www.westonmillwork.com

doors, garage

Amarr Garage Doors
165 Carriage Court
Winston-Salem, NC 27105
Tel: 800-503-DOOR
Fax: 336-767-3805
www.amarr.com

Architectural Garage Doors Mfg. Inc.
2420 Hiller Ridge
Johnsburg, IL 60050
Tel: 815-344-9910
Fax: 815-344-9912
customerservice@woodgaragedoor.com
www.woodgaragedoor.com

Artisan Custom Doorworks
975 Hemlock Road
Morgantown, PA 19543
Tel: 888-913-9170
Fax: 888-913-9179
info@artisandoorworks.com
www.artisandoorworks.com

Clopay Building Products
8585 Duke Boulevard
Mason, OH 45040-3101
Toll-free tel: 800-225-6729
www.clopaydoor.com

Garage Doors Incorporated
1001 South Fifth Street
San Jose, CA 95112
Toll-free tel: 800-223-9795
Tel: 408-293-7443
Fax: 408-293-7457
scott@garagedoorsinc.com
www.garagedoorsinc.com

Martin Door Manufacturing
PO Box 27437
Salt Lake City, UT 84127
Toll-free tel: 800-388-9310
Toll-free fax: 800-688-8182
Consumer@Martindoor.com
www.martindoor.com

Marvin Windows and Doors
PO Box 100
Warroad, MN 56763
Toll-free tel (US): 888-537-7828
Toll-free tel (Canada): 800-263-6161
www.marvin.com

MendocinoDoors.com
14660 Mitchell Creek Dr
Fort Bragg, CA 95437
Tel: 707-964-0635
leecab@mcn.org
www.mendocinodoors.com

Mon-Ray, Inc
801 Boone Avenue N
Minneapolis, MN 55427-4432
Toll-free tel: 800-544-3646
www.monray.com

Raynor Worldwide
PO Box 448
1101 E River Road
Dixon, IL 61021–0448
Toll-free tel: 800-472-9667
Tel: 815-288-1431
Fax: 815-288-7142
thegarage@raynor.com
www.raynor.com

Real Carriage Door Company, Inc
13417 82nd Avenue NW
Gig Harbor, WA 98329
Toll-free tel: 866-883-8021
Tel: 253-238-6908
Fax: 253-238-6231
info@realcarriagedoors.com
www.realcarriagedoors.com

Zeluck Incorporated LLC
5300 Kings Highway
Brooklyn, NY 11234
Toll-free tel: 800-233-0101
Tel: 718-251-8060
Fax: 718-531-2564
California
Tel: 310-456-6304
Fax: 310-456-6834
Florida
Tel: 561-833-0092
Fax: 561-832-9583
info@zeluck.com
www.zeluck.com

fireplaces, mantels

Adobelite
10031 Southern SE
Albuquerque, NM 87123
Tel: 505-291-0500
Fax: 505-291-9317
adobelite@adobelite.com
www.adobelite.com

A&M Victorian Decorations
2411 Chico Avenue
South El Monte, CA 91733
Toll-free tel: 800-671-0693
Tel: 626-575-0693
Fax: 626-575-1781
am-vic@pacbell.net
www.aandmvictorian.com

Architectural Stone Company
1900 Preston Road, Ste 267 PMB 100
Plano, TX 75093
Tel: 972-769-8379
Fax: 972-599-7638
contact@architecturalstone.net
www.architecturalstone.net

Walter S. Arnold, Stone Sculptor
Chicago, IL
Toll-free tel: 847-568-1188
Tel: 312-226-1141
www.stonecarver.com

Cantera Especial
343-15332 Antioch Street
Pacific Palisades, CA 90272
Toll-free tel: 800-564-8608
Tel: 818-907-7170
Fax: 818-907-0343
canteraespecial@aol.com
www.canteraespecial.com

Carroll's Mantels
421 CR 3321
Troy, AL 36079
Tel: 334-735-3217
Fax: 334-735-0877
carroll@carrollmantels.com
www.carrollmantels.com

Central Fireplace
20502 160th Street
Greenbush, MN 56726
Toll-free tel: 800-248-4681
Tel: 218-782-2575
Fax: 218-782-2580
www.centralboiler.com

Chesney's
D&D Building
979 Third Avenue, Ste 244, 2nd Floor
New York, NY 10022
Tel: 646-840-0609
Fax: 646-840-0602
newyorksales@chesneys-usa.com
www.chesneys.co.uk

Richard Cornelius
Cornelius Enterprises
4155 Jay Street
Wheatridge, CO 80033
Tel: 303-463-0236
Fax: 303-463-0237
rich@richdoor.com
www.richdoor.com

Cornerstone Masonry
PO Box 83
Pray, MT 59065
Tel/Fax: 406-333-4383
info@warmstone.com
www.warmstone.com

Da Vinci & Co. Ltd.
616 Merrick Road
Lynbrook, NY 11563
Tel: 516-596-6666
Fax: 516-596-6094
sales@davincimoldings.com
www.davincimoldings.com

DeSantana Handcarved Stone
PO Box 907
Higley, AZ 85236
Tel: 866-932-2783
Fax: 480-664-8682
info@desantanastone.com
www.desantanastone.com

Dracme Inc.
47 Avenue du Lac
Ste-Julie, PQ J3E 2Y6
Toll-free tel: 877-990-8635
Tel: 450-649-9525
Fax: 877-990-8636
sales@dracme.com
www.dracme.com

Eye of the Day Garden Design Center
4620 Carpenteria Avenue
Carpenteria, CA 93013
Tel: 805-566-0778
Fax: 805-566-0478
www.eyeofthedaygdc.com

Fairplay Stonecarvers
PO Box 0355
Oberlin, OH 44074-0355
Tel: 440-775-7878
Fax: 440-775-7979
meadows@fairplaystonecarvers.com
www.fairplaystonecarvers.com

Fine's Gallery
11400 South Cleveland Avenue
Fort Myers, FL 33907
Tel: 239-277-0009
Fax 239-418-0321
www.finesgallery.com

GrandRiverSupply.com
1421 Camino del Pueblo
Bernalillo, NM 87004
Toll-free tel: 877-477-8775
Tel: 505-867-4110
Fax: 505-867-9711
www.grandriverdoor.com

Haddonstone (USA) Ltd.
West
32207 United Avenue
Pueblo, CO 81001
Tel: 719-948-4554
Fax: 719-948-4285
East
201 Heller Place
Bellmawr, NJ 08031
Tel: 856-931-7011
Fax: 856-931-0040
info@haddonstone.com
www.haddonstone.com

Hawkeye Mantel Co.
16110 Asbury Road
Dubuque, IA 52002
Toll-free tel: 877-762-6835
Tel: 563-557-0927
Fax: 563-582-8667
hawkeyemantelco@mchsi.com
www.hawkeyemantels.com

Heat & Glo
20802 Kensington Boulevard
Lakeville, MN 55044
Toll-free tel: 888-427-3973
info@heatnglo.com
www.heatnglo.com

Homes By Marie
PO Box 2777
Corrales, NM 87048
Tel: 505-342-1532
Fax: 505-342-1579
marie@homesbymarie.com
www.homesbymarie.com

Hyde Park Architectural Decorating
130 Salt Point Turnpike
Poughkeepsie, NY 12603
Tel: 845-483-1340
Fax: 845-483-1343
info@hydeparksurrounds.com
www.hydeparksurrounds.com

Jim's Masonry
Albuquerque, NM
Tel: 505-237-9705 or 250-2046
respond@jimmasonry.com
www.jimsmasonry.com

La Puerta Originals
4523 State Road
Santa Fe, NM 87505
Tel: 505-984-8164
Fax: 505-986-5838
info@lapuertaoriginals.com
www.lapuertaoriginals.com

Masonry Heater Association of North America (MHA)
1252 Stock Farm Road
Randolph, VT 05060
Tel: 802-728-5896
Fax: 802-728-6004
bmarois@sovernet.com
www.mha-net.org

Molyneux Designs
PO Box 1637
Weaverville, CA 96093
Tel: 530-623-2268
Fax: 530-623-1671
md@CustomMade.com
www.molyneuxdesigns.com

New England Cast In Stone Manufacturing, Inc.
257 Clinton Street
PO Box 158, 132 Main Street
Springfield, VT 05156
Tel: 802-885-7866
Fax: 802-885-7870
info@newenglandcastinstone.com
www.newenglandcastinstone.com

Old Mill Mantels LLC
2522 Lansing Avenue
Jackson, MI 49202
Toll-free tel: 866-783-0707
Fax: 866-783-0708
www.oldmillmantels.com

Old World Stoneworks
5400 Miller Avenue
Dallas, TX 75206
Toll-free tel: 800-600-8336
Tel: 214-826-3645
Fax: 214-826-3227
info@oldworldstoneworks.com
www.oldworldstoneworks.com

Pearl Mantels
10846 E Shelby Drive
Collierville, TN 38017
Tel: 901-853-8237
Fax 901-861-1057
info@pearlmantels.com
www.pearlmantels.com

Pioneer Millworks
Toll-free tel: 800-951-9663
Eastern Office
1180 Commercial Drive
Farmington, NY 14425
Tel: 585-924-9970
Western Office
547 West 700 South
Salt Lake City, UT 84101
Tel: 801-328-9663
www.pioneermillworks.com

Poling Custom Homes
PO Box 747
Placitas, NM 87043
Tel: 505-867-3200
Mobile: 505-980-5280
rhpoling@polingcustomhomes.com
www.polingcustomhomes.com

Precision Development
5711 Clarewood
Houston, TX 77081
Tel: 713-667-1310
Fax: 713-667-0515
www.pdcaststone.com

Redwood Burl .Com
1834 Allard 34
Eureka, CA 95503
Tel: 707-441-1658
gbuckjr@redwoodburl.com
www.redwoodburl.com

Riley Brothers Inc.
10195 Southwest 186 Street
Miami, FL 33157
Tel: 305-252-4610
Fax: 305-252-5964
www.coralstone.com

Saint Tropez Stone Boutique
25 Evelyn Way
San Francisco, CA 94127
Tel: 415 2594820
Fax: 832-565-1100
sales@sainttropezstone.com
www.sainttropezstone.com

Santa Fe Kiva Fireplace Mfg. Inc.
Tel: 505-344-7555
pkincaid@santafekiva.com
www.santafekiva.com

Sierra Concrete Design, Inc.
2110 S. Anne Street
Santa Ana, CA 92704–4409
Tel: 714-557-8100
Fax: 714-557-8109
sales@sierraconcrete.com
www.sierraconcrete.com

Siteworks, Inc.
363 West Canino
Houston, TX 77037
Toll-free tel: 800-599-5463
Tel: 281-931-1000
Fax: 281-931-1044
tech@siteworkstone.com
www.chateaustone.com

Stone Age Designs
Atlanta
351 Peachtree Hills Avenue, Ste 501–A
Atlanta, GA 30305
Tel: 404-350-3333
Fax: 404-842-0936
Florida
935 Orange Avenue, Ste A
Winter Park, FL 32789
Tel: 407-628-5577
Fax: 407-628-5565
info@stoneagedesigns.net
www.stoneagedesigns.net

Stone Magic
301 Pleasant Drive
Dallas, TX 75217
Toll-free tel: 800-597-3606
Fax: 214-823-4503
info@stonemagic.com
www.stonemagic.com

Stonesculpt
Palo Alto, CA
Tel: 650-575-9683
Fax 650-322-5002
Lobykin@bigfoot.com
www.customstonecarving.com

Barbara Tattersfield Design Inc.
Palm Beach, FL Showroom
Tel: 561-833-3443
Fax: 561-833-3414
Corona del Mar, CA Showroom
Tel: 949-675-0600
Fax: 949-675-0601
info@btattersfielddesign.com
www.btattersfielddesign.com

Thunderstone LLC
3300 S 6th Street
Lincoln, NE 68502
Tel: 402-420-2322
Fax: 402-420-2542
cconiglio@castone.com
www.castone.com

Victorian Fireplace Shop
3121 W Broad Street
Richmond, VA 23230
Toll-free tel: 866-GASCOALS
Tel: 804-355-1688
Fax: 804-358-3728
karen@gascoals.com
www.thevictorianfireplace.com

Wausau Tile, Inc.
PO Box 1520
9001 Bus. Highway 51
Wausau, WI 54401
Toll-free tel: 800-388-8728
Tel: 715-359-3121
Fax: 715-355-4627
wtile@wausautile.com
www.wausautile.com

fireplaces, electric

American Chimney
201 N Maple Avenue
Purcellville, VA 20132
Tel: 540-338-2723
Fax: 540-338-2758
www.americanchimneyva.com

Dimplex North America
1367 Industrial Road
Cambridge, ON N1R 7G8
Toll-free tel: 800-668-6663
www.dimplex.com

Heat & Glo
20802 Kensington Boulevard
Lakeville, MN 55044
Toll-free tel: 888-427-3973
info@heatnglo.com
www.heatnglo.com

Heatilator
1915 W Saunders Street
Mount Pleasant, IA 52641
Toll-free tel: 800-927-6841
info@heatilator.com
www.heatilator.com

Lennox Hearth Products
1110 West Taft Avenue
Orange, CA 92865-4150
Toll-free tel: 800-953-6669
info@LennoxHP.com
www.lennoxhearthproducts.com

Napoleon Fireplaces
ask@napoleon.on.ca
www.napoleonfireplaces.com

Riley Brothers Inc.
10195 SW 186 Street
Miami, FL 33157
Tel: 305-252-4610
Fax: 305-252-5964
www.coralstone.com

Vermont Castings (CFM Corporation)
2695 Meadowvale Boulevard
Mississauga, ON L5N 8A3
Toll-free tel: 800-525-1898
Tel: 905-858-8010
Fax: 905-858-3966
www.vermontcastings.com

Victorian Fireplace Shop
3121 W. Broad Street
Richmond, VA 23230
Toll-Free: 866-GASCOALS
Tel: 804-355-1688
Fax: 804-358-3728
karen@gascoals.com
www.thevictorianfireplace.com

fireplaces, gas

American Chimney
201 N. Maple Avenue
Purcellville, VA 20132
Tel: 540-338-2723
Fax: 540-338-2758
www.americanchimneyva.com

Blaze King Industries Canada
1290 Commercial Way
Penticton, BC V2A 3H5
Tel: 250-493-7444
Blaze King Industries USA
146 A Street
Walla Walla, WA 99362
Tel: 509-522-2730
www.blazeking.com

Central Fireplace
20502 160th Street
Greenbush, MN 56726
Toll-free tel: 800-248-4681
Tel: 218-782-2575
Fax: 218-782-2580
www.centralboiler.com

ECS Inc.
918 Freeburg Avenue
Belleville, IL 62222–0529
Toll-free tel: 800-851-3153
Tel: 618-233-7420
Toll-free fax: 800-443-8648
Fax: 618-233-7097
info@empirecomfort.com
www.empirecomfort.com

Eiklor Flames
282 E Pivot Point Road
Paoli, IN 47454
Toll-free tel: 888-295-5647
Fax: 812-723-5708
info@eiklorflames.com
www.eiklorflames.com

European Home
10 Corey Street
Melrose, MA 02176
Tel: 781-662-1110
Fax: 413-832-4879
info@europeanhome.com
www.europeanhome.com

Heat & Glo
20802 Kensington Boulevard
Lakeville, MN 55044
Toll-free tel: 888-427-3973
info@heatnglo.com
www.heatnglo.com

Heatmaster, Inc.
PO Box 1717
Angier, NC 27501
Tel: 919-639-4568
Fax: 919-639-0101
info@heatmaster.com
www.heatmaster.com

Kingsman Fireplaces
2340 Logan Avenue
Winnipeg, MB R2R 2V3
Tel: 204-632-1962
Fax: 204-632-1960
contactus@kingsmanind.com
www.kingsmanind.com

Lennox Hearth Products
1110 West Taft Avenue
Orange, CA 92865–4150
Toll-free tel: 800-953-6669
info@LennoxHP.com
www.lennoxhearthproducts.com

Malm Fireplaces
368 Yolanda Avenue
Santa Rosa, CA 95404
Toll-free tel: 800-535-8955
Tel: 707-523-7755
Fax: 707-571-8036
info@malmfireplaces.com
www.malmfireplaces.com

Miles Industries Ltd.
190-2255 Dollarton Highway
North Vancouver, BC V7H 3B1
Toll-free tel: 800-468-2567
Tel: 604-984-3496
Toll-free fax: 800-268-0333
Fax: 604-984-0246
info@milesfireplaces.com
www.valorfireplaces.com

Monessen Hearth Systems
149 Cleveland Drive
Paris, KY 40361
Tel: 800-867-0454
Fax: 877-867-1875
www.monessenhearth.com

Moderustic
Rancho Cucamonga, CA 91730
Tel: 909-989-6129
Fax: 909-944-3811
Ed@Moderustic.com
www.moderustic.com

Napoleon Fireplaces
ask@napoleon.on.ca
www.napoleonfireplaces.com

Russo Products, Inc.
61 Pleasant Street
Randolph, MA 02368
Tel: 781-963-1182
sales@fireplaceseast.com
www.fireplaceseast.com

ThermArt / Off The Wall Fireplaces
106-20050 Stewart Crescent
Maple Ridge, BC V2X 0T4
Tel: 604-460-7101
Fax: 604-460-7102
infor@thermart.com
www.thermart.com

Vancouver Gas Fireplaces
235 West 7th Avenue
Vancouver, BC V5Y 1L9
Tel: 604-732-3470
Fax: 604-732-5729
salesvgf@vangasfireplaces.com
www.vangasfireplaces.com

Vermont Castings (CFM Corporation)
2695 Meadowvale Boulevard
Mississauga, ON L5N 8A3
Toll-free tel: 800-525-1898
Tel: 905-858-8010
Fax: 905-858-3966
www.vermontcastings.com

Victorian Fireplace Shop
3121 W Broad Street
Richmond, VA 23230
Toll-Free: 866-GASCOALS
Tel: 804-355-1688
Fax: 804-358-3728
karen@gascoals.com
www.thevictorianfireplace.com

fireplaces, wood

American Energy Systems, Inc
150 Michigan Street SE
Hutchinson, MN 55350
Toll-free tel: 800-495-3196
www.magnumfireplace.com

American Chimney
201 N. Maple Avenue
Purcellville, VA 20132
Tel: 540-338-2723
Fax: 540-338-2758
www.americanchimneyva.com

Blaze King Industries Canada
1290 Commercial Way
Penticton, BC V2A 3H5
Tel: 250-493-7444
Blaze King Industries USA
146 A Street
Walla Walla, WA 99362
Tel: 509-522-2730
www.blazeking.com

Buckley Rumford Co.
1035 Monroe Street
Port Townsend, WA 98368
Tel: 360-385-9974 or 360-385-9483
Fax: 360-385-1624
Tel (cell): 360-531-1081
buckley@rumford.com
www.rumford.com

Good Times Stove Company
PO Box 306
Goshen, MA 01032-0306
Tel: 413 -268-3677
stoveprincess@goodtimestove.com
www.goodtimestove.com

Heat & Glo
20802 Kensington Boulevard
Lakeville, MN 55044
Toll-free tel: 888-427-3973
info@heatnglo.com
www.heatnglo.com

Kozy Heat Fireplaces
204 Industrial Park Drive
Lakefield, MN 56150
Toll-free tel: 800-253-4904
Fax: 507-663-6644
www.kozyheat.com

Lennox Hearth Products
1110 West Taft Avenue
Orange, CA 92865–4150
Toll-free tel: 800-953-6669
info@LennoxHP.com
www.lennoxhearthproducts.com

Masonry Stove Builders
Tel: 819-647 5092
Fax: 819-647 6082
mheat@mha-net.org
www.mha-net.org

Malm Fireplaces
368 Yolanda Avenue
Santa Rosa, CA 95404
Toll-free tel: 800-535-8955
Tel: 707-523-7755
Fax: 707-571-8036
info@malmfireplaces.com
www.malmfireplaces.com

Monessen Hearth Systems
149 Cleveland Drive
Paris, KY 40361
Tel: 800-867-0454
Fax: 877-867-1875
www.monessenhearth.com

Napoleon Fireplaces
ask@napoleon.on.ca
www.napoleonfireplaces.com

Quadra-Fire, Hearth & Home Technologies
1445 North Highway
Colville, WA 99114–2008
Toll-free tel: 800-926-4356
info@quadrafire.com
www.quadrafire.com

Russo Products, Inc.
61 Pleasant Street
Randolph, MA 02368
Tel: 781-963-1182
sales@fireplaceseast.com
www.fireplaceseast.com

Temp Cast Masonry Heaters
3324 Yonge St
PO Box 94059
Toronto ON M4N 3R1
Toll-free tel: 800-561-8594
Fax: 416-486-3624
www.tempcast.com

Tulikivi U.S. Inc.
PO Box 7547
Charlottesville, VI 22906–7547
Toll-free tel: 800-843-3473
tulikivius@tulikivi.fi
www.tulikivi.com
Canadian Dealer
Soapstone Heating Systems Inc.
Box 97
Big Lake Ranch, BC V0L 1G0
Toll-free tel: 877-890-8770
Tel: 250-243-2100
boitier@wlake.com
www.soapstoneheating.com

Vermont Castings (CFM Corporation)
2695 Meadowvale Boulevard
Mississauga, ON L5N 8A3
Toll-free tel: 800-525-1898
Tel: 905-858-8010
Fax: 905-858-3966
www.vermontcastings.com

Vermont Marble, Granite, Slate & Soapstone Co.
Route 4, Killington, VT 05751
Tel: 802-747-7744
Fax: 802-217-1044
info@vermontwoodstove.com
www.vermontwoodstove.com

fireplaces, outdoor

Buckley Rumford Co.
1035 Monroe Street
Port Townsend, WA 98368
Tel: 360-385-9974 or 360-385- 9483
Fax: 360-385-1624
Tel (cell): 360 531 1081
buckley@rumford.com
www.rumford.com

G.I. Designs
700 Colorado Boulevard 120
Denver, CO 80206
Toll-free tel: 877-442-6773
Tel: 303-377-5323
Fax: 303-377-5332
Cs@gidesigns.net
www.gidesigns.net

Fire Science Inc.
2153 Niagara Falls Boulevard
Amherst, NY 14228
Tel: 716-568-2224
Fax: 716-568-2235
Sales@fire-science.com
www.fire-science.com

Masonry Stove Builders
Tel: 819-647-5092
Fax: 819-647-6082
mheat@mha-net.org
www.mha-net.org

Napoleon Fireplaces
ask@napoleon.on.ca
www.napoleonfireplaces.com

NEXO Fireplace
619 Dudley Way
Sacramento, CA 95818
Toll-free tel: 800-559-6396
Fax: 916-441-4915
info@nexofireplace.com
www.nexofireplace.com

fireplaces, firebrick

Rhodes Architectural Stone
2011 E Olive Street
Seattle, WA 98122
Tel: 206-709-3000
Fax: 206-709-3003
johnt@rhodes.org (West)
trevorb@rhodes.org (North, Midwest)
andrewg@rhodes.org (East)
www.rhodes.org

fireplaces, pellet stoves

American Chimney
201 N Maple Avenue
Purcellville, VA 20132
Tel: 540-338-2723
Fax: 540-338-2758
www.americanchimneyva.com

American Energy Systems, Inc
150 Michigan Street SE
Hutchinson, MN 55350
Toll-free tel: 800-495-3196
www.magnumfireplace.com

Quadra-Fire, Hearth & Home Technologies
1445 North Highway
Colville, WA 99114-2008
Toll-free tel: 800-926-4356
info@quadrafire.com
www.quadrafire.com

flooring, bamboo

123 Bamboo
311, New Universe, Yi Jin Street,
Lin'an City, Hangzhou, Zhejiang,
China – 311300
Tel: 0086-13216191755
chinabiz1@yahoo.com
www.123bamboo.com

Bamboard USA
13730 Lake City Way NE, Ste A
Seattle, WA 98125
Tel: 206-522-9789
Fax: 206-522-9758
charlesm@seanet.com
www.bamboardusa.com

Bamboo Direct Building Products Ltd.
151 West Second Street, Ste 2106
North Vancouver, BC V7M 3P1
Tel: 604-926-4257
Fax: 604-926-4280
info@bamboodirect.ca
www.bamboodirect.ca

Bamboo Flooring Hawaii, LLC
521 Ala Moana Boulevard, Ste 213
Honolulu, HI 96814
Toll-free tel: 877-502-2626
Tel: 808-550-8080
Fax: 808-550-8090
info@bambooflooringhawaii.com
www.bambooflooringhawaii.com

Bamboo Hardwoods
510 S. Industrial Way
Seattle, WA 98108
Tel: 206-264-2414
Toll-free tel: 800-783-0557
Fax: 206-264-9365
info@bamboohardwoods.com
www.bamboohardwoods.com

Bamboo Works Inc.
PO Box 1880
Kapaa, HI 96746
Toll-free tel: 808-632-0533
info@bambooworks.com
www.bambooworks.com

Bambooadvantage
PO Box 146617
Boston, MA 02114
Toll-free tel: 877-226-2728
Tel: 617-670-1288
Fax: 617-670-1837
info@bambooadvantage.com
www.bambooadvantage.com

Bameco Flooring
10579 Bancroft Lane
Frisco, TX 75035
Tel: 214-705-8338
Fax: 214-705-8338
sales@bameco.com
www.bameco.com

BamStar
4925 Galaxy Parkway, Ste O
Cleveland, OH 44128
Tel: 216-839-0900
Fax: 216-839-0890
www.bamstar.com

BuildDirect
1900–570 Granville Street
Vancouver, BC V6C 3P1
Toll-free tel: 877-631-2845
Tel: 604-662-8100
Fax: 604-662-8142
Toll-free fax: 877-664-2845
sales@builddirect.com
www.builddirect.com

BuildingGreen Inc.
122 Birge Street, Ste 30
Brattleboro, VT 05301
Tel: 802-257-7300
Fax: 802-257-7304
info@buildinggreen.com
www.buildinggreen.com

CFS Corporation
4485 Tench Road, Ste 330
Suwanee, GA 30024
Toll-free tel: 866-751-4893
Fax: 770-614-5833
Daphney@PremiumGreenBamboo.com
www.premiumgreenbamboo.com

DIYFlooring.com
1451 W. Cypress Creek Road, Ste 300
Ft. Lauderdale, FL 33309
Toll-free tel: 800-788-1098
info@diyflooring.com
www.diyflooring.com

D&M Bamboo
528 Congress Circle South
Roselle, IL 60172
Tel: 630-582-1600
Fax: 630-582-1700
sales@dmbamboo.com
www.dmbamboo.com

Duro-design Cork & Bamboo Flooring
2866 Daniel-Johnson Boulevard
Laval, QC H7P 5Z7
Toll-free tel: 888-528-8518
Tel: 450-978-3403
Fax: 450-978-2542
info@Duro-design.com
www.duro-design.com

Ecotimber
1611 Fourth Street
San Rafael, CA 94901
Tel: 415-258-8454
Fax: 415-258-8455
www.ecotimber.com

E.Z. Oriental Inc.
1455 Montery Pass, Ste 105
Montery Park, CA 91754
Toll-free tel: 888-395-8887
Tel: 323-526-1188
Fax: 323-526-9888
Jian@bamboofloor.net
www.bamboofloor.net

Fair Pacific Bamboo Flooring
8081 South Upham Street
Littleton, CO 80128
Toll-free tel: 877-633-5667
Tel: 720-227-9135
Fax: 720-554-8013
www.fairpacific.com

FloorCanada
89 Lindsay Road
Woodstock, ON N4V 1A8
Toll-free tel: 877-357-2632
Tel: 519-788-4948
tmckegney@floorcanada.ca
www.floorcanada.ca

Floor Fantasy Inc.
219-177 Main Street
Fort Lee, NJ 07024
Tel: 201-363-9956
Fax: 201-947-3016
sales@floorfantasy.com or
info@floorfantasy.com
www.floorfantasy.com

FloorShop.Com
18375 Olympic Ave S
Tukwila, WA 98188
Toll-free tel: 888-641-3566
Sales@floorshop.com
www.floorshop.com

Hardwood Brokers
2292 Natoma Drive
Virginia Beach, VA 23456
Toll-free tel: 800-313-1107
customerservice@hardwoodbrokers.com
www.hardwoodbrokers.com

iFLOOR.com
PO Box 3964
Bellevue, WA 98004
Tel: 800-454-3941
Fax: 425-455-7643
www.ifloor.com

Panda Bamboo Products Inc.
15-15 119th Street
College Point, NY 11356
Tel: 718-353-0700
Fax: 718-353-8899
sales@pandabamboo.com
www.pandabamboo.com

Prolex Flooring USA
200 Best Friend Court, Ste 215
Norcross, GA 30071
Toll-Free: 866-977-6539
Tel: 770-242-4080
Fax: 770-242-6338
sales@prolexflooring.com
www.prolexflooring.com

Robin-Reigi
48 W 21st Street, Ste 1002
New York, NY 10010
Tel: 212-924-5558
Fax: 212-924-2753
robin@robin-reigi.com
www.robin-reigi.com

Silkroad/K&M Bamboo Products Inc.
300 Esna Park Drive, Unit 26
Markham, ON L3R 1H3
Tel: 905-946-8128
Fax: 905-946-8126
kbow@silkroadflooring.com
www.silkroadflooring.com

Teragren LLC
12715 Miller Road NE, Ste 301
Bainbridge Island, WA 98110
Toll-free tel: 800-929-6333
Tel: 206-842-9477
Fax: 206-842-9456
info@teragren.com
www.teragren.com

TICO Bamboo
713 W Duarte Rd, G-318
Arcadia, CA 91007–7564
Toll-free tel: 800-690-8426
Fax: 626-628-3892
info@ticobamboo.com
www.ticobamboo.com

Triton International Woods LLC
PO Box 8767
Rocky Mount, NC 27804
Toll-free tel: 86-319-6637
Tel: 252-823-6675
Fax: 252-823-6491
info@tritonwoods.com
www.tritonwoods.com

Warner Bamboo Floors Inc.
701 E Linden Avenue
Linden, NJ 07036
Toll-free tel: 800-694-8030
Tel: 908-474-9666
sales@warnerbamboo.com.
www.warnerbamboo.com

Wood Flooring International
122 Kissel Road
Burlington, NJ 08016
Toll-free tel: 888-964-6832
Tel: 856-764-2501
Fax: 856-764-2503
info@bamtex.com
www.bamtex.com

Woodlist, Inc
277 Linden Street, Ste 204
Wellesley, MA 02482
Toll-free tel: 877-487-6504
Tel: 781-283-5757
Fax: 781-283-5707
info@woodlist.com
www.woodlist.com

flooring, ceramic, terra cotta, etc.

Almar Inc. Floor and Roof Tiles
6645 NW 77 Avenue
Miami, FL 33166
Toll-free tel: 800-54-TILE-1
Tel: 305-471-5830
Fax: 305-471-5833
www.altusa.com

Armstrong World Industries, Inc.
PO. Box 3001
Lancaster, PA 17604
Toll-free tel: 800-233-3823
www.armstrong.com

Classic Terra Cotta Company
15332 Antioch Ste 118
Pacific Palisades, CA 90272
Toll-free tel: 888-837-7286
Fax: 866-304-4082
www.terracottapavers.com

Complete Flooring
104-100 W Pflugerville Loop
Pflugerville, TX 78660
Tel: 512 989-1880
Fax: 512 989-1888
completeflooring1@netzero.com
www.completeflooring.net

Country Floors, Inc.
8735 Melrose Avenue
Los Angeles, CA 90069
Tel: 310-657-0510
Fax: 310- 659-6470
info@countryfloors.com
www.countryfloors.com

Crossville Inc.
PO Box 1168
Crossville, TN 38557
Toll-free tel: 800-221-9093
Tel: 931-484-2110
www.crossvilleinc.com

Design Tile Inc.
8455-B Tyco Road
Vienna, VA 22182
Tel: 703-734-8211
Fax: 703-749-7935
info@design-tile.com
www.design-tile.com

Imagine Tile
1515 Broad Street
Bloomfield, NJ 07003
Toll-free tel: 800-680-TILE
www.imaginetile.com

Karen Mack Tiles
2756-1 Capital Circle NE
Tallahassee, FL 32308
Tel: 850-942-6565
www.customtiles.com

Malibu Ceramic Works
PO Box 1406
Topanga, CA 90290
Tel: 310-455-2485
Fax: 310-455-4385
Mailub.ceramics@verizon.net
www.terracottapavers.com

Manet Tiles
16735 Roscoe Boulevard
North Hills, CA 91343
Tel: 818- 892-4154
Fax: 818-767-1752
manettiles@sbcglobal.net
www.manettiles.com

Paris Ceramics
3102 Roswell Road
Buckhead, Atlanta, GA 30305
Tel: 404-949-0294
Fax: 404-949-9435
atlanta@parisceramics.com
www.parisceramics.com

Pompei Mosaic Tile
11301 Olympic Boulevard, Ste 512
Los Angeles, CA 90064
Tel: 310-312-9893
Fax: 310-996-1929
info@pompei-mosaic.com
www.pompei-mosaic.com

Ann Sacks
8120 NE 33rd Drive
Portland, OR 97211
Toll-free tel: 800-278-8453
www.annsacks.com

Saint Tropez Stone Boutique
25 Evelyn Way
San Francisco, CA 94127
Tel: 415 2594820
Fax: 832-565-1100
sales@sainttropezstone.com
www.sainttropezstone.com

Savoia Canada Inc.
73 Samor Road
Toronto, ON M6A 1J2
Toll-free tel: 800-668-1537
Tel: 416-789-7778
Fax: 416-789-1009
www.savoia.com

Seneca Tiles
7100 S County Road 23
Attica, OH 44807-9796
Tel: 419-426-3561
Fax: 419-426-1735
info@senecatiles.com
www.handmold.com

Zion Tile Corp.
13450 SW 126 Street, Ste 10
Miami, FL 33186
Toll-free tel: 888- 986-9466
Tel: 305-252-9077
Fax: 305-252-9116
info@ziontile.com
www.ziontile.com

flooring, cork

American Cork Products Company
13406 Holston Hills Drive
Houston, TX 77069
Toll-free tel: 888-955-2675
Tel: 281-893-7033
Fax: 281-893-8313
sales@amcork.com
www.amcork.com

Bamboo Direct Building Products Ltd.
103 - 1075 West 1st Street
North Vancouver, BC V7P 3T4
Tel: 604-904-1224
Fax: 604-904-1229
info@bamboodirect.ca
www.bamboodirect.ca

Duro-design Cork & Bamboo Flooring
2866 Daniel-Johnson Boulevard
Laval, QC H7P 5Z7
Toll-free tel: 888-528-8518
Tel: 450-978-3403
Fax: 450-978-2542
info@Duro-design.com
www.duro-design.com

Eco Friendly Flooring
100 South Baldwin Street
Madison, WI 53703
Toll-free tel: 866-250-3273
Tel: 608-441-3265
Fax: 608-441-3264
ecofriendlyfloor@aol.com
www.ecofriendlyflooring.com

FloorShop.Com
18375 Olympic Avenue South
Tukwila, WA 98188
Toll-free tel: 888-641-3566
Sales@floorshop.com
www.floorshop.com

Globus Cork
741 E. 136 Street
Bronx New York, NY 10454
Tel: 718-742-7264
Fax: 718-742-7265
info@corkfloor.com
www.corkfloor.com

Hardwood Brokers
2292 Natoma Drive
Virginia Beach, VA 23456
Toll-free tel: 800-313-1107
customerservice@hardwoodbrokers.com
www.hardwoodbrokers.com

Jelinek Cork Group
U.S.
4500 Witmer Industrial Estates
PMB 167
Niagara Falls, NY 14305-1386
Tel: 716-439-4644
Fax: 716-439-4875
Canada
2260 Speers Road
Oakville, ON L6L 2X8
Tel: 905-827-4666
Fax: 905-827-6707
cork@jelinek.com
www.corkstore.com

Torlys
Head Office
1900 Derry Road E
Mississauga, ON L5S 1Y6
Toll-free tel: 800-461-2573
Tel: 905-612-8772
Fax: 905-612-9049
Western Canada
787 Clivedon Place, Unit 600
Delta, BC V3M 6C7
Tel: 604-777-9722
Fax: 604-777-9766
www.torlys.com

flooring, garage

American Garage Floor
1425 N Rangeline Road
Anderson, IN 46012
Tel: 800-401-4537
Fax: 765-642-0971
mike@americangaragefloor.com
www.americangaragefloor.com

Saint Tropez Stone Boutique

floors, hand-scraped, distressed

Ecotimber
1611 Fourth Street
San Rafael, CA 94901
Tel: 415-258-8454
Fax: 415-258-8455
www.ecotimber.com

HomerWood Hardwood Flooring Company
1026 Industrial Drive
Titusville, PA 16354
Tel: 814-827-3855
Fax: 814-827-3629
sales@homerwood.com
www.homerwood.com

Moreland Company
1617 South Tuttle Avenue, 3rd Floor
Sarasota, FL 34239
Toll-free tel: 800-397-7769
Fax: 941-953-5180
billm@morelandcompany.com
www.morelandcompany.com

Patina Old World Flooring
3820 North Ventura Avenue
Ventura, CA 9300
Toll-free tel: 800-501-1113
Tel: 805-648-7521
Fax: 805-648-7301
patina@patinawoodfloors.com
www.patinawoodfloors.com

Pennington Hardwoods
Tel: 812-248-4700
Fax: 812-248-4705
Quality@PenningtonHardwoods.com
www.penningtonhardwoods.com

Shaw Industries, Inc.
616 E Walnut Avenue
Dalton, GA 30722-2128
Toll-free tel: 800-441-7429
www.shawfloors.com

Urban Floor
6221 Randolph Street
Commerce, CA 90040
Toll-free tel: 866-758-7226
Tel: 323-890-0000
Fax: 323-890-0188
info@urbanfloor.com
www.urbanfloor.com

flooring, heated

Electro Industries Inc.
2150 West River Street
PO Box 538
Monticello, MN 55362
Toll-free tel: 800-922-4138
Tel: 763-295-4138
Fax: 763-295-4434
jbecke@electromn.com
www.electromn.com

Nuheat Industries Limited
1689 Cliveden Avenue
Delta, BC V3M 6V5
Toll-free tel: 800-778-9276
www.nuheat.com

Radiant Floor Company
PO Box 666
Barton, VT 05822
Toll-free tel: 866-927-6863
info@radiantcompany.com
www.radiantcompany.com

Warmup Inc.
32 Federal Road, Unit 1F
Danbury, CT 06810
Toll-free tel: 888-927-6333
Tel: 203-791-0072
Fax: 203-791-9272
Toll-free fax: 888-927-4721
us@warmup.com
www.warmup.com

Warmly Yours
2 Corporate Drive
Long Grove, IL 60047
Tel: 800-875-5285
Fax: 800-408-1100
sales@WarmlyYours.com
www.warmlyyours.com

Warm Tiles (EGS Easyheat Inc)
U.S.
2 Connecticut South Drive
East Granby, CT 06026
Tel: 860-653-1600
Fax: 860- 653-4938
Canada
99 Union Street
Elmira, ON N3B 3L7
Tel: 519-669-2444
Fax: 519-669-6419
www.easyheat.com

Warmzone
2056 South 1100 East
Salt Lake City, UT 84106
Toll-free tel: 888-488-9276
Tel: 801-326-5100
Fax: 801-326-5199
info@warmzone.com
www.warmzone.com

Watts Radiant Inc.
4500 E Progress Place
Springfield, MO 5803
Toll-free tel: 800-276-2419
www.wattsradiant.com

flooring, laminates, engineered, etc.

Armstrong World Industries, Inc.
PO. Box 3001
Lancaster, PA 17604
Toll-free tel: 800-233-3823
www.armstrong.com

Bameco Flooring
10579 Bancroft Lane
Frisco, TX 75035
Tel: 214-705-8338
Fax: 214-705-8338
sales@bameco.com
www.bameco.com

Eco Floors Inc.
3155 Pepper Mill Court, Unit 2
Mississauga, ON L5L 4X7
Tel: 905-828 9549
Fax: 905-828 2318
info@eco-floors.com
www.eco-floors.com

Formica Corporation
Toll-free tel: 800-367-6422
www.formica.com

Hardwood Brokers
2292 Natoma Drive
Virginia Beach, VA 23456
Toll-free tel: 800-313-1107
customerservice@hardwoodbrokers.com
www.hardwoodbrokers.com

Mountain Lumber Company
Toll-free tel: 800-445-2671
Tel: 434-985-3646
Fax: 434-985-4105
E-mail: sales@mountainlumber.com
Website: www.mountainlumber.com

Patina Old World Flooring
3820 North Ventura Avenue
Ventura, CA 9300
Toll-free tel: 800-501-1113
Tel: 805-648-7521
Fax: 805-648-7301
patina@patinawoodfloors.com
www.patinawoodfloors.com

Pergo
Toll-free tel: 800-337-3746
Tel: 919-773-6000
www.pergo.com

flooring, linoleum

Congoleum Corporation
Department C, PO Box 3127
Mercerville, NJ 08619-0127
Toll-free tel: 800-274-3266
Tel: 609-584-3000
www.congoleum.com

Forbo Linoleum Inc. (Marmoleum)
Humboldt Industrial Park
PO Box 667
Hazleton, PA 18201
Tel: 570-459-0771
Fax: 570-450-0277
info@fl.na.com
www.forbolinoleumna.com

flooring, salvaged and reclaimed wood

Aged Woods
2331 East Market Street, Ste 6
York, PA 17402
Toll-free tel: 800-233-9307
Tel: 717-840-0330
Fax: 717-840-1468
info@agedwoods.com
www.agedwoods.com

Sylvan Brandt
651 E Main Street
Lititz, PA 17543
Tel: 717-626-4520
Fax: 717-626-5867
dean@sylvanbrandt.com
www.sylvanbrandt.com

Carlson's Barnwood Company
8066 N 1200 Avenue
Cambridge, IL 61238
Tel: 309-522-5550
Fax: 309-522-5123
info@carlsonsbarnwood.com
www.carlsonsbarnwood.com

Country Road Associates
PO Box 885
63 Front Street
Millbrook, NY 12545
Tel: 845-677-6041
Fax: 845-677-6532
info@countryroadassociates.com
www.countryroadassociates.com

Elmwood Reclaimed Timber
PO Box 10750
Kansas City, MO 64188–0750
Tel: 816-532-0300
Toll-free tel: 800-705-0705
Fax: 816-532-0234
sales@elmwoodreclaimedtimber.com
www.elmwoodreclaimedtimber.com

Hill Country Woodworks
507 E Jackson Street
Burnet, TX 78611
Tel: 512-756-6950
lee@texaswoodwork.com
www.texaswoodwork.com

Longleaf Lumber
115 Fawcett Street
Cambridge, MA 02138
Toll-free tel: 866-653-3566
Tel: 617-871-6611
Fax: 617-871-6615
info@longleaflumber.com
www.longleaflumber.com

Maine Timber Works, LLC
823 Augusta Road
Rome, ME 04963
Tel: 207-397-3285
mainetimberworks@midmaine.net
www.mainetimberworks.com

Mountain Lumber Company
Toll-free tel: 800-445-2671
Tel: 434-985-3646
Fax: 434-985-4105
sales@mountainlumber.com
www.mountainlumber.com

Old Barn Wood Company
1911 Draper Street
Baraboo, WI 53913
Tel: 608-356-8849
david@old-barn-wood.com
www.old-barn-wood.com

Pioneer Millworks
Toll-free tel: 800-951-9663
Eastern Office
1180 Commercial Drive
Farmington, NY 14425
Tel: 585-924-9970
Western Office
547 West 700 South
Salt Lake City, UT 84101
Tel: 801-328-9663
www.pioneermillworks.com

Vintage Timberworks
47100 Rainbow Canyon
Temecula, CA 92592
Tel: 951-695-1003
Fax: 951-695-9003
wood@vintagetimber.com
www.vintagetimber.com

Vintage Log & Lumber, Inc.
Rt 1 Box 2F, Glen Ray Road
Alderson, WV 24910
Tel: 304-445-2300
Fax: 304-445 2249
Toll-free tel: 877-653-5647
sales@vintagelog.com
www.vintagelog.com

Wood Natural Restorations
3038 Woodlane Avenue
Orefield, PA 18069
Tel: 610-395-6451
Fax: 610-395-4662
ken@woodnatural.com
www.woodnatural.com

flooring, stone & brick

Atelier Jouvence Stonecutting and Tile
329 West 18th Street, Ste. 611
Chicago, IL 60016-1120
Tel: 312-492-7922
Cell: 773-251-1581
Fax: 312-492-7923
contact@atelierjouvence.com
www.atelierjouvence.com

Brick Floor Tile Inc.
Tel: 319-351-9733
info@brick-floor-tile.com
www.brick-floor-tile.com

Cleveland Quarries
230 West Main Street
S. Amherst, OH 44001
Toll-free tel: 800-248-0250
Tel: 440-986-4501
Toll-free fax: 800-649-8669
Fax: 440-986-4531
amst@amst.com
www.clevelandquarries.com

Country Floors Inc.
8735 Melrose Avenue
Los Angeles, CA 90069
Toll-free tel: 800-311-9995
Tel: 310-657-0510
Fax: 310-659-6470
info@countryfloors.com
www.countryfloors.com
Canada
Country Floors, Inc.
321 Davenport Road
Toronto, ON M5R 1K5
Tel: 416-922-9214
Fax: 416-922-3612
Country Floors, Inc.
5337 Rue Ferrier
Montreal, PQ H4P 1L9
Tel: 514-733-7596
Fax: 514-344-3755

Dixie Cut Stone and Marble Inc.
6128 Dixie Highway
Bridgeport, MI 48722
Tel: 888-450-2858
Fax: 989-777-8791
www.dixiestone.com

Gavin Historical Bricks
2050 Glendale Road
Iowa City, IA 52245
Tel: 319-354-5251
info@historicalbricks.com
www.historicalbricks.com

Echeguren Slate
1495 Illinois Street
San Francisco, CA 94107
Tel: 415-206-9343
Fax: 415-206-9353
slate@echeguren.com
www.echeguren.com

Endicott Clay Products Co. & Endicott Tile Ltd.
PO Box 17
Fairbury, NE 68352
Tel: 402-729-3323
Fax 402-729-5804
endicott@endicott.com
www.endicott.com

FloorCanada
89 Lindsay Road
Woodstock, ON N4V 1A8
Toll-free tel: 877-357-2632
Tel: 519-788-4948
tmckegney@floorcanada.ca
www.floorcanada.ca

Floor Tile and Slate Co.
1209 Carroll Avenue
Carrollton, TX 75006
Toll-free tel: 800-446-0220
Tel: 972-242-6647
Fax: 972-242-7253
sales@floortileandslate.com
www.floortileandslate.com

Intarsia, Inc.
9550 Satellite Boulevard, Ste 180
Orlando, FL 32837
Tel: 407-859-5800
Fax: 407-859-7555
www.intarsiainc.com

Monarch Stone
(European reproductions)
PO Box 2684
Carefree, AZ 85377
Tel: 480-220-8096
Fax: 480-595-0545
info@monarchstone.com
www.monarchstone.com

Ocean Stones
388 Carlaw Avenue
Toronto, ON M4M 2T4
Toll-free tel: 866-463-5832
Tel: 416-463-4805
Fax: 416-463-1205
www.kuda.com

Olympia Tile
1000 Lawrence Avenue W
Toronto, ON M6B 4A8
Toll-free tel U.S.: 800-667-8453
Toll-free tel Canada: 800-268-1613
Tel: 416-785-9555
Fax: 416-785-3204
info@olympiatile.com
www.olympiatile.com

Peacock Pavers
PO Box 519
Atmore, AL 36504-0519
Toll-free tel: 800-264-2072
Tel: 251-368-2072
Fax: 251-368-5080
rosaleigh@peacockpavers.com
www.peacockpavers.com

Pompei Mosaic Tile
11301 Olympic Boulevard, Ste. 512
Los Angeles, CA 90064
Tel: 310-312-9893
Fax: 310-996-1929
info@pompei-mosaic.com
www.pompei-mosaic.com

Porphyry USA, Inc.
220-7945 Mac Arthur Boulevard
Cabin John, MD 20818
Tel: 301-229-8725
Fax: 301-229-8739
pavers@porphyryusa.com
www.porphyryusa.com

Rhodes Architectural Stone
2011 East Olive Street
Seattle, WA 98122
Tel: 206-709-3000
Fax: 206-709-3003
johnt@rhodes.org (West)
trevorb@rhodes.org (North, Midwest)
andrewg@rhodes.org (East)
www.rhodes.org

Ring Brick and Stone
Albuquerque, NM
Tel: 505-891-0943

Ann Sacks
8120 NE 33rd Drive
Portland, OR 97211
Toll-free tel: 800-278-8453
www.annsacks.com

Saint Tropez Stone Boutique
25 Evelyn Way
San Francisco, CA 94127
Tel: 415 2594820
Fax: 832-565-1100
sales@sainttropezstone.com
www.sainttropezstone.com

Sheldon Slate
38 Farm Quarry Road
Monson, ME 04464
Tel: 207-997-3615
Fax: 207-997-2966
john@sheldonslate.com
Fox Road
Middle Granville, NY 12849
Tel: 518 642 1280
Fax: 518-642-9085
www.sheldonslate.com

StoneDecora
442-14622 Ventura Boulevard
Sherman Oaks, CA 91403
Tel: 818-986-1171
Fax: 818-907-0343
Info@StoneDecora.com
www.stonedecora.com

Stonehenge Slate Inc.
136 Lincoln Boulevard
Middlesex, NJ 08846
Tel: 732-748-0110
Fax: 732-748-0157
info@stonehengeslate.us
www.stonehengeslate.us

Stone Tile
1451 Castlefield Avenue
Toronto, ON M6M 1Y3
Tel: 416-515-9000
Fax: 534-1782
info@stone-tile.com
www.stone-tile.com

Stone Tile West
4040 - 7th Street SE
Calgary, AB T2G 2Y9
Tel: 402-234-7274
Fax: 402-214-0213
west@stone-tile.com
www.stone-tile.com

Topstone
3601 Range Road
Temple, TX 76504
Toll-free tel: 866-774-9197
Tel: 254-774-9197
Fax: 254-774-7631
Sales@PSITopstone.com
www.psitopstone.com

Universal Slate International Inc.
3821 - 9th Street SE
Calgary, AB T2G 3C7
Toll free tel: 888-677-5283
zimmer@universalslate.com
www.universalslate.com

Wausau Tile, Inc.
PO Box 1520
9001 Bus. Highway 51
Wausau, WI 54401
Toll-free tel: 800-388-8728
Tel: 715-359-3121
Fax: 715-355-4627
wtile@wausautile.com
www.wausautile.com

flooring, wood

Architectural Traditions
9280 E. Old Vail Road
Tucson, AZ 85747
Tel: 520-574-7374
www.architecturaltraditions.com

Armstrong World Industries, Inc.
PO Box 3001
Lancaster, PA 17604
Toll-free-tel: 800-233-3823
www.armstrong.com

Columbia Flooring
100 Maxine Road
Danville, VA 24541
Toll-free tel: 800-654-8796
Tel: 434-793-4647
Fax: 434-797-4005
info@columbiaflooring.
www.columbiaflooring.com

Ecotimber
1611 Fourth Street
San Rafael, CA 94901
Tel: 415-258-8454
Fax: 415-258 8455
www.ecotimber.com

FloorCanada
89 Lindsay Road
Woodstock, ON N4V 1A8
Toll-free tel: 877-357-2632
Tel: 519-788-4948
tmckegney@floorcanada.ca
www.floorcanada.ca

Gaetano Hardwood Floors
Seacliff Business Center
7071 Kearny Drive
Huntington Beach, CA 92648
Tel: 714-536-6942 or 949-376-9246
Fax: 714-596-6100
info@gaetanohardwoodfloors.com
www.gaetanohardwoodfloors.com

Global Specialty Products, Ltd.
976 Highway 212 E
Chaska, MN 55318
Toll-free tel: 800-964-1186
Tel: 952-448-6566
Fax: 952-448 6550
Toll-free fax: 800-964-0630
info@surfacingsolution.com
www.surfacingsolution.com

Hardwood Brokers
2292 Natoma Drive
Virginia Beach, VA 23456
Toll-free tel: 800-313-1107
customerservice@hardwoodbrokers.com
www.hardwoodbrokers.com

HomerWood Hardwood Flooring Company
1026 Industrial Drive
Titusville, PA 16354
Tel: 814-827-3855
Fax: 814-827-3629
sales@homerwood.com
www.homerwood.com

Hosking Hardwood Flooring
PO Box 163
Walpole, MA 02081
Toll-free tel: 877-356-6755
Tel: 508-643-0810
jeff@hoskinghardwood.com
www.hoskinghardwood.com

Launstein Hardwood Products
384 Every Road
Mason, Michigan 48854
Toll-free tel: 888-339-4639
Tel: 517- 676-1133
Fax: 517-676-6379
www.launstein.com

Patina Old World Flooring
3820 North Ventura Avenue
Ventura, CA 9300
Toll-free tel: 800-501-1113
Tel: 805-648-7521
Fax: 805-648-7301
patina@patinawoodfloors.com
www.patinawoodfloors.com

Pennington Hardwoods
Tel: 812-248-4700
Fax: 812-248-4705
Quality@PenningtonHardwoods.com
www.penningtonhardwoods.com

Redwood Burl .Com
34-1834 Allard
Eureka, CA 95503
Tel: 707-441-1658
gbuckjr@redwoodburl.com
www.redwoodburl.com

Richard Marshall Fine Floors Inc.
12520 Wilkie Avenue
Hawthorne, CA 90250
Toll-free tel: 800-689-5981
info@oldeboards.com
www.oldeboards.com

Shaw Industries, Inc.
616 E Walnut Avenue
Dalton, GA 30722-2128
Toll-free tel: 800-441-7429
www.shawfloors.com

Lydia's Forge

Southern Wood Floors
472 Flowing Wells Road
Augusta, GA 30907
Toll-free tel: 888-488-7463
Tel: 706-855-0779
Fax: 706-855-0383
products@southernwoodfloors.com
www.southernwoodfloors.com

Triton International Woods LLC
PO Box 8767
Rocky Mount, NC 27804
Toll-free tel: 866-319-6637
Tel: 252-823-6675
Fax: 252-823-6491
info@tritonwoods.com
www.tritonwoods.com

Urban Floor
6221 Randolph Street
Commerce, CA 90040
Toll-free tel: 866-758-7226
Tel: 323-890-0000
Fax: 323-890-0188
info@urbanfloor.com
www.urbanfloor.com

WoodFloorsOnline
info@woodfloorsonline.com
www.woodfloorsonline.com

Woodlist, Inc
277 Linden Street, Ste 204
Wellesley, MA 02482
Toll-free tel: 877-487-6504
Tel: 781-283-5757
Fax: 781-283-5707
info@woodlist.com
www.woodlist.com

green building

Center for Resource Conservation
1702 Walnut Street
Boulder, CO 80203
Tel: 303-441-3278
www.greenerbuilding.org

green building, pressed earth

Jim Hallock
Pagosa Springs, CO
earthblock@juno.com

Joaquim Karcher
Taos, NM
Tel: 505-758-9741
oneearth@laplaza.org

Todd Swanson
Hesperus, CO
Tel: 970-259-5985
biohab@gobrainstorm.net

green building, rammed earth

Earth and Sun Construction
5105 Cueva Mine Trail
Las Cruces, NM 88011
Tel: 505-644-5380
www.earthandsun.com

Huston Rammed Earth
Edgewood, NM
Tel: 505-281-9534

Rammed Earth Development
802 S. 8th Avenue
Tucson, AZ 85701
Tel: 520-623-2784
redinc@dakotacom.net
www.rammedearth.com

Rammed Earth Solar Homes
PO Box 654
Oracle, AZ 85623
Tel: 520-631-5807
info@rammedearthhomes.com
www.rammedearthhomes.com

Rammed Earth Works
4024 Hagen Road
Napa, CA 94558
Tel: 707-224-2532
easton@rammedearthworks.com
www.rammedearthworks.com

Soledad Canyon Earthbuilders
949 S. Melendres Street
Las Cruces, NM 88005
Tel: 505-527-9897
Fax: 505-647-8953
www.adobe-home.com

glass

Abraxis Art Glass, Inc.
212 West Laporte Avenue
Fort Collins, CO 80521
Tel: 970-493-7604
Fax: 970-493-7644
Info@Abraxisartglass.com
www.abraxisartglass.com

Architectural Glass Art, Inc.
815 West Market Street
Louisville, KY 40202
Toll-free tel: 800-795-9429
Fax: 502-585-2808
info@againc.com
www.againc.com

Architectural Glass, Inc.
71 Maple Street
Beacon, NY 12508
Tel: 845-733-4720
Fax: 845-733-4502
info@architecturalglassinc.com
www.architecturalglassinc.com

Armstrong Glass Company
1025 Cobb International Boulevard., Ste 250
Kennesaw, GA 30152
Tel: 770-919-9924
Fax: 770-919-9664
www.armstrongglass.com

Artglass & Metal
4033 Skyline Road
Carlsbad, CA 92008
Tel: 760-390-2994
jay@artglassandmetal.com
www.artglassandmetal.com

Artwork in Architectural Glass
East Coast
145 Highway 83
Good Hope, GA 30641
Tel: 949-251-0075
Fax: 678-623-3573
information@AAG-Glass.com
West Coast
20101 SW Birch, Ste 276
Newport Beach, CA 92660
Tel: 949-251-0075
Fax: 949-251-0126
westcoastinfo@AAG-Glass.com
www.artworkinglass.com

Art Zone
592 Markham Street (lower)
Toronto, ON M6G 2L8
Tel: 416-534-1892
artzone@bellnet.ca
www.artzone.ca

Bendheim
122 Hudson Street
New York, NY 10013
Toll-free tel: 800 606-7621
Tel: 212 226-6370
Fax: 212 431-3589
Bendheim West
3675 Alameda Avenue
Oakland, CA 94601
Toll-free tel: 800-900-3499
Tel: 510-535-6600
Fax: 510-535-6615
Bendheim East
61 Willett Street
Passaic, NJ 07055
Toll-free tel: 800-835-5304
Tel: 973 471-1778
Fax: 973 471-4202
www.bendheim.com

Ellenburg & Shaffer Glass Art Studio
344 S. Elm Street
Greensboro, NC 27401
Tel: 336-271-2811
Fax: 336-271-8000
info@stainedglassdomes.com
www.ellenburgandshaffer.com

Interstyle Ceramic and Glass Ltd.
3625 Brighton Avenue
Burnaby, BC V5A 3H5
Tel: 604-421-7229
Fax: 604-421-7544
info@interstyle.ca
www.interstyle.ca

Raynes & Co. Ltd.
255 Richmond Street E, Ste 412
Toronto ON M5A 4T7
Tel: 416-962-2100
Fax: 416-962-2175
mark@raynesandco.com
www.raynesandco.com

glass brick

Cleveland Glass Block
1225 E 222nd Street
Euclid, OH 44117
Toll-free tel: 800-245-1217
Tel: East: 216-531-6363
West: 440-886-0054
Fax: 216-531-2388
rtemple@clevelandglassblock.com
www.clevelandglassblock.com

IBP Glass Block
PO Box 425
Fort Worth, TX 76101-0425
Toll-free tel: 800-932-2263
Tel: 817-332-9124
Fax 817-332-1406
sweddle@ibpglassblock.com
www.ibpglassblock.com

Pittsburgh Corning Corporation
800 Presque Isle Drive
Pittsburgh, PA 15239
Toll-free tel: 800-624-2120
Fax: 724-325-9704
www.pittsburghcorning.com

Saint Tropez Stone Boutique
25 Evelyn Way
San Francisco, CA 94127
Tel: 415-259-4820
Fax: 832-565-1100
sales@sainttropezstone.com
www.sainttropezstone.com

Vetroarredo North America
PO Box 1812
Cranberry Township, PA 16066
Tel: 724-242-2121
Fax: 724-242-2323
tomgross@vanagb.com
www.vanagb.com

hardware

Acorn Manufacturing Company Inc.
Mansfield, MA 02048
Toll-free tel: 800-835-0121
acorninfo@acornmfg.com
www.acornmfg.com

Alamo Designs, Inc.
11765 West Avenue, 317
San Antonio TX 78216
Toll-free tel: 877-442-5266
Tel: 972-762-5472
Fax: 800-859-4716
mail@alamodesigns.com
www.alamodesigns.com

Albion Doors & Windows
PO Box 220
Albion, CA 95410
Tel: 707-937-0078
bysawyer@mcn.org
www.knobsession.com

Antique Hardware & Home
39771 SD Highway 34
PO Box 278
Woonsocket, SD 57385
Toll-free tel: 800-237-8833
Tel: 605-796-4425
Fax: 605-796-4085
antique.hardware@cabelas.com
www.antiquehardware.com

Architectural Traditions
9280 E. Old Vail Road
Tucson, AZ 85747
Tel: 520-574-7374
www.architecturaltraditions.com

Asselin U.S.A.
2870 Peachtree Road NW, 714
Atlanta, GA 30305–2918
Tel: 404-419-6114
Fax: 404-419-6116
contact@asselinusa.com
www.asselinusa.com

Atlas Homewares
326 Mira Loma Avenue
Glendale, CA 91204
Toll-free tel: 800-799-6755
Tel: 818-240-3500
atlasinformation@atlashomewares.com
www.atlashomewares.com

Ball & Ball Hardware Reproductions
463 W. Lincoln Highway
Exton, PA 19341
Toll-free tel: 800-257-3711
Tel: 610-363-7330
Fax: 610-363-7639
bill@ballandball.com
www.ballandball-us.com

Bouvet USA/Belvedere Hardware/Laforge/La Porcelaine
1060 Illinois Street
San Francisco, CA 94107
Tel: 415-864 0273
Fax: 415-864 2068
www.bouvet.com

The Bronze Craft Corporation
37 Will Street, PO Box 788
Nashua, NH 03061–0788
Toll-free tel: 800-488-7747
Tel: 603-883-7747
Fax: 603-883-0222
bronze@bronzecraft.com
www.bronzecraft.com

Bushere & Son Iron Studio
3968 E. Grand Avenue
Pomona, CA 91766
Tel: 909-469-0770
Fax: 909-469-0060
bushereiron@att.net
www.busihereandson.com

Charleston Hardware Co.
2143 Heriot Street
Charleston, SC 29402
Tel: 866-958-8626
Fax: 843-958-8446
www.charlestonhardwareco.com

Craftsmen In Wood
5441 West Hadley Street
Phoenix, AZ 85043
Tel: 602-296-1050
Fax: 602-296-1052
www.craftsmeninwood.com

Colleti Design
Tel: 866-922-3667
info@collettidesign.com
www.collettidesign.com

Crown Point Cabinetry
462 River Road
Claremont, NH 03743
Toll-free tel: 800-999-4994
Fax: 800-370-1218
info@crown-point.com
www.crown-point.com

Du Verre, The Hardware Co.
Tel: 416-593-0182
Fax: 416-593-5759
www.duverre.com

Freego Design
www.freegodesign.com

Susan Goldstick Inc.
200 Gate 5 Road, Ste 110
Sausalito, CA 94965
Toll-free tel: 888-566-2799
Tel: 415-332-6719
Fax: 415-332-6830
www.susangoldstick.com

GrandRiverSupply.com
1421 Camino del Pueblo
Bernalillo, NM 87004
Toll-free tel: 877-477-8775
Tel: 505-867-4110
Fax: 505-867-9711
www.grandriverdoor.com

HardwareSource
840 5th Avenue
San Diego, CA 92101
Toll-free tel: 877-944-6437
Fax: 619-232-6527
info@hardwaresource.com
www.hardwaresource.com

House of Antique Hardware
122 SE 27th Avenue
Portland, OR 97214
Toll-free tel: 888-223-2545
Tel: 503-231-4089
Fax: 503-233-1312
www.houseofantiquehardware.com

Kayne & Son Custom Hardware Inc
100 Daniel Ridge Road
Candler, NC 28715
Tel: 828-667-8868 or 828-665-1988
Fax: 828-665-8303
www.blacksmithsdepot.com

Lee Valley Tools
Canada
Lee Valley Tools Ltd.
PO Box 6295, Station J
Ottawa, ON K2A 1T4
Toll-free tel: 800-267-8767
U.S.
Lee Valley Tools Ltd.
PO Box 1780
Ogdensburg, NY 13669-6780
Toll-free tel: 800-871-8158
customerservice@leevalley.com
www.leevalley.com

Lydia's Forge
General Delivery
Cormac, ON K0J 1M0
Tel: 613-757-2574

The Modern Woodsmith, LLC
E9738 Gates Road
Munising, MI 49862
Tel: 906-387-5577
Fax: 906-387-5656
sales@themodernwoodsmith.com
www.themodernwoodsmith.com

Frank Morrow Company
129 Baker Street
Providence, RI 02905
Toll-free tel: 800 556 7688
Tel: 401-941-3900
Fax: 401-941-3810
sales@frankmorrow.com
www.frankmorrow.com

Nostalgic Warehouse Inc.
4661 Monaco Street
Denver, CO 80216-3304
Toll-free tel: 800-522-7336
Tel: 972-271-0319
Fax: 972-271-9726
www.nostalgicwarehouse.com

Martin Pierce
5437 Washington Boulevard
Los Angeles, CA 90016
Toll-free tel: 800-619-1521
Tel: 323-939-5929
www.martinpierce.com

Rejuvenation
2550 NW Nicolai Street
Portland, OR 97210
Toll-free tel: 888-401-1900
Toll-free fax: 800-526-7329
info@rejuvenation.com
www.rejuvenation.com

Restoration Hardware, Inc.
15 Koch Road, Ste J
Corte Madera, CA 94925
Toll-free tel: 800-910-9836
Tel: 415-924-1005
www.restorationhardware.com

Richelieu Hardware
7900, West Henri-Bourassa
Montreal, QC H4S 1V4
Toll-free tel (in Quebec): 866-832-4040
Toll-free tel (in Canada): 800-361-6000
Toll-free tel (from the US): 800-619-5446
info@richelieu.com
www.richelieu.com

Rocky Mountain Hardware
PO Box 4108
1030 Airport Way
Hailey, ID 83333
Toll-free tel: 888-788-2013
Tel: 208-788-2013
Fax: 208-788-2577
info@rockymountainhardware.com
www.rockymountainhardware.com

Saint-Gaudens Metal Arts
1784 La Costa Meadows Drive,
Ste 104
San Marcos, CA 92069
Tel: 760-891-0300
Fax 760-752-9914
Sales@VSGMetalArts.com
www.VSGmetalarts.com

Steptoe & Wife
90 Tycos Drive
Toronto, ON M6B 1V9
Toll-free tel: 800-461-0060
Tel: 416-780-1707
Toll-free fax: 877-256-4279
Fax: 416-780-1814
info@steptoewife.com
www.steptoewife.com

Talisman Handmade Tiles
Lowitz & Company
4401 N Ravenswood Avenue
Chicago, IL 60640
Tel: 773-784-2628
talisman@lowitzandcompany.com
www.lowitzandcompany.com

TRIMCO
PO Box 23277
3528 Emery Street
Los Angeles, CA 90023–0277
Tel: 323-262-4191
Fax: 323-264-7214 or 800-637-TRIM
info@trimcobbw.com
www.trimcobbw.com

Victoria Speciality Hardware
1990 Oak Bay Avenue
Victoria, BC V8R 1E2
Tel: 250-598-2966
Toll-free tel: 888-274-6779
vsh@shaw.ca
www.victoriaspecialityhardware.com

Whitechapel Ltd.
PO Box 11719
Jackson, WY 83002
Toll-free tel: 800-468-5534
Tel: 307-739-9478
Fax: 307-739-9458
info@whitechapel-ltd.com
www.whitechapel-ltd.com

heat registers

All American Wood Register
PO Box 348
7616 Hancock Drive
Wonder Lake, IL 60097
Tel: 815-728-8888
Fax: 815-728-9663
sales@allamericanwood.com
www.allamericanwood.com

Atlanta Supply
1333 Logan Circle NW
Atlanta, GA 30318
Toll-free: 800-972-5391
Tel: 404-815-9000
Fax: 404-876-7582
info@atlantasupply.com
www.atlantasupply.com

Beaux-Artes
1012 South Creek View Court
Churchton, MD 20733
Tel: 410-867-0790 or 301-855-4244
Fax: 410-867-8004
info@beaux-artes.com
www.beaux-artes.com

ClassicAire Wood Vents
9325 SW Barber Street
Wilsonville, OR 97070
Toll-free tel: 800-545-8368
Tel: 503-855-2000
Fax: 800-886-1667
classic@classicvents.com
www.classicvents.com

Hamilton Decorative
31 E 32nd Street, 11th Floor
New York, NY 10016
Toll-free tel: 866-900-3326
Tel: 212-760-3377
Fax: 212-760-3362
hamiltondeco@hamiltondeco.com
www.hamiltondeco.com

Handcraft Tile Inc.
1126 Yosemite Drive
Milpitas, CA 95035
Toll-free: 877-262-1140
Tel: 408-262-1140
Fax: 408-262-1441
info@handcrafttile.com
www.handcrafttile.com

Launstein Hardwood Products
384 Every Road
Mason, MI 48854
Toll-free tel: 888-339-4639
Tel: 517-676-1133
Fax: 517-676-6379
www.launstein.com

Pattern Cut, Inc.
2267 E Via Burton Street
Anaheim, CA. 92806-1222
Tel: 714-765-8138
Fax: 714-765-8139
info@patterncut.com
www.patterncut.com

The Reggio Register Co., Inc.
20 Central Avenue
PO Box 511
Ayer, MA 01432–0511
Toll-free tel: 800-880-3090
Tel: 978-772-3493
Fax: 978-772-5513
reggio@reggioregister.com
www.reggioregister.com

Rejuvenation
2550 NW Nicolai Street
Portland, OR 97210
Toll-free tel: 888-401-1900
Toll Free fax: 800-526-7329
info@rejuvenation.com
www.rejuvenation.com

Urban Registers
10595 Bloomfield Avenue
Los Alamitos, CA 90720
Tel: 562-795-0069
Fax: 562-598-5360
CherylD@socal.rr.com
www.shop.urbanregisters.com

kitchens

Cooper-Pacific Kitchens Inc.
G299-8687 Melrose Avenue
Los Angeles, CA 90069
Toll-free tel: 800-743-6284
Tel: 310-659-6147
Fax: 310-659-1835
www.cooperpacific.com

Craftsmen In Wood
5441 West Hadley Street
Phoenix, AZ 85043
Tel: 602-296-1050
Fax: 602-296-1052
www.craftsmeninwood.com

Crown Point Cabinetry
462 River Road
Claremont, NH 03743
Toll-free tel: 800-999-4994
Fax: 800-370-1218
info@crown-point.com
www.crown-point.com

DiaDot Custom Millwork
PO Box 113 Bldg 22
Sussex, NJ 07461
Tel: 973-875-5669
Fax: 973-875-5634
generalbox@diadot.com
www.diadot.com

Exotic Metal Interiors
Tel: 877-715-5483
Fax: 905 507-1044
exotic@primeliteusa.com
www.exoticmetalinteriors.com

Johnny Grey Inc. USA
Toll-free tel: 888-640-7879
Tel: 415-701-7701
www.johnnygrey.com

The Hat Factory Furniture Company
1000 N Division Street
Peekskill, NY 10566
Tel: 914-788-6288
Fax: 914-788-0019
info@hatfactoryfurniture.com
www.hatfactoryfurniture.com

Keystone Cabinetry Inc.
7333 Coldwater Canyon Boulevard
Building 38
North Hollywood, CA 91605
Tel: 818-503-0493
Fax: 818-503-1153
julian@keystonecabinetry.com
www.keystonecabinetry.com

Iripinia Kitchens
278 Newkirk Road
Richmond Hill, ON L4C 3G7
Tel: 905-780-7722
Fax: 905-780-0554
www.iripinia.com

Olympic Kitchens
1 Parnell Avenue
Scarborough, ON M1K 1B1
Tel: 416-266-8851
ikatsis@olympickitchens.ca
www.olympickitchens.ca

Christopher Peacock Cabinetry
148 The Merchandise Mart
Chicago, IL 60654
Tel: 312-321-9500
Fax: 312-321-9510
Chicago@peacockcabinetry.com
Greenwich Showroom
2 Deerfield Drive
Greenwich, CT 06830
Tel: 203-862-9333
Fax: 203-302-3411
Greenwich@peacockcabinetry.com
Boston Showroom
One Design Center Place
Boston, MA 02210
Tel: 617-204-9292
Fax: 617-204-9293
boston@peacockcabinetry.com
Naples Showroom
6320 Shirley Street, Ste 102
Naples, FL 34109
Tel: 239-596-8858
Fax: 239-596-0358
naples@peacockcabinetry.com
www.peacockcabinetry.com

Poliform USA, Inc.
150 E 58th Street
New York, NY 10155
Tel: 212-421-1220
Fax: 212-421-1290
www.poliformusa.com

Quality Custom Cabinetry, Inc.
PO Box 189
New Holland, PA 17557
Toll-free tel: 800-909-6006
www.qcc.com

Scheer's Cabinet & Doors Inc.
5315 Highway 2 E
Minot, ND 58701
Tel: 701-839-3384
Fax: 701-852-6090
info@scheers.com
www.scherrs.com

Snaidero US
20300 S. Vermont Avenue
Torrance, CA 90502
Toll-free tel: 877-SNAIDERO
Tel: 310-516-8499
Canada
Studio Snaidero Vancouver
1575 West Georgia Street
Vanvoucer, BC L6G 2V3
Tel: 604-669-4565
Fax: 604-669-4330
snaidero_vancouver@telus.net
www.snaidero-usa.com

Valcucine NY
66 Crosby Street
New York, NY 10012
Tel: 212-253-5969
Fax: 212-253-5889
valcuineSOHO@aol.com
www.valcucinena.com

Wood-Mode Fine Custom Cabinetry
One Second Street
Kreamer, PA 17833
Tel: 877-635-7500
Canada
C. I. K. Interiors
8118 Decarie Boulevard
Montreal, QC H4P 2S4
Tel: 514-737-4000
cikii@bellnet.ca
www.wood-mode.com

kitchen, ventilation hoods

Broan-NuTone, LLC
PO Box 140
Hartford, WI 53027
Toll-free tel: (U.S.): 800-558-1711
Toll-free tel: (Canada): 877-896-1119
www.broan.com

The Copper Shop
115 North First Avenue
Haubstadt, IN 47639
Tel: 812-768-5008
info@coppershop.net
www.coppershop.net

Euro Line Distribution, Inc.
PO Box 261066
Encino, CA 91426–1066
Toll-free tel: 800-706-1150
Tel: 818-205-1150
Fax: 818-205-1152
customerservice@eurolinedistribution.com
www.eurolinedistribution.com

Gaggenau (US and Canada)
BSH Home Appliances Corporation
5551 McFadden Avenue
Huntington Beach, CA 92649
Toll-free tel: 800-828-9165
Tel: 714-901-5360
questions@bshg.com
www.gaggenau.com

MetaStone, Inc.
Toll-free tel: 866-600-META,
Fax: 650-618-1917
www.metastone.com

MLD Hood Designs, Inc.
1107 Taylorsville Road
Washington Crossing, PA 18977
Tel: 215-493-2427
Fax: 215-493-7968
inquiries@mldhooddesigns.com
www.mldhooddesigns.com

Old World Stoneworks
5400 Miller Avenue
Dallas, TX 75206-6425
Toll-free tel: 800-600-8336
Tel: 214-826-3645
Fax: 214-826-3227
infor@oldworldstoneworks.com
www.oldworldstoneworks.com

Rangecraft Hoods
4–40 Banta Place
Fair Lawn, NJ 07410
Toll-free tel: 877-RC-HOODS
Tel: 201-791-0440
Fax: 201-791-4494
sales@rangecraft.com
www.rangecraft.com

Richelieu Hardware
7900, West Henri-Bourassa
Montreal, QC H4S 1V4
Toll-free tel (in Quebec): 866-832-4040
Toll-free tel (in Canada): 800-361-6000
Toll-free tel (from the US): 800-619-5446
info@richelieu.com
www.richelieu.com

Sirius Range Hoods (USA) Ltd.
1051 Clinton Street
Buffalo, NY 14206
Toll-free tel: 866-528-4987
Fax: 866-365-9204
info@siriushoods.com
www.siriushoods.com

Stanisci Design
14823 32 Mile Rd
Romeo, MI 48065
Tel: 586-752-3368
Fax: 586-752-3293
sales@wood-hood.com
www.wood-hood.com

Stone Age Designs
Atlanta
351 Peachtree Hills Ave, Ste 501–A
Atlanta, GA 30305
Tel: 404-350-3333
Fax: 404-842-0936
Florida
935 Orange Avenue, Ste A
Winter Park, FL 32789
Tel: 407-628-5577
Fax: 407-628-5565
info@stoneagedesigns.net
www.stoneagedesigns.net

Vent-A-Hood
PO Box 830426
Richardson, TX 75083–0426
Toll-free tel: 800-331-2492
www.ventahood.com

Mystic Sink, Elkay

Zephyr Corporation
395 Mendell Street
San Francisco, CA 94124
Toll-free tel: 888-880-8368
www.zephyronline.com

kitchens, sinks

Acorn Engineering Company
15125 Proctor Avenue
City of Industry, CA 91746
Toll-free tel: 800-591-9050
Fax: 626-937-4725
info@neo-metro.com
www.neo-metro.com

Allante International
Toll-free tel: 800-695-0805
Fax: 910-763-4213
info@allante.net
www.allante.net

American Standard
PO Box 6820
1 Centennial Plaza
Piscataway, NJ 08855-6820
Toll-free tel: 800-442-1902
www.americanstandard-us.com

Antique Hardware & Home
39771 SD Highway 34
PO Box 278
Woonsocket, SD 57385
Toll-free tel: 800-237-8833
Tel: 605-796-4425
Fax: 605-796-4085
antique.hardware@cabelas.com
www.antiquehardware.com

Bates & Bates
7310 Alondra Boulevard
Paramount, CA 90723
Toll-free tel: 800-726-7680
Tel: 562-808-2290
Fax: 562-808-2289
website@batesandbates.com
www.batesandbates.com

Blanco America, Inc.
110 Mount Holly By-Pass
Lumberton, NJ 08048
www.blancoamerica.com

Blanco Canada Inc.
3909 Nashua Drive, Units 6–9
Mississauga, ON L4V 1R3
Toll-free tel: 877-4BLANCO
Tel: 905-612-0554
Toll-free tel Fax: 877-8BLANCO
Fax: 905-612-9764
marketing@blancocanada.com
www.blancocanada.com

Elkay Sales, Inc.
2222 Camden Court
Oak Brook, IL 60523
Tel: 630-574-8484
Fax: 630-574-5012
www.elkayusa.com

Fine Crafts & Imports
3906 Ross Avenue
San Jose, CA 95124
Toll-free tel: 800-973-4099
Fax: 408-267-7536
www.finecraftsimports.com

Franke Kitchen Sinks
98 - 100 North Broadway, Route 28
Salem, NH 03079
Tel: 603-893-6777
Fax: 603-893-6555
info@frankekitchensinks.com
www.frankekitchensinks.com

Green Mountain Soapstone
680 E Hubbardton Road
PO Box 807
Castleton, VT 05735
Tel: 802-468-5636
vance@greenmountainsoapstone.com
www.greenmountainsoapstone.com

Iripinia Kitchens
278 Newkirk Road
Richmond Hill, ON L4C 3G7
Tel: 905-780-7722
Fax: 905-780-0554
www.iripinia.com

Kohler
Toll-free tel: 800-456-4537
www.us.kohler.com

MTI Whirlpools
670 North Price Road
Sugar Hill, GA 30518
Toll-free tel: 800-783-8827
info@mtiwhirlpools.com
www.mtiwhirlpools.com

Purple Sage Collections
1526B Howell Mill Road
Atlanta, GA 30318
Toll-free tel: 866-357-4657
Tel: 404-351-4445
Fax: 404-351-8250
info@purplesagecollections.com
www.purplesagecollections.com

Saint Tropez Stone Boutique
25 Evelyn Way
San Francisco, CA 94127
Tel: 415 2594820
Fax: 832-565-1100
sales@sainttropezstone.com
www.sainttropezstone.com

Sonoma Cast Stone Corporation
133 Copeland Street
Petaluma, CA 94952
Toll-free tel: 877-939-9929
Tel: 707-283-1888
Fax: 707-283-1899
sales@sonomastone.com
www.sonomastone.com

Specialty Stainless.com
Toll-free tel: 800-836-8015
Tel: 716-893-3100
Fax: 716-893-0443
info@specialtystainless.com
www.specialtystainless.com

Stone Forest
PO Box 2840
Santa Fe, NM 87504
Toll-free tel: 888-682 2987
Tel: 505-986-8883
Fax: 505-982-2712
info@stoneforest.com
www.stoneforest.com

The Swan Corporation
One City Centre, Ste 2300
St. Louis, MO 63101
Toll-free tel: 800-325-7008
Fax: 314-231-8165
infomail@swanstone.com
www.swanstone.com

kitchen, cabinetry

Advanced Commercial Contracting, Inc.
Architectural Wood Manufacturers
22473 Prat's Dairy Road
Abita Springs, LA 70420
Tel: 504-943-3368
Fax: 504-943-3369
adcomc@advancedmillwork.com
www.advancedmillwork.com

All Wood Creation, Inc.
11701 NW 102nd Road, Ste 9
Meldey, FL 33178
Tel: 305-887-6167
Fax: 305-887-6168
allwoodcreations@bellsouth.net
www.allwoodcreation.com

Architectural Millwork
3522 Lucy Road
Millington, TN 38053
Tel: 901-327-1384
Tel (cell): 901-218-0616
Fax: 901-353-1645
samtickle@aol.com
www.a-millwork.com

Architectural Traditions
9280 E. Old Vail Road
Tucson, AZ 85747
Tel: 520-574-7374
www.architecturaltraditions.com

Jack Arnold
7310 South Yale
Tulsa, OK 74136
Toll-free tel: 800-824-3565
Pat@jackarnold.com
www.jackarnold.com

Beaux-Artes
1012 South Creek View Court
Churchton, MD 20733
Tel: 410-867-0790 or 301-855-4244
Fax: 410-867-8004
info@beaux-artes.com
www.beaux-artes.com

Brentwood Kitchens
2007 Lancaster-Hutchins Road
Lancaster TX 75134
Tel: 972-227-6855
Fax: 972-227-5267
dcook@bentwoodkitchens.com
www.bentwoodkitchens.com

Byrne Custom Woodworking
17501 W 98th Street, 28–62
Lenexa, KS 66219
Tel: 913-894-4777
Fax: 913-894-4779
info@byrnecustomwood.com
www.byrnecustomwood.com

Caspar Designs
14660 Mitchell Creek Dr
Fort Bragg, CA 95437
Tel: 707-964-0635
leecab2@mcn.org
www.caspardesigns.com

Crown Point Cabinetry
PO Box 1560
Claremont, NH 03743
Toll-free tel: 800-999-4994
Fax: 800-370-1218
info@crown-point.com
www.crown-point.com

Crystal Cabinet Works, Inc.
Eleven Hundred Crystal Drive
Princeton, MN 55371
Toll-free tel: 800-347-5045
Fax: 763-389-5846
info@ccworks.com
www.ccworks.com

DiaDot Custom Millwork
PO Box 113 Bldg 22
Sussex, NJ 07461
Tel: 973-875-5669
Fax: 973-875-5634
generalbox@diadot.com
www.diadot.com

Iripinia Kitchens
278 Newkirk Road
Richmond Hill, ON L4C 3G7
Tel: 905-780-7722
Fax: 905-780-0554
www.iripinia.com

La Puerta Originals
4523 State Road
Santa Fe, NM 87505
Tel: 505-984-8164
Fax: 505-986-5838
info@lapuertaoriginals.com
www.lapuertaoriginals.com

McNulty Design Group
708 Vernon Avenue
Glencoe, IL 60022
Tel: 847-835-0868
Fax: 847-835-0867
www.mcnultydesign.com

Olympic Kitchens
1 Parnell Avenue
Scarborough, ON M1K 1B1
Tel: 416-266-8851
ikatsis@olympickitchens.ca
www.olympickitchens.ca

Regency Custom Cabinets
670 Old Highway 51
Nesbit, MS 38651
Tel: 662-429-5001
info@regencycabinets.com
www.regencycabinets.com

Wood-Mode Fine Custom Cabinetry
One Second Street
Kreamer, PA 17833
Toll-free tel: 877-635-7500
www.wood-mode.com

Stewart Wurtz Furniture
3410 Woodland Park Avenue N.
Seattle, WA 98103
Tel: 206-283-2586
Fax: 206-632-7188
lunaworks@earthlink.net
www.lunaworksdesign.com

lighting, antique

19th Century Lighting Co.
601 N Broadway Street
Union City, MI 49094
Tel: 517-741-4002
lampman@19thcenturylighting.com
www.19thcenturylighting.com

Allen's Antique Lighting
45 Elm Street
North Andover, MA 01845
Tel: 978-688-6466
Fax: 978-688-6466
antiquelight@comcast.net
www.antiquelight.com

Antique Hardware & Home
39771 SD Highway 34
PO Box 278
Woonsocket, SD 57385
Toll-free tel: 800-237-8833
Tel: 605-796-4425
Fax: 605-796-4085
antique.hardware@cabelas.com
www.antiquehardware.com

Arroyo Craftsman
4509 Little John Street
Baldwin Park, CA 91706
Tel: 626-960-9411
Fax: 626-960-9521
www.arroyo-craftsman.com

Architectural Emporium
207 Adams Avenue
Canonsburg, PA 15317
Toll-free tel: 724-746-4301
sales@architectural-emporium.com
www.architectural-emporium.com

Joan Bogart
PO Box 21
Rockville Centre, NY 11571
Tel: 516-764-5712
joanbogart@antiqueslighting.com
www.antiqueslighting.com

City Lights Antique Lighting
2226 Massachusetts Avenue
Cambridge, MA 02140
Tel: 617-547-1490
Fax: 617-497-2074
citylights2226@yahoo.com
www.citylights.nu

C.M.E. Antique Lighting
Tel: 304-422-5200
valleye@1st.net
www.cmeantiquelightingfixtures.com

The Copper Shop
115 N First Avenue
Haubstadt, IN 47639
Tel: 812-768-5008
info@coppershop.net
www.coppershop.net

Eagle Emporium
PO Box 282
Eagle, PA 19480
Tel: 610-458-7188
info@eagle-emporium.com
www.eagle-emporium.com

Genuine Antique Lighting
59a Wareham Street
Boston, MA 02118
Tel: 617-423-9790
info@genuineantiquelighting.com
www.genuineantiquelighting.com

Greg Davidson Antiques
1020 1st Avenue
Seattle, WA 98104-1008
Tel: 206-625-0406
greg@antiquelighting.biz
www.antiquelighting.biz

Howard's Antique Lighting
Tel: 413-528-1232
iromla@bcn.net
www.howardsantiquelighting.com

The Lampworks
435 Main Street
Hurleyville, NY 12747
Tel: 845-434-6155
info@thelampworks.com
www.thelampworks.com

Materials Unlimited
2 West Michigan Avenue
Ypsilanti, MI 48197
Toll-free tel: 800-299-9462
Fax: 734-482-3636
materials@materialsunlimited.com
www.materialsunlimited.com

C. Neri Antiques and Lighting
313 South Street
Philadelphia, PA 19147
Tel: 215-923-6669
Fax: 215-922-4189
neriantiquelites@aol.com
www.neriantiquelighting.com

Old World Lights
4 Inslee Street
Waterloo, NY 13165
Tel: 315-539-2747
admin@oldworldlights.com
www. Oldworldlights.com

Phoenix Antiques
RR 1 Indian River, ON
Tel: 705-872-8029
mccrea.ant@sympatico.ca
www.phoenixant.com

Rejuvenation
2550 NW Nicolai Street
Portland, OR 97210
Toll-free tel: 888-401-1900
Fax: 800-526-7329
info@rejuvenation.com
www.rejuvenation.com

Remember Yesterday
602 2nd Street
Petaluma, CA 94952
Tel: 707-778-1567
lamps@rememberyesterday.com
www.rememberyesterday.com

Renaissance Antique Lighting
Showroom
42 Spring Street
Newport, RI 02840
Toll-free tel: 800-850-8515
Tel: 401-849-8515
Fax: 401-849-8516
Wholesale and Restoration
Renaissance Antiques Inc.
657 Quarry Street Unit 4
Fall River, MA 02823
Toll-free tel: 800-891-8602
Fax: 401-849-8516
sales@antique-lighting.com.
www.antique-lighting.com

Sparrows Inc.
4115 Howard Avenue
Kensington, MD 20895
Toll-free tel: 888-800-1235
Tel: 301-530-0175
Fax: 301-530-0189
www.sparrows.com

St. Louis Antique Lighting Company
801 N Skinker Boulevard
St. Louis, MO 63130
Tel: 314-863-1414
Fax: 314-863-6702
htcstaff@traditional-building.com
www.Traditional-Building.com

Upton Studios LLC
145 Millville Ave
Milmay, NJ 08340
Tel: 609-261-2015
Fax: 609-249-3960
felixxus@comcast.net
www.uptonstudios.com

Vintage Lighting
Port Hope, ON L1A 1X4
905-231-1308
vintage@vintagelighting.com
www.vintagelighting.com

lighting, general

2nd Avenue Design
737 W 2nd Avenue
Mesa, AZ 85210
Toll-free tel: 800-843-1602
Fax: 800-826-2317
ms.click@2ndave.com
www.2ndave.com

Access Lighting
1200 Valencia
Tustin, CA 92780
Toll-free tel: 800-828-5483
Tel: 714-247-1270
info@accesslighting.com
www.accesslighting.com

Arroyo Craftsman
4509 Little John Street
Baldwin Park, CA 91706
Tel: 626-960-9411
Fax: 626-960-9521
www.arroyo-craftsman.com

Artcraft Lighting
8525 Jules Leger
Ville D'Anjou, QC H1J 1A8
Tel: 514-353-7200
Fax: 514-353-0013
info@artcraftlighting.com
www.artcraftlighting.com

The Basic Source, Inc.
655 Carlson Ct.
Rohnert Park, CA 94928
Tel: 707-586-5483
Fax: 707-586-5485
sales@basicsourcelighting.com
www.basicsourcelighting.com

Brass Light Gallery
PO Box 674
Milwaukee, WI 53201–0674
Toll-free tel: 800-243-9595
Toll-free fax: 800-505-9404
customerservice@brasslight.com
www.brasslight.com

Corbett Lighting
14625 E Clark Avenue
City of Industry, CA 91745
Tel: 626-336-4511
Fax: 626-330-4266
www.corbettlighting.com

Crystorama Lighting Manufacturer
95 Cantiague Rock Road
Westbury, NY 11590
Toll-free tel: 800-888-4470
Tel: 516-931-3179
crystallight95@aol.com
www.crystorama.com

Delightville Inc
22766 Ventura Blvd
Woodland Hills, CA 91364–1333
Tel: 818-225-9882

Designer's Fountain
20101 S. Santa Fe Avenue
Rancho Dominguez, CA 90221
Toll-free tel: 800-228-6197
Tel: 310-886-5143
Fax: 310-886-5161
Toll-free fax: 800-762-3401
email@designersftn.com
www.designersftn.com

Elk Lighting
101 West White Street
Summit Hill, PA 18250
Toll-free tel: 800-613-3261
Fax: 570-645-4574
elk@elklighting.com
www.elklighting.com

Eurofase Inc.
33 West Beaver Creek Rd
Richmond Hill, ON L4B 1L8
Tel: 905-695-2055
Toll-free tel: 800-660-5391
Fax: 905-695-2056
Toll-free fax: 800-660-5390
eurofase@eurofase.com
www.eurofase.com
Dallas Trade Mart
Eurofase Showroom, Ste 4002 & 3862
2100 Stemmons Freeway
Dallas, TX 75207
Tel: 214-741-4151

Eurolite
5 Lower Sherbourne, Ste 100
Toronto ON M5A 2P3
Tel: 416.203.1501
Fax: 416.203.6269
info@eurolite.com
www.eurolite.com

Fine Art Lamps
5770 Miami Lakes Drive
East Miami Lakes, FL 33014
Tel: 305-821-3850
Fax 305-821-1564
customerservice@fineartlamps.com.
www.fineartlamps.com

Fire Farm Inc.
PO Box 458
Elkader, IA 52043
Tel: 563-245-3515
Fax: 563-245-351
info@firefarm.com
www.firefarm.com

Forecast
1600 Fleetwood Drive
Elgin, IL 60123
Tel: 847-622-0416
Fax: 847-622-2542
www.forecastltg.com

Eurolite

H.A. Framburg & Company
941 Cernan Drive
Bellwood, IL 60104
Toll-free tel: 800-796-5514
Tel: 708-547-5757
Fax: 708-547-0064
framburg@framburg.com
www.framburg.com

G Lighting
9777 Reavis Park Drive
St. Louis, MO 63123
Toll-free tel: 800-331-2425
Tel: 314-631-6000
Fax: 314-631-7800
customerservice@glighting.com
www.glighting.com

Grand Light
580 Grand Ave
New Haven, CT 06511
Toll-free tel: 800-922-1469
Tel: 203-777-5781
Fax: 203-785-1184
info@grandlight.com
www.grandlight.com

GrandRiverSupply.com
1421 Camino del Pueblo
Bernalillo, NM 87004
Toll-free tel: 877-477-8775
Tel: 505-867-4110
Fax: 505-867-9711
www.grandriverdoor.com

Hampstead Lighting and Accessories
4505 Peachtree Industrial Boulevard, Ste E
Norcross, GA 30092
Toll-free tel: 888-WE-LIGHT
Tel: 770-447-1700
Fax: 770-447-1702
www.hampsteadlighting.com

Hinkley Lighting
12600 Berea Road
Cleveland, OH 44111
Tel: 216-671-3300
Fax: 216-671-4537
info@hinkleylighting.com
www.Hinkleylighting.com

Hubbardton Forge
154 Route 30 South
PO Box 827
Castleton, VT 05735
Tel: 802-468-3090
Fax: 802-468-3284
info@vtforge.com
www.vtforge.com

Hudson Valley Lighting, Inc.
106 Pierces Road
PO Box 7459
Newburgh, NY 12550
Tel: 845-561-0300
Fax: 845-561-6848
www.hudsonvalleylighting.com

International Association of Lighting Designers (IALD)
Merchandise Mart, Ste 9-104
200 World Trade Center
Chicago, IL 60654
Tel: 312-527-3677
Fax: 312-527-3680
www.iald.org

Kichler Lighting
7711 E Pleasant Valley Road
PO Box 318010
Cleveland, OH 44131–8010
Toll-free tel: 800-875-4216
Tel: 216-573-1003
dancer@kichler.com
www.kichler.com

Large Lighting.com
PO Box 12413
Charlotte, NC 28220
Tel: 800-650-1480
Sales@Chandeliers1.com
www.largelighting.com

LightTech Design
672 Fearrington Post
Pittsboro, NC 27312
Tel: 919-542-5577
Fax: 919-542-4965
stanpomeranz@earthlink.net
www.lighttechdesign.com

Magic Lite Ltd.
2526 Speers Road
Units 4–9
Oakville, ON L6L 5M2
Toll-free tel: 888-945-5483
TEL: 905-825-9592
Fax: 905-825-8334
lights@magiclite.com
www.magiclite.com

Maxim
253 N. Vineland Avenue
City of Industry, CA 91746
Tel: 626-956-4200
Fax: 626-956-4225
Info@Maximlighting.com
www.maximlighting.com

Minka Group
1151 W. Bradford Ct.
Corona, CA 92882
Tel: 951-735-9220
Fax: 951-735-9758
sales@minkagroup.net
www.georgekovacs.com

James R. Moder Crystal Chandelier Inc.
PO Box 420346
Dallas, TX 75342-0346
Tel: 800-663-1232
Fax: 800-886-5188
crystal@jamesrmoder.com
www.jamesrmoder.com

New Metal Crafts
812 N Wells Street
Chicago, IL 60610
Toll-free tel: 800-621-3907
Tel: 312-787-6991
Fax: 312-787-8692
inquiries@newmetalcrafts.com
www.newmetalcrafts.com

Norwell Manufacturing
82 Stevens Street
East Taunton, MA 02718
Toll-free: 800-822-2831
Tel: 508-823-1751
Fax: 508-823-9431
Service@Norwellinc.com
www.norwellinc.com

National Specialty Lighting
Toll-free tel: 800-527-2923
Tel: 303-926-1100
Fax: 800-527-4358 or 303-926-0011
sales@nslusa.com
www.nslusa.com

Nulco Lighting
30 Beecher Street
Pawtucket, RI 02860
Tel: 401-728-5200
Fax: 401-728-8210
www.nulcolighting.com

Pegasus Associates
PO Box 517
Beaver, PA 15009–0517
Toll-free tel: 800-392-4818
Tel: 724-846-5137
Fax: 724-847-7660
www.pegasusassociates.com

Progress Lighting
PO Box 5704
Spartanburg, SC 29304–5704
Tel: 864-599-6000
Fax: 864-599-6151
www.progresslighting.com

Fredrick Ramond Incorporated
16121 South Carmenita Road
Cerritos, CA 90703
Tel: 562-926-1361
Fax: 562-926-1015
info@fredrickramond.com
www.fredrickramond.com

Santa Fe Lights LLC
PO Box 1659
Espanola, NM 87532
Tel: 505-747-7744
Fax: 505-747-7733
info@santafelights.com
www.santafelights.com

Savoy House & Renaissance Guild
PO Box 491750
Lawrenceville, GA 30043
Toll-free tel: 800-801-1621
Tel: 770-407-4030
Fax: 770-407-4044
www.savoyhouse.com

Swarovski Canada Limited
3781 Victoria Park Avenue, Unit 8
Toronto, ON M1W 3K5
Tel: 416-492-8067
Fax: 416-496-1458
customer_relations.ca@swarovski.com
www.shop.swarovski.com

Seagull Lighting
301 W Washington Street
Riverside, NJ 08075
Toll-free: 800-347-5483
Fax: 800-877-4855
info@seagulllighting.com
www.seagulllighting.com

SPJ Lighting Inc.
2107 Chico Ave.
South El Monte, CA 91733
Toll-free tel: 800-469-3637
Tel: 626-433-4800
Fax: 626-433-4839
plestz@spjlighting.com
www.spjlighting.com

Thomas Lighting
10350 Ormsby Park Place, Ste 601
Louisville, KY 40223
Tel: 502-420-9600
Fax: 502-420-9640
www.thomaslighting.com

Top Brass Lighting
1719 Whitehead Road
Baltimore, MD 21207
Toll-free tel: 800-359-4135
Fax: 410-298-8042
sale@topbrasslighting.com
www.topbrasslighting.com

Troy-CSL Lighting Inc.
14625 E Clark Avenue
City of Industry, CA 91745
Tel: 626-336-4511
Fax: 626-330-4266
www.troy-lighting.com

Visionaire Lighting
19645 Rancho Way
Rancho Dominguez, CA 90220
Toll-free tel: 877-977-LITE
Tell: 310-512-6480
Fax: 310-512-6486
www.visionairelighting.com

Wilshire Manufacturing Company, USA
645 Myles Standish Boulevard
Taunton, MA 02780
Tel: 508-824-1970
Fax: 508-822-7046
cindyp@wilshiremfg.com
www.wilshiremfg.com

WPT Design
1881 Industrial Drive
Libertyville, IL 60048
Tel: 847-680-9043
Fax: 847-680-8140
www.matthewsfanco.com

lighting, designers

International Association of Lighting Designers (IALD)
Merchandise Mart, Ste 9–104
200 World Trade Center
Chicago, IL 60654
Tel: 312-527-3677
Fax: 312-527-3680
www.iald.org

Illuminations Landscape Lighting
Houston, TX 77007-5316
Toll-free tel: 800-863-1184
Tel: 713-863-1133
Fax: 713-863-0044
tom@illuminationslighting.com
www.illuminationslighting.com

lighting, recessed, track, specialty lighting

AmerTac
Saddle River Executive Centre
One Route 17 South
Saddle River, NJ 07458
Tel: 201-934-3224
Fax: 201-825-0802
www.amertac.com

Cooper Lighting
1121 Highway 74 South
Peachtree City, GA 30269
www.cooperlighting.com

Juno Lighting
1300 South Wolf Road
De Plaines, IL 60017
Tel: 847-827-9880
www.junolighting.com

Lightolier
631 Airport Road
Fall River, MA 02720
Tel: 508-679-8131
Fax: 508-674-4710
www.lightolier.com

Rejuvenation
2550 NW Nicolai Street
Portland, OR 97210
Toll-free tel: 888-401-1900
Fax: 800-526-7329
info@rejuvenation.com

Savoy House & Renaissance Guild
PO Box 491750
Lawrenceville, GA 30043
Toll-free tel: 800-801-1621
Tel: 770-407-4030
Fax: 770-407-4044
www.savoyhouse.com

Seagull Lighting
301 W Washington Street
Riverside, NJ 08075
Toll-free: 800-347-5483
Fax: 800-877-4855
info@seagulllighting.com
www.seagulllighting.com

Spectrum
Toll-free tel: 800-668-3899
spectrum@aol.com
www.greatchandelier.com

W.A.C. Lighting
New York
615 South Street
Garden City, NY 11530
California
168 Brea Canyon Road
City of Industry, CA 91789
Toll-free tel: 800-526-2588
Tel: 516-515-5000
Toll-free fax: 800-526-2585
Fax: 516-515-5050
sales@waclighting.com
www.waclighting.com

log building, associations

International Log Builders' Association
PO Box 775
Lumby, BC V0E 2G0
Tel: 250-547-8776
Fax: 250-547-8775
info@logassociation.org
www.logassociation.org

Log Builders Association of North America
Tel: 360-794-4469
info@loghomebuilders.org

Preservation Trades Network
PO Box 10236
Rockville, MD 20849
Tel: 301-315-8345
Fax: 301-315-8344
info@ptn.org
www.ptn.org

log home builders, canadian

ALBERTA

Langberg Log Homes Ltd.
Box 14, Site 9, RR 3
Rocky Mountain House, AB T4T 2A3
Tel: 403-729-5647
Fax: 403-729-5646
langbergloghomes@hotmail.com
www.langbergloghomes.com

Living Log and Timber Ltd.
James Maygard
Box 729
Breton, AB T0C 0P0
Tel: 780-696-3415
Fax: 780-696-3639
livinlog@telusplanet.net
www.livinglogandtimber.com

Moose Mountain Log Homes Inc.
PO Box 26
Bragg Creek, AB T0L 0K0
Tel: 403-932-3992
Fax: 403-932-9299
info@moosemountain.com
www.moosemountain.com

Twin Butte Log Homes Ltd.
Bruce Mackintosh
Box 474
Twin Butte, AB T0K 2J0
Tel: 403-627-2609
Fax: 403-627-2648
Tel (cell): 403-627-6497
tblh_ltd@telus.net

BRITISH COLUMBIA

Ram Creek Log Homes
Box 76
Wardner, BC V0B 2J0
Tel: 250-429-3445
Fax: 250-429-3449
ramcreekloghomes@telus.net
www.ramcreekloghomes.com

Big Foot Hand Hewn Log Homes Inc.
PO Box 309
Salmon Arm, BC V1E 4R8
Tel: 250-835-8885
Fax: 250-835-4732
henry@bigfoot-mfg.com
www.bigfootloghomes.com

Canada's Log People Inc.
Box 1981
100 Mile House, BC V0K 2E0
Tel: 250-791-5222
Fax: 250-791-5598
logpeople@bcinternet.net
www.canadaslogpeople.com

Cancedar Log Homes Ltd.
PO Box 143, 46229 Yale Road
Chilliwack, BC V2P 2P0
Tel: 604-703-1012
Fax: 604-703-1013
cancedar@shaw.ca
www.cancedarloghomes.com

Chinook Log Homes
RR 1, Site 16, Comp 34
Fort Street John, BC V1J 4M6
Tel: 250-261-4007
Fax: 250-261-6900
info@chinookloghomes.com
www.chinookloghomes.com

Creative Log Homes
Box 786
Summerland, BC V0H 1Z0
Tel: 250-494-9646
Fax: 250-494-9646
sales@creativeloghomes.net
www.creativeloghomes.net

Del Radomske's Okanagan School of Log Building International
1231 Philpott Road
Kelowna, BC V1P 1J7
Tel: 250-765-5166
Fax: 250-765-5167
info@okslb.ca
www.okslb.ca

ECO Log Homes
Peter Erbel
9702 Driftwood Road
Smithers, BC V0J 2N7
Tel: 250-847-3207
Fax: 250-847-3207
Peter@ecologhomes.com
www.ecologhomes.com

Island School of Building Arts
3199 Coast Road
Gabriola Island, BC V0R 1X7
Tel: 250-247-8922
Fax: 250-247-8978
info@logandtimberschool.com
www.logandtimberschool.com

Lake Country Log Homes Inc.
Box 885
Salmon Arm, BC V1T 4N9
Tel: 877-554-8881
Fax: 250-836-3874
info@lakecountrylog.com
www.lakecountrylog.com

The Log Connection
101-208 Ellis Street
Penticton, BC V2A 4L6
Toll-free tel: 888-207-0210
Tel: 250-770-9031
Toll-free fax: 877-564-5487
Fax: 250-770-9032
loghomes@thelogconnection.com
www.thelogconnection.com

Nicola Log Works Ltd.
Box 1027
Merritt, BC V1K 1B8
Tel: 250-378-4977
Fax: 250-378-4611
info@logworks.ca
www.logworks.ca

Nordic Spirit Timberworks, Inc.
831 Cascade Highway
PO Box 578
Rossland, BC V0G 1Y0
Tel: 250-362-5463
Fax: 250-362-5464
info@nstimber.com
www.nstimber.com

North American Log Crafters Ltd.
RR 1 S12 C21
Chase, BC V0E 1M0
Tel: 250-955-2485 or 877-955-2485
Fax: 250-955-0423
info@namericanlogcrafters.com
www.namericanlogcrafters.com

Original Log Homes Ltd.
811 Alder Street, Box 1301
100 Mile House, BC V0K 2E0
Tel: 250 395 3868
Fax: 250-395-2750
E-mail:original@originallog.com
www.originallog.com

Pacific Log Homes Ltd.
PO Box 64
Lone Butte, BC V0K 1X0
Tel: 250-395-4922
Fax: 250-395-3802
info@pacificloghomes.com
www.pacificloghomes.com

Pioneer Log Homes of British Columbia Ltd.
841A MacKenzie Avenue
Williams Lake, BC V2G 3X8
Toll-free tel: 877-822-5647
Tel: 250-392-5577
Fax: 250-392-5581
pioneerlog@telus.net
www.pioneerloghomesofbc.com

Red Willow Rustic Log Homes
14550 Fawn Road
Telkwa, BC V0J 2X2
Tel: 250-846-5699
Fax: 250-846-5680
info@redwillowrustic.ca
www.redwillowrustic.ca

Shuswap Log Homes International
PO Box 1178
Salmon Arm, BC V1E 4P3
Tel: 250-832-4003
Fax: 250-832-4099
prhoelz@telus.net
www.shuswaploghomes.com

Sitka Log Homes Inc.
5454 Tatton Road
100 Mile House, BC V0K 2E1
Tel: 250-791-6683
Fax: 250-791-6650
sitkalog@bcinternet.net
www.sitkaloghomes.com

Top Notch Log Construction
958 Finlayson Arm Road
Victoria, BC V9B 6E6
Tel: 250-478-0795
Fax: 250-478-4453
pat@topnotchlog.com
www.topnotchlog.com

West Coast Log Homes
PO Box 877
Gibsons, BC V0N 1V0
Tel: 604-740-8802
Fax: 604-886-0409
info@westcoastloghomes.com
www.westcoastloghomes.com

Whistler Valley Log Homes
Box 435
Garibaldi Highlands, BC V0N 1T0
Tel: 604-892-6328
Fax: 604-892-6328
info@loghomeswhistler.com
www.loghomeswhistler.com

NEW BRUNSWICK

Atlantic Log Works Ltd.
525 Mapleton Road
Moncton, NB E1G 2K5
Tel: 506-858-0048
atlanticlogworks@nb.aibn.com

NOVA SCOTIA

Heartwood Log Homes Ltd.
RR 1 Margaretville, NS B0S 1N0
Tel: 902-765-6596
Fax: 902-765-2414
roger@heartwood-log-homes.com
www.heartwood-log-homes.com

ONTARIO

Coyote Log Homes Inc.
RR 2 20594 Highway 60
Barry's Bay, ON K0J 1B0
Tel: 613-756-3394
Fax: 613-756-6186
danielalbert@coyoteloghomes.ca
www.coyoteloghomes.ca

Davidson Log & Timber Artisans Inc.
RR 1, 54 Rama-Dalton Boundary Road
Washago, ON, L0K 2B0
Tel: 705-833-1203
Fax: 705-833-1274
matt@davidsonloghomes.com
www.davidsonloghomes.com

John DeVries Log & Timber Homes 2000 Ltd.
RR 3 Tweed, ON K0K 3J0
Tel: 613-478-6830
Fax: 613-478-1400
info@jdvloghomes.com
www.jdvloghomes.com

Herwig Log Homes & Timber Framing Inc.
20513 Highway 17
RR 3, Cobden, ON K0J 1K0
Tel: 613-638-1500
Fax: 613-735-1724
herwig@herwigloghomes.com
www.herwigloghomes.com

New Frontier Logworks
70-C Mountjoy N, Ste 211
Timmins, ON P4N 4V7
Tel: 705-360-3090 or 705-465-1847
Fax: 705-267-1995
newfrontierlogworks@hotmail.com

Northern Comfort Log Homes
RR 1, 324 Kirby Road
Goulais River, ON P0S 1E0
Tel: 705-649-2780
Fax: 705-649-2780
northerncomfortloghomes@bellnet.ca
www.northerncomfortloghomes.com

Pioneer Logs Ltd.
RR 2 Singhampton, ON N0C 1M0
Tel: 519-922-2836
Fax: 519-922-2836
pioneerchr@bmts.com
www.pioneerlogsltd.com

Sunstream Log Homes
PO Box 176
Thorndale, ON N0M 2P0
Tel: 519-461-0114
Fax: 519-461-0117
info@sunstreamloghomes.com
www.suntreamloghomes.com

Winterwood Custom Builders
530 Greer Road
RR 3 Utterson, ON P0B 1M0
Toll-free tel: 800-814-5945
Tel: 705-385-8199
Fax: 705-385-8737
E-mail:winterwood@on.aibn.com
www.winterwood.ca

QUEBEC

Art Maison
161 Chemin du Grand Bois
St-Etienne de Bolton, QC J0E 2E0
Tel: 450-297-3513
Fax: 450-297-3513
artmaison@sympatico.ca

Douglas Lukian Inc.
785 Ch. Street Adolphe
Morin Heights, QC J0R 1H0
Tel: 450-226-6076
Fax: 450-226-2043
douglas_lukianinc@sympatico.ca

Flynn Log Homes
19 Chemin des Bois,
L'Ange-Gardien, QC J8L 2W8
Tel: 819-281-6956
Fax: 819-281-6956
E-mail:flh@flynnloghomes.com
www.flynnloghomes.com

Les habitations APEX
215, av. Richard
St-Boniface, QC G0X 2L0
Tel: 819-535-3200
Fax: 819-535-9370
info@apex-qc.ca
www.apex-qc.ca

Structures de bois rond Harkins Inc.
156 St-Andre
St-Faustin-Lac-Carre, QC J0T 1J1
Tel: 819-688-3824
Fax: 819-688-6600
info@boisrondharkins
www.boisrondharkins.com

SASKATCHEWAN

Back Country Logcrafters
310–16, RR 3
Saskatoon, SK S7K 3J6
Tel: 3064932448
Fax: 3064932481
backcountry@yourlink.ca

Canyon Hardwoods
RR 1
Carrot River, SK S0E 0L0
Tel: 306-768-2420
Fax: 306-768-2180
loghome@afo.net

YUKON

Woody's Log Homes
Box 183
Teslin, YK Y0A 1B0
Tel: 867-390-2739
Fax: 867-390-2424
woodys.loghomes@northwestel.net

Yukon Alaska Log Homes
207 Tlinget Street
Whitehorse, YK Y1A 2Z1
Tel: 867-668-2206
Fax: 867-393-3498
mikkelsen@yt.sympatico.ca
www.ykakloghomes.com

log home builders, u.s.

ALASKA

Husky Logwork
PO Box 873148
Wasilla, AK 99687
Tel: 907-357-6006
huskylogwork@hotmail.com
www.huskylogwork.com

Top Notch Log Builders, Inc.
Box 401
Talkeetna, AK 99676
Tel: 907-733-2427
Fax: 907-733-5647
topnotch@gci.net

ARIZONA

Canavest Builders Inc.
1210 Metate Lane
Prescott, AZ 86303
Tel: 928-717-0742
Fax: 928-717-0761
bev@canavest.com
www.canavest.com

CALIFORNIA

Gaudet Log Homes
701 Day Valley Road
Aptos, CA 95003
Tel: 831-662-8801
Fax: 831-662-8806
sales@gaudetloghomes.com
www.gaudetloghomes.com

COLORADO

Ackerman Handcrafted Log Homes
PO Box 1318
Carbondale, CO 81623
Tel: 970-963-0119
Fax: 970-963-9180
ackerman@sopris.net
www.ackermanloghomes.com

Blue Ox Logcrafters
PO Box 644
Carbondale, CO 81623
Tel: 970-963-3689
Fax: 970-963-3689
blueox@rof.net
www.blueoxlogcrafters.com

Mountain State Log Homes, Inc.
PO Box 547
Montrose, CO 81402
Tel: 970-249-3738
Fax: 970-323-5050
info@msloghomes.com
www.msloghomes.com

Timmerhus
Main Office
Tel: 303- 449-1336
Fax 303-449-9170
ed@timmerhusinc.com
East Coast Office
Tel: 603-643-2002
Fax: 603- 643-5651
elevin@valley.net
www.timmerhusinc.com

GEORGIA

Dream Crafters
PO Box 2161
Elijay, GA 30540
Tel: 706-636-4559
gifford@ellijay.com

IDAHO

Caribou Creek Log Homes, Inc.
HCR 85 Box 3 (Highway 95 N)
Bonners Ferry, ID 83805
Tel: 800-619-1156
Fax: 208-267-7215
cariboulinfo@cariboucreekloghomes.com
www.cariboucreekloghomes.com

Edgewood Log Structures
PO Box 1030
Coeur D'Alene, ID 83816–1030
Tel: 208-683-3330
Fax: 208-683-3331
brian@edgewoodlog.com
www.edgewoodlog.com

Precision Craft Log Homes
Jim Young
711 E Broadway Avenue
Meridian, ID 83642
Tel: 208-887-1020
Fax: 208-887-1253
jyoung@precisioncraft.com
www.precisioncraft.com

IOWA

Gabriel's Carpentry
4025 Evergreen Avenue
Joice, Iowa 50446
Tel: 641-588-3854
Fax: 641-588-3855
beal4us@yahoo.com

Rustic Home Builders
2046 Cleveland Ave
Inwood, IA 51240
Tel: 712-753-2666
Fax: 712-753-2722
rusthome@alliancecom.net
www.rustichomebuilders.com

MAINE

BG Stadig Handcrafted Log Homes
898 Oxbow Road
Oxbow, ME 04764
Tel: 207-435-6237
kiki@mfx.net

MICHIGAN

Keweenaw Bay Log Homes
Route 1, Box 22A
Covington, MI 49919
Tel: 906-355-2154
Fax: 906-355-2154
kblh@up.net
www.keweenawbayloghomes.com

Maple Island Log Homes
Eric Gordon
2387 Bayne Road
Twin Lake, MI 49457
Tel: 800-748-0137
Fax: 231-821-0485
eric@mapleisland.com
www.mapleisland.com

MINNESOTA

Andersen Log Homes Company
6425 State 371 NW
Walker, MN 56484
Tel: 218-547-3433
Fax: 218-547-3425
alhloghomes@arvig.net
www.andersenloghomes.com

Senty Log Homes
Mike Senty
PO Box 969
Grand Marais, MN 55604
Tel: 217-387-2644 Fax: 218-387-2740
logs@senty.com
www.senty.com

MONTANA

Artisan Log Works
6585 Farm to Market Road
Whitefish, MT 59937
Tel: 406 250 3664
Fax: 406-862-5647
tomnixon@centurytel.net
www.artisanlogworks.com

Bitterroot Timber Frames
567 Three Mile Creek Road,
Stevensville, MT 59870
Tel: 406-777-5546
Fax: 406-777-5547
bitterroot@bitterroottimberframes.com
www.bitterroottimberframes.com

Montana Dry Log & Lumber
93-5th Lane
Fort Shaw, MT 59443
Tel: 406-467-3199
Fax: 406-467-3106
mdl@3rivers.net
www.logsandlumber.com

Shady Grove Log & Timber Builders, LLC
PO Box 232
Whitefish, MT 59937
Tel: Pat: 406-212-0388; Paul: 406-261-0256
Fax: 406-862-0504
info@shadygrovelog.com
www.shadygrovelog.com

NEW YORK

Beaver Creek Log Homes
35 Territory Road
Oneida, NY 13421
Tel: 315-245-4112
Fax: 315-245-5787
robbin@beavercreeklog.com
www.beavercreekloghomes.com

NORTH CAROLINA

Atali Log Homes
PO Box 949
Skyland, NC 28776
Tel: 828-681-5585
Fax: 828-681-5848
atalihomes@aol.com
www.atalihomes.com

Farrell Structures, LLC
22 Cimmaron Drive
Pisgah Forest, NC 28768
Tel: 828-885-7969
Fax: 828-885-7967
sfarrell@citcom.net
www.farrellstructures.com

OREGON

Homestead Log Homes Inc.
6301 Crater Lake Highway
Medford, OR 97502
Tel: 541-826-6888
Fax: 541-826-1797
info@homesteadloghomes.com
www.homesteadloghomes.com

Swiss Mountain Log Homes
PO Box 2012
Sisters, OR 97759
Tel: 541-385-6006
Fax: 541-385-6006
info@swissmtloghomes.com
www.swissmtloghomes.com

Treehouse Log Homes of the NW, LLC
1126 Edgewater Street, NW
Salem, OR 97304
Tel: 503-370-7284
Fax: 503-581-1948
amber@treehouseloghomes.com
www.treehouseloghomes.com

PENNSYLVANIA

Home Field Advantage Ltd.
PO Box 500
St. Peter's Village, PA 19470
Tel: 610-323-4283
Fax: 610-327-1018
hfaltd@msn.com
www.home-field-advantage.com

Lukcik's Log Homes
75 Hemlock Drive
Blairsville, PA 15717
Tel: 724-248-9433
Fax: 724-248-9433
lukcikloghomes@yourinter.net
www.lukciksloghomes.com

TEXAS

Southwest Log Homes, Inc.
6821 Nob Hill Drive, N.
Richland Hills, TX 76180
Tel: 800-733-4228
Fax: 817-656-8543
swlog@flash.net
www.southwestloghomes.com

VERMONT

Mitchell Mountain Company
PO Box 15
Granby, VT 05840
Tel: 802-328-3886
Fax: 802-328-9800
pcci@sover.net

Mountain Logworks
1892 Daniels Farm Road
Waterford, VT 05819
Tel: 802-748-5929
mountainlogworks@yahoo.com
www.mountainlogworks.com

The Wooden House Co.
3714 North Road
Newbury, VT 05085
Tel: 802-429-2490
Fax: 802-429-2890
john@woodenhousecompany.com
www.woodenhousecompany.com

VIRGINIA

Highlands Log Structures, Inc.
PO Box 1747
Abingdon, VA 24212
Tel: 276-623-1580
Fax: 276-623-1798
highlandslog@msn.com
www.highlandslogstructures.com

WASHINGTON

North Region Log Homes
PO Box 95
Curlew, WA 99118
Tel: 509-779-4260 ext. 22
Fax: 509-779-4260
jsghome@msn.com
www.northregion.org

True Log Homes
4208 Mount Baker Highway
Everson, WA 98247
Tel: 360-592-2322
Fax: 360-592-2719
jim@truelog.com
www.truelog.com

WISCONSIN

Natural Log Homes Ltd.
PO Box 283
River Falls, WI 54022
Tel: 715-425-1739
Tel (cell): 612-804-2300
Fax: 715-425-1746
robert@logbuilding.org
www.naturalloghomes.com

Ojibwa Log Homes
N3881 County Road J
Winter, WI 54896
Tel: 715-266-3435
Fax: 715-266-3435
ojibwaloghomes@pctcnet.net
www.ojibwa-loghomes.com

Wild Wood Custom Builders LLC
E3290 Baggs Hill Road
Waupaca, WI 54981
Tel: 715-256-0871
Fax: 715-256-1965
wildwood@wolfnet.net
www.wildwoodloghomes.net

WYOMING

Bromley Log Homes
81 Whitney Drive
Cody, WY 82414
Tel: 307-587-5010
Fax: 307-587-9301
mbromley@tritel.net
www.bromleyloghomes.com

Legend Log Crafters LLC
PO Box 54
Meeteetse, WY 82433
Tel: 307-868-9293
Fax: 307-868-9293
Legendlog@yahoo.com
www.legendlog.com

metal, decorative

Alloy Castings Co. Inc.
151 West Union Street
PO Box 473
East Bridgewater, MA 02333-0473
Tel: 508-378-2541
Fax: 508-378-1240
www.alloycastings.com

Artistic Iron Works
805 Winnipeg Street
Regina, SK S4R 1J1
Toll-free tel: 800-667-4766
www.artisticironworks.com

Chelsea Decorative Metal Company
8212 Braewick Drive
Houston, TX 77074
Tel: 713-721-9200
Fax: 713-776-8661
www.metals.about.com

D.J.A. Imports Ltd.
1672 E. 233rd Street
Bronx, NY 10466
Tel: 718-324-6871
Fax: 718-324-0726
info@djaimports.com
www.djaimports.com

King Architectural Metals
Toll-free tel: 800-542-2379
itemrequest@kingmetals.com
www.kingmetals.com

McNichols
5505 West Gray Street
PO Box 30300
Tampa, FL 33630–3300
Tel: 813-286-2771
Fax: 813-282-8620
sales@mcnichols.com
www.mcnichols.com

National Association of Architectural Metal Manufacturers (NAAMM)
8 South Michigan Avenue, Ste 1000
Chicago, IL 60603
Tel: 312-332-0405
Fax: 312-332-0706
www.naamm.org

NIKO Contracting Co., Inc.
3434 Parkview Avenue
Pittsburgh, PA 15213
Tel: 412-687-1517
Fax: 412-687-7969
info@nikocontracting.com
www.nikocontracting.com

Novel Architectural Products
3425 NW 167th Street
Miami, FL 33056
Toll-free tel: 866-623-0563
Tel: 305-623-0563
Fax: 305-663-1332
sales@novelamerica.com
www.novelamerica.com

Pacific Crest Industries, Inc. (Spectra Metals)
2411 Pomona Road
Corona, CA 92880
Toll-free tel: 877-674-3026
Tel: 951-520-0517
Fax: 951-520-0618
www.pacificcrestind.com

Roof Drainage Components & Accessories Inc.
60 Don Westbrook Avenue N
Jasper, GA 30143
Tel: 706-692-7333
Fax: 706-692-7335
www.roofdrainagecomp.com

Stainless Steel and Metal of Florida Inc.
5025 N Hiatus Road
Sunrise, FL 33351
Toll-free tel: 866-881-8335
Tel: 954-746-9155
Fax: 954-746-9252
info@ssmfinc.com
www.ssmfinc.com

Steptoe & Wife
90 Tycos Drive
Toronto, ON M6B 1V9
Toll-free tel: 800-461-0060
Tel: 416-780-1707
Toll-free fax: 877-256-4279
Fax: 416-780-1814
info@steptoewife.com
www.steptoewife.com

TWP Inc.
2831 Tenth Street
Berkeley, CA 94710
Toll-free tel: 800-227-1570
Tel: 510-548-4434
Fax: 510-548-3073
sales@twpinc.com
www.twpinc.com

Uniquities Architectural Antiques
5240 – 1a Street, SE
Calgary, AB T2H 1J1
Tel: 403-228-9221
Fax: 403-283-9226
info@uniquities-archant.com
www.uniquities-archant.com

Universal Wire Company
16 North Steel Road
Morrisville, PA 19067-3699
Toll-free tel: 888-523-0575
Tel: 215-736-8981
Fax: 215-736-8994
uwcman@localnet.com
www.universalwirecloth.com

moldings, millwork, columns etc.

Alloy Casting Co. Inc.
3900 S. Peachtree Road
Mesquite, TX 75180-2724
Toll-free tel: 800-527-1318
Tel: 972-286-2368
Fax: 972-557-4727
jon@alloynet.com
www.alloynet.com

Architectural Adornments
PO Box 726
Canby, OR 97013-0726
Toll-free tel: 877-835-0023
Tel: 503-655-0023
Fax: 503-655-2536
info@architecturaladornments.com
www.architecturaladronments.com

Architectural Products by Outwater LLC
East Coast
4 Passaic Street PO Drawer 403
Wood-Ridge, NJ 07075
Toll-free tel: 800-631-8375
Toll-free fax: 800-888-3315
West Coast
4720 West Van Buren
PO Box 18190
Phoenix, AZ 85043
Toll-free tel: 800-248-2067
info@balmer.com
www.archpro.com

Architectural Millwork
3522 Lucy Road
Millington, TN 38053
Tel: 901-327-1384
Tel (cell): 901-218-0616
Fax: 353-1645
samtickle@aol.com
www.a-millwork.com

Balmer Architectural Moldings
271 Yorkland Boulevard
Toronto, ON M2J 1S5
Toll-free tel: 800-665-3454
Tel: 416-491-6425
Fax: 416-491-7023
www.balmer.com

Burton Moldings (MDF)
7 – 12320 Trites Road
Richmond, BC V7E 3R7
Toll-free tel: 888-323-8926
Tel: 604-241-2606
Fax: 604-241-2625
sales@burtonmoldings.com
www.burtonmoldings.com

Chadsworth's 1.800 Columns
277 North Front Street
Historic Wilmington, NC 28401
Toll-free tel: 800-265-8667
Tel: 910-763-7600
Fax: 910-763-3191
catalog@columns.com
www.columns.com

Cumberland Woodcraft Co.
10 Stover Drive
PO Drawer 609
Carlisle, PA 17013
Toll-free tel: 800-367-1884
Tel: 717-243-0063
info@cumberlandwoodcraft.com
www.cumberlandwoodcraft.com

The Design-Build Store
U.S.
255 Sunrise Avenue, Ste 200
Palm Beach, FL 33480
Canada
1046 Yonge Street
Toronto, ON M4W 2L1
Toll-free tel: 866-889-9919
info@designbuildstore.com
www.designbuildstore.com

First Class Building Products Inc.
1418 Fenwick Drive
Marietta, GA 30064
Toll-free tel: 888-514-8141
Tel: 770-514-8141
Fax: 770-514-0731
firstclassbp@mindspring.com
www.firstclassbp.com

Fretworks International
150 - 1623 Military Road
Niagara Falls, NY 14304–1745
Toll-free tel: 800-465-7598
Tel: 905-892-6697
Fax: 905-892-7791
info@fretworks.com
www.fretworks.com

Fypon, Ltd.
960 West Barre Road
Archbold, OH 43502
Toll-free tel.: 800-446-3040
Toll-free fax: 800-446-9373
www.fypon.com

Hill Country Woodworks
507 E Jackson Street
Burnet, TX 78611
Tel: 512-756-6950
lee@texaswoodwork.com
www.texaswoodwork.com

Historical Arts & Casting, Inc.
5580 W Bagley Park Road
West Jordan, UT 84088
Toll-free tel: 800-225-1414
Tel: 801-280-2400
Fax: 801-280-2493
info@historicalarts.com
www.historicalarts.com

Hyde Park Fine Art of Moldings Inc.
29-16 40th Avenue
Long Island City, NY 11101
Tel: 718-706-0504
Fax: 718-706-0507
info@hyde-park.com
www.hyde-park.com

Lutes Custom Woodworking
PO Box 726
Canby, Oregon 97013
Toll-free tel: 877-835-0023
Tel: 503-655-0023
Fax: 503-655-2536
BLutes@compuserve.com
www.lutescw.com

Maine Timber Works, LLC
823 Augusta Road
Rome, ME 04963
Tel: 207-397-3285
mainetimberworks@midmaine.net
www.mainetimberworks.com

Pioneer Millworks
Toll-free tel: 800-951-9663
Eastern Office
1180 Commercial Drive
Farmington, NY 14425
Tel: 585-924-9970
Western Office
547 West 700 South
Salt Lake City, UT 84101
Tel: 801-328-9663
www.pioneermillworks.com

Richelieu Hardware Ltd
7900 Henri Bourassa Boulevard West
Montreal, QC H4S 1V4
Toll-free tel: (Quebec) 866-832-4040
(Canada): 800-361-6000
(U.S.): 800-619-5446
Tel: 514-832-4010
info@richelieu.com
www.richelieu.com

Royal Corinthian, Inc
603 Fenton Lane
West Chicago, IL 60185
Toll-free tel: 888-265-8661
Tel: 630-876-8899
Fax: 630-876-3098
royalcor@dls.net
www.royalcorinthian.com

Shanker Industries Inc.
Oceanside, NY 11572
Tel: 516-766-4477
Fax: 516-766-6655
sales@shanko.com
www.shanko.com

Michael Shea Woodcarving
PO Box 4014
Mount Forest, ON NOG 2LO
Tel/Fax: 519-323-1179
theresa@michaelsheawoodcarving.com
www.michaelsheawoodcarving.com

Southern Rose Architectural Specialties
PO Box 280144
Columbia, SC 29228
Tel: 803-356-4545
mail@classicdetails.com
www.classicdetails.com

Trimwork By Design
148 Creditstone Road, Unit 3
Vaughan, ON L4K 5V8
Toll-free tel: 866-248-7776
Tel: 905-417-6611
Fax: 905-417-4069
info@trimworkbydesign.com
www.trimworkbydesign.com

Vintage Woodworks
Highway 34 S.
PO Box 39
MSC 4141
Quinlan, TX 75474–0039
Tel: 903-356-2158
mail@vintagewoodworks.com
www.vintagewoodworks.com

Worthington Millworks
6950 Phillips Highway, Ste 20
Jacksonville, FL 32216
Toll-free tel: 800-872-1608
Tel: 904-281-1485
Fax: 904-281-1488
sales@WorthingtonMillwork.com
www.worthingtonmillwork.com

prefab houses

Alchemy Architects
856 Raymond Avenue
St. Paul, MN 55114
Tel: 651-647-6650
geo@alchemyarch.com
www.alchemyarch.com

Anderson Anderson Architecture
83 Columbia Street, Ste 300
Seattle, WA 98104
Tel: 206-332-9500
Fax: 415-520-9522
aaa@andersonanderson.com
www.andersonanderson.com

Clever Homes, LLC
665 Third Street, Ste 400
San Francisco, CA 94107
Tel: 415-344-0806
Fax: 415-344-0807
www.cleverhomes.net

Fabprefab
www.fabprefab.com

Hive Modular
1330 Quincy Street N.E.
Minneapolis, MN 55413
Tel: 612-379-4382
Fax: 612-331-4638
info@hivemodular.com
www.hivemodular.com

Marmol Radziner Prefab
12210 Nebraska Avenue
Los Angeles, CA 90025
Tel: 310-689-0089
Fax: 310-826-6226
info@marmolradzinerprefab.com
www.marmol-radziner.com

Myson Inc.
Colchester, VT 05446
Toll-free tel: 800-698-9690
info@MysonInc.com
www.mysoninc.com

Rocio Romero, LLC
PO Box 30
Perryville, MO 63775
Tel: 573-547-9078
sales@rocioromero.com
www.rocioromero.com

radiators

Antique Plumbing & Radiators
30 Prospect Street
Somerville, MA 02143
Tel: 617-625-6140
a1plumbing@rcn.com
www.antiqueplumbingandradiators.com

Beautiful Radioators
3330 E Kemper Rd
Cincinnati, OH 45241
Toll-free tel: 800-543-7040
Tel: 513-385-0555
Fax: 513-741-6292
info@beautifulradiators.com
www.beautifulradiators.com

Burnham Hydronics
PO Box 3079
Lancaster, PA 17604
Tel: 717-397-4701
Fax: 717-481-8409
www.burnham.com

R & D Energy Savers
2861Sherwood Heights Drive
Unit 21
Oakville, ON L6J 7K1
Tel: 905-829-4941
Fax: 905-829-4942
www.rdes.ca

Runtal North America, Inc.
187 Neck Road, PO Box 8278
Ward Hill, MA 01835
Toll-free tel (US): 800-526-2621
Toll-free tel (Canada): 888-829-4901
Fax: 978-372-7140
info@runtalnorthamerica.com
www.runtalnorthamerica.com

rain chains

Japanese Style Inc.
16159 320th Street
New Prague, MN 56071
Toll-free tel: 877-226-4387
Fax: 952-758-1922
customercare@cherryblossomgardens.com
www.cherryblossomgardens.com

Kinetic Fountains
Toll-free tel: 877-271-1112
Fax: 828-651-9268
sales@kineticfountains.com
www.kineticfountains.com

Park City Rain Gutter
6421 Business Park Loop Road
Park City, UT 84098–6215
Tel: 435-649-2805
Fax: 435-649-2605
info@pcraingutter.com
www.pcraingutter.com

Rutland Gutter Supply
10895 Rocket Boulevard
Orlando, FL 32824
Tel: 407-859-1119
Fax: 407-859-1123
rutlandguttersupply@yahoo.com
www.rutlandguttersupply.com

restorations

The Association for Preservation Technology International
1224 Centre West, Ste 400B
Springfield, IL 62704
Tel: 217-793-7874
Toll-free fax: 888-723-4242
information@apti.org
www.apti.org

Clearheart
Tel: 415-571-5338
clearheart@clearheart.org
www.clearheart.org

Historic Plaster Conservation Services Limited
26 Barrett Street
Port Hope, ON L1A 1M7
Tel: 905-885-8764
Fax: 905-885-8330
info@historicplaster.com
www.historicplaster.com

Preservation Trades Network
Bryan Blundell
PO Box 10236
Rockville, MD 20849
Tel: 301-315-8345
Fax: 301-315-8344
info@ptn.org
www.ptn.org

roof cresting

Architectural Iron Company, Inc.
104 Ironwood Court
PO Box 126
Milford, PA 18337–0126
Toll-free tel: 800-442-4766
Tel: 570- 296-7722 766
Fax: 570-296-4766
info@architecturaliron.com
www.architecturaliron.com

Franklin Weathervanes
PMB 552, 250 H Street
Blaine, WA 98230-4033
Toll-free tel.: 800-561-4355
Fax: 604-533-9802
sales@franklinvanes.com
www.franklinvanes.com

Heather & Little Limited
3205 14th Avenue
Markham, ON L3R 0H1
Toll-free tel: 800-450-0659
Tel: 905-475-9763
Fax: 905-475-9764
info@heatherandlittle.com
www.heatherandlittle.com

Historical Arts & Casting, Inc.
5580 West Bagley Park Road
West Jordan, UT 84088
Toll-Free: 800-225-1414
Tel: 801-280-2400
Fax: 801-280-2493
info@historicalarts.com
www.historicalarts.com

roofing tiles (clay, slate, concrete, etc.)

Boston Valley Terra Cotta
6860 South Abbott Road
Orchard Park, NY 14127
Toll-free tel: 888-214-3655
Tel: 716-649-7490
Fax: 716-649-7688
info@bostonvalley.com
www.bostonvalley.com

Camara Slate Products
PO Box 8, Rt 22A
Fair Haven, VT 05743
Tel: 802-265-3200
Fax 802-265-2211
info@camaraslate.com
www.camaraslate.com

CertainTeed Corporation
PO Box 860
750 East Swedesford Road
Valley Forge, PA 19482
Toll-free tel (building professional):
800-233-8990
Toll-free tel (consumer): 800-782-8777
Tel: 610-341-7000
Fax: 610-341-7777
www.certainteed.com

Columbia Concrete Products Limited
8650 - 130 Street
Surrey, B.C. V3W 1G1
Toll-free tel: 877-388-8453
Tel: 604-596-3388
Fax: 604-599-5972
rooftile@crooftile.com
www.columbiarooftile.com

Decra Roofing Systems
1230 Railroad Street
Corona, CA 92882
Tel: 951-272-8180
Fax: 951-272-4476
marketing@decra.com
www.decra.com

Dura-Loc Roofing Systems Limited
PO Box 220
Courtland, ON N0J 1E0
Toll-free tel: 800-265-9357
Tel: 519-688-2200
Fax: 519-688-2201
www.duraloc.com

The Durable Slate Co.
1050 N Fourth Street
Columbus, OH 43201
Toll-free tel: 800-666-7445
Tel: 614-299-5522
Fax: 614-299-7100
www.durableslate.com

Echeguren Slate
1495 Illinois Street
San Francisco, CA 94107
Tel: 415-206-9343
Fax: 415-206-9353
slate@echeguren.com
www.echeguren.com

EcoStar, a division of Carlisle SynTec
PO Box 7000
Carlisle, PA 17013
Toll-free tel: 800-211-7170
Fax: 888-780-9870
www.ecostarinc.com

Evergreen Slate Company
68 East Granville NY 12832
Toll-free tel: 866-815-2900
Tel: 518-642-2530
Fax: 518-642-9313
sales@evergreenslate.com
www.evergreenslate.com

Vande Hey Raleigh

Floor Tile and Slate Co.
1209 Carroll Ave.
Carrollton, TX 75006
Toll-free tel: 800-446-0220
Tel: 972-242-6647
Fax: 972-242-7253
sales@floortileandslate.com
www.floortileandslate.com

Gladding McBean
601 7th Street
Lincoln, CA 95648-1828
Toll-free tel: 800-776-1133
lynn.haines@paccoast.com
www.gladdingmcbean.paccoast.com

Greenstone Slate Company
Upper Road
PO Box 134
Poultney, Vermont 05764
Toll-free tel: 800-619-4333
Tel: 802-287-4333
Fax: 802-287-5720

Ludowici Roof Tile
4757 Tile Plant Road
New Lexington, OH 43764
Tel: 800-945-8453
Fax: 740-342-0025
info@ludowici.com
www.ludowici.com

MCA Tile
1985 Sampson Avenue
Corona, CA 92879
Toll-free tel: 800-736-6221
Fax: 909-736-6052
sales@mca-tile.com
www.mca-tile.com

North Country Slate
8800 Sheppard Avenue E
Toronto, ON M1B 5R4
Toll-free tel: 800-975-2835
Tel: 416-724-4666
Fax: 416-281-8842
info@ncslate.com
www.ncslate.com

Northern Roof Tiles
50 Dundas Street E, Ste 2
Dundas, ON L9H 7K6
Toll-free tel: 888-678-6866
Tel: 905- 689-4035
Fax: 905-689-7099
sales@northernrooftiles.com
www.northernrooftiles.com

Penn Big Bed Slate Co. Inc.
PO Box 184
8450 Brown Street
Slatington, PA 18080
Tel: 610-767-4601
Fax: 610-767-9252
pennbbs@aol.com
www.pennbigbedslate.com

Premier Roofing Specialist, Inc.
PO Box 2298
Lake City, FL 32056–2298
Toll-free tel: 888-492-4789
Fax: 386-719-9905
www.premierroofs.com

Sheldon Slate
Maine
38 Farm Quarry Road
Monson, ME 04464
Tel: 207-997-3615
Fax: 207-997-2966
john@sheldonslate.com
New York
Fox Road
Middle Granville, NY 12849
Tel: 518-642-1280
Fax: 518-642-9085
www.sheldonslate.com

SouthSide Roofing & Sheet Metal Co.
290 Hanley Industrial Ct.
St. Louis, MO 63144
314-968-4800
general.mail.ssr@SouthSideRoofing.com
www.southsideroofing.com

The Tile Man Inc.
PO Box 329
LouisBurg, NC 27549-6923
Toll-free tel: 888-263-0077
Tel: 866-686-9394
Fax: 919-853-6634
Fax Vintage Tile: 919-556-2072
info@thetileman.com
E-mail Vintage Tile:
loriak7@earthlink.net
www.thetileman.com

TileSearch
PO Box 580
Roanoke, TX 76262
Tel: 817-491-2444
Fax: 817-491-2457
ts@tilesearch.net
www.tilesearch.net

Universal Slate International Inc.
3821-9th Street SE
Calgary, AB T2G 3C7
Toll-free tel.: 888-677-5283
zimmer@universalslate.com
www.universalslate.com

U.S. Tile
909 West Railroad Street
Corona, CA 92882–1906
Toll-free tel: 800-252-9548
Tel: 909-737-0200
Fax: 909-734-9591
clayinfo@ustile.com
www.ustile.com

Vande Hey Raleigh Mfg., Inc.
1665 Bohm Drive
Little Chute, WI 54140–2529
Toll-free tel: 800-236-8453
Tel: 920-766-0156
Fax: 920-766-0776
www.vhr-roof-tile.com

Vermont Specialty Slate, Inc.
PO Box 4
Brandon, VT 05733
Toll-free tel: 866-US-SLATE
Tel: 802-247-6615
Fax: 802-247-4209
slate@vtslate.com
www.vtslate.com

Vermont Structural Slate Company, Inc.
Box 98, 3 Prospect Street
Fair Haven, VT 05743
Toll-free tel: 800-343-1900
Tel: 802-265-4933
Fax: 802-265-3865
info@vermontstructuralslate.com
www.vermontstructuralslate.com

roofing, metal

Baschnagel Brothers
150–25 14th Avenue
Whitestone, NY 11357
Tel: 718-767-1919
Fax: 718-767-5141
sales@baschnagel.com
www.baschnagel.com

Charleston Metalworks
2256 Wren Drive
North Charleston, SC 29406
Tel: 888-235-4452
Fax: 843-569-0902
mail@charlestonmetalworks.com
www.charlestonmetalworks.com

Classic Products Canada
3505 Laird Road, Unit 16
Mississauga, ON L5L 5Y7
Toll-free tel: 866-969-7663
Tel: 905-608-8800
info@classicproducts.ca
www.classicproducts.ca

Classic Metal Roofing
8510 Industry Park Drive
Piqua, OH 45356
Toll-free tel: 800-543-8938
Fax: 937-778-5116
www.classicroof.com

Classic Metal Roofs LLC
Toll-free tel: 866-660-6668
www.classicproducts.ca

Copper-Inc.
PO Box 244
Dickinson, TX 77539
Toll-free tel: 888-499-1962
Toll-free fax: 888-499-1963
sales@copper-inc.com
www.copper-inc.com

The Copper Shop
115 N First Avenue
Haubstadt, IN 47639
Tel: 812-768-5008
info@coppershop.net
www.coppershop.net

Custom-Bilt Metals
13940 Magnolia Avenue
Chino, CA 91710–7029
Toll-free tel: 800-826-7813
Tel: 909-664-1500
Fax: 909-664-1520
www.custombiltmetals.com

Drexel Metals Corporation
204 Railroad Drive
Ivyland, PA 18974
Toll-free tel: 888-321-9630
Tel: 215-396-4470
Fax: 877-321-9638
www.drexmet.com

Follansbee
Follansbee, WV 26037
Toll-free tel: 800-624-6909
Tel: 304-527-1260
Fax: 304-527-1269
www.follansbeeroofing.com

Gerard Roofing Technologies
955 Columbia Street
Brea, CA 92821
Tel: 714-529-0407
Fax: 714-529-6643
gerardusa@metalsusa.com
www.gerardusa.com

Hans Liebscher Custom Copperworks & Sheet Metal
PO Box 38
San Marcos, CA 92979
Tel: 760-471-5114
Fax: 760-471-7884
www.hansliebschercopperwks.com

Heather & Little Ltd.
3205 14th Avenue
Markham, ON L3R 0H1
Toll-free tel: 800-450-0659
Tel: 905-475-9763
Fax: 905-475-9764
info@heatherandlittle.com
www.heatherandlittle.com

Interlock Industries (Ont.) Inc.
230 Admiral Boulevard
Mississauga, ON L5T 2N6
Toll-free tel: 888-766-3661
Fax: 905-564-9980
info@ontariosbestroof.com
www.ontariosbestroof.com

KBS Metal Roofing
PO Box 1711
La Porte, IN 46352
Toll-free tel: 800-837-2897
Tel: 219-324-2083
Fax: 219-324-3531
www.kbsmetalroofing.com

Metro Roof Products
3093 "A" Industry Street
Oceanside, CA 920054
Toll-free tel: 866-638-7648
www.smartroofs.com

Met-Tile
1745 E. Monticello Court
Ontario, CA 91761
Tel: 909-947-0311
Fax: 909-947-1510
met-tile@met-tile.com
www.met-tile.com

Nicholson & Galloway
261 Glen Head Road
Glen Head, NY 11545
Tel: 516-671-3900
Fax: 516-759-3569
tomc@nicholsonandgalloway.com
www.nicholsonandgalloway.com

NIKO Contracting Co., Inc.
3434 Parkview Avenue
Pittsburgh, PA 15213
Tel: 412-687-1517
Fax: 412-687-7969
info@nikocontracting.com
www.nikocontracting.com

W.F. Norman Corp.
214 N. Cedar
PO Box 323
Nevada, MO 64772
Toll-free tel: 800-641-4038
Tel: 417-667-5552
Fax: 417-667-2708
ceilings@wfnorman.com
www.wfnorman.com

Park City Rain Gutter
6421 Business Park Loop Road
Park City, UT 84098-6215
Tel: 435-649-2805
Fax: 435-649-2605
info@pcraingutter.com
www.pcraingutter.com

Premier Roofing Specialist, Inc.
PO Box 2298
Lake City, FL 32056–2298
Toll-free tel: 888-492-4789
Fax: 386-719-9905
www.premierroofs.com

Preservation Products, Inc.
221 Brooke Street
Media, PA 19063
Toll-free tel: 800-553-0523
Tel: 610-565-5755
Fax: 610-891-0834
info@preservationproducts.com
www.preservationproducts.com

Revere Copper Products, Inc.
One Revere Park
Rome, NY 13440-5561
Toll-free tel: 800-448-1776
Tel: 315-338-2022
Fax: 315-338-2105
revere@reverecopper.com
www.reverecopper.com

Roof Drainage Components & Accessories Inc.
60 Don Westbrook Ave, N.
Jasper, GA 30143
Tel: 706-692-7333
Fax: 706-692-7335
www.roofdrainagecomp.com

Rutland Gutter Supply
10895 Rocket Boulevard
Orlando, FL 32824
Tel: 407-859-1119
Fax: 407-859-1123
rutlandguttersupply@yahoo.com
www.rutlandguttersupply.com

Standing Seam Roofing Company, LLC
PO Box 31
599 Island Lane
West Haven, CT 06516
Toll-free tel: 800-545-8726
Tel: 203-931-0945
Fax: 203-934-7404
info@standing-seamroofing.com
www.standing-seamroofing.com

Threadgill Sheet Metal Works Inc.
17515 A Huffmeister
Cypress, TX 77429
Tel: 281-373-0016
Fax: 281-373-0010
staff@threadgillsheetmetal.com
www.tsmwinc.com

Triple-S Chemical Products, Inc.
3464 Union Pacific Avenue.
Los Angeles, CA 90023
Tel: 323-261-7301
Fax: 323-261-5567
info@ssschemical.com
www.ssschemical.com

Vail Metal Systems LLC
PO Box 2030
Edwards, CO 81632
Toll-free tel: 800-358-8245
Fax: 800-358-3469
roofing@vailmetal.com
www.vailmetal.com

Vulcan Supply Corporation
PO Box 100
Westford, VT 05494
Toll-free tel: 800-659-4732
Fax: 802-893-0534
info@vulcansupply.com
www.vulcansupply.com

Zappone Manufacturing
2928 N Pittsburg Street
Spokane, WA 99207
Toll-free tel: 800-285-2677
www.zappone.com

roofing, rubber

EcoStar
PO Box 7000
Carlisle, PA 17013
Toll-free tel: 800-211-7170
Fax: 888-780-9870
www.ecostarinc.com

roofing, wood shingles/shakes

Anbrook Industries Ltd.
17830 Fraser Dyke Road,
Pitt Meadows, BC V3Y 1Z1
US Mailing Address
PO Box 3044
Sumas, WA 98295-3044
Tel: 604-465-5657
Fax: 604-465-3657
brooke@anbrook.com
www.anbrook.com

BCF Shake Mill Ltd.
Site 25, C-1, RR 1
Fanny Bay, BC V0R 1W0
Toll-free tel: 877-707-4253
Tel: 250-335-2969
Fax: 250-335-1425
www.bcfshake.com

Cedar Plus
PO Box 515
Sumas, WA 98295
Toll-free tel: 800-963-3388
www.clarkegroup.com

Cedar Valley Shingle Systems
943 San Felipe Road
Hollister, CA 95023
Tel: 800-521-9523
www.cedar-valley.com

C.J. Cedar Ltd.
Box 156
Merville, BC V0R 2M0E
Tel: 250-337-8333
sales@cjcedar.com
www.cjcedar.com

Coppin Ridge Company
411 US Highway 101
Hoquiam, WA 98550
Tel: 360-532-8479
info@coppinridge.com
www.coppinridge.com

Decra Roofing Systems
1230 Railroad Street
Corona, CA 92882
Tel: 951-272-8180
Fax: 951-272-4476
marketing@decra.com
www.decra.com

EcoStar, a division of Carlisle SynTec
PO Box 7000
Carlisle, PA 17013
Toll-free tel: 800-211-7170
Fax: 888-780-9870
www.ecostarinc.com

Fraser Cedar Products Ltd.
Canadian address
27400 Lougheed Highway
Maple Ridge, BC V2W 1L1
U.S. address (Mail Only)
PO Box 713
Sumas, WA 98295
Tel: 604-462-7335
Toll-free tel: 800-388-7022
Fax: 604-462-7246
bsweet@frasercedarproducts.com
www.frasercedarproducts.com

Goat Lake Forest Products
RR 3, Weldwood Road
Powell River, BC V8A 5C1
Tel: 604-487-4266
Fax: 604-487-4432
sales@goatlake.com
www.goatlake.com

Shakertown Inc.
PO Box 400
1200 Kerron Street
Winlock, WA 98596
Tel: 360-785-3501
Toll-free tel: 800-426-8970
Fax: 360-785-3076
Mill Tel: 604-462-8422 or 604-462-8425

S&K Cedar Products Ltd.
31121 Silverhill Avenue
Mission, BC V4S 1G8
Toll-free tel: 877-977-8445
Sales-Cell: 604-302-2473
Mill Fax: 604-462-8427
info@shakes.ca
www.cedarshake.com

S&W Forest Products Ltd.
9486 288th Street
Maple Ridge, BC V2X 8Y6
Tel: 604-462-0045
Toll-free tel: 800-806-9663
Fax: 604-462-0258
swforest@dowco.com
www.swforest.com

Silver Creek Premium Products
Canadian Address
PO Box 261
Matsqui, BC V4X 3R2
U.S. Address
Silver Creek Premium Products
PO Box 1445
Sumas, WA 98295
Toll-free tel: 877-292-3327 (U.S. and Canada)
Tel: (Sales) 604-826-1499,
(Main Office) 604-826-5971
Fax: 604-820-0072
contact@silvercreek.bc.ca
www.silvercreek.bc.ca

Star Cedar Sales, Inc.
PO Box 1027
Kamiah, ID 83536
Tel: 208-935-2566
Fax: 208-935-2745
yocedarman@yahoo.com
www.starcedar.com

Stave Lake Cedar
Canada
Box 16, RR 5
Maple Ridge, BC V2X 8Y6
U.S.
PO Box 96
Sumas, WA 98295
Tel: (Canadian Sales) 604-462-8266
(U.S. Sales) 800-492-5386
Fax: 604-462-8264
sales@stavelake.com
www.stavelake.com

roofing, thatched

McGhee & Co.
PO Box 2098
Staunton, VA 24401
Toll-free tel: 888-842-8241
Tel: 845-721-0443
Fax: 540-886-0007
thatchit@aol.com.
www.thatching.com

Premier Roofing Specialist, Inc.
PO Box 2298
Lake City, FL 32056–2298
Toll-free tel: 888-492-4789
Fax: 386-719-9905
www.premierroofs.com

roofing tiles, recycled

Authentic Roof
Crowe Building Products
116 Burris Street
Hamilton, ON L8M 2J5
Tel: 905-529-6818
Fax: 905-529-1755
webmaster@authentic-roof.com
www.authentic-roof.com

The Tile Man Inc.
PO Box 329
LouisBurg, NC 27549–6923
Toll-free tel: 888-263-0077
Fax: 919-853-6634
Fax (Vintage Tile): 919-556-2072
info@thetileman.com
E-mail (Vintage Tile):
loriak7@earthlink.net
www.thetileman.com

shutters

Architectural Millwork
3522 Lucy Road
Millington, TN 38053
Tel: 901-327-1384
Tel (cell): 901-218-0616
Fax: 901-353-1645
samtickle@aol.com
www.a-millwork.com

La Puerta Originals
4523 State Road
Santa Fe, NM 87505
Tel: 505-984-8164
Fax: 505-986-5838
info@lapuertaoriginals.com
www.lapuertaoriginals.com

skylights

Aluplex Skylights Inc.
34 Martin Ross Avenue
Toronto, ON M3J 2K8
Toll-free tel: 877- 258-7539
Tel: 416-665-4482
Fax: 416-665-2826
www.aluplex.com

Architectural Glazing Technologies
661 Main Street
Waterboro, ME 04087
Toll-free tel: 800-345-7899
Tel: 207-247-6747
www.agtglazing.com

Bristolite Skylights
PO Box 2515
Santa Ana, CA 92707
Toll-free tel: 800-854-8618
Tel: 714-540-5415
sales@bristolite.com
www.bristolite.com

Homes By Marie
PO Box 2777
Corrales, NM 87048
Tel: 505-342-1532
Fax: 505-342-1579
marie@homesbymarie.com
www.homesbymarie.com

New England Conservatories, Inc.
Ware, MA
Tel: 413-967-9093
sales@newenglandconservatories.net
www.newenglandconservatories.net

Solatube International, Inc
2210 Oak Ridge Way
Vista, CA 92081
Toll-free tel: 888-765-2882
Tel: 760-477-1120
Fax: 760-599-5181
www.solatube.com

J. Sussman Inc.
109-10 180th Street
Jamaica, NY 11433
Tel: 718-297-0228
Fax: 718-297-3090
webmaster@jsussmaninc.com
www.jsussmaninc.com

Velux
Toronto Office
2740 Sherwood Heights Drive,
Oakville, ON L6J 7V5
Toll-free tel: 800-888-3589
Vancouver Office
Mary Hill Business Park
305-1515 Broadway Street
Port Coquitlam, BC V3C 6M2
Velux-cdn@celux.com
www.velux.ca
www.veluxusa.com

stone

A & A Natural Stone Ltd.
381297 Concession 17
Keppel Township
R.R. 1 PO Box 524
Wiarton, ON N0H 2T0
Tel: 519-534-5966 or 534-5540
Fax: 519-534-2435

All Granite & Marble
1A Mount Vernon Street
Ridgefield Park, NJ 07660
Tel: 201-440-6779
Fax: 201-440-6855
www.marble.com

Architectural Stone Corp.
65 Nelson Road
Lively, ON N2P 2L2
Toll-free tel: 800-204-1189
Tel: 705-682-4303
Fax: 705-682-4305
www.architecturalstone.ca

Arc Stone
9020 Edgeworth Dr
Capitol Heights, MD 20743
Toll-free tel: 800-825-9086
Fax: 301-499-8875
www.arc-stone.com

Atelier Jouvence Stonecutting and Tile
329 West 18th Street, Ste 611
Chicago, IL 60016–1120
Tel: 312-492-7922
Tel (cell): 773-251-1581
Fax: 312-492-7923
contact@atelierjouvence.com
www.atelierjouvence.com

Bedrosians Ceramic Tile and Stone
4285 N Golden State Boulevard
Fresno, CA 93722
Tel: 559-275-5000
Fax: 559-275-1753
bedrosians@aol.com
www.bedrosians.com

Belmont Rose Granite Corporation
7225 Woodbine Avenue, Ste 116A
Markham, ON L3R 1A3
Toll-free tel: 866-621-1281
Tel: 905-940-0700
Fax: 905-940-0522
belmontrose@rogers.com
www.proud.com/belmontrose

Canada Stone
7 - 3871 North Fraser Way
Burnaby, BC V5J 5G6
Tel: 604-430-3077
Fax: 604-266-3006
canadastone@canadastone.net
www.canadastone.net

CanAmerican Granite Corp.
1561 Erin Street
Winnipeg, MN R3E 2T2
Tel: 204-774-3378
Fax: 204-480-4175
info@canamericangranite.com
www.canamericangranite.com

Caruso Marble & Stone Corporation
169 Lodi Street
Hackensack, NJ 07601
Tel: 201-343-2840 or 201-343-2843
Fax: 201-343-2845
ccmarble59@yahoo.com
www.carusomarble.com

Cleveland Quarries
230 West Main Street
S. Amherst, OH 44001
Toll-free tel: 800-248-0250
Tel: 440-986-4501
Toll-free fax: 800-649-8669
Fax: 440-986-4531
amst@amst.com
www.clevelandquarries.com

Champlain Stone Ltd.
PO Box 650
Warrensburg, NY 12885
Tel: 518-623-2902
Fax:518-623-3088
info@champlainstone.com
www.champlainstone.com

Chen & Ragen
2100 E Union Street
Seattle, WA 98122
Tel: 206-325-2456
Fax: 206-325-7779
info@chenragen.com
www.chenragen.com

DeSantana Handcarved Stone
PO Box 907
Higley, AZ 85236
Tel: 866-932-2783
Fax: 480-664-8682
info@desantanastone.com
www.desantanastone.com

Dixie Cut Stone and Marble Inc.
6128 Dixie Highway
Bridgeport, MI 48722
Tel: 888-450-2858
Fax: 989-777-8791
www.dixiestone.com

Earth Products Inc.
5601 S. Madison Avenue
Indianapolis, IN 46227
Tel: 317-786-7980
Fax: 317-781-9044
gonzo@earthproductsinc.com
www.earthproductsinc.onsmartpages.com

Endless Mountain Stone Co.
Box 273 Brushville Road
Susquehanna, PA 18847
Tel: 570-465-7200
Fax: 800-672-3524
sales@endlessmountainstone.com
www.endlessmountainstone.com

Fieldstone Center, Inc.
990 Green Street
Conyers, GA 30012
Tel: 770-483-6770
Fax: 770-483-9255
poynter@mindspring.com
www.fieldstonecenter.com

Floor Tile and Slate Co.
1209 Carroll Avenue
Carrollton, TX 75006
Toll-free tel: 800-446-0220
Tel: 972-242-6647
Fax: 972-242-7253
sales@floortileandslate.com
www.floortileandslate.com

Genesee Cut Stone & Marble
5276 South Saginaw Road
Flint, MI 48507
Tel: 810-743-1800
Fax: 810-694-7901
stone@gcsm.com
www.gcsm.com

Green Mountain Soapstone Corporation
680 East Hubbardton Road
Castleton, VT 05735
802-468-5636
www.greenmountainsoapstone.com

Krukowski Stone
3781 City Road C.
Mosinee, WI 54455
Toll-free tel: 800-628-0314
Tel: 715-693-6300
Fax: 715-693-7223
JoanieW@krukowskistone.com
www.krukowskistone.com

Kuhlman Corporation
1845 Indian Wood Circle
Maumee, OH 43537
Toll-Free: 800-669-3309
Tel: 419-897-6000
Fax: 419-897-6061
info@kuhlman-corp.com
www.kuhlman-corp.com

Monarch Stone International
647 Camino de Los Mares,
Ste 108–230
San Clemente, CA 92673
Tel: 949-498-0971
Fax: 949-498-0941
info@monarchstone.net
www.monarchstone.net

Onyx Marble and Granite Ltd.
157 Toryork Drive
North York, ON M9L 1X9
Tel: 416-739-9339
Fax: 416-739-9449
info@onyxmarble.ca
www.onyxmarble.ca

Penn Big Bed Slate Co. Inc.
PO Box 184
8450 Brown Street
Slatington, PA 18080
Tel: 610-767-4601
Fax: 610-767-9252
pennbbs@aol.com
www.pennbigbedslate.com

Porphyry USA, Inc.
7945 Mac Arthur Boulevard 220
Cabin John, MD 20818
Tel: 301-229-8725
Fax: 301-229-8739
pavers@porphyryusa.com
www.porphyryusa.com

Rhodes Architectural Stone
2011 E Olive Street
Seattle, WA 98122
Tel: 206-709-3000
Fax: 206-709-3003
johnt@rhodes.org (West)
trevorb@rhodes.org (North, Midwest)
andrewg@rhodes.org (East)
www.rhodes.org

RMG Stone Products Inc.
PO Box 807
E Hubbardton Road
Castleton VT 05735
Tel: 802-468-5636
Fax: 802-468-8968
sales@rmgstone.com
www.rmgstone.com

Robinson Brick
Toll-free tel: 800-477-9002
Tel: 303-738-3000
Fax: 303-781-1818
www.robinsonbrick.com

Sheldon Slate
38 Farm Quarry Road
Monson, ME 04464
Tel: 207-997-3615
Fax: 207-997-2966
john@sheldonslate.com
Fox Road
Middle Granville, NY 12849
Tel: 518-642-1280
Fax: 518-642-9085
www.sheldonslate.com

StoneDecora
14622 Ventura Blvd, 442
Sherman Oaks, CA 91403
Tel: 818-986-1171
Fax: 818-907-0343
Info@StoneDecora.com
www.stonedecora.com

Barbara Tattersfield Design Inc.
Palm Beach, FL Showroom
Tel: 561-833-3443
Fax: 561-833-3414
Corona del Mar, CA Showroom
Tel: 949-675-0600
Fax: 949-675-0601
info@btattersfielddesign.com
www.btattersfielddesign.com

Topstone
3601 Range Road
Temple, TX 76504
Toll-free tel: 866-774-9197
Tel: 254-774-9197
Fax: 254-774-7631
Sales@PSITopstone.com
www.psitopstone.com

Universal Slate International Inc.
3821-9th Street S.E.
Calgary, AB T2G 3C7
Toll-free tel: 888-677-5283
zimmer@universalslate.com
www.universalslate.com

Vermont Quarries
Tel: 802-775-1065
Fax: 802-775-1369
vtquarries@aol.com
www.vermontquarries.com

Weser Brownstone (U.S.)
PO Box 323
New Haven, CT 06513
Tel: 203-467-2354
Fax: 203-469-2352
info@brownstone.us
www.brownstone.us

stone, precast & concrete

Advanced Rock Technologies
879 Brickyard Circle, Unit B9
PO Box 941
Golden, CO 80402
Toll-free tel: 800-898-7991
Tel: 303-278-9913
Fax: 303-278-9914
info@stonewallpanels.com
www.advancedrock.net

A&M Victorian Decorations, Inc.
2411 Chico Avenue
So. El Monte, CA 91733
Toll-free tel: 800-671-0693
Tel: 626-575-0693
Fax: 626-575-1781
am-vic@pacbell.net
www.aandmvictorian.com

Architectural Facades Unlimited
600 E Luchessa Avenue
Gilroy, CA 95020
Tel: 408-846-5350
Fax: 408-846-6911
www.architecturalfacades.com

Architectural Reproductions Inc.
525 N. Tillamook Street
Portland, OR 97227
Toll-free tel: 888-440-8007
Tel: 503 284 8007
Fax: 503-281.6926
admin@archrepro.com
www.archrepro.com

Architectural Stone Company
1900 Preston Road, Ste 267 PMB 100
Plano, TX 75093
Tel: 972-769-8379
Fax: 972-599-7638
contact@architecturalstone.net
www.architecturalstone.net

Belden Brick Sales & Service, Inc.
386 Park Avenue South
New York, NY 10016
Tel: 212-686-3939
Fax: 212-686-4387
david@NYNJBrick.com
www.nynjbrick.com

CaesarStone Canada
9151 Boulevard Saint-Laurent
Montreal, QC H2N 1N2
Tel: 514-382- 5180
Fax: 514-382 5990
tecnica@ciot.com
www.ciot.com
8899 Jane Street
Toronto, L4K 2M6
Tel: 416-739-800
Fax: 905-660-3818
totecnica@ciot.com
www.ciot.com

CaesarStone USA - U.S. Quartz Products
11830 Sheldon Street
Sun Valley CA 91352
Tel: 818-394-6000
Fax: 818-394-6006
Toll-free tel: 877-978-2789
info@caesarstoneus.com
New York
36–16 19th Ave
Astoria, NY 11105
Tel: 718-777-9780
Fax: 718-777-9784
ny@caesarstoneus.com
www.caesarstoneus.com

Cantera Especial
15332 Antioch Street 343
Pacific Palisades, CA 90272
Tel: 818-907-7170 or 800-564-8608
canteraespecial@aol.com
www.cantera-especial.com

Concrete Designs Inc.
3650 S. Broadmont Drive
Tucson, AZ 85713
Toll-free tel: 800-279-2278
Tel: 520-624-6653
Fax: 520-624-3420
concrete.designs@oldcastleapg.com
www.concrete-designs.com

Continental Cast Stone (East)
400 Cooper Road
W. Berlin, NJ 08091
Tel: 856-753-4000
Fax: 856-753-8700
Estimating Fax: 856-753-8711
sales@caststone.net
www.caststone.net

Continental Cast Stone of Kansas Inc.
22001 West 83rd Street
Shawnee, KS 66227
Tel: 800-989-7866
Fax: 913-422-7272
Estimating Fax: 913-422-3680
info@continentalcaststone.com
www.caststone.net

Continental Cast Stone of Texas Inc.
101 E Shady Grove Road
Grand Prairie, TX 75050
Tel: 866-871-7866
Fax: 972-871-1251
info@caststone-texas.com
www.caststone.net

Continental Cast Stone South
508 Stella Avenue
Savannah, GA 31415
Tel: 912-447-0207
Fax: 912-447-0338
info@caststone-south.com
www.caststone.net

Dixie Cut Stone and Marble Inc.
6128 Dixie Highway
Bridgeport, MI 48722
Tel: 888-450-2858
Fax: 989-777-8791
www.dixiestone.com

Doty & Sons Concrete Products Inc.
1275 E State Street
Sycamore, IL 60178
Toll-free tel: 800-233-3907
Tel: 815-895-2884
Fax: 815-895-8035
info@dotyconcrete.com
www.dotyconcrete.com

Dura Art Stone
11010 Live Oak
Fontana, CA 92337
Toll-free tel: 800-821-1120
Tel: 909-350-9000
Fax: 909-350-9632
duraartstone@duraartstone.com
www.duraartstone.com

GenStone
1075 S. Yukon Street, Ste 250
Lakewood, CO 80226
Toll-free tel: 800-955-7866
Fax: 303-854-0238
info@genstoneproducts.com
www.genstoneproducts.com

Haddonstone (USA) Ltd.
West
32207 United Avenue
Pueblo, CO 81001
Tel: 719-948-4554
Fax: 719-948-4285
East
201 Heller Place
Bellmawr, NJ 08031
Tel: 856-931-7011
Fax: 856-931-0040
info@haddonstone.com
www.haddonstone.com

Melton Classics Inc.
PO Box 465020
Lawrenceville, GA 30042
Toll-free tel: 800-963-3060
Tel: 800-963-3060
Fax: 770-962-6988
mclassics@aol.com
www.meltonclassics.com

Owens Corning Cultured Stone
One Owens Corning Parkway
Toledo, OH 43659
Tel: 800-255-1727
Fax: 866-213-3037
answers@answers.owenscorning.com
www.culturedstone.com

Precision Development
5711 Clarewood
Houston, TX 77081
Tel: 713-667-1310
Fax: 713-667-0515
www.pdcaststone.com

Stonecraft Industries Inc.
24777 Fraser Highway
Langley, BC L2Z 2L2
info@stonecraftindustries.com
www.stonecraftindustries.com

Stone Legends
301 Pleasant Drive
Dallas, TX 75217
Toll-free tel: 800-398-1199
Fax: 214-398-1293
sales@stonelegends.com
www.stonelegends.com

Stonex Cast Products Inc.
127 Squankum – Yellowbrook Road
Farmingdale, NJ 07727
Tel: 732-938-2334
Fax: 732-919-0918
info@stonexonline.com
www.stonexonline.com

Stonetile
4055 96th Avenue SE
Calgary, AB T2C 4T7
Tel: 403-279-2214
Fax: 403- 236-8979
www.stonetile.com

Sun Precast Co. Inc.
PO Box 423, Ridge Road
McClure, PA 17841
Tel: 570-658-8000
Fax: 570-658-8008
sales@sunprecast.com
www.sunprecast.com

Topstone
3601 Range Road
Temple, TX 76504
Toll-free tel: 866-774-9197
Tel: 254-774-9197
Fax: 254-774-7631
Sales@PSITopstone.com
www.psitopstone.com

Towne House Restorations Inc.
592 Johnson Avenue
Brooklyn, NY 11237
Tel: 718-497-9200
Fax: 718-497-3556
info@THRcaststone.com
www.thrcaststone.com

Thunderstone LLC
3300 South 6th Street
Lincoln, NE 68502
Tel: 402-420-2322
Fax: 402-420-2542
cconiglio@castone.com
www.castone.com

TriLite Stone Company
PO Box 308
Howard Lake, MN 55349
Toll-free tel: 888-786-6626
Tel: 320-543-2254
Fax: 320-543-2294
info@trilitestone.com
www.trilitestone.com

soapstone

Arc Stone
9020 Edgeworth Dr
Capitol Heights, MD 20743
Toll-free tel: 800-825-9086
Fax: 301-499-8875
www.arc-stone.com

Canadian Soapstone
3423 Torbolton Ridge Road
Woodlawn, ON K0A 3M0
Tel: 613-832-4256
Fax: 613-832-0539
soapstonecounters@operamail.com
www.soapstonecounters.com

Cornerstone Masonry
PO Box 83
Pray, MT 59065
Tel/Fax: 406-333-4383
info@warmstone.com
www.warmstone.com

Green Mountain Soapstone
680 E Hubbardton Road
PO Box 807
Castleton, VT 05735
Tel: 802-468-5636
vance@greenmountainsoapstone.com
www.greenmountainsoapstone.com

Soapstone Canada/
H.A. Ness & Co. Inc.
PO Box 11, Station U
Toronto, ON M8Z 5M4
Toll-free tel: 800-668-6377
Tel: 416-231-1645
Fax: 416-231-0231
glen.ness@soapstonecanada.com
www.soapstonecanada.com

alternative energy, solar, off grid, etc.

American Solar Energy Society
2400 Central Avenue, Ste A
Boulder, CO 80301
Tel: 303-443-3130
Fax: 303-443-3212
ases@ases.org
www.ases.org

Big Frog Mountain Corporation
100 Cherokee Boulevard, Ste 321
Chattanooga, TN 37405
Toll-free tel: 877-232-1580
Tel: 423-265-0307
Fax: 423-265-9030
www.bigfrogmountain.com

BP Solar
630 Solarex Court
Frederick, MD 21703
Tel: 301-698-4200
Fax: 301-698-4201
www.bpsolar.us

Carmanah Power Systems Group
Building 4, 203 Harbour Road
Victoria, BC V9A 3S2
Toll-free tel: 877-722-8877
Tel: 250-380-0052
Fax: 250-380-0062
info@carmanah.com
www.carmanah.com

Creative Energy Technologies Inc
2872 State Rt 10
Summit, NY 12175
Tel: 518-287-1428
info@cetsolar.com
www.cetsolar.com

Energy Outfitters, Ltd.
543 Northeast E Street
Grants Pass, OR 97526
Toll-free tel: 800-467-6527
Fax: 541-476-7480
Toll-free fax: 888-597-5357
Canada
Energy Outfitters, Ltd
17–220 Bayview Drive
Barrie, ON L4N 4Y8
Contact: Dan Lampkin
Eastern Canada Regional Manager
Toll-free tel: 877-357-6527
Fax: 519-632-7865
Tel (cell): 519-716-9777
info@energyoutfitters.com
www.energyoutfitters.com

EV Solar Products
2655 N Highway 89
Chino Valley, AZ 86323
Tel: 928-636-2201
info@evsolar.com
www.evsolar.com

Florida Alternative Energy Corp.
dba Healey & Associates
120 Venetian Way, Ste 16
Merritt Island, FL 32953
Tel: 321-452-2173
www.flaenergy.com

Heliodyne, Inc.
4910 Seaport Avenue
Richmond, CA
Tel: 510-237-9614
Fax: 510-237-7018
sales@heliodyne.com
www.heliodyne.com

Innovative Power Systems
(IPS Solar)
1153 Sixteenth Avenue SE
Minneapolis, MN 55414
Tel: 612-623-3246
www.ips-solar.com

Interstate Renewable Energy
Council
POB 1156
Latham, NY 12110–1156
Tel: 518-458-6059
info@irecusa.org
www.irecusas.org

Power Solutions
PO Box 607
Los Gatos, CA 95031–0607
CSL 833338
Tel: 408-656-9113
Fax 408-354-6951
info@solutionsforpower.com
www.solutionsforpower.com

SBT Designs
25581 IH-10 West
San Antonio, TX 78257
Toll-free tel: 800-895-9808
Tel: 210-698-7109
Fax: 210-698-7147
info@sbtdesigns.com
www.sbtdesigns.com

Sierra Solar Systems
563 C Idaho Maryland Road
Grass Valley, CA 95945
Toll-free tel: 888-667-6527
Tel: 530-273-6754
Fax: 530-273-1760
www.sierrasolar.com

Solar Energy Power Association
805 15th St, NW, Ste 510
Washington, DC 20005
Tel: 202-857-0898
Fax: 202-682-0559
info@solarelectricpower.org
www.solarelectricpower.org

Solar Services Inc.
1364 London Bridge Road, Ste 102
Virginia Beach, VA 23453
Tel: 757-427-6300
solserv@solarservices.com
www.solarservices.com

The Solar Store
2833 N Country Club Rd
Tucson, AZ 85716
Toll-free tel: 877-264-6374
Tel: 520-322-5180
Fax: 520-322-9531
www.solarstore.com

Sonideft Solar
217 Beaverbank Crossroad
Sackville, NS B4E 2E7
Tel: 902-865-2189
Fax: 902-484-7897
customerservice@heatwithsolar.com.
www.heatwithsolar.com

Windtech
Tel: 914-232-2354
Fax: 914-232-2356
info@windmillpower.com.
www.windmillpower.com

Zomeworks Corporation
PO Box 25805
1011A Sawmill Road
Albuquerque, NM 87125
Toll-free tel: 800-279-6342
Tel: 505-242-5354
Fax: 505-243-5187
zomework@zomeworks.com
www.zomeworks.com

staircases

Architectural Millwork
3522 Lucy Road
Millington, TN 38053
Tel: 901-327-1384
Tel (cell): 901-218-0616
Fax: 353-1645
samtickle@aol.com
www.a-millwork.com

Arcways
PO Box 763
Neenah, WI 54957
Toll-free tel: 800-558-5096
Fax: 920-725-2053
info@arcways.com
www.arcways.com

The Design-Build Store
U.S.
255 Sunrise Avenue, Ste 200
Palm Beach, FL 33480
Canada
1046 Yonge Street
Toronto, ON M4W 2L1
Toll-free tel 866-889-9919
info@designbuildstore.com
www.designbuildstore.com

Goddard Spiral Stairs
PO Box 502
Logan, KS 67646
Toll-free tel: 800-536-4341
spiralstaircases@ruraltel.net
www.spiral-staircases.com

Historical Arts & Casting, Inc.
5580 West Bagley Park Road
West Jordan, UT 84088
Toll-Free: 800-225-1414
Tel: 801-280-2400
Fax: 801-280-2493
info@historicalarts.com
www.historicalarts.com

Log Stairs by Premium Woodworks
605 Ridge Street
Sault Ste. Marie, MI 4 9783
Toll-free tel: 800 509 6871
Tel: 705 843 0609
info@spirallogstairs.com
www.spirallogstairs.com

Mylen Stairs
650 Washington Street
Peekskill, NY 10566
Toll-free tel: 800-431-2155
Tel: 914 739-8486
Fax: 914 739-9744
Info@MylenStairs.com
www.mylenstairs.com

New England Stair Company
www.newenglandstair.com

Rintal Canada
701 Brook Road N
PO Box 545
Cobourg, ON K9A 4L3
Rintal Inc.
171 Cooper Ave., Unit 110
Tonawanda, NY 14150
Toll-free tel: 877-816-2113
Tel: 905-373-0600
Fax: 905-373-0612
info@modularstairs.com
www.modularstairs.com

Sierra Stair Company, Inc.
3432 Swetzer Road
Loomis, CA 95650
Tel: 916-652-2800
info@sierrastair.com
www.sierrastair.com

Slabaugh Custom Stairs Ltd.
357 Fairview Avenue
Quakertown, PA 18951
Tel: 215-538-9000
Fax: 215-536-2291
sales@slabaughstairs.com
www.slabaughstairs.com

Stairways, Inc.
4166 Pinemont
Houston, TX 77018
Toll-free tel: 800-231-0793
Tel: 713-680-3110
Fax: 713-680-2571
swinfo@stairwaysinc.com
www.stairwaysinc.com

Stairsmiths.com
108 Commercial Avenue
Carrollton, GA 30117
Toll-free tel: 888-830-6880
Fax: 770-830-6885
sales@stairsmiths.com
www.stairsmiths.com

Steptoe and Wife Antiques, Ltd.
90 Tycos Drive
Toronto, ON M6B1V9
Toll-free tel: 800-461-0060
Tel: 416-780-1707
Fax: 416-780-1814
info@steptoewife.com
www.steptoewife.com

Unique Spiral Stairs
117 Benton Road
Albion, ME 04910
Toll-free tel: 800-924-2985
Tel: 207-437-2415
Fax 207-437-2196
usstairs@uniquespiralstairs.com
www.uniquespiralstairs.com

siding, wood

Aker Woods Company
14347 Mahaffey Drive
Piedmont, SD 57769
Tel: 605-786-1127
www.akerwoods.com

Buffalo Lumber
Toll-free tel: 877-960-9663
Tel: 615-563-8680
Chris@BuffaloLumber.com
www.buffalo-lumber.com

Cape Cod Finished Wood Siding
Toll-free tel: 800-565-7577
capecod@marwoodltd.com
www.capecod.ca

Carolina Colortones
10 Industrial Drive
Arden, North Carolina 28704
Toll-free tel: 800-948-4349
Tel: 828-687-9510
Fax: 828-687-9532
info@carolinacolortones.com
www.carolinacolortones.com

Granville Manufacturing Company
Route 100
Granville, VT 05747
Tel: 802-767-4747
Fax: 802-767-3107
www.woodsiding.com

The Knotty Pine and Cedar Company
2337 W Houghton Lake Drive
Houghton Lake, MI 48629
Tel: 989-366-8811
contact@logsidinghomes.com
www.logsidinghomes.com

Kootenay Innovative Wood Ltd.
PO Box 130
3020 South Slocan Station Road
South Slocan, BC V0G 2G0
Tel: 250-359-8050
Fax: 250-359-8052
www.kiwood.com

Maibec
660 Lenoir Street
Sainte-Foy, PQ G1X 3W3
Tel: 418-659-3323
info-prod@maibec.com
www.maibec.com

Michigan Prestain
1701 Clyde Park SW
Grand Rapids, MI 49509
Toll-free tel: 800-641-9663
Tel: 616-241-1440
Fax: 616-241-6661
siding@michiganprestain.com
www.michiganprestain.com

Midwest Cyprus Products
9195 22nd Street
Perry, KS 66073
Toll-free tel: 800-545-8884
Fax: 785-597-5637
www.midwestcypress.com

Redwood Products
2727 W Main
Oklahoma City, OK 73107
Toll-free tel: 866-318-3325
Tel: 405-235-DECK
redwoodproducts_@excite.com
www.redwoodproducts.com

Western Red Cedar Lumber Association
Seattle Office
PMB 1705
914 - 164th Street, SE, B12
Mill Creek, WA 98012-6339
Toll-free tel: 877-316-8845
Telephone: 425-316 8845
Fax: 425-316-3979
mackie@wrcla.org
New York Office
PO Box 952
Riverhead, NY 11901-2136
Toll-free tel: 800-266-1910
Telephone: 631-643-9725
Fax: 631-643-7252
burke@wrcla.org
www.cedar-siding.org

White Cedar Shingles
PO Box 3039
Gaylord, MI 49734
Tel: 989-731-1757
Fax: 989-732-9151
kcskarl@charterinternet.com
www.whitecedarshingles.com

siding, fiber cement

BuildDirect
1900-570 Granville Street
Vancouver, BC V6C 3P1
Toll-free tel: 877-631-2845
Tel: 604-662-8100
Fax: 604-662-8142
Toll Free Fax: 877-631-2845
sales@builddirect.com
www.builddirect.com

CertainTeed Corporation
PO Box 860
750 East Swedesford Road
Valley Forge, PA 19482
Toll Free (building professional): 800-233-8990
Toll Free (consumer): 800-782-8777
Tel: 610-341-7000
Fax: 610-341-7777
www.certainteed.com

James Hardie
26300 La Alameda, Ste. 250
Mission Viejo, CA 92691
Toll-free tel: 888 542-7343
info@JamesHardie.com
www.jameshardie.com

Michigan Prestain
1701 Clyde Park SW
Grand Rapids, MI 49509
Toll-free tel: 800-641-9663
Tel: 616-241-1440
Fax: 616-241-6661
siding@michiganprestain.com
www.michiganprestain.com

tiles, metal

Aji Tiles
601 Montrose Avenue
South Plainfield, NJ 07080
Tel: 877-658-4537
www.ajitiles.com

American Marazzi Tile, Inc.
359 Clay Road
Sunnyvale, TX 75182
Tel: 972-232-3801
Fax: 972-226-5629
contact@marazzitile.com
www.marazzitile.com

Jeffrey Court, Inc.
620 Parkridge Avenue
Norco, CA 92860
Tel: 951-340-3383
Fax: 951-340-2429
www.jeffreycourt.com

Crossville, Inc.
PO Box 1168
Crossville, TN 38557
Tel: 931-484-2110
www.crossvilleinc.com

Metaphor Bronze
PO Box 176
Belfast, ME 04915
Toll-free tel: 800-907-8200
inpho@metaphorbronze.com
www.metaphorbronze.com

Ridalco Industires Inc.
1551 Michael Street
Ottawa, ON K1B 3T4
Tel: 613-745-9161
Fax: 613-745-6452
Toll-free fax: 800-268-6526
info@ridalco.com
www.ridalco.com

Rigidized Metals Corporation
658 Ohio Street
Buffalo, NY 14203
Toll-free tel: 800- 836-2580
Tel: 716-849-4760
Fax: 716-849-0401
www.rigidized.com

Saint-Gaudens Metal Arts
1784 La Costa Meadows Drive, Ste 104
San Marcos, CA 92069
Tel: 760-891-0300
Fax 760-752-9914
Sales@VSGMetalArts.com
www.vsgmetalarts.com

Talisman Handmade Tiles
Lowitz & Company
4401 N. Ravenswood Avenue
Chicago, IL 60640
Tel: 773-784-2628
talisman@lowitzandcompany.com
www.lowitzandcompany.com

tiles, general

Accent Glass Tile
PO Box 230322
Tigard, OR 97223–0322
Fax: 503-213-5973
info@accentglasstile.com
accentglasstile.com

Agape Tile
531 E Main Street
Albemarle, NC 28001
Tel: 704-982-7673
agapetile@agapetile.com
www.agapetile.com

American Restoration Tile
11416 Otter Creek South Road
Mabelvale, AR 72103
Tel: 501-455-1000
Fax: 501-455-1004
bebyrd@restorationtile.com
www.restorationtile.com

Artisan Tile and Marble Company of New Jersey, Inc.
468 Elizabeth Avenue
Somerset, NJ 08873-5200
Tel: 732-764-6700
Fax: 732-764-6767
info@artisannj.com
www.artisannj.com

Arts and Crafts Tile.Com
702 Camino De Los Mares
San Clemente CA 92673
Tel: 949-292-0552
info@artsandcraftstile.com
www.artsandcraftstile.com

Lorna Auerbach Associates
2416 Wilshire Boulevard
Santa Monica, CA 90403
Tel: 310-453-5400
www.auerbachassociates.com

Bambino Ceramics
1441 NW Pioneer Hill Road
Poulsbo, WA 98370
Tel: 360-779-2553
eb@bambinoceramics.com
www.bambinoceramics.com

Batik Tile
Toll-free tel: 888-692-2845
joanne@batiktile.com
www.batiktile.com

Bedrosians Ceramic Tile and Stone
4285 N Golden State Boulevard
Fresno, CA 93722
Tel: 559-275-5000
Fax: 559-275-1753
bedrosians@aol.com
www.bedrosians.com

California Pottery and Tile Works
859 E 60th Street
Los Angeles, CA 90001
Tel: 323-235-4151
Fax: 323-235-4161
califpot@aol.com
www.malibutile.com

Clay Squared to Infinity
34 13th Avenue NE
Minneapolis, MN 55413
Tel: 612-781-6409
josh@claysquared.com

Cornerstone Ceramics
5076 Alhambra Avenue
Los Angeles, CA 90032
Tel: 323-221-6818
info@cornerstoneceramics.com
www.cornerstoneceramics.com

Country Floors, Inc.
8735 Melrose Avenue
Los Angeles, CA 90069
Tel: 310-657-0510
Fax: 310- 659-6470
info@countryfloors.com
www.countryfloors.com

Jeffrey Court, Inc.
369 Meyer Circle
Corona, CA 92879
Tel: 951-340-3383
Fax: 951-340-2429
www.jeffreycourt.com

CR STUDIO 4
Tel: 760-731-9040
tilestudio@aol.com

Designs in Tile
PO Box 358
Mount Shasta, CA 96067
Tel: 530-926-2629
Fax: 530-926-6467
info@designsintile.com
www.designsintile.com

Dodge Lane Potters' Group
96 E Dodge Lane
Sonora, CA 95370
Tel: 209-532-9124
Fax: 209-532-7713
Toll-free tel: 877-511-9124
tiles@dlpg.com

Diamond Tech Glass Tiles
5600 Airport Boulevard, Ste C
Tampa, FL 33634
Toll-free tel: 800-937-9593
Tel: 813-806-2923
info@dtglasstiles.com
www.dtglasstiles.com

DuQuella Tile & ClayWorks
PO Box 90065
Portland, OR 97290
Toll-free tel: 866-218-8221
Tel: 503-256-8330
Fax: 503-257-4773
www.tiledecorative.com

Dy's Art Tiles
14200 SE 42nd Terrace
Summerfield, FL 34491
Tel: 352-347-9809
www.dyztilz.com

Earth Marks
7885 North Bradburn Boulevard
Westminster, CO 80030
Tel: 303-487-9222
jennifer@earth-marks.com
www.earth-marks.com

Bettina Elsner Artistic Tiles
PO Box 1185
Indian Rocks Beach, FL 33785
Tel: 727-596-3038
bettina1@elsnertile.com
www.elsnertile.com

European Reclamation
4520 Brazil Street
Los Angeles, CA 90039
Tel: 818-241-2152
Fax: 818-547-2734
htc@wgn.net
www.historictile.com

Fly on the Wall Tile
Michele Rock
1735 Wellesley Avenue
St. Paul, MN 55105
Tel: 651-699-5792
rockpane@comcast.com

Fraser Clay Works, Inc.
64 Myrtlewood Drive
Mountain Home, AR 72653
Tel: 870-492-5031
Fax: 870-492-4754
fraserclayworks@cox-internet.com
www.fraserclayworks.com

Hand Carved Clay
Deb LeAir
1047 Earl Street
St. Paul, MN 55106
Tel: 651-793-3426
debleair@yahoo.com

Handcraft Tile Inc.
1126 Yosemite Drive
Milpitas, CA 95035
Toll-free tel: 877-262-1140
Tel: 408-262-1140
Fax: 408-262-1441
info@handcrafttile.com
www.handcrafttile.com

Hand Made Mexican Tiles
California Office
7595 Caroll Road
San Diego, CA 92121
Tel: 858-689-9596
Fax: 858-689-9597
Connecticut Office
21 Bernhard Road
North Haven, CT 06473
Tel: 203-946-0861
Fax: 203-946-0862
www.mexicanhandcraftedtile.com

Joan Rothchild Hardin
4-393 West Broadway
New York, NY 10012
Tel: 212-966-9433
Fax: 212-431-9196
joan@hardintiles.com
www.hardintiles.com

Hastings Tile & Bath Inc.
230 Park Avenue South
New York, NY 10003
Tel: 212-674-9700
Fax: 212-674-8083
nycsales@hastings30.com
Chicago
120 Merchandise Mart
Chicago, IL 60654
Tel: 312-527-0565
chicagosales@hastings30.com
www.hastingstilebath.com

Hispanic Designe'
Showroom and Warehouse
6125 N Cicero Avenue
Chicago, IL 60646
Tel: 773-725-3100
Fax: 773-725-0167
hispdesn@hispanicdesigne.com
www.hispanicdesigne.com

imagewares inc.
Box 990
Nobleton, ON L0G 1N0
Tel: 905-859-7643
Fax: 905-859-7742
info@imagewares.com
www.imagewares.com

Imagine Tile
1515 Broad Street
Bloomfield, NJ 07003
Toll-free tel: 800-680-TILE
www.imaginetile.com

Interstyle Ceramic and Glass Ltd.
3625 Brighton Avenue
Burnaby, BC V5A 3H5
Tel 604-421-7229
Fax 604-421-7544
info@interstyle.ca
www.interstyle.ca

Jacobs Tile Studios
1472 67th Street
Brooklyn, NY 11219
Tel: 917-734-6034
info@jacobstiles.com
www.jacobstiles.com

Lavabo
lavabo@giant.co.uk
www.giant.co.uk

Levy Laroque
One Cottage Street
Easthamptom, MA 01027
Tel: 413-527-5040
marcia@levylarocque.com
www.tileandwood.com

London Tile Co.
65 Walnut Street
New London, OH 44851
Tel: 419-929-1551
Toll-free tel: 888-757-1551
info@londontile.com
www.londontile.com

Lowitz and Company
4401 N Ravenswood Avenue
Chicago, IL USA
Tel: 773-784-2628
info@lowitzandcompany.com

Majolica Mosaics
Tel: 207-772-0802
akearney@gwi.net
www.majolicamosaics.com

Manet Tiles
7595 San Fernando Road
Burbank, CA 91505
Toll-free tel: 818-767-1657
Fax: 818-767-1752
manettiles@sbcglobal.net
www.manettiles.com

The Maya Romanoff Corporation
1730 W Greenleaf
Chicago, IL 60626
Tel: 773-465-6909
Fax: 773-465-7089
customerservice@mayaromanoff.com
www.mayaromanoff.com

McIntyre Tile Company, Inc.
PO Box 14
Healdsburg, CA 95448
Tel: 707-433-8866
www.mcintyre-tile.com

Mexican Handcrafted Tile/ MC Designs
7595 Carroll Road
San Diego, CA 92121
Tel: 858-689-9596
andi@mexicanhandcraftedtile.com
www.mexicanhandcraftedtile.com

Mercury Mosaics
Mercedes Mattila
125 West Broadway
Minneapolis, MN 55411
Tel: 612-250-3299
www.mercurymosaics.com

Mission Guild Studio
222 Main Street
Worcester, NY 12197
Tel/Fax: 607-397-1808
www.angelfire.com/ny5/missionguild/

Mission Tile West
853 Mission Street
South Pasadena, CA 91030
Tel: 626-799-4595
Fax: 626-799-8769
southpasadena@missiontilewest.com
www.missiontilewest.com

Mission Tile West Design Studio
1207 4th Street, Ste 100
Santa Monica, CA 90401
Tel: 310-434-9697
Fax: 310-434-9795
santamonica@missiontilewest.com
www.missiontilewest.com

Monterey Ceramics
3130 N Del Mar Avenue
Rosemead, CA 91770
Tel: 626-288-8693
Fax: 626-288-8105
info@montereyceramictile.com
www.montereyceramictile.com

Motawi Tile Works
170 Enterprise Drive
Ann Arbor, MI 48103
Tel: 734-213-0017
Fax: 734-213-2569

Moving Color Tiles
2351 Sunset Boulevard, Ste 170-425
Rocklin, CA 95765
Tel: 916-337-6296
sales@movingcolor.net
www.movingcolor.net

North Prairie Tileworks
2845 Harriet Avenue S
Minneapolis, MI 55408
Tel: 612-871-3421
Fax: 612-871-2923
Information@handmadetile.com
www.handmadetile.com

Ocean Stones
388 Carlaw Avenue
Toronto, ON M4M 2T4
Tel: 416-463-4805
Toll-free tel: 866-463-5832
Fax: 416-463-1208

Olympia Tile
1000 Lawrence Ave West
Toronto, ON M6B 4A8
Tel: 416-785-9555
Fax: 416-785-3204

Pave Tile & Stone Inc.
10 West Street
West Hatfield, MA 01088
Toll-free tel: 800-239-6437
Toll-free fax: 800-560-5787
Tel: 413-247-7677
Fax: 413-247-8383
info@pavetile.com
www.pavetile.com

Peace Valley Tile
64 Beulah Road
New Britain, PA 18901
Tel: 215-340-0888
info@peacevalleytile.com
www.peacevalleytile.com

Eric Pilhofer/PilhoferWerks
1618 Central Avenue NE, Ste 25
Minneapolis, MN 55413
Tel: 952-484-6990
Fax: 952-945-0451
info@pilhoferwerks.com
www.pilhoferwerks.com

Purple Sage Collections
1526B Howell Mill Road
Atlanta, GA 30318
Toll-free tel: 866-357-4657
Tel: 404-351-4445
Fax: 404-351-8250
info@purplesagecollections.com
www.purplesagecollections.com

Quarry Tile Company
6328 E Utah Avenue
Spokane, WA 99212
Tel: 509-536-2812
Fax: 509-536-4072
sales@quarrytile.com
www.quarrytile.com

Revival Tileworks
PO Box 230191
Encinitas, CA 92023
Tel: 760-730-9141
inquiries@revivaltileworks.com
www.revivaltileworks.com

Rocheford Handmade Tile
3315 Garfield Avenue S
Minneapolis, MN 55408
Tel: 612-824-6216
Fax: 612-821-8825
sales@housenumbertiles.com
www.housenumbertiles.com

Charles Rupert: The Shop
2005 Oak Bay Avenue
Victoria, BC V8R 1E5
Tel: 250-592-4916
Fax: 250-592-4999
theshop@charles-rupert.com
www.charles-rupert.com

Ann Sacks
8120 NE 33rd Drive
Portland, OR 97211
Toll-free tel: 800-278-8453
www.annsacks.com

Saint-Gaudens Metal Arts
1784 La Costa Meadows Drive, Ste 104
San Marcos, CA 92069
Tel: 760-891-0300
Fax 760-752-9914
Sales@VSGMetalArts.com
www.vsgmetalarts.com

Solar Antique Tiles
306 E. 61st
New York, NY 10021
Tel: 212-755-2403
Fax: 212-980-2649
PLEITAO@aol.com
www.solarantiquetiles.com

Sonoma Cast Stone
133 Copeland Street
Petaluma, CA 94952
Toll-free tel: 877-939-9929
Tel: 707-283-1888
Fax: 707-283-1899
sales@sonomastone.com
www.sonomastone.com

Motawi Tile

Sumon Company
Maple Valley, WA
Tel: 425-432-6492
Fax: 425-432-3921
sumon@nwlink.com
www.sumon.com

Surving Studios
17 Millsburg Road
Middletown, NY 10940
Tel: 845-355-1430
info@surving.com
Toll-free tel: 800-768-4954
Fax: 845-355-1517
www.surving.com

Swan Tile, Inc.
4737 Abbott Avenue South
Minneapolis, MN 55410
Tel: 612-929-9720
Fax 612-922-3054
jjjswan@msn.com

Tactile Geometrics
310 7th Avenue
Brooklyn, NY 11215
Tel: 718-768-5486
info@tactilegeometrics.com
www.tactilegeometrics.com

Talisman Handmade Tiles
Lowitz & Company
4401 N. Ravenswood Avenue
Chicago, IL 60640
Tel: 773-784-2628
talisman@lowitzandcompany.com
www.lowitzandcompany.com

Terrapin Tile
PO Box 337115
Greeley, CO 80633-0619
Tel: 970-402-5067
terrapintile@comcast.net
www.terrapintile.com

Trikeenan Tileworks, Inc.
PO Box 22
Keene, NH 03431
Sales and Design tel: 603-355-2961
Factory tel: 603-352-4299
Fax: 603-352-9843
showroom@trikeenan.com
www.trikeenan.com

United States Ceramic Tile Company
4244 Mount Pleasant Street NW, Ste 100
North Canton, OH 44720
Toll-free tel: 800-321-0684
Tel: 330-649-5000
Fax: 330-649-5055
info@usctco.com
www.usctco.com

Urban Jungle Art & Design
San Diego, CA
Tel: 619-299-1644
Fax: 619-299-6644
deirdre@UrbanJungleArt.com
www.urbanjungleart.com

Heather Weisz
1775 Hillsdale Road
Southhapton, PA 18966
Tel: 215-322-5128
Fax: 215-322-5062
HFWeisz@comcast.net
www.members.aol.com/hfweisz

Whistling Frog Tile Company
311 E Maplehurst
Ferndale, MI 48220
Tel: 248 542 1112
Fax: 810 892 9549
whistlefrog@comcast.net
www.whistlefrog.com

Wiseman & Spaulding Designs
12 Shaw Hill Road
Hampden, ME 04444
Tel: 207-862-3513
Fax: 207-862-4513
www.antiquitytile.com

timberframe homes

1867 Confederation Log Homes
PO Box 9
Bobcaygeon, ON K0M 1A0
Tel: 705-738-5131
Fax: 705-738-5283
rkinsman@confederationloghomes.com
www.confederationloghomes.com

Artisan Log Works
6585 Farm to Market Road
Whitefish, MT 59937
Tel: 406-250-3664
Fax: 406-862-5647
tomnixon@centurytel.net
www.artisanlogworks.com

Art Maison
Sylvain Metivier
161 Chemin du Grand Bois
St-Etienne de Bolton, PQ J0E 2E0
Tel: 450-297-3513
Fax: 450-297-3513
artmaison@sympatico.ca

Bitterroot Timber Frames
567 Three Mile Creek Road,
Stevensville, MT 59870
Tel: 406-777-5546
Fax: 406-777-5547
bitterroot@bitterroottimberframes.com
www.bitterroottimberframes.com

Blue Ox Logcrafters
PO Box 644
Carbondale, CO 81623
Tel: 970-963-3689
Fax: 970-963-3689
blueox@rof.net
www.blueoxlogcrafters.com

Bromley Log Homes
81 Whitney Drive
Cody, WY 82414
Tel: 307-587-5010
Fax: 307-587-9301
mbromley@tritel.net
www.bromleyloghomes.com

Continental Log Homes
Box 185
Pemberton, BC V0N 2L0
Tel: 604-894-6449
Fax: 604-894-5444
sales@continentalloghomes.com
www.continentalloghomes.com

Davidson Log & Timber Artisans Inc.
RR 1, 54 Rama-Dalton Boundary Road
Washago, ON L0K 2B0
Tel: 705-833-1203
Fax: 705-833-1274
matt@davidsonloghomes.com
www.davidsonloghomes.com

Douglas Lukian Inc.
785 Ch. St. Adolphe
Morin Heights, QC J0R 1H0
Tel: 450-226-6076
Fax: 450-226-2043
douglas_lukianinc@sympatico.ca

Farrell Structures, LLC
22 Cimmaron Drive
Pisgah Forest, NC 28768
Tel: 828-885-7969
Fax: 828-885-7967
sfarrell@citcom.net
www.farrellstructures.com

Forbes Landing Log Homes
Box 807
Campbell River, BC V9W 6Y4
Tel: 250-286-3725
Fax: 250-286-3720
info@forbeslandingloghomes.ca
www.forbeslandingloghomes.bc.ca

Herwig Log Homes & Timber Framing Inc.
20513 Hwy 17
RR3, Cobden ON K0J 1K0
Tel: 613-638-1500
Fax: 613-735-1724
herwig@herwigloghomes.com
www.herwigloghomes.com

Laverty Log Homes
3343 Heron Road
New Hamburg, ON N3A 3C4
Tel: 519-662-4479
earl@lavertyloghomes.com
www.lavertyloghomes.com

Living Log and Timber Ltd.
Box 729 Breton, AB T0C 0P0
Tel: 780-696-3415
Fax: 780-696-3639
livinlog@telusplanet.net
www.livinglogandtimber.com

Minde Log Construction Inc.
2112 E Pioneer Road
Duluth, MN 55804
Tel: 218-525-1070
Fax: 218-525-3181
mindelog@cpinternet.com
www.mindelog.com

Mountain Spring Log Homes
130 Mcalpine Road
Maynooth, ON K0L 2S0
Tel: 613-338-3008
Fax: 613-321-1229
mountainspring@bellnet.ca
www.mountain-spring.com

Pioneer Log Homes of British Columbia Ltd.
841A MacKenzie Avenue
Williams Lake, BC V2G 3X8
Toll-free tel: 877-822-5647
Tel: 250-392-5577
Fax: 250-392-5581
pioneerlog@telus.net
www.pioneerloghomesofbc.com

Shady Grove Log & Timber Builders, LLC
PO Box 232 Whitefish, MT 59937
Tel: (Pat) 406-212-0388,
(Paul) 406-261-0256
Fax: 406-862-0504
info@shadygrovelog.com
www.shadygrovelog.com

Silver Plume Log & Timberworks L.L.C.
PO Box 1382
Idaho Springs, CO 80452
Tel: 303-567-4207
Fax: 303-567-4305
fishingwolf@worldnet.att.net

Timber Framers Guild
176 Pratt Road
Alstead, NH 03602-9722
Tel: 888-453-0879
Fax: 888-453-0879
joel@tfguild.org
www.tfguild.org

Timberline Builders Inc.
435 Mountain View Drive
Rollinsville, CO 80474
Tel: 303-258-1887
Fax: 303-258-1887
mark@tbiloghomes.com
www.tbiloghomes.com

Treehouse Log Homes Ltd
Box 866
Lumby, BC V0E 2G0
Tel: 250-547-9444
Fax: 250-547-9444
treehouse@loghome.bc.ca
www.loghome.bc.ca

Winterwood Custom Builders
530 Greer Road, RR3
Utterson, ON P0B 1M0
Toll-free tel: 800-814-5945
Tel: 705-385-8199
Fax: 705-385-8737
winterwood@on.aibn.com
www.winterwood.ca

vitrolite

Timothy J. Dunn
2402 Bredell Street
Louis, MO 63143
Tel: 314-645-4317
Fax: 314-647-6276
vitrolite@earthlink.com.
www.vitrolitespecialist.com

walls, panels, composites, etc.

3form
2300 South 2300 W, Ste B
Salt Lake City, UT 84119
Toll-free tel: 800-726-0126
Tel: 801-649-2500
Fax: 801-649-2699
info@3-form.com
www.3-form.com

American Acrylic Corporation
400 Sheffield Avenue
West Babylon, NY 11704
Toll-free tel: 800-627-9025
Tel: 631-422-2200
Fax: 631-422-2811
www.thomasregister.com
www.americanacrylic.com

Richelieu Hardware
7900, West Henri-Bourassa
Montreal, PQ H4S 1V4
Toll-free tel in Quebec: 866-832-4040
Toll-free tel Canada: 800-361-6000
Toll-free tel from the US: 800-619-5446
info@richelieu.com
www.richelieu.com

Robin-Reigi
48 W 21st Street, Ste 1002
New York, NY 10010
Tel: 212-924-5558
Fax: 212-924-2753
robin@robin-reigi.com
www.robin-reigi.com

weathervanes, cupolas

American Weathervane
333 Mamaroneck Avenue, Ste 368
White Plains, NY 10605
Toll-free tel: 800-310-0043
Tel: 914-428-6564
service@americanweathervane.com
www.americanweathervane.com

Cape Cod Cupola Co.
78 State Road, Rt. 6
North Dartmouth, MA 02747
Tel: 508-994-2119
jebernier@capecodcupola.com
www.capecodcupola.com

SkyArt Studio Weathervanes
290 Pratt Street
Meriden, CT 06450
Tel: 203.630.9171
Fax: 203.886.0029
info@skyartstudio.com
www.skyartstudio.com

Vintage Copper
3215 N 448 Road
Salina, OK 74365
Toll-free tel: 877-220-5355
Tel: 918-434-6675
Fax: 918-434-6676
info@vintagecopper.com
www.vintagecopper.com

Weathervanes of Maine
Searsport, ME 04974
Tel: 207-548-0050
weathervanesofmaine.com
West Coast Weather Vanes
377 Westdale Drive
Santa Cruz, CA 95060
Toll-free tel: 800-762-8736
Tel: 831-425-5505
info@westcoastweathervanes.com
www.weathervanesofmaine.com
www.westcoastweathervanes.com

West Coast Weather Vanes
377 Westdale Drive
Santa Cruz, CA 95060-9446
Toll-free tel: 800-762-8736
Tel: 831-425-5505
Fax: 831-425-5514
info@westcoastweathervanes.com
www.westcoastweathervanes.com

windows

Adams Architectural Wood Products
300 Trails Road
Eldridge, IA 52748
Toll-free tel: 888-285-8120
Tel: 563-285-8000
Fax: 563-285-8003
www.adamsarch.com

Allied Window Inc.
11111 Canal Road
Cincinnati, OH 45241-1861
Toll-free tel: 800-445-5411
Fax: 513-559-1883
info@invisiblestorms.com
www.alliedwindow.com

Architectural Traditions
9280 E. Old Vail Road
Tucson, AZ 85747
Tel: 520-574-7374
www.architecturaltraditions.com

Artistic Windows & Doors
10 South Inman Avenue
Avenel, NJ 07001
Toll-free tel: 800-278-3667
Tel: 732-726-9400
Fax: 732-726-9494
info@artisticdoorsandwindows.com
www.artisticdoorsandwindows.com

Asselin U.S.A., Asselin Old World Craftsmen Inc.
2870 Peachtree Road NW 714
Atlanta, GA 30305–2918
Tel: 404-419-6114
Fax: 404-419-6116
contact@asselinusa.com
www.asselinusa.com

Cityproof Corporation
10–11 43rd Avenue
Long Island City, NY 11101
Tel: 718-786-1600
Fax: 718-786-2713
Cityproof@aol.com
www.cityproof.com

Dazzle Glazz
6605 Jefferson Highway
Baton Rouge, LA 70806–8104
Toll-free tel: 888-717-9484
Tel: 225-216-9484
duncan@dazzleglazz.com
www.dazzleglazz.com

Enjo Architectural Millwork
16 Park Avenue
Staten Island, NY 10302
Toll-free tel: 800-437-3656
sales@enjo.com
www.enjo.com

Grabill Windows & Doors Incorporated
7463 Research Drive
Almont, MI 48003
Tel: 810-798-2817
Fax: 810-798-2809
tgrabill@bignet.net
www.grabillwindow.com

GreenWood Workshop, LLC
4045 Richland Avenue
Louisville, KY 40207
Tel: 502-894-0501
Sales@GreenWoodWorkshop.com
www.greenwoodworkshop.com

JELD-WEN Inc.
PO Box 1329
Klamath Falls, OR 97601
Toll-free tel: 800-535-3936
www.jeld-wen.com

Joseph Millworks
37123 Hansen Ln.
Baker City, OR 97814
Tel/Fax: 541-894-2347
randy@josephmillworks.com
www.josephmillworks.com

Loewen Windows and Doors
77 Highway 52 West
Box 2260
Steinbach, MB R5G 1B2
Toll-free tel: 800-563-9367
Tel: 204-326-6446
www.loewen.com

Marvin Windows and Doors
PO Box 100 Warroad, MN 56763
Toll-free tel (U.S.): 888-537-7828
Toll-free tel (Canada): 800-263-6161
www.marvin.com

Molyneux Designs
PO Box 1637 Weaverville, CA 96093
Tel: 530-623-2268
Fax: 530-623-1671
md@CustomMade.com
www.molyneuxdesigns.com

Mon-Ray, Inc
801 Boone Avenue N
Minneapolis, MN 55427–4432
Toll-free tel: 800-544-3646
www.monray.com

North Star Manufacturing
40684 Talbot Line
St. Thomas, ON N5P 3T2
Toll-free tel: 800-265-5701
Tel: 519-637-7899
Toll-free fax: 800-681-1064
Fax: 519-637-3403
www.northstarwindows.com

Parrett Windows
810 Second Avenue E
PO Box 440
Dorchester, WI 54425–0440
Toll-free tel: 800-541-9527
Fax: 715-654-6555
www.parrettwindows.com

Pella Corporation
102 Main Street
Pella, IA 50219
Toll-free tel: 800-374-4758
www.pella.com

Restoration Works
1345 Stanford Drive
Kankakee, IL 60901
Tel: 815-937-0556
Fax: 815-937-4072
gwallace@restorationworksinc.com
www.restorationworksinc.com

Re-View
1235 Saline
North Kansas City, MO 64116
Tel: 816-746-9339
Fax: 816-746-9331
www.re-view.biz

Smith Restoration Sash
122 Manton Avenue, Unit 714
Providence, RI 02909
Tel: 401-351-1222
Fax: 401-351-1245
windows@smithrestorationsash.com
www.smithrestorationsash.com

Sperlich Lighting Art Glass and Doors
7005 N Waterway Drive
Miami, FL 33155
Tel: 786-388-7522
Fax: 786-388-7523
sales@sperlich.com
www.sperlich.com

J. Sussman Inc.
109-10 180th Street
Jamaica, NY 11433
Tel: 718-297-0228
Fax: 718-297-3090
webmaster@jsussmaninc.com
www.jsussmaninc.com

Weston Millwork Company
722 Washington Street
Weston, MO 64098
Tel: 816-640-5555
Fax: 816-386-5555
info@westonmillwork.com
www.westonmillwork.com

Wood Window Workshop
839 Broad Street
Utica, NY 13501
Tel: 800-724-3081
Fax: 315-733-0933
mike@woodwindowworkshop.com
www.woodwindowworkshop.com

Zeluck Incorporated LLC
5300 Kings Highway
Brooklyn, NY 11234
Toll-free tel: 800-233-0101
Tel: 718-251-8060
Fax: 718-531-2564
California
Tel: 310-456-6304
Fax: 310-456-6834
Florida
Tel: 561-833-0092
Fax: 561-832-9583
info@zeluck.com
www.zeluck.com

woodwork, custom

Advanced Commercial Contracting, Inc.
22473 Prat's Dairy Road
Abita Springs, LA 70420
Tel: 504-943-3368
Fax: 504-943-3369
adcomc@advancedmillwork.com
www.advancedmillwork.com

Architectural Woodworking & Design
4341 Roslyn Road
Downers Grove, IL 60515
Tel: 630-810-1604
Fax: 630-960-0055
info@poccidesign.com
www.poccidesign.com

Bramms Custom Cabinetry
10 Ontario Road
St. Thomas, ON N5P 3N4
Toll-free tel: 877-663-2537
Tel: 519-631-8138
Fax: 519-631-8195
Braams@on.aibn.com
Michigan Showroom
114 S. Old Woodward Avenue
Birmingham, MI 48009
Tel: 248-646-3395
Fax: 248-646-3403
www.braams.ca

Byrne Custom Woodworking
17501 W. 98th St 28–62
Lenexa, KS 66219
Tel: 913-894-4777
Fax: 913-894-4779
info@byrnecustomwood.com
www.byrnecustomwood.com

Crown Point Cabinetry
PO Box 1560
Claremont, NH 03743
Toll-free tel: 800-999-4994
Fax: 800-370-1218
info@crown-point.com
www.crown-point.com

Cumberland Woodcraft Co.
PO Drawer 609
Carlisle, PA 17013–0609
Toll-free tel: 800-367-1884
Tel: 717-243-0063
www.cumberlandwoodcraft.com

DiaDot Custom Millwork
PO Box 113 Bldg 22
Sussex, NJ 07461
Tel: 973-875-5669
Fax: 973-875-5634
generalbox@diadot.com
www.diadot.com

The Hat Factory Furniture Company
1000 N Division Street
Peekskill, NY 10566
Tel: 914-788-6288
Fax: 914-788-0019
info@hatfactoryfurniture.com
www.hatfactoryfurniture.com

Lutes Custom Woodworking and Design
PO Box 726
Canby, OR 97013
Toll-free tel: 877-835-0023
Tel: 503-655-0023
Fax: 503-655-2536
BLutes@compuserve.com
www.lutescw.com

Quality Custom Cabinetry, Inc.
PO Box 189
New Holland, PA 17557
Toll-free tel: 800-909-6006
www.qcc.com

woodwork, carving

Edward Falkenberg
RR5 Claremont, ON L1Y 1A2
edwardfalkenberg@hotmail.com
www.falkenberg.ca

Molyneux Designs
PO Box 1637
Weaverville, CA 96093
Tel: 530-623-2268
Fax: 530-623-1671
md@CustomMade.com
www.molyneuxdesigns.com

Ron Ramsey
Lake Tahoe, NV
Tel: 530-546-9394
ron@carvedbyramsey.com
www.carvedbyramsey.com

Redwood Burl .Com
34 - 1834 Allard
Eureka, CA 95503
Tel: 707-441-1658
gbuckjr@redwoodburl.com
www.redwoodburl.com

Simmonds Woodturning
PO Box 884
Durham ON N0G 1R0
Tel: 519-855-6013
Toll-free tel: 877-247-8408
Fax: 519-855-4633
simmondswt@sympatico.ca
www.simmondswoodturning.com

Whitehead Carvings
Tel: 250-675-2334
Fax: 250-675-2646
info@whiteheadcarvings.com
www.whiteheadcarvings.com

glossary

A-frame
A low-cost form of building in which rafters that are spaced apart on the ground meet at the roof ridge-pole in an inverted V. This building style sacrifices usable interior space for simplicity of construction.

Alabaster
A very fine variety of gypsum (hydrous calcium sulfate) found in nature. The epitome of white.

Ambient lighting
General lighting used to illuminate an entire space rather than certain areas or tasks.

American Foursquare (1895–1930)
These boxlike, simple two-story homes feature low-hipped roofs with wide overhangs, big porches and often a large central dormer. A cousin of the Prairie Style, they are often dressed up with decorative elements taken from other house styles.

American Streamline
The North American adaptation of Art Deco, with a visual emphasis on speed and rounded corners, while focusing on industrial materials such as nickel, concrete and glass block.

Apex stone
The top stone in a gable, pediment, vault or dome.

Art Deco
Coined from the Exposition des Arts Decoratifs held in Paris in 1925, Art Deco is a masculine decorative style that has many influences, from Ancient Egypt and African art to industrial design. Art Deco homes are known for their use of cubic forms, concrete or block projections (especially over doors), and pastel coloring.

Art glass
A type of colored glass used in windows in America during the late 19th and early 20th centuries; characterized by cubic forms, combinations of hues and special effects in transparencies and opaqueness. Art glass was used to great effect by Frank Lloyd Wright.

Art Moderne (1930–1945)
Reflecting North America's love affair with all things fast and technological, Art Moderne put a streamlined spin on Art Deco to make homes that looked much like a sleek, rounded diesel train. The style is characterized by smooth white walls, often of stuccoed block or painted bricks, cubic forms, flat roofs, metal sash or glass block windows set close to the wall or even wrapped around a rounded corner. Horizontal blocks or indentations are used to symbolize speed and movement.

Baluster
Regularly spaced miniature columns or posts that support a railing.

Balustrade
A railing with supporting balusters. Also the low, ornamental railing used on the roofs of Georgian, Federal and some other 18th- and 19th-century houses — sometimes enclosing a rooftop space known as a "widow's walk."

Baseboard
A molded board placed against a wall around a room on the floor to conceal a joint between the floor and the wall finish.

Batten
A narrow strip of wood used to cover joints between boards or panels.

Beam
A wooden, metal or other material that is a main horizontal supporting piece.

Beaux Arts Style (1885–1925)
A historical style stemming from the Ecole des Beaux Arts in Paris during the 19th century, it combines traditional Greek, Roman and Renaissance ideas. Used mostly for large public buildings.

Bell-cast or bell-curve roof
A bell-shaped eave or roof that flares out. French influenced.

Bisque
The bottom layer of a ceramic tile that lies under the glaze.

Bluestone
A dense, hard, fine-grained sandstone of bluish-gray color that splits readily into thin slabs. A variety of flagstone.

Board and batten
A type of usually vertical wooden cladding for wood frame buildings in which battens cover the joints between the boards.

Bonneted dormer
A dormer with a curved roofline, normally containing an arched window.

Boomtown architecture
Associated with North American frontier towns in the 19th century that went up quickly and cheaply using local materials, characterized by a false front to make humble buildings look more grand.

Border tiles
Tiles that outline the perimeter of a wall, floor or a field of tiles.

Boss
A decorative piece (of wood, stone or plastic) that covers the intersection of beams or the center of a panel or coffer.

Box beam
A simulated, purely decorative beam that has no supportive capabilities.

Bracket
A horizontally projecting ornamental support for an arch, roof cornice or entablature. Most often seen beneath the roof overhangs of Italianate houses.

Brick, clinker
A now highly prized antique brick that was discarded for its imperfections. They make a "clinking" sound when knocked together.

Brick, facing
Highly finished bricks for the façade wall of a house or building.

Brick, veneer
A facing of brick tied to a wood frame or masonry wall, serving as a wall covering only and carrying no structural loads.

Brownstone
A brown or reddish-brown sandstone.

Bull nose
Convex rounding of a stone or wood piece such as a stair tread, border tile or window sill.

Bungalow
A one- or one-and-a-half-story residential structure. Bungalows often feature gable or hip roofs with a large central dormer. Popular in North America for its simple plan and low cost, the bungalow had many proponents, including the California firm of Greene & Greene.

Butler's pantry
A small service room off of the dining room, often with a sink, warmer and cupboards.

Cape Cod cottage
A simple wood-frame house with a gabled roof and shingle siding. Originating in the coastal areas of eastern North America, its simplicity of design and inexpensive cost make it popular throughout the continent.

Capital
The decorative top of a column.

Carpenter Gothic
Frame homes in the Gothic Revival style that make up for their lack of stone and brickwork by including large amounts of wooden gingerbread. (See Gothic Revival)

Cedar shakes
Overlapping roofing material hand-split on one side and sawn on the other.

Cedar shingles
Roofing material similar to cedar shakes, only sawn on both sides and therefore thinner.

Ceramic tile
Formed pieces of clay that have been fired in an oven at high temperatures, either with a decorative, glossy, baked-on surface called a glaze, or unglazed and therefore the color of the clay they are made of.

Chair rail
A molding at chair back height to prevent damage to walls.

Chateau Style
A housing style based on French Medieval castles, characterized by steep roofs and turrets.

Chimney pot
An often elaborate cap that sits on top of a chimney. Can be made of brick, clay or metal. Sometimes designed to increase the draw of the chimney but can also be purely decorative.

Chinking
The material used to fill the gap between the logs in the wall of a log building. Refers to the process of filling the gaps as well as the material used.

Clapboard
Overlapping wooden boards applied horizontally, used as siding on buildings. Sometimes called "shiplap." Most aluminum, vinyl and fiber cement siding mimics the look of clapboard.

Classical Revival (1770–1830)
The Classical Revival style was a transitional style between the Federal and Greek Revival styles, employing a symmetrical facade featuring a portico or entrance porch supported by columns of Greek or Roman design.

Clerestory
The upper level of a room that is above the first story, often containing a window.

Coffered ceiling
An ornamental ceiling composed of a network of beams that form square or polygonal panels.

Colonial Revival Style (1876–1955)
A simplified version of the Georgian style, the Colonial Revival incorporates columns and pillars, Greek-like porticos with pediment, gabled roofs and doorways with sidelights and transoms.

Colonial windows
Windows with small rectangular panes, or lights, in even series of eight, ten, twelve and more.

Colonnade
A row of columns holding up a roof, often with arches separating them.

Column
Vertical, round supports, usually with a base at the bottom and a capital at top. Can either be load bearing or purely decorative. Columns are important design elements in Classical Revival and Neoclassical styles and usually fall into the traditional Greek and Roman orders Doric, Ionic, Corinthian, Tuscan. The distinguishing elements of the column are almost always in the capitals.

Composite
A manufactured product in which ground stone is permanently bonded to other material. Often used for countertops, sinks and bathtubs.

Concrete
A mixture of cement, sand, gravel and water that hardens into a stone-like substance. Because it is mixed wet, concrete can be poured into molds of any shape, and has therefore been used to make everything from concrete blocks to entire walls, countertops, floors and ceilings.

Conservatory
A greenhouse or other glassed-in room attached to a house.

Contemporary Style (1965–present)
A stripped-down, unornamented style known for its use of stone and brick, with many large windows, and open-plan interiors. Often with flat roofs but can also have roofs incorporating cathedral ceilings.

Coquina
A coarse and highly porous limestone composed predominantly of seashells cemented together by calcite, mainly quarried in Florida.

Corbel
An architectural bracket that projects to act as a support (but can be purely decorative) for pilasters, mantels, shelves and more.

Cornice
In interiors, a strip of decorative molding at the top of a wall. For exteriors, a part of the roof that sticks out past the wall), used most commonly in classical architecture. Often made of two or more moldings.

Covered Loggia
A roofed and columned gallery that is open to the air on one side. An architectural detail found on many Italian villas.

Craftsman Style (1905–1930)
The Craftsman style was the architectural result of the Arts and Crafts movement, a reaction to the mass production of the time. It celebrated and encouraged good design and handcrafted details in every aspect of life, including houses. Craftsman houses have strong architectural details such as open plan interiors, built-in components, extra wide eaves, rafters exposed at the eaves, stone porch supports, and beamed ceilings. Wherever possible the materials used are natural, such as unpainted wood, stone, ceramics and copper. Frank Lloyd Wright was one American architect influenced by the Craftsman approach.

Crown molding
Decorative moldings used to cover the unfinished areas where wall and ceilings meet and to provide visual interest. Sometimes used above kitchen cabinets.

Cupola
A structure, usually domed, on top of a roof or dome. Can be lantern-shaped.

Curved staircase
A staircase that climbs in a circular direction. Has an inside radius less than twice the width of the individual treads.

Delft tiles
A distinctive form of blue-and-white glazed pictorial or patterned tiles, originally from the Dutch town of the same name.

Dentil molding
A decorative molding attached below a crown molding that is made of regular blocks that make a tooth-like pattern. Inside the house they are found most often in crown moldings. Outside they appear most frequently on Georgian houses but are also seen on Federal and Early Classical Revival houses.

Dormer
A roof extension that contains a vertical window. From the French *dormir,* "to sleep."

Dormer window
A window in a wall that either projects from a sloping roof, or is recessed (inset dormer) into the roof, or a combination of both.

Drip chain
A long line of linked objects that channel water from the gutters to the ground.

Drip ledge or drip molding
A projecting molding, usually above a window, that allows rainwater to "drip" onto the ground.

Dry press method
A common way of making ceramic and porcelain tiles by pressing powdered material under pressure.

Eave
The lower part of a roof that projects beyond the face of the walls.

Eavestrough
A trough fixed to an eave designed to collect and carry away the water runoff from the roof. Also known as a gutter.

Engineered floor
A flooring product made of hardwood veneer bonded onto a core of multiple layers of hardwood, plywood or High Density Fiberboard (HDF).

Entablature
A decorative horizontal consisting of classical elements such as the architrave, frieze and cornice, and lies directly above a column or other support.

Facade
The front of a building.

Face-frame cabinets
Cabinets with a front frame with rails and stiles around the cabinet opening. This results in a door and drawer space that is usually less than frameless cabinets.

Fanlight
A fixed, fan-shaped (semicircular or elliptical), multi-mullioned window that sits above a door or window unit. A common element of Georgian architecture.

Fascia board
A finish board attached to the faces of eaves and other roof projections.

Federal Style (1780–1820)
One of the first truly American styles that came into being after independence in 1776. Similar to the English Adams style of house, the Federal style incorporates more delicate detailing as well as fanlights and sidelights in the entry door.

Feng Shui
Chinese for "wind-water," the ancient Chinese art of placement. The philosophy behind Feng Shui is that harmony and balance is achieved first in your environment — starting with your home — and then in your life. Complex rules governing the arrangement and relationship of every aspect of house design, from where the door goes to the shape of columns, form the basis of this philosophy.

Fiber cement siding
The combination of portland cement mixed with ground sand, cellulose fiber, and other additives (as manufactured by James Hardie Building Products) has resulted in durable, low-maintenance siding products that resemble traditional cladding, such as shingles and wood.

Field tile
The group of tiles making up the largest area of a tiled surface, exclusive of the border.

Finial
A decorative piece used to terminate a gable, spire or piece of furniture.

Flagstone
Thin slabs of stone used for flagging or paving walks, driveways, patios and more.

Fluted rail
Decorative molding, usually between cabinets.

Frameless cabinets
A more modern, European style in which doors cover the entire cabinet face.

French window
Two casement windows hinged on the side to open in the middle. When the windows go all the way the floor it becomes a French door.

Frieze
A decorative belt of molding that runs underneath a cornice.

Gable
The upper triangular-shaped portion of the end wall of a house.

Gable end
The entire end wall of a house having a gable roof.

Gambrel roof
A doubled sloped (or hipped) gable roof where the upper slope is of a lesser pitch than the lower and both slopes are straight.

Georgian Style (1714–1830)
The dominant style in British architecture during the time of the Kings George. This style is characterized by simple, elegant buildings with classical features such as a symmetrical front facade and usually two windows on either side of a central doorway (often including a fanlight), with a row of five windows above. Georgian homes often have a central chimney in the northern United States, and a pair of chimneys at either end in the South. Early Georgian homes had dormered gable roofs, while later homes often have hipped roofs and are more ornate overall.

Gingerbread
Decoratively sawn woodwork, often around the eaves, which became popular with the invention of the steam-powered saw. Also known as fretwork. Gingerbread-cut boards are known as "bargeboards" or "vergeboards."

Glazed tiles
Ceramic tiles coated with glass-forming minerals and ceramic stains that are fused under high heat. This produces either a matte, semi-gloss or high-gloss finish.

Gothic-head window
A window topped with a pointed arch.

Gothic Revival (1840–1880)
Based on Gothic cathedrals and castles. Because of their complexity and expense, most Gothic Revival homes were built for the rich. The style is characterized by pointed Gothic windows and arches, steeply pitched roofs, and castle or cathedral elements such as crenellated battlements, turrets and parapets, large verandahs and oriel windows.

Granite
A very hard, fine- to coarse-grained igneous rock formed by volcanic action that gets its characteristic mineral-flecked appearance from the presence of particles of quartz, feldspar, mica, and other trace minerals.

Greek Revival (1825–1860)
The Greek Revival style became popular when archeologists first began to excavate ancient Greek and Roman buildings. Homes of this style are designed to resemble classic Greek temples. This is accomplished by the use of free-standing columns supporting a portico and shallow, gabled roofs and a deep entablature (representing a heavy stone lintel), also supported by classical Grecian columns. Simpler adaptations of the style have pilasters instead of full-blown columns

Green roofs
Using flat or slightly pitched roofs to grow plants in a lightweight medium. The benefits include evaporative cooling, protection from destructive UV light, the conversion of carbon dioxide to oxygen, and the reduction of storm water runoff.

Grout
Cement-like material used to fill the joints between tiles.

Gutters
Channels hanging below the eves of a roof that direct water into downspouts and away from the foundation of the building.

Half timber
A post-and-beam frame structure whose external cladding exposes the timberwork. Popular in Queen Anne and Tudor style homes.

Hearth
The floor of a fireplace and the area immediately in front of it.

Heat sink
A medium (usually stone, brick or concrete) that, when exposed to the sun's rays inside a room, will become heated and radiate the heat back into the room during the sunless night hours. A key element of passive solar heating.

Hip
The sloping ridge of a roof formed by two intersecting roof slopes.

Hipped roof
A roof that slopes on four sides.

Honed finish
Material polished to a dull, matte finish. Not as fine or reflective as a polished finish.

HVAC
An acronym for Heat, Ventilation and Air Conditioning systems.

I-beam
A steel beam with a cross section resembling the letter I.

Inglenook
Two facing benches that form a nook in front of a fireplace.

Insulated Concrete Forms (ICFs)
A system of building walls in which hollow foam blocks are filled with concrete. Extremely strong and well-insulated but requiring skill to put up.

Ipe
An extremely dense tropical wood. Pronounced *ee-pay*.

International Style
The American interpretation of German Bauhaus architecture, the International Style is the forbear of Contemporary architectural style. Like Bauhaus, it is a style stripped of ornamentation, concentrating on simple, clean lines, flat roofs, and regular, cubic shapes.

Italianate (1840–1885)
The dominant American house style of the Civil War period, the Italianate style has low-pitched roofs with wide overhangs and ornamental brackets and tall, arched windows. They are usually square-shaped two-story homes and can be topped by a cupola.

Jalousie windows
A window composed of overlapping narrow glass, metal, or wooden louvers, whose angle is adjustable. Also known as louvered windows.

Jamb
The side post or lining of a doorway, window or other opening.

Jarrah
A dense, dark wood from Australia.

Joist
One of a series of horizontal wood members used to support a floor, ceiling or roof.

Kiva fireplace
An adobe (or concrete block and stucco) fireplace with rounded contours that resemble a beehive, hence its familiar name. Often set into a corner.

Laminate flooring
A flooring product made from a top layer that is a photographic image of a flooring type (wood, ceramic tile, and so on) bonded onto a core of multiple layers of High Density Fiber (HDF).

Lancet window
Tall, narrow window with a pointed-arch top. One of the characteristics of Gothic architecture.

Limestone
A sedimentary rock composed of calcium carbonate. Marble and onyx are forms of limestone.

Linoleum
A manufactured flooring material made of oxidized linseed oil, mixed with cork or wood flour, mineral filler and pigments and bonded to jute or another suitable backing. Comes in rolls or individual tiles.

Lintel
A horizontal structural member (beam) that supports the load over an opening such as a door or window. Can be made of stone, concrete or other hard material.

Long log (Scandinavian scribe)
A style of log building employing round logs in which the full log lengths in each wall are interlocked at their corners. Each log is grooved out on its underside to fit tightly over the log below and prevent air and water from passing between them.

Majolica tile
A form of decorative tile with an opaque, textured glaze.

Mansard roof
A double sloped roof with the lower slope being longer and steeper, with a concave curve. First designed by 17th-century French architect François Mansart.

Mantel
Commonly, a shelf over a fireplace opening that frames the fireplace, but the term mantel also includes the fireplace surround.

Manufactured stone
Ground stone (often quartz) bonded with resin into a homogenous and durable surface. Used for countertops, sinks and more.

Marble
Limestone and other sedimentary carbonate rocks transformed through metamorphic forces and capable of taking a high polish. Occurs in a wide range of colors and variations.

Masonry fireplace
A fireplace made of masonry brick or stone with channels within that circulate heated air. The heated masonry releases the captured heat for many hours, even after the fire has become low or gone out.

Medallions
Ornate round accents used on ceilings. Can act as chandelier focal points.

Medium Density Fiberboard (MDF)
An engineered wood.

Molding
A decorative piece of wood, metal or other material applied to walls, doors and other surfaces.

Mortar
A mixture of cement, lime, sand, and water used to bond bricks or blocks.

Mortise and tenon
A joint made between two pieces of wood in which a protruding piece (the tenon) fits into the corresponding hole (the mortise) of the other.

Mullion
A vertical rail or molding that separates window panes.

Neoclassical Architecture
An approach to architecture that is used in many different styles — Greek Revival, American Foursquare, and Beaux Arts — and relies on ancient Greek and Roman principles of proportion and design. Many important public buildings, including the White House, and private residences have been influenced by Neoclassical architecture since the 1700s.

Neo-eclectic (1985–present)
A new form of residential house design that freely incorporates any and all previous housing styles. Exteriors may include faux stone and veneer brick. Interiors are marked by semi-open floor plans, large kitchen/family room complexes, and other innovations for busy families, such as second-floor laundry rooms. Some critics frown upon neo-eclectic architecture for playing with traditional forms too freely.

Neo Mediterranean (1960–present)
Similar to Spanish Colonial and other Mediterranean styles, characterized by white stucco walls, red tile roofs, arches, elaborately carved wooden doors, and floor plans that are open to the outside.

Newel post
The structural and decorative terminus of a staircase balustrade.

Non-vitreous tiles
Indoor tiles with a water absorption of 7 percent or greater.

Nosing
The rounded and projecting edge of a stair tread, windowsill, and more.

Onyx
A form of marble characterized by a dense, crystalline look that is generally translucent and layered.

Oriented Strand Board (OSB)
Strong panels made of different layers of aspen, poplar or southern yellow pine, bonded together under heat and pressure using a waterproof phenolic resin adhesive. Also known as Waferboard. Used in place of plywood.

Palladian window
Three windows usually separated by pilasters, with a fanlight over the taller, middle window. Originated by 16th-century Italian architect Andrea Palladio. Often seen in high-style Georgian and Federal houses, usually over the front entranceway so it can illuminate an upstairs hall or landing.

Pantiles
Clay roofing tiles.

Parapet
Often the highest point of a roof, a low wall that extends upward above the height of the building, useful for hiding the mechanicals.

Parquet floor
Small rectangular pieces of wood inlaid in geometric forms, sometimes using contrasting woods, to make flooring.

Parquet de Versailles
A parquet floor that is bordered either with wood or some other material such as brick or stone.

Particleboard
An engineered board made from large particles of wood held together with a binder.

Passive solar
Heating and cooling a building without relying on manmade energy sources through the use of energy efficient materials and proper site placement.

Pavers
Glazed or unglazed or porcelain tiles made from the dust-pressed method. A paver has a surface area of more than 6 inches square.

Pediment
An ornamental accent found above entranceways and windows.

Physical Vapor Deposition (PVD)
A clear protective coating for hardware such as faucets and taps.

Picture / plate rail
A wall shelf for pictures or plates.

Picture window
A large-paned fixed window, usually on the front of a house.

Pilaster
A decorative, flat, rather than rounded, column (complete with capital, shaft and base) that projects only slightly from a wall.

Piece-on-piece
A style of log building in which short lengths of logs are used to form wall panels in sections between the vertical support beams that form the main structural frame of the building.

Pitched roof
A roof that has one or more surfaces sloping at angles greater than necessary for drainage.

Plinths
The square, lower part of the base of a column.

Plywood
Manufactured wood, usually in 4-by-8-foot sheets, made of three or more layers of veneer wood bonded with glue. The middle layer usually has an alternating grain to the layers above and below it for added strength.

Pointing
The filling of mortar joints with mortar or caulking compounds.

Polystyrene
A lightweight plastic that comes in two forms — solid and expanded or extruded.

Polyvinyl chloride (PVC)
A hard plastic used on everything from vinyl siding to resilient flooring.

Porcelain tiles
Ceramic tiles made by the dust-pressed method. Hard tiles with very low water-absorption rates, porcelain tiles are often used for flooring and other high-use areas because of their durability and the fact that they keep their color all the way through, even when chipped or scratched.

Porte cochere
A covered projection out from the main door of the house covering the driveway to protect visitors from the weather when getting out of vehicles. Can also lead into an interior courtyard. From the French for "coach door."

Portico
An entranceway or porch, with columns and pediment. Often the central piece of a façade. Very common on Federal, Early Classical Revival and Greek Revival houses.

Post and beam
A traditional way of building a frame for a building by combining large horizontal beams and vertical posts. The first stage of a timber frame house.

Postmodern Style (1925–1980)
A reaction to the purity and stark look of Modernism, Postmodern Style is characterized by often humorous juxtaposition of different styles or the exaggeration of classical architectural details.

Prairie Style (1893–1920)
A unique American style, Prairie is characterized by its accent on horizontal lines, low-pitched roofs, big eaves and an open floor plan. Its initial and biggest proponent was architect Frank Lloyd Wright. Prairie Style is compatible with similar Arts and Crafts designs.

Pueblo Style
Sometimes called Adobe for the prevalent building material, Pueblo Style homes are based on the structures built by the Native American in the southwest hundreds of years ago. They are characterized by flat roofs, timbers extending through the walls, rain spouts, rounded contours, and deep-set window and door openings.

Quarry tile
Extruded clay tiles that can be glazed or unglazed. Used predominantly for flooring.

Quartz
Essentially silicon dioxide, the most common mineral on earth. Because of quartz's hardness and resistance to scratches, it is used for countertops, often manufactured in a process in which ground quartz is bound with a resin.

Quartzite
A compact granular rock composed of quartz crystals, occurring in a wide range of colors and with very high tensile strength.

Queen Anne Style (1880–1910)
Mass-production techniques enabled this elaborate, fairytale style to reign in the late 19th century. Characterized by turrets, porches (usually done in gingerbread bargeboards), bay and oriel windows, ornamental brackets, and siding that mixes clapboard and wooden shingles in patterns. Some strikingly painted examples of the style are known as "Painted Ladies."

Queen Anne window
A window with small glass lights on the upper sash. From the late 1800s.

Queen post
A vertical structural member used to transfer the weight of roof purlins (rafters) onto the gable ends or onto horizontal cross beams.

Quoin
Stone or brick accent on an exterior corner, sometimes simulated on frame structures to look like stone. A mainstay of Georgian architecture.

Radiant heating
A method of heating, usually consisting of coils or pipe or electric heating elements embedded in the floor, wall or ceiling.

Rafter
One of the series of structural members of a roof, usually of two-inch nominal thickness, that slopes up from the top of the wall to the ridge, designed to support roof loads but not as a ceiling finish.

Rail
Horizontal member of a window sash.

Rammed earth
A building method that mixes soil with bonding and waterproofing additives that is rammed or forced into forms to produce walls. Resembles adobe in appearance and is considered an ecologically "green" building technique.

Ranch Style (1932–present)
The ubiquitous and very simple one-story house that was the staple of early North American subdivisions. Characterized by a low gable roof, attached garage, L-shaped floor plans that open to the outdoors, and large windows. Some variations are known as California Ramblers.

Reclaimed wood
Wood used for an architectural purpose (often flooring) that was once used for another purpose, such as railroad bridges, warehouse beams or flooring, and so on.

Resilient flooring
A generic term for vinyl and linoleum flooring.

Return
A pattern that continues, or returns, around a corner. The most classic is the "Greek return."

Reveal
The decorative highlighting of the joint between the meeting of two planes — such as a wall and a ceiling. The reveal's intent is the opposite of a molding's.

Richardsonian Romanesque (1880–1900)
Copying the rounded arches of the buildings of ancient Rome, Henry Hobson Richardson popularized the look of his robust, rough-hewn style by building famous public buildings. Homes built in this style are characterized by rounded and patterned arches over windows and doors, rough-cut stone cladding, medieval-looking towers and spiral-cut columns.

Riser
A vertical board separating the treads on a staircase.

Roman arch
A semi-circular arch.

Romantic
All of the 19th-century Revival styles (Greek, Gothic, Egyptian), as well as the Italianate and Italian Villa styles, are considered Romantic.

Rosette
A smaller version of a medallion used as an accent on a ceiling, mantel, or any other form of woodwork.

Roundel
A decorative circular window or panel molding.

Rubblestone
A decorative or veneer cladding stone that is irregularly shaped with a rough finish.

Saltbox roof
Shed roof built as an extension of a gable roof of the same pitch and width, originating on the Eastern seaboard of North America.

Saltillo tile
Tile made of unglazed clay, originating in Mexico.

Santa Fe Style
A derivation of Pueblo style, originating in New Mexico and inspired by Spanish Colonial and Native Pueblo forms.

Sash
The framework of a window. A double-hung window has an upper and lower sash.

Scribing
Fitting woodwork to an irregular surface. A term used in long log building that refers to the process of transferring the contours of one log onto another so that they may be cut to fit tightly together.

Second Empire Style (1852–1870)
An ornate style that originated during the Second Empire in France. It is characterized by a Mansard roof and curved (bonneted) dormers.

Serpentine
A form of marble characterized by a dull green, mottled appearance.

Shake
A shingle split (not sawn) from a block of wood and used for roofing or siding.

Shed roof
A sloping roof having its surface on one plane.

Shingle Style (1880–1900)
Originally the elaborately gabled and porched summer homes of the wealthy in the New England area, but now referring to any home that is clad in wooden shingles.

Shiplap
See clapboard.

Shoji screen
A sliding rice-paper and wood room divider popular in Japan.

Sidelight
A window beside the door that forms an integral part of the door unit.

Siding
In wood-frame construction, the material other than stucco or masonry used as an exterior wall covering.

Sill
The horizontal bottom shelf of a window frame.

Slate
A very fine-grained metamorphic rock derived from sedimentary rock shale, easily split into relatively thin slabs. Used mainly for roofing, flooring and countertops.

Soffit
The underside of elements of a building, such as staircases and roof overhangs.

Spanish Mission Style (1890–1920)
Modeled after the mission churches built by the Hispanic monks in the southwest United States, these homes are known for their deeply shaded verandahs, arched dormers and roof parapets. They are usually faced with stucco and have red tile roofs.

Stackwall
A style of log building in which logs of one to three feet in length are laid in a bed of mortar with their end grain exposed, to form the walls of the building. Also called cord wood construction.

Stair landing
A platform between flights of stairs.

Stile
A vertical structural member of a window frame, at the outer edge.

Stockade Style
A way of building with logs in which the wall logs are laid side by side and vertical, joined at the top and bottom with horizontal sill and ceiling beams.

Straw bale house
A form of construction using bales of straw covered with lath or chicken wire and then plastered. Straw bale is a renewable building material.

Strike plate
The part of a door lock set which is fastened to the jamb.

Stucco
A cladding material used since ancient times consisting of various ratios of cement, sand lime, applied wet to a web of chicken wire, masonry wall, or wooden lath. Stucco can be tinted various colors.

Tempered glass
Glass that has been strengthened by treating it with high heat. Also known as safety glass.

Terra cotta
Latin for "baked earth." A glazed or unglazed fired clay tile that is usually the color of the clay it was made of.

Terrazzo
Stone chips — usually marble — bonded with cement and ground to a flat surface that exposes the chips, which are then polished. Used for flooring.

Toe kick
Also, toe plate. Molding used to cover the open space under kitchen cabinets.

Tongue and groove
Boards that are milled so that there is a groove on one edge and a corresponding tongue on the other so that they fit together.

Transom
A horizontal window above a doorway or tall window, either fixed or able to open.

Travertine
A form of limestone precipitated from ground waters often found in caves or around springs. Has a characteristic pitted appearance.

Tread
The horizontal part of a step or staircase.

Tread rise
The vertical measurement from the top of one tread to the top of the next.

Tread run
The horizontal distance measured from the front of one tread to the front of the next.

Turret
An ornamental tower emerging from the roof of a structure.

Ultraviolet
A type of radiation in wavelengths shorter than those of visible light and longer than those of x-rays. (UV.)

Unglazed tiles
Tiles with no coat of fired-on glazing, which therefore don't show wear as much as glazed tiles. Common unglazed tiles are porcelain tiles and quarry tiles.

Usonian
An abbreviation of United States of North America coined by architect Frank Lloyd Wright in 1936 to describe a type of stylish but inexpensive home he hoped every family could afford. Lacking basements and with carports instead of garages, the homes used inexpensive materials such as plywood and concrete.

Vault
A masonry arched ceiling.

Veneer
A thin piece of solid wood laminated or glued onto a piece of backing. Used for furniture and flooring.

Venetian plaster
A multi-layered, tinted plaster.

Verandah
A covered porch, open at the sides.

Vergeboards
Decorative trim cut out of boards along the gable edges of a roof or dormer. Sometimes called "bargeboards" or "gingerbread."

Vernacular architecture
A building style that is governed more by local needs and materials than by any recognized architectural style.

Victorian Style (1850–1900)
A highly ornamental residential style that was popular during the reign of Queen Victoria, characterized by vertical motifs, two or three stories, balconies and porches, gingerbread, steep roofs and tall, narrow windows. Interiors were cut up into many small, purpose-built rooms.

Viga
A heavy wooden beam used to support the roof in adobe homes. Usually the ends are left exposed.

Vitreous tiles
Tiles with a water absorption less than 3 percent moisture, but more than 0.5 percent. Cannot be used in exterior areas where freeze-thaw conditions could cause tile cracking.

Vitrine
A glass display cabinet.

Vitrolite
A pigmented plate glass, mechanically ground to a mirror finish. Popular in Art Deco details.

Wall niche
A recessed enclosure in a wall used to showcase decorative elements such as statuary or vases.

Wear layer
The surface layer of a manufactured floor that protects the colored or patterned layer underneath.

Western red cedar
A species of cedar whose natural oils act as a preservative. That and its fine grain and strength make it ideal of making cedar roof shakes and shingles.

Window, awning
A window with a top-hinged sash.

Window, bay
A window that projects outside the main wall of a building, in three or more segments.

Window, bow
A bay window that projects out from the wall in a semicircle, usually in five segments.

Window, casement
A single window sash hinged at one side. English casements swing out and French casements swing in.

Window, Chicago
A large fixed sash flanked by a narrow, often movable sash on either side.

Window, double-hung
A window having two vertically sliding sashes, one on top of the other, designed to open either the top or bottom half at one time. The most common North American window design.

Window, eyebrow
Small, curved-topped windows often found on roofs or built into the top molding of the house. Found in Greek Revival and Italianate houses.

Window, jalousie
Overlapping rectangular slats of glass, that pivot open and closed.

Window, oriel
A built-up window that projects from the wall and does not extend to the ground, often supported by brackets or corbels. Used in the Gothic Revival style.

Window, single-hung
A window with a fixed top sash and a lower sash that slides vertically.

bibliography

Ackerman, James S. *Palladio.* New York: Penguin, 1974

Blumenson, John. *Identifying American Architecture: A Pictorial Guide to Styles and Terms: 1600-1945*. New York: Norton & Co., 1981

Calloway, Stephen. *The Elements of Style: A Practical Encyclopedia of Interior Architectural Details from 1485 to the Present.* Toronto: Firefly Books, 2005

Christopher, Peter and Richard Skinulis. *Log Homes: Classics of the North.* Toronto: Boston Mills Press, 2004

Hess, Alan with Kenneth Frampton, Thomas S. Hines, Bruce Brooks Pfeiffer and Alan Weintraub. *Frank Lloyd Wright: The Houses*. New York: Rizzoli, 2005

Langdon, Philip. *American Houses.* New York: Stewart, Tabori & Chang, 1987

Larkin, David, June Sprigg and James Johnson. *Colonial Design in the New World,* New York: Stewart, Tabori & Chang, 1988

Lawrence, Richard Russell. and Teresa Chris. *The Period House: Style, Detail & Decoration 1774-1914*. London: Weidenfeld & Nicolson, 1998

Lee, Vinny and Roy Main. *Recycled Spaces.* San Francisco: Soma Books, 2000

Mann, Dale and Richard Skinulis. *The Complete Log House Book: A Canadian Guide to Building With Logs.* Toronto: McGraw-Hill Ryerson, 1979

Masey, James C. and Shirley Maxwell. *House Styles in America: The Old-House Journal Guide to the Architecture of American Homes.* New York: House Styles in America Studio, 1999

McAlester, Virginia, Lee McAlester, Juan Rodriguez-Arnaiz, and Lauren Jarr. *A Field Guide to American Houses.* New York: Knopf, 1984

Miller, Judith. *The Style Source Book: The Definitive Illustrated Directory of Fabrics, Wallpapers, Paints, Flooring, Tiles.* Toronto: Firefly Books, 2003

Myers, Barton. *3 Steel Houses.* Mulgrave, Australia: The Images Publishing Group, 2005

Paul, Linda Leigh. *Desert Retreats Sedona Style*. New York: Universe Publishing, 2003

Pavlovits, Daniel. *Interiors Now.* Mulgrave, Australia: Images Publishing, 2004

Reiss, Marcia. *Architectural Details.* San Diego, CA: Thunder Bay Press, 2004

Rybczynski, Witold. *Home: A Short History of an Idea.* New York: Viking Penguin, 1986

Schmidt, Philip and Jessie Walker. *Decorating With Architectural Details.* Upper Saddle River, NJ: Creative Homeowner, 2004

Scully, Vincent. *The Shingle Style Today.* New York: George Braziller Publishing, 2000

acknowledgments

This book would not have been possible without the help and cooperation of all of the people who participated — the architects and designers who created the environments we photographed; the owners who let us into their homes; and the artists, artisans, manufacturers and distributors who supplied the images of and information on products that are the physical details that combine to make a house a home. We would also like to give special thanks to those who went out of their way to give us extra help: Lynn Appleby, Dee Chenier, Paul Froncek, Tony Jenkins, Mary and Peter Noetzel, Barton Phelps, Ed Sure, Tish Cohen and Nancy Shanoff. And finally, we would like to thank our editor, Kathleen Fraser, and our designer, Joseph Gisini of PageWave Graphics Inc., for working so hard to pull a very complex book together, and also Boston Mills Press and Firefly Books for having the vision to commission the book in the first place.

photo credits

All photography in this book is by Peter Christopher unless otherwise noted.

Additional photographs by:

Eric Brandt Architect: (page 18).

Kuwabara Payne McKenna Blumberg Architects: exterior of home, front (page 9); exterior of home, rear (page 14); entrance to wine cellar (page 238).

Joel Puliatti: Vetrazzo product shots (page 404).

Nancy Shanoff: Frank Lloyd Wright exterior (page 34); interior (page 46).

Alexander Vertikoff Photography: Gamble House (page 31).

Paul Warchol Photography: metal bed chambers (page 100); lift kitchen, with close-ups (page 240).

Jim Westphalen: Hubbardton Forge light fixtures (Cover, pages 356, 358, 359, 360, 361).

Ted Yarwood Photographer: interior (page 249).

We are grateful to those suppliers, manufacturers, artisans and artists who allowed us to use images of their work and wares for the Product sections of this book. (No payment or other recompense was required or received from any of the companies or individuals whose products are shown.) Obviously, the photos shown here represent just a tiny fraction of the products and providers available. In a book of this scope there is no way we could show you the infinite number of selections available, but we have enjoyed choosing many of our particular favorites to help spark your ideas. For more choices and possibilities, consult the Sources guide beginning at page 472. It will lead you to contact information and web addresses and more inspirational ideas multiplied by the tens of thousands.

index